PROJECT

MANAGEMENT
BEST PRACTICES

PROJECT

MANAGEMENT
BEST PRACTICES

Achieving Global Excellence

SECOND EDITION

HAROLD KERZNER, PH. D.

WILEY

John Wiley & Sons, Inc.

INTERNATIONAL
Institute for Learning, Inc.

Library of Congress Cataloging-in-Publication Data:
Kerzner, Harold.
    Project management : best practices : achieving global excellence / Harold Kerzner.—2nd ed.
        p.   cm.
    Includes bibliographical references and index.
    ISBN 978-0-470-52829-7 (cloth : alk. paper)   1. Project management—Case studies.   I. Title.
    HD69.P75K472 2010
    658.4'04—dc22

                                                                                    2009042592

Printed in the United States of America
10   9   8   7   6   5   4   3

*To
my wife, Jo Ellyn,
who showed me that excellence
can be achieved in
marriage, family, and life
as well as at work*

Contents

Preface

For almost 40 years, project management was viewed as a process that might be nice to have but not one that was necessary for the survival of the firm. Companies reluctantly invested in some training courses simply to provide their personnel with basic knowledge on planning and scheduling. Project management was viewed as a threat to established lines of authority, and in many companies only partial project management was used. This half-hearted implementation occurred simply to placate lower- and middle-level personnel as well as selected customers.

During this 40-year period, we did everything possible to prevent excellence in project management from occurring. We provided only lip service to empowerment, teamwork, and trust. We hoarded information because the control of information was viewed as power. We placed personal and functional interests ahead of the best interest of the company in the hierarchy of priorities. And we maintained the faulty belief that time was a luxury rather than a constraint.

By the mid-1990s, this mentality began to subside, largely due to two recessions. Companies were under severe competitive pressure to create quality products in a shorter period of time. The importance of developing a long-term trusting relationship with the customers had come to the forefront. Businesses were being forced by the stakeholders to change for the better. The survival of the firm was now at stake.

Today, businesses have changed for the better. Trust between the customer and contractor is at an all-time high. New products are being developed at a faster rate than ever before. Project management has become a competitive weapon during competitive bidding. Some companies are receiving sole-source contracts because of the faith that the customer has in the contractor's ability to deliver a continuous stream of successful projects using a project management methodology. All of these factors have allowed a multitude of companies to achieve some degree of excellence in project management. Business decisions are now being emphasized ahead of personal decisions.

Words that were commonplace six years ago have taken on new meanings today. Change is no longer being viewed as being entirely bad. Today, change implies continuous improvement. Conflicts are no longer seen as detrimental. Conflicts managed well can be beneficial. Project management is no longer viewed as a system entirely internal to the organization. It is now a competitive weapon that brings higher levels of quality and increased value-added opportunities for the customer.

Companies that were considered excellent in management in the past may no longer be regarded as excellent today, especially with regard to project management. Consider the book entitled In *Search of Excellence*, written by Tom Peters and Robert Waterman in 1982 (published by Harper & Row, New York). How many of those companies identified in their book are still considered excellent today? How many of those companies have won the prestigious Malcolm Baldrige Award? How many of those companies that have won the award are excellent in project management today? Excellence in project management is a never-ending journey. Companies that are reluctant to invest in continuous improvements in project management soon find themselves with low customer satisfaction ratings.

The differentiation between the first 40 years of project management and the last 10 years is in the implementation of project management on a companywide basis. For more than three decades, we emphasized the quantitative and behavioral tools of project management. Basic knowledge and primary skills were emphasized, and education on project management was provided only to a relatively small group of people. However, within the past 10 years, emphasis has been on implementation across the entire company. What was now strategically important was how to put 30 years of basic project management theory in the hands of a few into practice. Today it is the implementation of companywide project management applications that constitutes advanced project management. Subjects such as earned-value analysis, situational leadership, and cost and change control are part of basic project management courses today whereas 15 years ago they were considered advanced topics in project management. So, what constitutes applied project management today? Topics related to project management implementation, enterprise project management methodologies, project management offices, and working with stakeholders are advanced project management concepts.

This book covers the advanced project management topics necessary for implementation of and excellence in project management. The book contains numerous quotes from people in the field who have benchmarked best practices in project management and are currently implementing these processes within their own firms. Quotes in this book were provided by 10 CEOs, 5 Presidents, several COOs, CIOs, CFOs, senior VPs, VPs, global VPs, general managers, PMO directors, and others. The quotes are invaluable because they show the thought process of these leaders and the direction in which their firms are heading. These companies have obtained some degree of excellence in project management, and what is truly remarkable is the fact that this happened in less than five or six years. Best practices in implementation will be the future of project management well into the twenty-first century. Companies have created best practices libraries for project management. Many of the libraries are used during competitive bidding for differentiation from other competitors. Best practices in project management are now viewed as intellectual property.

Companies that are discussed in this book include:

ABB	McElroy Translation
Alcatel-Lucent	Microsoft
American Greetings	Motorola

Antares	NASA
AT&T	Nortel
Boeing	NXP
Computer Associates	Perot Systems
Convergent Computing	Roadway Express
Churchill Downs, Inc.	Rockwell Automation
Comau	Satyam
Computer Sciences Corp.	SENTEL
Deloitte	Sherwin-Williams
Department of Defense	Siemens
DFCU	Slalom
Diebold	Star Alliance
DTE Energy	Synovus
EDS	Sypris Electronics
Enakta	Teradyne
Ericsson	Texas Instruments
EXEL	Tyco
General Motors	Visteon
Harris	Vitalize Consulting
Holcim	Westfield
Hewlett-Packard	WWF
IBM	Zurich America Insurance Co.
Indra	
ITC	
Johnson Controls	
Jefferson County	
Lilly	

Seminars and webinar courses on project management principles and best practices in project management are available using this text and my text *Project Management: A Systems Approach to Planning, Scheduling, and Controlling*, 10th edition (Wiley, Hoboken, New Jersey, 2009). Seminars on advanced project management are also available using this text. Information on these courses, E-learning courses, and in-house and public seminars can be obtained by contacting:

Lori Milhaven, Executive Vice President, IIL:

Phone:	800-325-1533 or 212-515-5121
Fax:	212-755-0777
E-mail:	lori.milhaven@iil.com

Harold Kerzner
International Institute for Learning, Inc.
2010

INTERNATIONAL INSTITUTE FOR LEARNING, INC. (IIL)

International Institute for Learning, Inc. (IIL) specializes in professional training and comprehensive consulting services that improve the effectiveness and productivity of individuals and organizations.

As a recognized global leader, IIL offers comprehensive learning solutions in hard and soft skills for individuals, as well as training in enterprise-wide Project, Program, and Portfolio Management.

PRINCE2™,[1] Lean Six Sigma; Microsoft® Office Project and Project Server,[2] and Business Analysis.

After you have completed *Project Management Best Practices: Achieving Global Excellence, Second Edition*, IIL invites you to explore our supplementary course offerings. Through an interactive, instructor-led environment, these virtual courses will provide you with even more tools and skills for delivering the value that your customers and stakeholders have come to expect.

For more information, visit http://www.iil.com or call +1-212-758-0177.

---

1. PRINCE2 is a trademark of the Office of Government Commerce in the United Kingdom and other countries.
2. Microsoft Office Project and Microsoft Office Project Server are registered trademarks of the Microsoft Corporation.

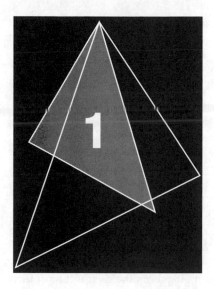

Understanding Best Practices

1.0 INTRODUCTION

Project management has evolved from a set of processes that were once considered "nice" to have to a structured methodology that is considered mandatory for the survival of the firm. Companies are now realizing that their entire business, including most of the routine activities, can be regarded as a series of projects. Simply stated, we are managing our business by projects.

Project management is now regarded as both a project management process and a business process. As such, project managers are expected to make business decisions as well as project decisions. The necessity for achieving project management excellence is now readily apparent to almost all businesses. Steven Deffley, Project Management Professional (PMP), Global Product Manager at Tyco Electronics, believes that:

> Achieving Project Management Excellence addresses how Project Management has evolved into a business process, providing concepts that can be employed to improve the effectiveness and financial contribution of an organization. Excellence is driven by a focus on critical success factors and key performance indicators as it relates to a project. Excellence in Project Management illustrates how the intellectual value of lessons learned can lead to a proprietary competitive advantage. Achieving Project Management Excellence demonstrates how Project Management has matured in encouraging and supporting an organization to perform at a higher level.

As the relative importance of project management permeates each facet of the business, knowledge is captured on best practices in project management. Some companies view this knowledge as intellectual property to be closely guarded in the vaults of the company. Others share this knowledge in hope of discovering other best practices. Companies are now performing strategic planning for project management.

One of the benefits of performing strategic planning for project management is that it usually identifies the need for capturing and retaining best practices. Unfortunately this is easier said than done. One of the

1

reasons for this difficulty, as will be seen later in the chapter, is that companies today are not in agreement on the definition of a best practice, nor do they understand that best practices lead to continuous improvement, which in turn leads to the capturing of more best practices.

1.1 PROJECT MANAGEMENT BEST PRACTICES: 1945–1960

During the 1940s, line managers functioned as project managers and used the concept of over-the-fence management to manage projects. Each line manager, wearing the hat of a project manager, would perform the work necessitated by his or her line organization and, when completed, would throw the "ball" over the fence in hopes that someone would catch it. Once the ball was thrown over the fence, the line managers would wash their hands of any responsibility for the project because the ball was no longer in their yard. If a project failed, blame was placed on whichever line manager had the ball at that time.

The problem with over-the-fence management was that the customer had no single contact point for questions. The filtering of information wasted precious time for both the customer and the contractor. Customers who wanted first-hand information had to seek out the manager in possession of the ball. For small projects, this was easy. But as projects grew in size and complexity, this became more difficult.

During this time period, very few best practices were identified. If there were best practices, then they would stay within a given functional area never to be shared with the remainder of the company. Suboptimal project management decision making was the norm.

Following Word War II, the United States entered into the Cold War. To win a Cold War, one must compete in the arms race and rapidly build weapons of mass destruction. The victor in a Cold War is the one who can retaliate with such force as to obliterate the enemy. Development of weapons of mass destruction was comprised of very large projects involving potentially thousands of contractors.

The arms race made it clear that the traditional use of over-the-fence management would not be acceptable to the Department of Defense (DoD) for projects such as the B52 bomber, the Minuteman Intercontinental Ballistic Missile, and the Polaris submarine. The government wanted a single point of contact, namely, a project manager who had total accountability through all project phases. In addition, the government wanted the project manager to possess a command of technology rather than just an understanding of technology, which mandated that the project manager be an engineer preferably with an advanced degree in some branch of technology. The use of project management was then mandated for some of the smaller weapon systems such as jet fighters and tanks. The National Aeronautics and Space Administration (NASA) mandated the use of project management for all activities related to the space program.

Projects in the aerospace and defense industries were having cost overruns in excess of 200–300 percent. Blame was erroneously placed upon improper implementation of project management when, in fact, the real problem was the inability to forecast technology, resulting in numerous scope changes occurring. Forecasting technology is extremely difficult for projects that could last 10–20 years.

By the late 1950s and early 1960s, the aerospace and defense industries were using project management on virtually all projects, and they were pressuring their suppliers to use it as well. Project management was growing but at a relatively slow rate except for aerospace and defense.

Because of the vast number of contractors and subcontractors, the government needed standardization, especially in the planning process and the reporting of information. The government established a life-cycle planning and control model and a cost-monitoring system and created a group of project management auditors to make sure that the government's money was being spent as planned. These practices were to be used on all government programs above a certain dollar value. Private industry viewed these practices as an overmanagement cost and saw no practical value in project management.

In the early years of project management, because many firms saw no practical value in project management, there were misconceptions concerning project management. Some of the misconceptions included:

- Project management is a scheduling tool such as PERT/CPM (program evaluation and review technique/critical-path method) scheduling.
- Project management applies to large projects only.
- Project management is designed for government projects only.
- Project managers must be engineers and preferably with advanced degrees.
- Project managers need a "command of technology" to be successful.
- Project success is measured in technical terms only. (Did it work?)

1.2 PROJECT MANAGEMENT BEST PRACTICES: 1960–1985

During this time period, with a better understanding of project management, the growth of project management had come about more through necessity than through desire, but at a very slow rate. Its slow growth can be attributed mainly to lack of acceptance of the new management techniques necessary for its successful implementation. An inherent fear of the unknown acted as a deterrent for both managers and executives.

Other than aerospace, defense, and construction, the majority of companies in the 1960s maintained an informal method for managing projects. In informal project management, just as the words imply, the projects were handled on an informal basis whereby the authority of the project manager was minimized. Most projects were handled by functional managers and stayed in one or two functional lines, and formal communications were either unnecessary or handled informally because of the good working relationships between line managers. Those individuals that were assigned as project managers soon found that they were functioning more as project leaders or project monitors than as real project managers. Many organizations today, such as low-technology manufacturing, have line managers who have been working side by side for 10 or more years. In such situations, informal project management may be effective on capital equipment or facility development projects and project management is not regarded as a profession.

By 1970 and through the early 1980s, more companies departed from informal project management and restructured to formalize the project management process, mainly because the size and complexity of their activities had grown to a point where they were unmanageable within the current structure.

Not all industries need project management, and executives must determine whether there is an actual need before making a commitment. Several industries with simple tasks, whether in a static or a dynamic environment, do not need project management. Manufacturing industries with slowly changing technology do not need project management, unless of course they have a requirement for several special projects, such as capital equipment activities, that could interrupt the normal flow of work in the routine manufacturing operations. The slow growth rate and acceptance of project management were related to the fact that the limitations of project management were readily apparent yet the advantages were not completely recognizable. Project management requires organizational restructuring. The question, of course, is "How much restructuring?" Executives have avoided the subject of project management for fear that "revolutionary" changes must be made in the organization.

Project management restructuring has permitted companies to:

- Accomplish tasks that could not be effectively handled by the traditional structure
- Accomplish one-time activities with minimum disruption of routine business

The second item implies that project management is a "temporary" management structure and, therefore, causes minimum organizational disruption. The major problems identified by those managers who endeavored to adapt to the new system all revolved around conflicts in authority and resources.

Another major concern was that project management required upper level managers to relinquish some of their authority through delegation to middle managers. In several situations, middle managers soon occupied the power positions, even more so than upper level managers.

Project management became a necessity for many companies as they expanded into multiple product lines, many of which were dissimilar, and organizational complexities grew. This growth can be attributed to:

- Technology increasing at an astounding rate
- More money invested in research and development (R&D)
- More information available
- Shortening of project life cycles

To satisfy the requirements imposed by these four factors, management was "forced" into organizational restructuring; the traditional organizational form that had survived for decades was inadequate for integrating activities across functional "empires."

By 1970, the environment began to change rapidly. Companies in aerospace, defense, and construction pioneered in implementing project management, and other industries soon followed, some with great reluctance. NASA and the DoD "forced" subcontractors into accepting project management.

Because current organizational structures are unable to accommodate the wide variety of interrelated tasks necessary for successful project completion, the need for project management has become apparent. It is usually first identified by those lower level and middle managers who find it impossible to control their resources effectively for the diverse activities within their line organization. Quite often middle managers feel the impact of changing environment more than upper level executives.

Once the need for change is identified, middle management must convince upper level management that such a change is actually warranted. If top-level executives cannot recognize the problems with resource control, then project management will not be adopted, at least formally. Informal acceptance, however, is another story.

As project management developed, some essential factors in its successful implementation were recognized. The major factor was the role of the project manager, which became the focal point for integrative responsibility. The need for integrative responsibility was first identified in complex R&D projects.

The R&D technology has broken down the boundaries that used to exist between industries. Once-stable markets and distribution channels are now in a state of flux. The industrial environment is turbulent and increasingly hard to predict. Many complex facts about markets, production methods, costs, and scientific potentials are related to investment decisions in R&D.

All of these factors have combined to produce a king-size managerial headache. There are just too many crucial decisions to have them all processed and resolved at the top of the organization through regular line hierarchy. They must be integrated in some other way.

Providing the project manager with integrative responsibility resulted in:

- Total project accountability assumed by a single person
- Project rather than functional dedication
- A requirement for coordination across functional interfaces
- Proper utilization of integrated planning and control

Without project management, these four elements have to be accomplished by executives, and it is questionable whether these activities should be part of an executive's job description. An executive in a Fortune 500 corporation stated that he was spending 70 hours each week working as both an executive and a project manager, and he did not feel that he was performing either job to the best of his abilities. During a presentation to the staff, the executive stated what he expected of the organization after project management implementation:

- Push decision making down in the organization.
- Eliminate the need for committee solutions.
- Trust the decisions of peers.

Those executives who chose to accept project management soon found the advantages of the new technique:

- Easy adaptation to an ever-changing environment
- Ability to handle a multidisciplinary activity within a specified period of time

- Horizontal as well as vertical work flow
- Better orientation toward customer problems
- Easier identification of activity responsibilities
- A multidisciplinary decision-making process
- Innovation in organizational design

As project management evolved, best practices became important. Best practices were learned from both successes and failures. In the early years of project management, private industry focused on learning best practices from successes. The government, however, focused on learning about best practices from failures. When the government finally focused on learning from successes, the knowledge of best practices came from their relationships with both their prime contractors and the subcontractors. Some of these best practices that came out of the government included:

- Use of life-cycle phases
- Standardization and consistency
- Use of templates [e.g., for statement of work (SOW), work breakdown structure (WBS), and risk management]
- Providing military personnel in project management positions with extended tours of duty at the same location
- Use of integrated project teams (IPTs)
- Control of contractor-generated scope changes
- Use of earned-value measurement

1.3 PROJECT MANAGEMENT BEST PRACTICES: 1985–2010

By the 1990s, companies had begun to realize that implementing project management was a necessity, not a choice. By 2010, project management had spread to virtually every industry and best practices were being captured. In the author's opinion, the appearance of best practices from an industry perspective might be:

- 1960–1985: Aerospace, defense, and construction
- 1986–1993: Automotive suppliers
- 1994–1999: Telecommunications
- 2000–2003: Information technology
- 2004–2006: Health care
- 2007–2008: Marketing and sales
- 2009–Present: Government agencies

The question now was not how to implement project management, but how fast could it be done? How quickly can we become mature in project management? Can we use the best practices to accelerate the implementation of project management?

Table 1–1 shows the typical life-cycle phases that an organization goes through to implement project management. In the first phase, the embryonic phase, the organization

TABLE 1–1.   FIVE PHASES OF PROJECT MANAGEMENT LIFE CYCLE

Embryonic	Executive Management Acceptance	Line Management Acceptance	Growth	Maturity
Recognize need	Get visible executive support	Get line management support	Recognize use of life-cycle phases	Develop a management cost/schedule control system
Recognize benefits	Achieve executive understanding of project management	Achieve line management commitment	Develop a project management methodology	Integrate cost and schedule control
Recognize applications	Establish project sponsorship at executive levels	Provide line management education	Make the commitment to planning	Develop an educational program to enhance project management skills
Recognize what must be done	Become willing to change way of doing business	Become willing to release employees for project management training	Minimize creeping scope Select a project tracking system	

recognizes the apparent need for project management. This recognition normally takes place at the lower and middle levels of management where the project activities actually take place. The executives are then informed of the need and assess the situation.

There are six driving forces that lead executives to recognize the need for project management:

- Capital projects
- Customer expectations
- Competitiveness
- Executive understanding
- New project development
- Efficiency and effectiveness

Manufacturing companies are driven to project management because of large capital projects or a multitude of simultaneous projects. Executives soon realize the impact on cash flow and that slippages in the schedule could end up idling workers.

Companies that sell products or services, including installation, to their clients must have good project management practices. These companies are usually non-project-driven but function as though they were project-driven. These companies now sell solutions to their customers rather than products. It is almost impossible to sell complete solutions to customers without having superior project management practices because what you are actually selling is your project management expertise.

There are two situations where competitiveness becomes the driving force: internal projects and external (outside customer) projects. Internally, companies get into trouble when they realize that much of the work can be outsourced for less than it would cost to perform the work themselves. Externally, companies get into trouble when they are no longer competitive on price or quality or simply cannot increase their market share.

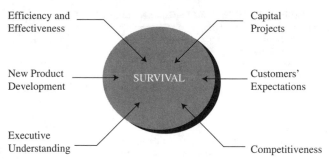

FIGURE 1–1. The components of survival. *Source:* Reprinted from H. Kerzner, *In Search of Excellence in Project Management,* New York: Wiley, 1998, p. 51.

Executive understanding is the driving force in those organizations that have a rigid traditional structure that performs routine, repetitive activities. These organizations are quite resistant to change unless driven by the executives. This driving force can exist in conjunction with any of the other driving forces.

New product development is the driving force for those organizations that are heavily invested in R&D activities. Given that only a small percentage of R&D projects ever make it into commercialization, where the R&D costs can be recovered, project management becomes a necessity. Project management can also be used as an early-warning system that a project should be canceled.

Efficiency and effectiveness, as driving forces, can exist in conjunction with any other driving forces. Efficiency and effectiveness take on paramount importance for small companies experiencing growing pains. Project management can be used to help such companies remain competitive during periods of growth and to assist in determining capacity constraints.

Because of the interrelatedness of these driving forces, some people contend that the only true driving force is survival. This is illustrated in Figure 1–1. When the company recognizes that survival of the firm is at stake, the implementation of project management becomes easier.

Enrique Sevilla Molina, PMP, Corporate PMO Director, discusses the driving forces at Indra that necessitated the need for excellence in project management:

> The internal forces were based on our own history and business experience. We soon found out that the better the project managers, the better the project results. This realization came together with the need to demonstrate in national and international contracts, with both US and European customers, our real capabilities to handle big projects. These big projects required world class project management, and for us managing the project was a greater challenge than just being able to technically execute the project. Summarizing, these big projects set the pace to define precise procedures on how handling stakeholders, big subcontractors and becoming a reliable main point of contact for all issues related with the project.

Sandra Kumorowski, Marketing and Operations Consultant, discusses the driving forces at Enakta:

The company was a project-based company and it made sense to turn to project management as a tool for continuous improvement. The main issues that also drove the company to use project management were reoccurring time/cost/quality management issues, team productivity issues and client satisfaction issues. The table shown below [Table 1–2] illustrates the necessity:

The speed by which companies reach some degree of maturity in project management is most often based upon how important they perceive the driving forces to be. This is illustrated generically in Figure 1–2. Non-project-driven and hybrid organizations move quickly to maturity if increased internal efficiencies and effectiveness are needed. Competitiveness is the slowest path because these types of organizations do not recognize that project management affects their competitive position directly. For project-driven organizations, the path is reversed. Competitiveness is the name of the game and the vehicle used is project management.

TABLE 1–2.   THE NECESSITY FOR PROJECT MANAGEMENT

Why?	Benefits
We are a project-based company	1. We know how to deliver projects 2. PM is a <u>tool for successful delivery of actionable insights</u>
To build credibility, grow, and compete: We should be perceived as a systematic and organized organization. We have to prevent mistakes	1. Earned reputation for systematic work 2. Focused business strategy 3. PM could be one of our <u>competitive advantages</u>
To control cost, time, resources: We should establish effective project control system	1. Decreased uncertainty 2. Increased product quality 3. Happy people 4. More effective planning (project & company level)
To learn as an organization & individuals:	1. PM concepts as part of our <u>continuous education</u> 2. Learning organization is always ahead of time and its competitors

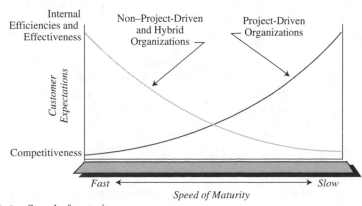

FIGURE 1–2.   Speed of maturity.

Once the organization perceives the need for project management, it enters the second life-cycle phase of Table 1–1, executive acceptance. Project management cannot be implemented rapidly in the near term without executive support. Furthermore, the support must be visible to all.

The third life-cycle phase is line management acceptance. It is highly unlikely that any line manager would actively support the implementation of project management without first recognizing the same support coming from above. Even minimal line management support will still cause project management to struggle.

The fourth life-cycle phase is the growth phase, where the organization becomes committed to the development of the corporate tools for project management. This includes the processes and project management methodology for planning, scheduling, and controlling as well as selection of the appropriate supporting software. Portions of this phase can begin during earlier phases.

The fifth life-cycle phase is maturity. In this phase, the organization begins using the tools developed in the previous phase. Here, the organization must be totally dedicated to project management. The organization must develop a reasonable project management curriculum to provide the appropriate training and education in support of the tools as well as the expected organizational behavior.

By the 1990s, companies finally began to recognize the benefits of project management. Table 1–3 shows the benefits of project management and how our view of project

TABLE 1–3. CRITICAL FACTORS IN PROJECT MANAGEMENT LIFE CYCLE

Critical Success Factors	Critical Failure Factors
Executive Management Acceptance Phase	
Consider employee recommendations	Refuse to consider ideas of associates
Recognize that change is necessary	Unwilling to admit that change may be necessary
Understand the executive role in project management	Believe that project management control belongs at executive levels
Line Management Acceptance Phase	
Willing to place company interest before personal interest	Reluctant to share information
Willing to accept accountability	Refuse to accept accountability
Willing to see associates advance	Not willing to see associates advance
Growth Phase	
Recognize the need for a corporate-wide methodology	View a standard methodology as a threat rather than as a benefit
Support uniform status monitoring/reporting	Fail to understand the benefits of project management
Recognize the importance of effective planning	Provide only lip service to planning
Maturity Phase	
Recognize that cost and schedule are inseparable	Believe that project status can be determined from schedule alone
Track actual costs	See no need to track actual costs
Develop project management training	Believe that growth and success in project management are the same

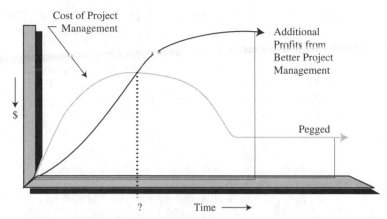

FIGURE 1–3. Project management costs versus benefits.

management has changed. Many of these benefits were identified through the discovery and implementation of best practices.

Recognizing that the organization can benefit from the implementation of project management is just the starting point. The question now becomes, "How long will it take us to achieve these benefits?" This can be partially answered from Figure 1–3. In the beginning of the implementation process, there will be added expenses to develop the project management methodology and establish the support systems for planning, scheduling, and control. Eventually, the cost will level off and become pegged. The question mark in Figure 1–3 is the point at which the benefits equal the cost of implementation. This point can be pushed to the left through training and education.

1.4 AN EXECUTIVE'S VIEW OF PROJECT MANAGEMENT

Today's executives have a much better understanding and appreciation for project management than did their predecessors. Early on, project management was seen as simply scheduling a project and then managing the project using network-based software. Today, this parochial view has changed significantly. It is now a necessity for survival.

Although there are several drivers for this, three significant reasons seem to stand out. First, as businesses downsize because of poor economic conditions or stiffening competition, the employees remaining in the company are expected to do more with less. Executives expect the employees to become more efficient and more effective when carrying out their duties. Second, business growth today requires the acceptance of significant risks, specifically in the development of new products and services for which there may not be reasonable estimating techniques or standards. Simply stated, we are undertaking more jobs that are neither routine nor predictable. Third, and perhaps most important, is that we believe we are managing our business as though it is a series of projects. Projects now make up a significant part of one's jobs. As such, all employees are actually project managers to some degree and are expected to make business as well as project decisions.

The new breed of executive seems to have a much broader view of the value of project management, ranging from the benefits of project management, to the selection criteria for project managers, to organizational structures that can make companies more effective. This is apparent from the four comments below, which were provided by Tom Lucas, Chief Information Officer for the Sherwin-Williams Company:

- We have all managed projects at one time or another, but few of us are capable of being Project Managers.
- The difference between managing projects and professional project management is like the difference between getting across the lake in a rowboat versus a racing boat. Both will get you across the lake but the rowboat is a long and painful process. But how do people know until you give them a ride?
- Don't be misguided into thinking professional project management is about process. It is about delivering business results.
- If you don't appreciate that implementing a PMO is a cultural transition, you are destined to fail.

The comments below from other executives clearly indicate their understanding and appreciation of project management:

Managing projects successfully is the only way we can be effective for our customers. As our government contracts must stand up to high levels of scrutiny in tight budgetary times and limited resources, it is critical that SENTEL's project manager is empowered and accountable for reporting their status, in good times and bad. SENTEL continues to invest in the development of . . . sustainable and flexible project management methodologies that we support from . . . all parts of our company and we share best practices so that all groups can benefit from the successes of others. (Darrell Crapps, Chief Executive Officer [CEO], SENTEL)

Our Customers, which are multinational industrial groups, expect from Comau Project Managers an international, multicultural and global approach. In the meantime our Shareholder is asking us for high projects governance obtained though a global Project Management effective framework. In 2006 we have adopted a world-class Project Management approach (i.e. PMI) which, together with the implementation of the best practices on the global Comau footprint, allowed us to demonstrate that both Customers and Shareholder goals can be fulfilled. I am sure that we are on the right tracks and that this continuous improvement strategy has to be pursued in the next years with motivation and perseverance. (Riccardo Tarantini, COMAU CEO, Fiat Group)

Program and Project Management expertise is core to our success. It is with these professions as a core competency that we can execute in delivering value to our customers. Whether it is deploying internal IT projects, new product development, cost saving projects or integration services, these are more successful when they are staffed with experienced Program and Project Management personnel. We have invested over the years and we will continue to invest to ensure that we maintain this expertise throughout Rockwell Automation. (Keith Nosbusch, CEO, Rockwell Automation)

Over the past 15 years, ongoing transformation has become a defining characteristic of IBM—and a key factor in our success. Effective change in process and IT transformation doesn't just happen, it must be enabled by highly skilled Project Managers. Our Project Managers analyze processes, enabled by IT, in a way that allows us to innovate and eliminate unnecessary steps, simplify and automate. They help us become more efficient and effective by pulling together the right resources to get things done—on time and on budget. They are invaluable as we continue to make progress in our transformation journey. (Linda S. Sanford, Senior Vice President, Enterprise on Demand Transformation and Information Technology, IBM Corporation)

Project managers are a critical element of our end-to-end development and business execution model. Our goal is to have sound project management practices in place to provide better predictability in support of our products and offerings. As a team, you help us see challenges before they become gating issues and ensure we meet our commitments to STG and clients. . . . We continue to focus on project management as a career path for high-potential employees and we strongly encourage our project managers to become certified, not only PMI, but ultimately IBM certified. . . . End-to-end project management must become ingrained in the fabric of our business. (Rod Adkins, Senior Vice President of Development & Manufacturing, IBM's System and Technology Group [STG])

At leading IT software services providers, project management has evolved and matured from a complex process of identifying and meeting a customer's unique requirements to applying a core set of proven, second-generation best practices captured and packaged in standard offerings. The standard offerings deliver repeatable success and accelerated time to value for the customer. They also give the customer and the project manager the ability to take a phased approach to building the customer's comprehensive IT management vision. (Dave Yusuf, SVP Global PMO, Computer Associates Services)

Project Management is a core process at Johnson Controls. The Project Manager is ultimately responsible for the execution, profitability and the quality of our new product launches. We expect high performance from our Project Managers, and we are careful to select the best people to fill these roles. I believe that companies need a profound understanding of this critical discipline to be successful in the century to come. With a steady stream of high quality, profitable projects and products, we can be assured to maintain our competitive advantage. (Dr. Beda Bolzenius, Vice President and President, Automotive Experience, Johnson Controls Inc.)

Successful project management is mission critical to us from two points of view:

- First, as we define and implement PLM (Product Lifecycle Management) solutions, we help customers to streamline their entire product lifecycle across all functional units. This can make any large PLM project an intricate and even complex undertaking. To live up to our company mantra of "we never let a customer fail", robust and reliable project management is often the most critical component we provide aside from the PLM platform itself; the combination of the two enables our customers to achieve the business benefits they strive for by investing in PLM.
- Second, Siemens itself is one of our largest customers. This is a great opportunity and, at the same time, a great challenge. Keeping a project's objectives and scope under control with our "internal" customer is at least as challenging as with external customers; yet it is critical in order to keep our development roadmaps and deployment schedules on track. Our job is to continue to successfully develop and deploy the first and only true

end-to-end industry software platform. This comprehensive platform covers the entire product lifecycle from initial requirements, through product development, manufacturing planning, controlling the shop floor and even managing the maintenance, repair and overhaul of the product in question. As a result, effective project management is vital to our success. (Dr. Helmuth Ludwig, President, Siemens PLM Software)

Project Management is vital to the success of any organization. Whether projects are focused on customer acquisition, loyalty and insight or driven by the need to increase enterprise efficiency, excellence in Project Management ensures that tangible, meaningful results are achieved on time and on budget. (Brad Jackson, CEO, Slalom Consulting)

Projects and Project Management play a vital role in our business of IT Services. While being a key enabler for delighting customers, Project Management also helps in setting the right expectations of stakeholders and more importantly, maintaining a balance between their expectations. Effective Project Management becomes a strong competitive advantage or differentiator for our delivery capabilities. Excelling in Project Management has allowed us not only to increase the quality of our services, reduce our time-to-market, decrease rework costs and increase staff motivation, but also to create a more integrated and agile organization. (A. S. Murthy, CEO, Satyam Computer Services Limited)

Solid project management is the glue that binds a successful implementation together. Each of our project managers is [a] knowledgeable technical experts, but more often than not, our customers need a soft skilled project manager to take control of an unorganized project and turn it around. The challenge is to get everyone motivated and moving in sync like an orchestra, by respecting the complicated and subtle aspects of human dynamics. Turning a project around is no easy feat, but given the correct methodology and executive support, it can be done. (Bruce Cerullo, CEO, Vitalize Consulting Solutions, Inc.)

I believe that operational excellence is achieved when we have the right people, processes, and technology deployed every day in the most efficient and effective way to achieve client satisfaction. One of my top priorities as COO is to establish a single, common culture across the organization—a culture based on excellence. By raising the maturity of our project management people, processes, and tools, and by increasing our focus on adoption of new practices, we have improved consistency in project delivery across Perot Systems and made significant progress toward producing predictable, repeatable, high-quality results for our clients. (Russell Freeman, Chief Operating Officer [COO], Perot Systems Corporation)

In this age of instant communications and rapidly evolving networks, Nortel continues to maximize use of its project management discipline to ensure the successful deployment of increasingly complex projects. We foster an environment that maintains a focus on sharing best practices and leveraging lessons learned across the organization, largely driven by our project managers. We are also striving to further integrate project management capabilities with supply chain management through the introduction of SAP business management software. Project management remains an integral part of Nortel's business and strategy as it moves forward in a more services- and solutions-oriented environment. (Sue Spradley, previously President, Global Operations, Nortel Networks)[1]

---

1. H. Kerzner, *Best Practices in Project Management: Achieving Global Excellence*, Wiley, Hoboken, NJ, 2006, p. 17

The PMO process has been essential to the success of several major IS projects within Our Lady of Lourdes Regional Medical Center. This was especially true of our recent conversion from MedCath IS support to Franciscan Missionaries of our Lady Health System (FMOLHS) IS support at our newest physician joint venture: The Heart Hospital of Lafayette. PMO built trust through transparency, accountability and a framework for real-time project assessment. Without this structure I seriously doubt we could have succeeded in bringing the conversion on time and under budget. (W. F. "Bud" Barrow, President and CEO, Our Lady of Lourdes Regional Medical Center)

Through project management, we've learned how to make fact-based decisions. Too often in the past we based our decisions on what we thought could happen or what we hoped would happen. Now we can look at the facts, interpret the facts honestly and make sound decisions and set realistic goals based on this information. (Zev Weiss, CEO, American Greetings)[2]

The program management office provides the structure and discipline to complete the work that needs to get done. From launch to completion, each project has a roadmap for meeting the objectives that were set. (Jeff Weiss, President and CEO, American Greetings)[3]

Through project management, we learned the value of defining specific projects and empowering teams to make them happen. We've embraced the program management philosophy and now we can use it again and again to reach our goals. (Jim Spira, Retired President and CEO, American Greetings)[4]

In the services industry, how we deliver (i.e. the project management methodology) is as important as what we deliver (i.e. the deliverable). Customers expect to maximize their return on IT investments from our collective knowledge and experience when we deliver best-in-class solutions. The collective knowledge and experience of HP (Hewlett-Packard) Services is easily accessible in HP Global Method. This integrated set of methodologies is a first step in enabling HPS to optimize our efficiency in delivering value to our customers. The next step is to know what is available and learn how and when to apply it when delivering to your customers. HP Global Method is the first step toward a set of best-in-class methodologies to increase the credibility as a trusted partner, reflecting the collective knowledge and expertise of HP Services. This also improves our cost structures by customizing pre-defined proven approaches, using existing checklists to ensure all the bases are covered and share experiences and learning to improve Global Method. (Mike Rigodanzo, formerly Senior Vice President, HP Services Operations and Information Technology)[5]

In 1996, we began looking at our business from the viewpoint of its core processes. . . . As you might expect, project management made the short list as one of the vital, core processes to which quality principles needed to be applied. (Martin O'Sullivan, retired Vice President, Motorola)[6]

The disciplines of project management constitute an essential foundation for all initiatives toward business or indeed human advancement. I can't conceive crossing the vision/reality chasm without them. (Keith Thomas, Chairman, ITC [Information Technology & Communications] Business Unit, Neal & Massy Holdings, Ltd.)

---

2. H. Kerzner, *Advanced Project Management: Best Practices on Implementation*, Wiley, Hoboken, NJ, 2004, p.273.
3. Ibid.
4. Ibid.
5. Ibid., p. 67.
6. Ibid., p. 184.

The comments by Keith Thomas clearly indicate that today's executives recognize that project management is a strategic or core competency needed for survival because it interfaces with perhaps all other business processes, including quality initiatives. The following comments by John Walsh, President of Sypris Electronics, are indicative of executives that recognize the broad applications of project management, especially the integration of project management with other processes. Also from John Walsh's comments, you can see that today's executives are taking the lead role in spearheading these initiatives rather than delegating them to subordinates.

- Proper project management is the cornerstone to any successful company. Breaking the paradigm that project management is for customer initiatives is essential. Everything that we do at Sypris Electronics can be classified as a project, whether it is internally or externally focused. Furthermore, the success of our internal policy deployment of hoshin kanri hinges upon proper project management fundamentals.[7]
- Project management is essentially the training ground for presidents and CEO's for tomorrow due to the breadth and depth of experiences that are dealt with. Early on in my career I was fortunate enough to work with some of the best names in project management, including Dr. Harold Kerzner. This early mentoring not only helped from a tactical execution standpoint, but it also stressed the importance of proper project management and the benefits of the right implementation.
- In 1999 I was faced with the opportunity of turning around an automotive occupant safety components business. Challenged with having to implement Lean Manufacturing principles and Six Sigma as the methodology to transform operations, I fell back to my project management training and skills. The end result was co-chairing that year's Six Sigma conference as a key note speaker to describe how essential project management is in deploying Six Sigma and turning around operations.
- In these tough economic times, there is even more scrutiny on project success. So, with less tolerance for project failure it is imperative that the right people, practices, and infrastructure are in place to manage the project for value maximization.

1.5 BEST PRACTICES PROCESS

"Why capture best practices?" The reasons or objectives for capturing best practices might include:

- Continuous improvements (efficiencies, accuracy of estimates, waste reduction, etc.)
- Enhanced reputation

---

7. *Hoshin kanri* can mean many things to an organization. It can be used as a method of strategic planning and a tool for managing complex projects, a quality operating system geared to ensuring that the organization faithfully translates the voice of the customer into new products, or a business operating system that ensures reliable profit growth. It is also a method for cross-functional management and for integrating the lean supply chain. But, most of all, it is an organizational learning method and competitive resource development system. For additional information, see T. L. Jackson, *Hoshin Kanri for the Lean Enterprise*, Productivity Press, New York, 2006.

● Winning new business
● Survival of the firm

Survival of the firm has become the most important reason today for capturing best practices. In the last few years, customers have put pressure on contractors in requests for proposals (RFPs) by requesting:

● A listing of the number of PMPs in the company and how many will be assigned to this project
● A demonstration that the contractor has an enterprise project management methodology that is acceptable to the customer or else the contractor must use some other methodology approved by the customer
● Supporting documentation identifying the contractor's maturity level in project management, possibly using a project management maturity model for assessments
● A willingness to share lessons learned and best practices discovered on this project and perhaps previous projects for other customers

Recognizing the need for capturing best practices is a lot easier than actually doing it. Companies are developing processes for identifying, evaluating, storing, and disseminating information on best practices. There are nine best practices activities as shown in Figure 1–4, and most companies that recognize the value of capturing best practices accomplish all of these steps.

The processes answer the following nine questions:

● What is the definition of a best practice?
● Who is responsible for identifying the best practice and where do we look?
● How do we validate that something is a best practice?
● Are there levels or categories of best practices?

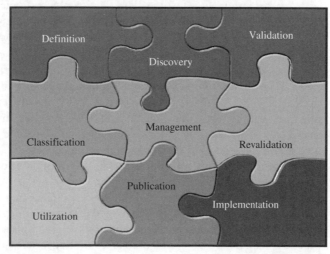

FIGURE 1–4. Best practices processes.

- Who is responsible for the administration of the best practice once approved?
- How often do we re-evaluate that something is still a best practice?
- How do companies use best practices once they are validated?
- How do large companies make sure that everyone knows about the existence of the best practices?
- How do we make sure that the employees are using the best practices and using them properly?

Each of these questions will be addressed in the next several sections.

1.6 STEP 1: DEFINITION OF A BEST PRACTICE

For more than a decade, companies have become fascinated with the expression "best practices." But now, after a decade or more of use, we are beginning to scrutinize the term and perhaps better expressions exist.

A best practice begins with an idea that there is a technique, process, method, or activity that can be more effective at delivering an outcome than any other approach and provides us with the desired outcome with fewer problems and unforeseen complications. As a result, we supposedly end up with the most efficient and effective way of accomplishing a task based upon a repeatable process that has been proven over time for a large number of people and/or projects.

But once this idea has been proven to be effective, we normally integrate the best practice into our processes so that it becomes a standard way of doing business. Therefore, after acceptance and proven use of the idea, the better expression possibly should be a "proven practice" rather than a best practice. This is just one argument why a best practice may be just a buzzword and should be replaced by proven practice.

Another argument is that the identification of a best practice may lead some to believe that we were performing some activities incorrectly in the past, and that may not have been the case. This may simply be a more efficient and effective way of achieving a deliverable. Another issue is that some people believe that best practices imply that there is one and only one way of accomplishing a task. This also may be a faulty interpretation.

Perhaps in the future the expression best practices will be replaced by proven practices. However, for the remainder of this text, we will refer to the expression as best practices, but the reader must understand that other terms may be more appropriate. This interpretation is necessary in this book because most of the companies that have contributed to this book still use the expression best practices.

As project management evolved, so did the definitions of a best practice. Some definitions of a best practice are highly complex while others are relatively simplistic. Yet, they both achieve the same purpose of promoting excellence in project management throughout the company. Companies must decide on the amount of depth to go into the best practice. Should it be generic and at a high level or detailed and at a low level? High-level best practices may not achieve the efficiencies desired whereas highly detailed best practices may have limited applicability.

Every company can have its own definition of a best practice and there might even be industry standards on the definition of a best practice. Typical definitions of a best practice might be:

- Something that works
- Something that works well
- Something that works well on a repetitive basis
- Something that leads to a competitive advantage
- Something that can be identified in a proposal to generate business
- Something that keeps the company out of trouble and, if trouble occurs, the best practice will assist in getting the company out of trouble

Every company has its own definition of a best practice. There appear to be four primary reasons for capturing best practices:

- Improve efficiency
- Improve effectiveness
- Standardization
- Consistency

In each of the following definitions, you should be able to identify which of the four, or combination thereof, the company targets:

- At Orange Switzerland, a best practice is defined as an experience based, proven, and published way of proceeding to achieve an objective.[8]
- We do have best practices that are detailed in our policies/procedures and work-flows. These are guidelines and templates as well as processes that we all (members of the EPMO—enterprise project management office) have agreed to abide by as well as that they are effective and efficient methods for all parties involved. In addition, when we wrap up (conclude) a project, we conduct a formal lessons learned session (involving the project manager, sponsors, core team, and other parties impacted by the project) which is stored in a collective database and reviewed with the entire team. These lessons learned are in effect what create our best practices. We share these with other health care organizations for those vendors for which we are reference sites. All of our templates, policies/procedures, and work-flows are accessible by request and, when necessary, we set meetings to review as well as explain them in detail.[9]
- Any tool, template or activity used by a project manager that has had a positive impact on the *PMBOK® Guide* Knowledge and/or process areas and/or the triple constraint. An example of a best practice would be: Performing customer satisfaction

8. H. Kerzner, *Best Practices in Project Management: Achieving Global Excellence*, Wiley, Hoboken, NJ, 2006, p.12.
9. Ibid.., p.13.

assessments during each phase of a project allows adjustments during the project life cycle, which improves deliverables to the client, and improves overall project management. [This would be accompanied by a template for a customer satisfaction survey.] (Spokesperson for AT&T)

- Generally we view a best practice as any activity or process that improves a given situation, eliminates the need of other more cumbersome methods, or significantly enhances an existing process. Each best practice is a living entity and subject to review, amendments, or removal.[10]
- For Churchill Downs Incorporated, a best practice is any method or process that has been proven to produce the desired results through practical application. We do not accept "industry" or "professional standards" as best practices until we have validated that the method or process works in our corporate environment.

Examples of some of our best practices include:

- *Charter Signatures:* One of our best practices is requiring stakeholder signatures on project and program charters. This seems basic, but my experience is that a formal review and approval of a project's business objectives and goals is rarely documented. By documenting business objectives and their associated metrics, we have been able to proactively manage expectations and ensure alignment between various stakeholders.
- *Process Definition:* In addition to defining the organization's project, program and portfolio management processes, the PMO has also taken an active role in mapping all of the financial processes for Churchill Downs Incorporated, from check requests and employee reimbursement requests to procedures for requesting capital expenses and purchase orders. This practice has increased corporate-wide awareness of how standardizing processes can enhance efficiency.
- *Access to Information:* The PMO developed process maps, procedures and policies for the end-to-end budgeting processes, associated workflows and templates. These have been made available company-wide via CCN, the company's intranet site.[11]

- At Indra, we consider a "best practice" in project management as a management action or activity that usually generates a positive outcome. As such, it is accepted by the management community and eventually becomes a recommended or required way of performing the task. We also consider as a "best practice", the use of predefined indicators, thresholds or metrics to make or facilitate decisions with regard to project management processes.[12]
- In the PMO, a best practice is a process, methodology or procedure that is followed in order to ensure a consistent approach and standard is utilized. Best practices within VCS are evaluated for efficiency (internally and externally)

10. Ibid.
11. Comments by Chuck Millhollan, Director of Program Management, Churchill Downs Inc.
12. Comments by Enrique Sevilla Molina, PMP, Corporate PMO Director, Indra.

and updated to reflect the lessons learned and the practical, real world experience from our project management consulting base working in the field with our customers.[13]

● Sandra Kumorowski believes that a best practice is . . . either a method, tactic, or process that has been proven through implementation and tested use to add specific and measurable benefits and long-term value in terms of increased project performance outcome like decreased project cost, increased employee productivity, improved client experience (rate of retention), and an increased number of new projects. Best practices add both short-term and long-term value to the organization.

I was responsible for providing Best Practices updates to all employees. Right after each Post Mortem session and after an idea was confirmed as a best practice, I would send an e-mail update to all employees. Typical best practices included:

● *Workload Distribution and Team Leadership:* Within a team will be equalized among the project lead (senior consultant) and junior consultant who would usually do the hard work of analysis without participating in the composition of the actual strategy. [Many times, junior people felt left out and did not feel they had contributed to the success of the project. They often felt inferior and that became a big issue that greatly decreased team productivity. The issue was part of the senior strategist job description (they were responsible for 80% of strategic thinking on a project) that could have been interpreted differently by different people.]

● *Project Kick-Off Meeting:* All project stakeholders must be in that meeting to discuss project scope and objectives. A specific scope statement format must be followed to clearly define objectives. During the meeting, all milestones dates must be determined and agreed on by all stakeholders.

● *Project Milestones Meetings:* Must be scheduled (some using GoToMeeting software) and included on all stakeholders' calendars right after the Kick-Off Meeting.

● *Project Post Mortem Meeting:* All team members must evaluate the project performance, fill in the Post Mortem questionnaire and discuss it with other team members. Post Mortem meetings must be scheduled no later than one week after the final client presentation to ensure the project issues and/or successes stay fresh in everyone's mind.[14]

Best practices can be implemented on either a formal or an informal basis. For example, Doug Bolzman, Consultant Architect, PMP®, ITIL Service Manager at EDS,

---

13. Comments by Marc Hirshfield, PMP, Director, Project Management Office, Vitalize Consulting Solutions, Inc. Vitalize Consulting Solutions, Inc. (VCS) is a health care IT consulting company in the hospital and ambulatory (physician practice) marketspace. The PMO at VCS serves our project management consultants who work on behalf of our customer base in support of implementing their health care IT initiatives. Maybe best stated, we act and service in the role of a virtual PMO to our consultants and customer base.
14. Comments by Sandra Kumorowski, Marketing and Operations Consultant, Enakta.

discusses the approach at EDS for defining a best practice and how the final decision is made[15]:

Within EDS, specific work activities are identified, promoted and leveraged as best practices. With thousands of clients, many innovative methods are discovered and utilized to support specific client needs. These innovations that can be leveraged are identified and promoted as part of an existing best practice or as a new best practice. By leveraging these efforts, all innovations can benefit all clients, thus improving the maturity of service provision by the organization.

A best practice is a work package comprised of a process, tool (or templates), and people that when used together can enable a project team to produce a consistent and stable deliverable for a client with increased accuracy and efficiency. Best practices are project management–based work types that at the start of a project can be identified during the planning stage and leveraged by the project team.

In order to manage each best practice consistently, each practice is documented following the best practice profile template. Some of the information contained within the profile includes the description of the practice, the type, the value to the company, and a list of practitioners to use the practice. Each practice documents all of the assets and asset status, and finally all of the practices document the business drivers that have been used to develop the practice.

EDS established a Client Facing Best Practice Design Board (comprised of best practice owners) to define and oversee the framework used to manage best practices. EDS has registered and has received a patent for the process and tools used to manage its best practices.

Once a practice has been nominated and approved to be a best practice, it is only sanctioned until the next yearly review cycle. Over time, best practices have the tendency to lose value and become ineffective if they are allowed to age. To allow for continuous improvement, a new level of maturity is assigned to all of the sanctioned best practices for a yearly resanctioning. By continually moving the bar higher, it is necessary to prove the value of each best practice and demonstrate improvements from the previous version of the process, tools, and people designs.

Being part of the integrated set of client facing best practices involves an assessment of a "candidate" best practice against a standard set of maturity criteria. Three levels of best practice maturity have been defined with minimum requirements for an "associate" best practice, "best practice," and "mastery" best practice. The requirements are centered on the following sections:

- Value
- Client
- User community
- Training
- Assets

15. Doug Bolzman has been with EDS more than 20 years and is currently a member of the EDS Project Management Delivery-IT Enterprise Management/ITIL capability. EDS has been awarded a patent on behalf of Doug's processes titled, "System and Method for Identifying and Monitoring Best Practices of an Enterprise."

TABLE 1–4.   BEST PRACTICES INCENTIVES

Private Sector	Public Sector
Profit	Minimization of cost
Competitiveness	On-time delivery
Efficiency	Efficiency
Effectiveness	Effectiveness
Customer satisfaction	Stakeholder satisfaction
Partnerships	Sole-source procurement

- Governance
- Release management
- Integration with internal structures
- Integration with external structures, if applicable

Everyone seems to have his or her own definition of a best practice. In the author's opinion, for simplicity sake, *best practices are those actions or activities undertaken by the company or individuals that lead to a sustained competitive advantage in project management while providing value for the company, the client, and the stakeholders.* What is important in this definition is the term "sustained competitive advantage." In other words, best practices are what differentiate you from your competitors.

Another important word in this definition is "value." There is no point in identifying a best practice unless people, specifically the users of the best practice, can see the value in using it. This is critical in getting acceptance in the use of a best practice.

This definition of a best practice discussed above focuses more on the private sector than on the public sector. A comparison of possible incentives for discovery and implementation of best practices in the public and private sectors is shown in Table 1–4.

1.7 STEP 2: SEEKING OUT BEST PRACTICES

Best practices can be captured either within your organization or external to your organization. Benchmarking is one way to capture external best practices possibly by using the project management office as the lead for external benchmarking activities. However, there are external sources other than benchmarking for identifying best practices:

- Project Management Institute (PMI) publications
- Forms, guidelines, templates, and checklists that can impact the execution of the project
- Forms, guidelines, templates, and checklists that can impact our definition of success on a project
- Each of the *PMBOK® Guide* areas of knowledge or domain areas
- Within companywide or isolated business units
- Seminars and symposiums on general project management concepts

- Seminars and symposiums specializing on project management best practices
- Relationships with other professional societies
- Graduate-level theses

With more universities offering masters- and doctorate-level work in project management, graduate-level theses can provide up-to-date research on best practices.

The problem with external benchmarking is that best practices discovered in one company may not be transferable to another company. In the author's opinion, most of the best practices are discovered internally and are specifically related to the company's use of its project management methodology and processes. Good project management methodologies allow for the identification and extraction of best practices. However, good ideas can come from benchmarking as well.

Sometimes, the identification of the drivers or metrics that affect each best practice is more readily apparent than the best practice itself. Metrics and drivers can be treated as early indicators that a best practice may have been found. It is possible to have several drivers for each best practice. It is also possible to establish a universal set of drivers for each best practice, such as:

- Reduction in risk by a certain percentage, cost, or time
- Improve estimating accuracy by a certain percentage or dollar value
- Cost savings of a certain percentage or dollar value
- Efficiency increase by a certain percentage
- Reduction in waste, paperwork, or time by a certain percentage

There are several advantages of this approach for searching for drivers. First, the drivers can change over time and new drivers can emerge rapidly. Second, the best practices process is more of a science than an art. And third, we can establish levels of best practices such as shown in Figure 1–5. In this figure, a level 4 best practice, which is the best, would satisfy 60 percent or more of the list of drivers or characteristics of the ideal best practice.

Best practices may not be transferable from company to company, nor will they always be transferable from division to division within the same company. As an example, consider the following best practice discovered by a telecommunications company:

- A company institutionalized a set of values that professed that quality was everything. The result was that employees were focusing so much on quality that there was a degradation of customer satisfaction. The company then reprioritized its values with customer satisfaction being the most important, and quality actually improved.

In this company, customer satisfaction emphasis led to improved quality. However, in another company, emphasis on quality could just as easily have led to an improvement in customer satisfaction. Care must be taken during benchmarking activities to make sure that whatever best practices are discovered are in fact directly applicable to your company.

Best practices need not be overly complex. As an example, the following list of best practices is taken from companies discussed in this textbook, and as you can see, some of the best practices were learned from failures rather than successes:

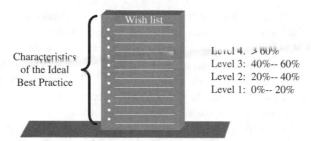

FIGURE 1–5. Best practices levels. Each level contains a percentage of the ideal characteristics.

- Changing project managers in midstream is bad even if the project is in trouble. Changing project managers inevitably elongates the project and can make it worse.
- Standardization yields excellent results. The more standardization placed in a project management methodology, usually the better are the results.
- Maximization of benefits occurs with a methodology based upon templates, forms, guidelines, and checklists rather than policies and procedures.
- Methodologies must be updated to include the results of discovering best practices. The more frequently the methodology is updated, the quicker the benefits are realized.

As stated previously, best practices need not be complex. Even though some best practices seem simplistic and common sense, the constant reminder and use of these best practices lead to excellence and customer satisfaction. As an example, Antares Management Solutions developed a small handout listing nine best practices. This handout was given to all employees, inserted into proposals during competitive bidding, and also shown to customers. The nine best practices in its brochure are:

- Making use of project management concepts and terminology throughout the enterprise, not just in information systems
- Providing ongoing project management training throughout the enterprise
- Structuring every major project in a consistent manner, including scope, responsibilities, risks, high-level milestones, and project planning
- Communicating and updating project plans on an ongoing basis using online tools when appropriate
- Maintaining an official project issue list, including who is responsible, what are the potential impacts, and how the issues are being resolved
- Using a formal change control process, including an executive steering committee to resolve major change issues
- Concluding every major project with an open presentation of results to share knowledge gained, demonstrate new technology, and gain official closure
- Periodically auditing and/or benchmarking the project management process and selected projects to determine how well the methodology is working and to identify opportunities for improvement
- Adapt to meet the business needs of the client

There is nothing proprietary or classified about these nine best practices. But at Antares Management Solutions, these best practices serve as constant reminders that project management is seen as a strategic competency and continuous improvement is expected.

Another way to identify sources of best practices is from the definition of project success, critical success factors (CSFs), and key performance indicators (KPIs). Extracting best practices from the definition of success on a project may be difficult and misleading, especially if we have a poor definition of success.

Over the years, a lot of the changes that have taken place in project management have been the result of the way we define project success. As an example, consider the following chronological events that took place over the past several decades:

- *Success is measured by the triple constraint:* The triple constraint is time, cost, and performance (which include quality, scope, and technical performance). This was the basis for defining success during the birth of project management.
- *Customer satisfaction must be considered as well:* Managing a project within the triple constraint is always a good idea, but the customer must be satisfied with the end result. A contractor can complete a project within the triple constraint and still find that the customer is unhappy with the end result.
- *Other (or secondary) factors must be considered as well:* These include using the customer's name as a reference, corporate reputation and image, compliance with government regulations, strategic alignment, technical superiority, ethical conduct, and other such factors. The secondary factors may end up being more important than the primary factors of the triple constraint.
- *Success must include a business component:* Project managers are managing part of a business rather than merely a project and are expected to make sound business decisions as well as project decisions. There must be a business purpose for each project. Each project is considered as a contribution of business value to the company when completed.
- *Prioritization of constraints must occur:* Not all project constraints are equal. The prioritization of constraints is on a project-by-project basis. Sponsorship involvement in this decision is essential.
- *The definition of success must be agreed upon between the customer and the contractor.* Each project can have a different definition of success. There must be up-front agreement between the customer and the contractor at project initiation or even at the first meeting between them on what constitutes success.
- *The definition of success must include a "value" component:* Why work on a project that does not provide the correct expected value at completion?

The problem with defining success as on time, within cost, and at the desired quality or performance level is that this is an internal definition of success only. Bad things can happen on projects when the contractor, customer, and various stakeholders are all focusing on different definitions of project success. There must be an upfront agreement on what constitutes project success. The ultimate customer or stakeholder should have some say in the definition of success, and ultimately there may be numerous best practices discovered that relate to customer/stakeholder interfacing.

TABLE 1–5. COMPARING PROJECT SUCCESS VERSUS FAILURE

MEASURE OF PM PROJECT SUCCESS

Successful =	ORGANIZATIONAL LEVEL

❑ More business from the client
❑ Clients contacting us
❑ Client satisfaction during/after the project
❑ Project team happy
	● Ratio of successful to unsuccessful projects per year
	● Increased number of successful projects per period (ROI)
	● Everybody is on board accepting changes, no resistance
	● Effective project portfolio management=balanced use of resources
	● Increased client satisfaction
	    Number of returning clients
	    Number of recommended clients

Unsuccessful =	**PROJECT LEVEL**
❑ No more business from client ❑ We have to call clients ❑ Client dissatisfaction during/after the project ❑ Project team not happy	● Reduced project costs (working on more than one project at a time effectively, etc.) ● Well-distributed project time, no nights/weekends per project ● Reduced number of unexpected events/changes throughout the project ● Good team dynamics, met expectations ● Reduced number of negative issues per project

Today, we recognize that the customer rather than the contractor defines quality. The same holds true for project success. There must be customer and stakeholder acceptance included in any definition of project success. You can complete a project internally within your company within time, within cost, and within quality or specification limits and yet the project is not fully accepted by the customer or stakeholders.

At Enakta, the definition of project success is compared against the definition of project failure. According to Sandra Kumorowski, Marketing and Operations Consultant, the definition can appear as shown in Table 1–5.

Although companies may maintain a definition of project success (and even project failure), they may not have a clear definition of excellence in project management. This occurs when project management either is fully embedded into all of the company's work flow processes or is seen as a supportive role. Sandra Kumorowski believes:

> At Enakta, we do not have a formal definition of what is excellence in project management. However, our company wanted to view project management as a supportive role in creativity. Excellence is then achieved by complete customization to the current organizational structure and full alignment with long-term organizational goals, as shown in Figure 1–6.

Although some definitions of project success seem quite simple, many companies have elaborated on the primary definition of project success. Consider the comments below provided by Colin Spence, Project Manager/Partner at Convergent Computing (CCO):

> General guidelines for a successful project are as follows:
>
> ● Meeting the technology and business goals of the client on time, on budget, and on scope
> ● Setting the resource or team up for success, so that all participants have the best chance to succeed and have positive experiences in the process

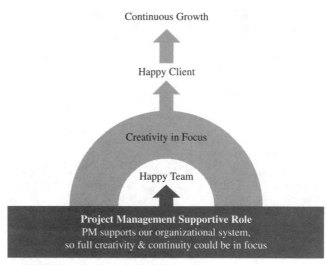

FIGURE 1–6.   Project management as a supportive role.

- Exceeding the client's expectations in terms of abilities, teamwork, and professionalism and generating the highest level of customer satisfaction
- Winning additional business from the client and being able to use the client as a reference account and/or agree to a case study
- Creating or fine-tuning processes, documentation, and deliverables that can be shared with the organization and leveraged in other engagements

Another critical question today is, "Who is defining success on a project?" Colin Spence responded to this question by stating:

> Both the client stakeholders and CCO team members ultimately define whether a project is successful.
>
> Success criteria for the project are defined during the initial steps in the engagement by the client and the project team in designing the proposal/scope-of-work (SOW) document. The creation of the scope of work typically involves stakeholders from the client side, a technical consultant from CCO, and an account manager from CCO. Depending on the size and complexity of the project, this process can be completed in one meeting with the client or may involve a more complete discovery and design process. CCO consultants are trained to focus on the business concerns of the client as well as their technology goals and to ensure that the solution recommended meets the full set of requirements.
>
> Once the information has been gathered, the CCO team will create the SOW document that may involve assistance from additional resources (such as a technical writer or other expert on the technologies involved). The draft of the SOW is reviewed by the CCO team, delivered to the client, and then presented to the client. Often a project plan accompanies the SOW, but if not, one is created prior to the commencement of the work.

Once the project starts, regular checkpoint meetings are critical to ensure the project is successful on all fronts and for change management purposes. These meetings involve client stakeholders and members of the CCO team as appropriate. If a project manager is not assigned to the project full time, one will be allocated to attend in an advisory role. Team members are encouraged to raise flags at any time during the project.

Once the project is complete, a satisfaction review meeting is scheduled, where the results of the project are discussed and the client can freely report on what worked and what did not and make suggestions for perfecting the relationship. A project can be successful in that it met the goals set forth in the SOW but still be a failure if the client is not satisfied and chooses to not engage in other projects with CCO. Additionally the client may be satisfied with the results, but the CCO team may not be, so it is important to assess success from both the external point of view and the internal one.

At Churchill Downs Incorporated (CDI), success is defined differently. According to Chuck Millhollan, Director of Program Management:

Project success is defined in our PMO charter as follows;

Based on input from CDI's executive management, the PMO considers a project to be a success when the following are true:

a. Pre-defined business objectives and project goals were achieved or exceeded.
b. A high-quality product is fully implemented and utilized.
c. Project delivery met or beat schedule and budget targets.
d. There are multiple winners:

   i. Project participants have pride of ownership and feel good about their work.
   ii. The customer's (internal and/or external) expectations are met.
   iii. Management has met its goals.

e. Project results helped build a good reputation.
f. Methods are in place for continual monitoring and evaluation (benefit realization).

We do not use project management "process" indicators to define project success. While schedule and budget targets are part of the criteria, sponsor acceptance, project completion, and ultimately project success, is based on meeting defined business objectives.

Enrique Sevilla Molina, PMP, Corporate PMO Director at Indra, provides us with his company's definition of project success and program success:

● Project success is based on achieving the proposed project targets in budget, scope, performance and schedule. Many times, the economic criteria appears as the main driving factor to measure project success, but there are other factors just as important such as building a durable relationship with the customer and building strong alliances with selected partners. Another significant criteria for project success measurement is the reliability of the project data forecast. It may be the case that, when the economic results of the project are not as good as they should be, if the fact is pointed out and reported soon enough, the success of the project is equally achieved.
● Program success is based on achieving the Program's overall strategic targets defined during Program definition and, at this level, the success is measured not only by

achieving the expected economic outcomes but, most of all, reaching the expected position in the market with regard to a product or a line of products, and establishing a more advantageous position with regard to our competitors. Leadership in a product line constitutes the ultimate measure of success in a Program. It is worthwhile to mention that, quite often, the success of a Program is based on the partnership concept developed with our major subcontractors at the Project level.

- Project success is defined at a business unit level by the responsible director, in accordance with the strategic goals assigned to the project.
- Program success is defined at the company level by the Chief Operations management in accordance with the program's defined mission.

AT&T defines project and program success in a similar manner. According to a spokesperson for AT&T:

- Project success is defined as a Client Satisfaction rating of "Very Satisfied" and On-Time Performance of Project Delivery of 98% or greater. The Project Management Organizational Leadership Team sets the objectives, which are tracked to determine project success. Program success is defined and tracked the same way as project success.

- Excellence [in project management] is defined as a consistent Project Management Methodology applied to all projects across the organization, continued recognition by our customers, and high customer satisfaction. Also our project management excellence is a key selling factor for our sales teams. This results in repeat business from our customers. In addition there is internal acknowledgement that project management is value-added and a must have.

Project success can be measured intermittently throughout the phase or gate review meetings that are part of the project management methodology. This allows a company to establish interim metrics for measuring success. An example of this will appear in a later chapter on project management methodologies.

Another element that is becoming important in the definition of success is the word *value*. Doug Bolzman, Consultant Architect, PMP®, ITIL Service Manager at EDS, believes:

At one point, customers were measuring project success as being on time and under budget. But if the project provided no real business value, what good is it being on time or under budget? Value for projects are being transformed within the planning of the project to depict the value to the user or the client of the project.

The users or the customers of the project define success. This can be difficult to identify at the start of the project (especially if not the norm). The executives can determine the overall value of how the projects map to the success of a program or initiative, but the users or customers will be the entity to receive the value of the project.

The comments by Doug Bolzman indicate that perhaps the single most important criterion for defining a potential best practice is that it must add value to the company and/or the client. Hewlett-Packard also sees the necessity for understanding the importance of value. According to a program manager at Hewlett-Packard, the following three best practices are added-value best practices:

- Project Collaboration Portals with standardized PM templates and Integrated tool kits with ability to request additional features by a Support staff.
- Project Retrospectives—very helpful for group learning and eliciting/recognizing/ documenting "best practices" but indeed communication beyond the immediate team is the challenge.
- Virtual Projects—given sufficient infrastructure I feel virtual projects are more productive and effective than burning up time and money on travel. I think HP utilizes these capabilities internally very well.

The ultimate definition of success might very well be when the customer is so pleased with the project that the customer allows you to use his or her name as a reference. This occurred in one company that bid on a project at 40 percent below its cost of doing the work. When asked why the bid was so low, company representatives responded that they knew they were losing money but what was really important was getting the customer's name on the corporate resume of clients. Therefore, the secondary factors may be more important than the primary factors.

The definition of success can also change based upon whether you are project- or non-project-driven. In a project-driven firm, the entire business of the company is projects. But in a non-project-driven firm, projects exist to support the ongoing business of production or services. In a non-project-driven firm, the definition of success also includes completion of the project *without* disturbing the ongoing business of this firm. It is possible to complete a project within time, within cost, and within quality and at the same time cause irrevocable damage to the organization. This occurs when the project manager does not realize that the project is *secondary* in importance to the ongoing business.

Some companies define success in terms of CSFs and KPIs. Critical success factors identify those factors necessary to meet the desired deliverables of the customer. CSFs and KPIs do not need to be elaborate or sophisticated metrics. Simple metrics, possibly based upon the triple constraint, can be quite effective. According to a spokesperson from AT&T:

> The critical success factors include Time, Scope, Budget and Customer Satisfaction. Key performance indicators include on-time performance for key deliverables. These include customer installation, customer satisfaction and cycle-time for common milestones.

Typical CSFs for most companies include:

- Adherence to schedules
- Adherence to budgets
- Adherence to quality
- Appropriateness and timing of signoffs
- Adherence to the change control process
- Add-ons to the contract

Critical success factors measure the end result usually as seen through the eyes of the customer. KPIs measure the quality of the process to achieve the end results. KPIs are internal measures and can be reviewed on a periodic basis throughout the life cycle of a project. Typical KPIs include:

- Use of the project management methodology
- Establish control processes
- Use of interim metrics
- Quality of resources assigned versus planned for
- Client involvement

Key performance indicators answer such questions as: Did we use the methodology correctly? Did we keep management informed and how frequently? Were the proper resources assigned and were they used effectively? Were there lessons learned which could necessitate updating the methodology or its use? Companies that are excellent in project management measure success both internally and externally using KPIs and CSFs. As an example, consider the following remarks provided by a spokesperson from Nortel Networks[16]:

> Nortel defines project success based on schedule, cost, and quality measurements, as mutually agreed-upon by the customer, the project team, and key stakeholders. Examples of key performance indicators may include completion of key project milestones, product installation/integration results, change management results, completion within budget, and so on. Project status and results are closely monitored and jointly reviewed with the customer and project team on a regular basis throughout a project to ensure consistent expectations and overall success. Project success is ultimately measured by customer satisfaction.

Shown below are additional definitions of CSFs and KPIs:

- CSFs:
  - Typically projects either improve something or reduce something. These improvements come in the form of capability or functionality of the company (through the employees/users). These produce additional productivity, new products and services, or more efficiency for existing products. Critical success factors are mapped to the overall business objectives. (Provided by Doug Bolzman, Consultant Architect, PMP®, ITIL Service Manager, EDS)
  - Success factors are defined at the initial stages of the project or program, even before they become actual contracts, and are a direct consequence of the strategic goals allocated to the project or program. Many times these factors are associated with expanding the market share in a product line or developing new markets, both technically and geographically. (Provided by Enrique Sevilla Molina, PMP, Corporate PMO Director, Indra)

---

16. H. Kerzner, *Best Practices in Project Management: Achieving Global Excellence*, Wiley, Hoboken, NJ, 2006, p. 26.

- Obviously, CSFs vary with projects and intent. Below are some that apply over a large variety of projects:
  - Early customer involvement
  - High-quality standards
  - Defined processes and formalized gate reviews
  - Cross-functional team organizational structure
  - Control of requirements, prevention of scope creep
  - Commitment to schedules—disciplined planning to appropriate level of detail and objective and frequent tracking
  - Commitment of resources—right skill level at necessary time
  - Communication among internal teams and with customer
  - Early risk identification, management, and mitigation—no surprises
  - Unequaled technical execution based on rigorous Engineering. (Comments provided by a spokesperson at Motorola)[17]
- CCO has identified a number of CSFs involved in delivering outstanding technology services:
  - Have experienced and well-rounded technical resources. These resources need to not only have outstanding technical skills but also be good communicators, work well in challenging environments, and thrive in a team environment.
  - Make sure we understand the full range of the clients' needs, including both technical and business needs, and document a plan of action (the scope of work) for meeting these needs.
  - Have well-defined policies and processes for delivering technology services that leverage "best practice" project management concepts and practices.
  - Have carefully crafted teams, with well-defined roles and responsibilities for the team members, designed to suit the specific needs of the client.
  - Enhance collaborations and communications both internally (within the team and from the team to the CCO) and externally with our clients.
  - Leverage our experience and knowledge base as much as possible to enhance our efficiency and the quality of our deliverables. (Provided by Colin Spence, Project Manager/Partner, Convergent Computing)

- KPIs:
  - Key performance indicators allow the customer to make a series of measurements to ensure the performance is within the stated thresholds (success factors). This is called "keeping the pulse of the company" by the executives. KPIs are determined, measured, and communicated through mechanisms such as dashboards or metrics.(Provided by Doug Bolzman, Consultant Architect, PMP®, ITIL Service Manager, EDS)

---

17. Ibid., p. 27.

- Vitalize Consulting Solutions (VCS) holds our PMs accountable to a quarterly assignment quality assessment (AQA) score. This survey asks our customers to rate our PMs on a scale of 1–5 using a list of 15 questions regarding performance, customer satisfaction, and overall customer expectations. We measure performance based on our project manager's ability to meet these indicators and our commitment to providing high-quality project management resources. VCS also reviews the typical "on-time and on-budget" metrics through our project review process but avoids penalizing PMs for delays that are outside their control (for example, delays in software being delivered). (Marc Hirshfield, PMP, Director, Project Management Office)
- Our most common KPIs are associated to the financial projects results, for instance, project margin compliance with the allocated strategic target, new contracts figure for the business development area goals, etc. Success factors are translated into performance indicators so they are periodically checked.
- By default, a first indication of projects health is provided by the schedule and cost performance indices (SPI and CPI) embedded into the PM tools. They are monthly provided by the project management information system and they are also available for historical analysis and review. These indicators are also calculated for each department, so they constitute an indicator of the overall cost and schedule performance of the department or business unit. (Provided by Enrique Sevilla Molina, PMP, Corporate PMO Director, Indra)
- Postship acceptance indicators:
  - Profit and loss
  - Warranty returns
  - Customer reported unique defects
  - Satisfaction metrics
- In-process indicators:
  - Defect trends against plan
  - Stability for each build (part count changes) against plan
  - Feature completion against plan
  - Schedule plan versus actual performance
  - Effort plan versus actual performance
  - Manufacturing costs and quality metrics
  - Conformance to quality processes and results of quality audits
  - System test completion rate and pass/fail against plan
  - Defect/issue resolution closure rate
  - Accelerated life-testing failure rates against plan
  - Prototype defects per hundred units (DPHU) during development against plan (Provided by a spokesperson at Motorola)[18]
- The SOW provides a checklist of basic indicators for the success of the project, but client satisfaction is also important. The SOW will indicate what the

---

18. Ibid.

deliverables are and will provide information on costs and timelines that are easily tracked.

However, it is also critical that the SOW identify not only the goals of the project but also which are the most important to the client. For example, one client may not be overly concerned about the budget but must have the project meet certain deadlines. Part of the project manager's job is to understand which are the key criteria to meet and manage the project accordingly. If no project manager is assigned, this task is assigned to one of the other team members, typically the consultant.

So the project manager needs to periodically assess whether the project is under/over/on budget, ahead/behind/on schedule, and fulfilling the other goals of the project and to assess whether the client is satisfied with the work and deliverables. Additionally, the project manager needs to assess the internal functions of the team and make adjustments if needed. (Provided by Colin Spence, Project Manager/Partner, Convergent Computing)

Most people seem to understand that CSFs and KPIs can be different from project to project. However, there is a common misbelief that CSFs and KPIs, once established, must not change throughout the project. As projects go through various life-cycle phases, these indicators can change. Carl Manello, PMP, Solutions Lead—Program and Project Management, Slalom Consulting, believes that:

> Establishing the right Critical Success Factors and Key Performance Indicators is crucial. A project's defined success (i.e., defined success means that . . . all stakeholders are in agreement at the earliest stages of the project as to what the end-state will look like) is identified through the Critical Success Factors. A project's selection of the right Key Performance Indicators establishes the measurement metrics for tracking to determine whether the Critical Success Factors will be met. As a starting point, On Time, On Budget and with the agreed upon specifications (or within a tolerance for all three metrics) are good basic KPIs. As projects mature in their ability to deliver results, more sophisticated performance indicators may be implemented. For example, instead of using the loosely defined "agreed upon specifications," projects may choose to use the formality of a requirements volatility measure and some acceptable variation around a baseline.

In the author's experience, more than 90 percent of the best practices that companies identify come from analysis of the KPIs during the debriefing sessions at the completion of a project or at selected gate review meetings. Because of the importance of extracting these best practices, some companies are now training professional facilitators capable of debriefing project teams and capturing the best practices.

Before leaving this section, it is necessary to understand who discovers the best practice. Best practices are discovered by the people performing the work, namely the project manager, project team, and possibly the line manager. According to a spokesperson from Motorola[19]:

> The decision as to what is termed a best practice is made within the community that performs the practice. Process capabilities are generally known and baselined. To claim best practice

---

19. Ibid., p. 14.

status, the practice or process must quantitatively demonstrate significant improvements in quality, efficiency, cost, and/or cycle time. The management of the organization affected as well as process management must approve the new practice prior to institutionalization.

Generally, the process of identification begins with the appropriate team member. If the team member believes that he or she has discovered a best practice, they then approach their respective line manager and possibly project manager for confirmation. Once confirmation is agreed upon, the material is sent to the Project Management Office (PMO) for validation. After validation, the person that identified the best practice is given the title of "Best Practice Owner" and has the responsibility of nurturing and cultivating the best practice.

Some companies use professional facilitators to debrief project teams in order to extract best practices. These facilitators may be assigned to the PMO and are professionally trained in how to extract lessons learned and best practices from both successes and failures. Checklists and templates may be used as part of the facilitation process. As an example, consider the following statements from Sandra Kumorowski, Marketing and Operations Consultant, Enakta:

> After each project, we conduct Post Mortem meetings in order to evaluate the project performance. I have created a standard format—list of questions—and all team members had to prepare their answers ahead of time. If there was an idea, process, or tactic that has added measurable benefit or long-term value to the organization, the top management would then add it to our Library. I was responsible for an update to all employees when a new best practice has been added to our Best Practices Library.

1.8 DASHBOARDS AND SCORECARDS

In our attempt to go to paperless project management, emphasis is being given to visual displays such as dashboard and scorecards utilizing and displaying CSFs and KPIs. Executives and customers desire a visual display of the most critical project performance information in the least amount of space. Simple dashboard techniques, such as traffic light reporting, can convey critical performance information. As an example:

- Red traffic light: A problem exists which may affect time, cost, quality, or scope. Sponsorship involvement is necessary.
- Yellow or amber light: This is a caution. A potential problem may exist, perhaps in the future if not monitored. The sponsor is informed but no action by the sponsor is necessary at this time.
- Green light: Work is progressing as planned. No involvement by the sponsor is necessary.

While a traffic light dashboard with just three colors is most common, some companies use many more colors. The information technology (IT) group of a retailer had an eight-color dashboard for IT projects. An amber color meant that the targeted end date had past and the project was still not complete. A purple color meant that this work package was undergoing a scope change that could have an impact on the triple constraint.

Some people confuse dashboards with scorecards. There is a difference between dashboards and scorecards. According to Eckerson[20]:

- Dashboards are visual display mechanisms used in an *operationally* oriented performance measurement system that measure performance against targets and thresholds using right-time data.
- Scorecards are visual displays used in a *strategically* oriented performance measurement system that chart progress towards achieving strategic goals and objectives by comparing performance against targets and thresholds.

Both dashboards and scorecards are visual display mechanisms within a performance measurement system that convey critical information. The primary difference between dashboards and scorecards is that dashboards monitor operational processes such as those used in project management, whereas scorecards chart the progress of tactical goals. Table 1–6 and the description following it show how Eckerson compares the features of dashboards and scorecards.[21]

Dashboards. Dashboards are more like automobile dashboards. They let operational specialists and their supervisors monitor events generated by key business processes. But unlike automobiles, most business dashboards do not display events in "real time," as they occur; they display them in "right time," as users need to view them. This could be every second, minute, hour, day, week, or month depending on the business process, its volatility, and how critical it is to the business. However, most elements on a dashboard are updated on an intraday basis, with latency measured in either minutes or hours.

Dashboards often display performance visually, using charts or simple graphs, such as gauges and meters. However, dashboard graphs are often updated in place, causing the graph to "flicker" or change dynamically. Ironically, people who monitor operational processes often find the visual glitz distracting and prefer to view the data in the original form, as numbers or text, perhaps accompanied by visual graphs.

TABLE 1–6. COMPARING FEATURES

Feature	Dashboard	Scorecard
Purpose	Measures performance	Charts progress
Users	Supervisors, specialists	Executives, managers, and staff
Updates	Right-time feeds	Periodic snapshots
Data	Events	Summaries
Display	Visual graphs, raw data	Visual graphs, comments

20. W. W. Eckerson, *Performance Dashboards: Measuring, Monitoring and Managing Your Business*, Wiley, Hoboken, NJ, 2006, pp.293, 295. Chapter 12 provides an excellent approach to designing dashboard screens.
21. Ibid., p. 13.

Scorecards. Scorecards, on the other hand, look more like performance charts used to track progress toward achieving goals. Scorecards usually display monthly snapshots of summarized data for business executives who track strategic and long-term objectives, or daily and weekly snapshots of data for managers who need to chart the progress of their group of project toward achieving goals. In both cases, the data are fairly summarized so users can view their performance status at a glance.

Like dashboards, scorecards also make use of charts and visual graphs to indicate performance state, trends, and variance against goals. The higher up the users are in the organization, the more they prefer to see performance encoded visually. However, most scorecards also contain (or should contain) a great deal of textual commentary that interprets performance results, describes action taken, and forecasts future results.

Summary. In the end, it does not really matter whether you use the term dashboard or scorecard as long as the tool helps to focus users and organizations on what really matters. Both dashboards and scorecards need to display critical performance information on a single screen so users can monitor results at a glance.

Although the terms are used interchangeably, most project managers prefer to use dashboards and/or dashboard reporting. Eckerson defines three types of dashboards, as shown in Table 1–7 and the description that follows.[22]

- *Operational dashboards* monitor core operational processes and are used primarily by front-line workers and their supervisors who deal directly with customers or manage the creation or delivery of organizational products and services. Operational dashboards primarily deliver detailed information that is only lightly summarized. For example, an online Web merchant may track transactions at the product level rather than the customer level. In addition, most metrics in an operational dashboard are updated on an intraday basis, ranging from minutes to hours depending on the application. As a result, operational dashboards emphasize monitoring more than analysis and management.
- *Tactical dashboards* track departmental processes and projects that are of interest to a segment of the organization or a limited group of people. Managers and business analysts use tactical dashboards to compare performance of their areas or projects, to budget plans, forecasts, or last period's results. For example, a project to reduce the number

TABLE 1–7.   THREE TYPES OF PERFORMANCE DASHBOARDS

	Operational	Tactical	Strategic
Purpose	Monitor operations	Measure progress	Execute strategy
Users	Supervisors, specialists	Managers, analysts	Executives, managers, staff
Scope	Operational	Departmental	Enterprise
Information	Detailed	Detailed/summary	Detailed/summary
Updates	Intraday	Daily/weekly	Monthly/quarterly
Emphasis	Monitoring	Analysis	Management

---

22. Ibid., pp. 17–18.

of errors in a customer database might use a tactical dashboard to display, monitor, and analyze progress during the previous 12 months toward achieving 99.9 percent defect-free customer data by 2007.

- *Strategic dashboards* monitor the execution of strategic objectives and are frequently implemented using a balanced scorecard approach, although total quality management, Six Sigma, and other methodologies are used as well. The goal of a strategic dashboard is to align the organization around strategic objectives and get every group marching in the same direction. To do this, organizations roll out customized scorecards to every group in the organization and sometimes to every individual as well. These "cascading" scorecards, which are usually updated weekly or monthly, give executives a powerful tool to communicate strategy, gain visibility into operations, and identify the key drivers of performance and business value. Strategic dashboards emphasize management more than monitoring and analysis.

There are three critical steps that must be considered when using dashboards: (1) the target audience for the dashboard, (2) the type of dashboard to be used, and (3) the frequency in which the data will be updated. Some project dashboards focus on the key performance indicators that are part of earned-value measurement. These dashboards may need to be updated daily or weekly. Dashboards related to the financial health of the company may be updated weekly or quarterly. Figures 1–7 and 1–8 show the type of information that would be tracked weekly or quarterly to view corporate financial health.[23]

1.9 KEY PERFORMANCE INDICATORS

Most often, the items that appear in the dashboards are elements that both customers and project managers track. These items are referred to as key performance indicators (KPIs) that were discussed previously. According to Eckerson[24]:

> A KPI is a metric measuring how well the organization or individual performs an operational, tactical or strategic activity that is critical for the current and future success of the organization.

Some people confuse a KPI with leading indicators. A leading indicator is actually a KPI that measures how the work you are doing now will affect the future.

KPIs are critical components of all earned-value measurement systems. Terms such as cost variance, schedule variance, schedule performance index, cost performance index, and time/cost at completion are actually KPIs but are not referred to as such. The need for these KPIs is simple: What gets measured gets done! If the goal of a performance measurement system is to improve efficiency and effectiveness, then the KPI must reflect

23. J. Alexander, *Performance Dashboards and Analysis for Value Creation*, Wiley, Hoboken, NJ, 2007, pp. 87—88. Reproduced by permission of John Wiley & Sons.
24. W. Eckerson, *Performance Dashboards: Measuring, Monitoring and Managing Your Business*, Wiley, Hoboken, NJ, 2006, p. 294.

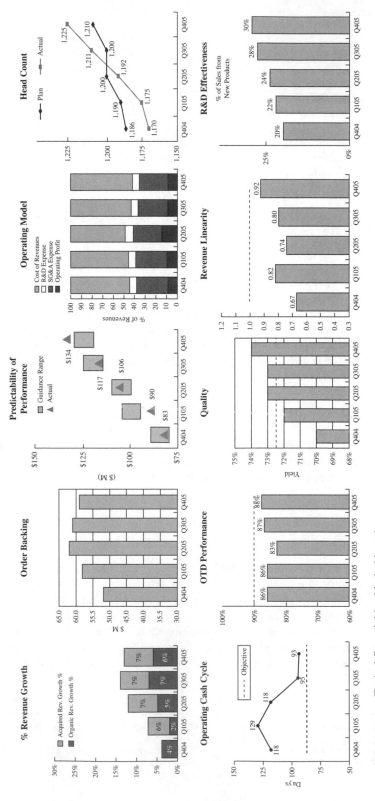

FIGURE 1-7. Typical financial health dashboards.

XYZ Company
Q4' 05 Week #7 of 13/ 54% of Q4
($ in Millions)

Bookings	Week	Unit	QTD	Forecast	% Achieved	$ Required
	0.7	BU 1	15.0	30.0	50%	15.0
	—	BU 2	0.9	1.0	89	0.1
	0.5	BU 3	4.0	6.0	67	2.0
	0.4	BU 4	1.7	4.7	37	2.9
	0.0	Other	0.1	—		(0.1)
	1.6	Totals	21.7	41.7	52%	$20.0

Revenue	Week	Unit	QTD	Forecast	% Achieved	Backlog	Req'd Fill
	2.0	BU 1	13.0	28.0	46%	5.0	10.00
	0.4	BU 2	3.0	5.0	60	1.0	1.00
	0.0	BU 3	3.0	6.0	50	2.0	1.00
	2.6	BU 4	3.0	7.0	43	1.0	3.00
	—	Other	—	—			
	5.0	Totals	22.0	46.0	48%	9.0	15.0

Receivable Collections (Cumulative)	Week	1	2	3	4	5
	Actual	1.0	5.0	19.0		
	Target	4.0	9.0	17.0	28.0	35.0

	Day	1	2	3	4	5
Process Yield		77%	80%	81%	68%	82%

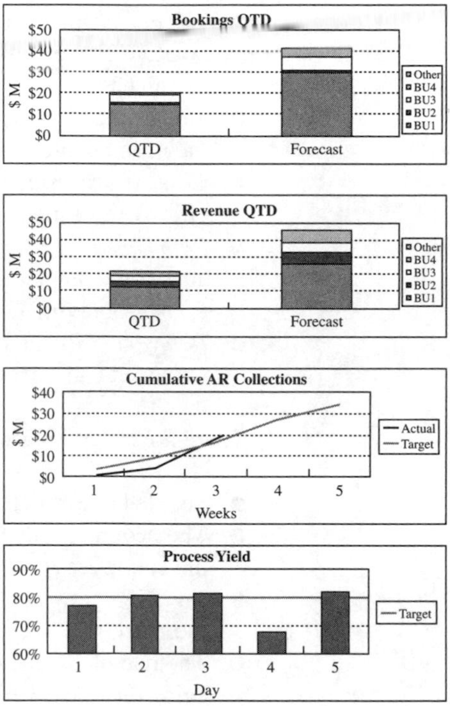

FIGURE 1–8. Typical financial health dashboards.

controllable factors. There is no point in measuring an activity if the users cannot change the outcome.

Eckerson identifies 12 characteristics of effective KPIs[25]:

● *Aligned.* KPIs are always aligned with corporate [or project] strategy and objectives.
● *Owned.* Every KPI is "owned" by an individual or group on the business [or project] side who is accountable for its outcome.
● *Predictive.* KPIs measure drivers of business [or project] value. Thus, they are "leading" indicators of performance desired by the organization.
● *Actionable.* KPIs are populated with timely, actionable data so users can intervene to improve performance before it is too late.
● *Few in number.* KPIs should focus users on a few high-value tasks, not scatter their attention and energy on too many things.
● *Easy to understand.* KPIs should be straightforward and easy to understand, not based upon complex indices that users do not know how to influence directly.

25. Ibid., p. 201.

- *Balanced and linked.* KPIs should balance and reinforce each other, not undermine each other and suboptimize processes.
- *Trigger changes.* The act of measuring a KPI should trigger a chain reaction of positive changes in the organization [or project], especially when it is monitored by the CEO [or customers or sponsors].
- *Standardized.* KPIs are based upon standard definitions, rules, and calculations so they can be integrated across dashboards throughout the organization.
- *Context driven.* KPIs put performance in context by applying targets and thresholds to performance so users can gauge their progress over time.
- *Reinforced with incentives.* Organizations can magnify the impact of KPIs by attaching compensation or incentives to them. However, they should do this cautiously, applying incentives only to well-understood and stable KPIs.
- *Relevant.* KPIs gradually lose their impact over time, so they must be periodically reviewed and refreshed.

There are several reasons why the use of KPIs often fails on projects, including:

- People believe that the tracking of a KPI ends at the first line manager level.
- The actions needed to regulate unfavorable indications are beyond the control of the employees doing the monitoring or tracking.
- The KPIs are not related to the actions or work of the employees doing the monitoring.
- The rate of change of the KPIs is too slow, thus making them unsuitable for managing the daily work of the employees.
- Actions needed to correct unfavorable KPIs take too long.
- Measurement of the KPIs does not provide enough meaning or data to make them useful.
- The company identifies too many KPIs to the point where confusion reigns among the people doing the measurements.

Years ago, the only metrics that some companies used were those identified as part of the earned-value measurement system. The metrics generally focused only on time and cost and neglected metrics related to business success as opposed to project success. As such, the measurement metrics were the same on each project and the same for each life-cycle phase. Today, metrics can change from phase to phase and from project to project. The hard part is obviously deciding upon which metrics to use. Care must be taken that whatever metrics are established do not end up comparing apples and oranges. Fortunately, there are several good books in the marketplace that can assist in identifying proper or meaningful metrics.[26]

Selecting the right KPIs is critical. Since a KPI is a form of measurement, some people believe that KPIs should be assigned only to those elements that are tangible.

---

26. Three books that provide examples of metric identification are P. F. Rad and G. Levin, *Metrics for Project Management,* Management Concepts, Vienna, VA, 2006; M. Schnapper and S. Rollins, *Value-Based Metric For Improving Results*, J. Ross Publishing, Ft. Lauderdale, FL, 2006; and D. W. Hubbard, *How To Measure Anything*; Wiley, Hoboken, NJ, 2007.

Therefore, many intangible elements that should be tracked by KPIs never get looked at because someone believes that measurement is impossible. Anything can be measured regardless of what some people think. According to Hubbard[27]:

- Measurement is a set of observations that reduces uncertainty where the results are expressed as a quantity.
- A mere reduction, not necessarily elimination, of uncertainty will suffice for a measurement.

Therefore, KPIs can be established even for intangibles like those discussed later in this book in the chapter on value-driven project management.

Hubbard believes that five questions should be asked before we establish KPIs for measurement[28]:

- What is the decision this [KPI] is supposed to support?
- What really is the thing being measured [by the KPI]?
- Why does this thing [and the KPI] matter to the decision being asked?
- What do you know about it now?
- What is the value to measuring it further?

Hubbard also identifies four useful measurement assumptions that should be considered when selecting KPIs[29]:

- Your problem [in selecting a KPI] is not as unique as you think
- You have more data than you think
- You need less data than you think
- There is a useful measurement that is much simpler than you think

Selecting the right KPIs is essential. On most projects, only a few KPIs are needed. Sometimes we seem to select too many KPIs and end up with some KPIs that provide us with little or no information value, and the KPI ends up being unnecessary or useless in assisting us in making project decisions.

Sometimes, companies believe that the measures that they have selected are KPIs when, in fact, they are forms of performance measures but not necessarily KPIs. David Parmenter discusses three types of performance measures[30]:

- *Key results indicators* (KRIs) tell you how you have done in a perspective
- *Performance indicators* (PIs) tell you what to do
- *KPIs* tell you what to do to increase performance drastically

Parmenter believes that[31]:

27. D. W. Hubbard, *How To Measure Anything*, Wiley, Hoboken, NJ, 2007, p. 21.
28. Ibid, p. 43.
29. Ibid, p. 31.
30. D. Parmenter, *Key Performance Indicators*, Wiley, Hoboken, NJ, 2007, p. 1.
31. Ibid., p. 19.

The ultimate success of a change strategy depends greatly on how the change is introduced and implemented, rather than on the merit of the strategy itself. Successful development and utilization of key performance indicators (KPIs) in the workplace is determined by the presence or absence of four foundation stones:

- Partnership with the staff, unions, key suppliers, and key customers
- Transfer of power to the front line
- Integration of measurement, reporting, and improvement of performance
- Linkage of performance measures to strategy

In a project environment, the performance measures can change from project to project and phase to phase. The identification of these measures is performed by the project team, including the project sponsor. Project stakeholders may have an input as well. Corporate performance measures are heavily financially oriented and may undergo very little change over time. The measurements indicate the financial health of the corporation.

Establishing corporate performance measures related to strategic initiatives or other such activities must be treated as a project in itself, and supported by the senior management team (SMT).

The SMT attitude is critical—any lack of understanding, commitment, and prioritizing of this important process will prevent success. It is common for the project team and the SMT to fit a KPI project around other competing, less important firefighting activities.

The SMT must be committed to the KPI project, to driving it down through the organization. Properly implemented, the KPI project will create a dynamic environment. Before it can do this, the SMT must be sold on the concept. This will lead to the KPI project's being treated as the top priority, which may mean the SMT's allowing some of those distracting fires to "burn themselves out."[32]

1.10 STEP 3: VALIDATING THE BEST PRACTICE

Previously we stated that seeking out of a best practice is done by the project manager, project team, functional manager, and/or possibly a professional facilitator trained in how to debrief a project team and extract best practices. Any or all of these people must believe that what they have discovered is, in fact, a best practice. When project managers are quite active in a project, emphasis is placed upon the project manager for the final decision on what constitutes a best practice. According to a spokesperson for AT&T, the responsibility for determining what is a best practice rests with:

> The individual project manager that shows how it had a positive impact on their project.

However, although this is quite common, there are other validation methods that may involve a significant number of people. Sometimes, project managers may be removed from

---

32. Ibid., p. 27. Chapter 5 of this book has excellent templates for reporting KPIs.

where the work is taking place and may not be familiar with activities that could lead to the identification of a best practice. Companies that have a PMO place a heavy reliance on the PMO for support because the approved best practices are later incorporated into the methodology, and the PMO is usually the custodian of the methodology. According to Marc Hirshfield, PMP, Director, Project Management Office, Vitalize Consulting Solutions (VCS):

> In the PMO at VCS, we have a change management process that is used to evaluate updates to our methodology. The ultimate decision rests with the change management team. Updates are then made to the methodology and re-distributed to the original author and project management practice so everyone is aware of the updates. This way the original author of the request views their direct impact on improving the PMO methodology.

Once the management of the organization affected initially approves the new best practice, it is forwarded to the PMO or process management for validation and then institutionalization. The PMO may have a separate set of checklists to validate the proposed best practice. The PMO must also determine whether or not the best practice is company proprietary because that will determine where the best practice is stored and whether the best practice will be shared with customers.

The best practice may be placed in the company's best practice library or, if appropriate, incorporated directly into the company's stage gate checklist. Based upon the complexity of the company's stage gate checklist process and enterprise project management methodology, the incorporation process may occur immediately or on a quarterly basis.

According to Chuck Millhollan, Director of Program Management at Churchill Downs, Incorporated:

> We do not label our processes or methods as "best practices." We simply learn from our lessons and ensure that learning is incorporated into our methodology, processes, templates, etc.

Some organizations have committees not affiliated with the PMO that have as their primary function the evaluation of potential best practices. Anyone in the company can provide potential best practices data to the committee and the committee in turn does the analysis. Project managers may be members of the committee. Other organizations use the PMO to perform this work. These committees and the PMO most often report to the senior levels of management.

The fourth edition of the *PMBOK® Guide* emphasizes the importance of stakeholder involvement in projects. This involvement may also include the final decision on whether or not a discovery is a best practice. According to Chuck Millhollan, Director of Program Management, Churchill Downs, Inc.:

> Ultimately, the final decision resides with our stakeholders, both internal and external. Another way of putting this is that the PMO does not make the decision if a method or process works. We actively seek feedback from our project stakeholders and use their inputs to determine if our processes are "best practices" for Churchill Downs Incorporated. The specific best practices identified previously, among others, have even been accepted outside of the PMO as generally accepted practices.

Another example of stakeholder involvement is provided by Enrique Sevilla Molina, PMP, Corporate PMO Director, Indra:

> The decision is taken by the corporate PMO responsible, the business unit manager, the local PMO authority, or even the cognizant authority, if it is the case. It depends on the subject and the scope of the task. Some of the management best practices have been established at corporate level, and they have been incorporated into the PM methodology. Many of them have also been incorporated into the Project Management Information Systems and the corporate PM tooling.

Evaluating whether or not something is a best practice is not time consuming, but it is complex. Simply because someone believes that what he or she is doing is a best practice does not mean that it is in fact a best practice. Some PMOs are currently developing templates and criteria for determining that an activity may qualify as a best practice. Some items that are included in the template might be:

- Is transferable to many projects
- Enables efficient and effective performance that can be measured (i.e., can serve as a metric)
- Enables measurement of possible profitability using the best practice
- Allows an activity to be completed in less time and at a lower cost
- Adds value to both the company and the client
- Can differentiate us from everyone else

One company had two unique characteristics in its best practices template:

- Helps to avoid failure
- If a crisis exists, helps us to get out of a critical situation

Executives must realize that these best practices are, in fact, intellectual property to benefit the entire organization. If the best practice can be quantified, then it is usually easier to convince senior management of its value.

1.11 STEP 4: LEVELS OF BEST PRACTICES

As stated previously, best practices come from knowledge transfer and can be discovered anywhere within or outside of your organization. This is shown in Figure 1–9.

Companies that maintain best practices libraries that contain a large number of best practices may create levels of best practices. Figure 1–10 shows various levels of best practices. Each level can have categories within the level. The bottom level is the professional standards level, which would include professional standards as defined by PMI. The professional standards level contains the greatest number of best practices, but they are more of a general nature than specific and have a low level of complexity.

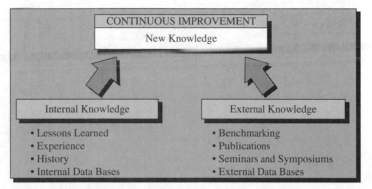

FIGURE 1–9. Knowledge transfer.

The industry standards level would identify best practices related to performance within the industry. The automotive industry has established standards and best practices specific to the auto industry.

As we progress to the individual best practices in Figure 1–10, the complexity of the best practices goes from general to very specific applications and, as expected, the quantity of best practices is less. An example of a best practice at each level might be (from general to specific):

- *Professional Standards:* Preparation and use of a risk management plan, including templates, guidelines, forms, and checklists for risk management.
- *Industry Specific:* The risk management plan includes industry best practices such as the best way to transition from engineering to manufacturing.
- *Company Specific:* The risk management plan identifies the roles and interactions of engineering, manufacturing, and quality assurance groups during transition.
- *Project Specific:* The risk management plan identifies the roles and interactions of affected groups as they relate to a specific product/service for a customer.
- *Individual:* The risk management plan identifies the roles and interactions of affected groups based upon their personal tolerance for risk, possibly through the use of a responsibility assignment matrix prepared by the project manager.

FIGURE 1–10. Levels of best practices.

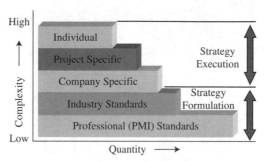

FIGURE 1–11. Usefulness of best practices.

Best practices can be extremely useful during strategic planning activities. As shown in Figure 1–11, the bottom two levels may be more useful for project management strategy formulation whereas the top three levels are more appropriate for the execution or implementation of a strategy.

Not all companies maintain a formal best practices library. In some companies, when a best practice is identified and validated, it is immediately placed into the stage gate process or the project management methodology. In such a case, the methodology itself becomes the best practice. Enrique Sevilla Molina, PMP, Corporate PMO Director at Indra, states:

> In fact, our Project Management methodology constitutes our established library of best practices applicable to every project in the company. There are additional best practices libraries in different business units. There are, for instance, detailed instructions for proposal preparation or for cost and schedule estimation purposes, which are appropriate for the specific business or operations area.

When asked how many best practices they maintain at Indra, Enrique commented:

- It is hard to say because of the subject itself and the multiplicity of business areas in the company. If we consider our PM methodology as a set of "best practices", it would be difficult to count every best practice included.
- Besides our internally published Indra Project Management Methodology Manual, we have for instance specific guides at corporate level for WBS elaboration, project risk management, and the project's performance measurement based on earned value techniques. We have also specific instructions published for proposal preparation, costs estimation, and even detailed WBS preparation rules and formats for different business unit levels.

1.12 STEP 5: MANAGEMENT OF BEST PRACTICES

There are three players involved in the management of the best practices:

- The best practice's owner
- The PMO
- The best practices' library administrator who may reside in the PMO

The best practice's owner, who usually resides in the functional area, has the responsibility of maintaining the integrity of the best practice. Being a best practice owner is usually a noncompensated, unofficial title but does appear as a symbol of prestige. Therefore, the owner of the best practice tries to enhance it and keep the best practice alive as long as possible.

The PMO usually has the final authority over best practices and makes the final decision on where to place the best practice, who should be allowed to see it, how often it should be reviewed or revalidated, and when it should be removed from service.

The library administrator is merely the caretaker of the best practice and may keep track of how often people review the best practice assuming it is readily accessible in the best practices library. The library administrator may not have a good understanding of each of the best practices and may not have any voting rights on when to terminate a best practice.

1.13 STEP 6: REVALIDATING BEST PRACTICES

Best practices do not remain best practices forever. Because best practices are directly related to the company's definition of project success, the definition of a best practice can change and age as the definition of success changes. Therefore, best practices must be periodically reviewed. The critical question is, "How often should they be reviewed?" The answer to this question is based upon how many best practices are in the library. Some companies maintain just a few best practices whereas large, multinational companies may have thousands of clients and maintain hundreds of best practices in their libraries. If the company sells products as well as services, then there can be both product-related and process-related best practices in the library.

The following two examples illustrate the need for reviewing best practices.

- Once a practice has been nominated and approved to be a best practice, it is only sanctioned until the next yearly review cycle. Over time, best practices have the tendency to lose value and become ineffective if they are allowed to age. (EDS)
- Best practices are reviewed every four months. Input into the review process includes:

  - Lessons learned documents from project completed within the past four months
  - Feedback from project managers, architects, and consultants
  - Knowledge that subject matter experts (i.e., best practices owners) bring to the table; this includes information gathered externally as well as internally
  - Best practices library reporting and activity data (Computer Associates)

There are usually three types of decisions that can be made during the review process:

- Keep the best practice as is until the next review process.
- Update the best practice and continue using it until the next review process.
- Retire the best practice from service.

1.14 STEP 7: WHAT TO DO WITH A BEST PRACTICE

With the definition that a best practice is an activity that leads to a sustained competitive advantage, it is no wonder that some companies have been reluctant to make their best practices known to the general public. Therefore, what should a company do with its best practices if not publicize them? The most common options available include:

- *Sharing Knowledge Internally Only:* This is accomplished using the company intranet to share information to employees. There may be a separate group within the company responsible for control of the information, perhaps even the PMO. Not all best practices are available to every employee. Some best practices may be password protected, as discussed below.
- *Hidden from All But a Selected Few:* Some companies spend vast amounts of money on the preparation of forms, guidelines, templates, and checklists for project management. These documents are viewed as both company-proprietary information and best practices and are provided to only a select few on a need-to-know basis. An example of a "restricted" best practice might be specialized forms and templates for project approval where information contained within may be company-sensitive financial data or the company's position on profitability and market share.
- *Advertise to Your Customers:* In this approach, companies may develop a best practices brochure to market their achievements and may also maintain an extensive best practices library that is shared with their customers after contract award. In this case, best practices are viewed as competitive weapons.

Most companies today utilize some form of best practices library. According to a spokesperson from AT&T:

> The best practices library is Sharepoint based and very easy to use both from a submission and a search perspective. Any Project Manager can submit a best practice at any time and can search for best practices submitted by others.

Even though companies collect best practices, not all best practices are shared outside of the company even during benchmarking studies where all parties are expected to share information. Students often ask why textbooks do not include more information on detailed best practices such as forms and templates. One company commented to the author:

> We must have spent at least $1 million over the last several years developing an extensive template on how to evaluate the risks associated with transitioning a project from engineering to manufacturing. Our company would not be happy giving this template to everyone who wants to purchase a book for $85. Some best practices templates are common knowledge and we would certainly share this information. But we view the transitioning risk template as proprietary knowledge not to be shared.

1.15 STEP 8: COMMUNICATING BEST PRACTICES ACROSS THE COMPANY _____

Knowledge transfer is one of the greatest challenges facing corporations. The larger the corporation, the greater the challenge with knowledge transfer. The situation is further complicated when corporate locations are dispersed over several continents. Without a structured approach for knowledge transfer, corporations can repeat mistakes as well as missing valuable opportunities. Corporate collaboration methods must be developed. NXP has found a way to overcome several of these barriers. Mark Gray, MBA, PMP, Ph.D. candidate, Senior Project Manager at NXP Semiconductor, discusses this approach. Mark calls this: To grow oaks, you need to start with nuts. . . :

One of the biggest problems facing project managers and their organizations today is how best to get knowledge transferred from the experts (or at least experienced) project managers to the rest of the organization. Much has been written on lessons *ignored* and equally as much has been written on having lessons learned as a value-added component in the toolbox of project management. What seems to be missing is a sound approach for getting the knowledge transmitted (the identified, captured, stored parts are all pretty obvious).

At NXP we noted that there are many lessons that are captured during projects or during their review but very little evidence that these lessons were even being seen by other project managers. One solution we put in place was to use the model of a Community of Practice (CoP) as described by Wenger[33] in his work on the subject. Using the same basic idea as used by Shell in their off-shore drilling platforms, we established local forums of "experts" with the specific mandate to create an arena in which project managers would feel comfortable sharing their findings and learning's from their projects.

The process itself is very simple; Lessons are identified by the project managers either from project debriefs or from peer reviews and these are then presented to the forum as a type of "war story". It's important to note here that we look for both good and "less good" incidents to learn from. In general this leads to a good (sometimes spirited) debate on the topic from which the participants can take away a genuine learning experience. The results are of course somewhat qualitative in nature so to say we have clear measurable improvements as a direct result of these sessions would be difficult, however we have seen a general improvement in the overall performance of projects since we started this initiative.

Now for the hard part—building, and maintaining the CoP. Wenger gives some very good basics of how to construct CoP's, including the need to have a good core team, involvement of executive sponsors, clear outcomes, general house rules etc. He also talks about the need for a high level of energy input from the core team to create and maintain the momentum of the CoP. We have seen this in real life and add our own particular elements to the recipe:

● Start with a nut—You need to have at least one very extroverted and charismatic lead figure. We seek out people that are not just experts in the field but are almost fanatical in their dedication to developing and maturing project management.

---

33. E. Wenger, R. McDermott, and W. M. Snyder, *Cultivating Communities of Practice: A Guide to Managing Knowledge*, Harvard Business School Press, Cambridge, MA, 2002.

- Plant it in the right place—You need to ensure the CoP does not interfere too much with either normal work or personal time. We generally hold our sessions around the lunch hour with food provided.
- You need occasionally to prune the CoP branches to promote strong growth. Groups can develop in directions which do not really promote the central theme of the CoP, in which case it's a good idea to spin them off to avoid dilution.
- When the members leave, don't panic, this is just seasonal. People come and go as their jobs change, their project pressure changes, etc. They'll be back . . .
- The CoP needs to produce more nuts which can be planted in other sites and thus grow an interlinked community of communities. In NXP's case we created a solid core team over the last few years with active communities sharing and learning in 10 different sites around the world.

There is no point in capturing best practices unless the workers know about it. The problem, as identified above, is how to communicate this information to the workers, especially in large, multinational companies. Some of the techniques include:

- Websites
- Best practices libraries
- Community of practice
- Newsletters
- E-mailings
- Internal seminars
- Transferring people
- Case studies
- Other techniques

Nortel Networks strives to ensure timely and consistent communications to all project managers worldwide to help drive continued success in the application of the global project management process. Examples of the various communication methods used by Nortel include:

- The *PM Newsflash* is published on a monthly basis to facilitate communications across the project management organization and related functions.
- Project management communications sessions are held regularly, with a strong focus on providing training, metrics reviews, process and template updates, and so on.
- Broadcast bulletins are utilized to communicate time-sensitive information.
- A centralized repository has been established for project managers to facilitate easy access to and sharing of project management–related information.[34]

The comments by Nortel make it clear that best practices in project management can now permeate all business units of a company, especially those companies that are multinational.

---

34. H. Kerzner, *Best Practices in Project Management: Achieving Global Excellence* Wiley, Hoboken, NJ, 2006, p. 18.

One of the reasons for this is that we now view all activities in a company as a series of projects. Therefore, we are managing our business by projects. Given this fact, best practices in project management are now appearing throughout the company.

Publishing best practices in some form seems to be the preferred method of communications. At Indra, Enrique Sevilla Molina, PMP, Corporate PMO Director, states:

> They are published at corporate level and at the corresponding level inside the affected business unit. Regular courses and training is also provided for newly appointed Project Managers, and their use is periodically reviewed and verified by the internal audit teams. Moreover, the PM corporate tools automate the applications of best practices in projects, as PM best practices become requirements to the PM information systems.

According to a spokesperson from AT&T:

> We have defined a best practice as any tool, template or activity that has had a positive impact on the triple constraint and/or any of the *PMBOK® Guide* Process or Knowledge areas. We allow the individual project manager to determine if it is a best practice based on these criteria. We communicate this through a monthly project management newsletter and highlight a best practice of the month for our project management community.

Another strategic importance of best practices in project management can be seen from the comments below by Suzanne Zale, Global Program Manager at EDS:

> Driven by the world economy, there is a tendency toward an increasing number of large-scale global or international projects. Project managers who do not have global experience tend to treat these global projects as large national projects. However, they are completely different. A more robust project management framework will become more important for such projects. Planning up front with a global perspective becomes extremely important. As an example, establishing a team that has knowledge about geographic regions relevant to the project will be critical to the success of the projects. Project managers must also know how to operate in those geographic areas. It is also essential that all project team members are trained and understand the same overall project management methodology.
> Globalization and technology will make sound project management practice even more important.

Suzanne Zale's comments illustrate the importance of extracting best practices from global projects. This could very well be the future of best practices by the end of this decade.

1.16 STEP 9: ENSURING USAGE OF THE BEST PRACTICES

Why go through the complex process of capturing best practices if people are not going to use them? When companies advertise to their clients that they have best practices, it is understood that tracking of the best practices and how they are used must be done. This is normally part of the responsibility of the PMO. The PMO may have the authority to regularly audit projects to ensure the usage of a best practice but may not have the authority to enforce the usage. The PMO may need to seek out assistance from the head of the PMO, the project sponsor, or various stakeholders for enforcement.

When best practices are used as competitive weapons and advertised to potential customers as part of competitive bidding, the marketing and sales force must understand the best practices and explain this usage to the customers. Unlike 10 years ago, the marketing and sales force today has a good understanding of project management and the accompanying best practices.

1.17 COMMON BELIEFS

There are several common beliefs concerning best practices that companies have found to be valid. A partial list is:

- Because best practices can be interrelated, the identification of one best practice can lead to the discovery of another best practice, especially in the same category or level of best practices. Best practices may be self-perpetuating.
- Because of the dependencies that can exist between best practices, it is often easier to identify categories for best practices rather than individual best practices.
- Best practices may not be transferable. What works well for one company may not work for another company.
- Even though some best practices seem simplistic and common sense in most companies, the constant reminder and use of these best practices lead to excellence and customer satisfaction.
- Best practices are not limited exclusively to companies in good financial health. Companies that are cash rich can make a $10 million mistake and write it off. But companies that are cash poor are very careful in how they approve projects, monitor performance, and evaluate whether or not to cancel the project.

Care must be taken that the implementation of a best practice does not lead to detrimental results. One company decided that the organization must recognize project management as a profession in order to maximize performance and retain qualified people. A project management career path was created and integrated into the corporate reward system.

Unfortunately the company made a severe mistake. Project managers were given significantly larger salary increases than line managers and workers. People became jealous of the project managers and applied for transfer into project management thinking that the "grass was greener." The company's technical prowess diminished and some people resigned when not given the opportunity to become a project manager.

Sometimes, the implementation of a best practice is done with the best of intentions but the final result either does not meet management's expectations or may even produce an undesirable effect. The undesirable effect may not be apparent for some time. As an example, consider the first best practice in Table 1–8. Several companies are now using traffic light reporting for their projects. One company streamlined its intranet project

TABLE 1–8.   IMPROPER APPLICATION OF BEST PRACTICES

Type of Best Practice	Expected Advantage	Potential Disadvantage
Use of traffic light reporting	Speed and simplicity	Poor accuracy of information
Use of a risk management template/form	Forward looking and accurate	Inability to see all possible risks
Highly detailed WBS	Control, accuracy, and completeness	More control and cost of reporting
Using enterprise project management on all projects	Standardization and consistency	Too expensive on certain projects
Using specialized software	Better decision making	Too much reliance on tools

management methodology to include "traffic light" status reporting. Beside every work package in the work breakdown was a traffic light capable of turning red, yellow, or green. Status reporting was simplified and easy for management to follow. The time spent by executives in status review meetings was significantly reduced and significant cost savings were realized.

Initially, this best practice appeared to be beneficial for the company. However, after a few months, it became apparent that the status of a work package, as seen by a traffic light, was not as accurate as the more expensive written reports. There was also some concern as to who would make the decision on the color of the traffic light. Eventually, the traffic light system was enlarged to include eight colors, and guidelines were established for the decision on the color of the lights. In this case, the company was fortunate enough to identify the disadvantage of the best practice and correct it. Not all disadvantages are easily identified, and those that are may not always be correctable.

There are other reasons why best practices can fail or provide unsatisfactory results. These include:

- Lack of stability, clarity, or understanding of the best practice
- Failure to use best practices correctly
- Identifying a best practice that lacks rigor
- Identifying a best practice based upon erroneous judgment
- Failing to provide value

1.18 BEST PRACTICES LIBRARY

With the premise that project management knowledge and best practices are intellectual properties, how does a company retain this information? The solution is usually the creation of a best practices library. Figure 1–12 shows the three levels of best practices that seem most appropriate for storage in a best practices library.

Figure 1–13 shows the process of creating a best practices library. The bottom level is the discovery and understanding of what is or is not a "potential" best practice. The sources for potential best practices can originate anywhere within the organization.

FIGURE 1–12. Levels of best practices.

The next level is the evaluation level to confirm that it is a best practice. The evaluation process can be done by the PMO or a committee but should have involvement by the senior levels of management. The evaluation process is very difficult because a one-time positive occurrence may not reflect a best practice that will be repetitive. There must exist established criteria for the evaluation of a best practice.

Once a best practice is established, most companies provide a more detailed explanation of the best practice as well as providing a means for answering questions concerning its use. However, each company may have a different approach on how to disseminate this critical intellectual property. Most companies prefer to make maximum utilization out of the company's intranet websites. However, some companies simply consider their current forms and templates as the ongoing best practices library. Consider the following example:

> EDS has a website that lists all of the sanctioned best practices. There are currently 12 best practices listed in the library. The library provides a high-level graphic to depict the

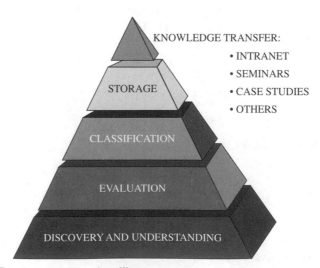

FIGURE 1–13. Creating a best practices library.

relationships of the best practices, a high-level profile of each best practice, and a link to each of the individual best practice websites. All of the best practices are included in a single glossary of terms.

The Client Facing Best Practice website identifies best practices by EDS roles, the situation, problem, or opportunity that one is facing, and from the perspectives of the client relationship life cycle. A best practices grid is provided to the employees with a profile of each best practice, an overview presentation to explain the purpose and uses of the best practice, a link to the training materials, and finally a link to the individual best practice web site. (Provided by Doug Bolzman, Consultant Architect, PMP®, ITIL Service Manager, EDS)

Figure 1–12 showed the levels of best practices, but the classification system for storage purposes can be significantly different. Figure 1–14 shows a typical classification system for a best practices library.

The purpose for creating a best practices library is to transfer knowledge to the employees. The knowledge can be transferred through the company intranet, seminars on best practices, and case studies. Some companies require that the project team prepare case studies on lessons learned and best practices before the team is disbanded. These companies then use the case studies in company-sponsored seminars. Best practices and lessons learned must be communicated to the entire organization. The problem is determining how to do it effectively.

Another critical problem is best practices overload. One company started up a best practices library and, after a few years, had amassed what it considered to be hundreds of best practices. Nobody bothered to reevaluate whether or not all of these were still best practices. After reevaluation had taken place, it was determined that less

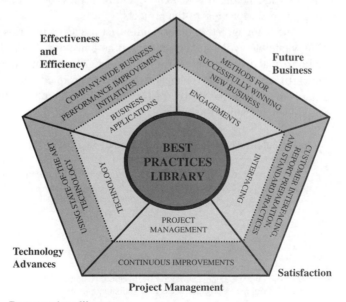

FIGURE 1–14. Best practices library.

than one-third of these were still regarded as best practices. Some were no longer best practices, others needed to be updated, and others had to be replaced with newer best practices.

1.19 DTE ENERGY[35]

In 2002, the Information Technology Services (ITS) Organization at DTE Energy initiated an effort to collect and document best practices for project management. Our intent was to publish, communicate, and ensure these best practices were adopted across the culture. We believed this approach would lead to continuous improvement opportunities resulting in higher quality, more timely, and less expensive software-based business solutions.

Rather than describe the ideal state of project management as we could envision it, we decided to describe the current state as it was being practiced by project managers in the organization. The goal was to establish a baseline set of standards that we knew project managers could meet because we were already doing them.

We formed a small team of our most experienced project managers and IT leaders. This team drew upon their recent project experiences to identify a set of best practices. While not every project manager uniformly followed them, the best practices represented the highest common denominator rather than the lowest. We knew that these were feasible since they represented practical experiences of our most successful project managers, and they also characterized the practices we wanted applied consistently across all projects. In this way, the bar was low enough to be achievable but high enough to be a meaningful improvement for most of the projects.

The team agreed to describe the best practices in terms of "what" rather than "how." We wanted to avoid the difficult and time-consuming task of defining detailed procedures with formal documentation. Rather, we described what project managers needed to do and the artifacts that project managers needed to produce. This provided the practicing project managers with some degree of flexibility in the methods (the "how") they employed to produce the results (the "what").

> We published the best practices in a hundred-page "Project Management Standards and Guidelines" manual (Table of Contents, Table 1–9) and also posted them on our Solution Delivery Process Center (SDPC) intranet website (Screen Shot, Figure 1–15). We included references to other resources such as standard forms, templates, and procedures that already existed.

---

35. Material on DTE Energy provided by Joseph C. Thomas, PMP, Senior Project Manager, and Steve Baker, PMP, CCP, Principal Analyst, Process and Skills Organization.

TABLE 1–9. DFCU FINANCIAL CORPORATE PROJECTS LIST REPORT HEADERS

Priority	1 = Board reported and/or top priority 2 = High priority 3 = Corporate priority, but can be delayed 4 = Business unit focused or completed as time permits
Project	Project name
Description	Brief entry, especially for new initiatives
Requirements document status	R = Required Y = Received N/A = Not needed
Status	Phase (discovery, development, implementation) and percentage completed for current phase
Business owner	Business unit manager who owns the project
Project manager	Person assigned to this role
Projected delivery time	Year/quarter targeted for delivery
Resources	Functional areas or specific staff involved
Project notes	Brief narrative on major upcoming milestones or issues

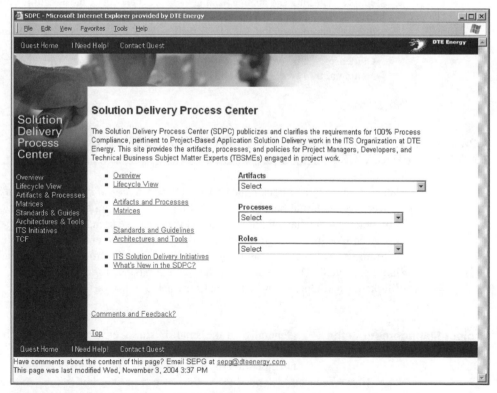

FIGURE 1–15. Solution delivery process center screen shot.

The SDPC is our process asset library containing scores of high-level role and process descriptions, easy-to-access templates and examples, and links to various resources from other departments across ITS. We designed our digital library to be usable for a variety of perspectives, including role-based ("I am a . . ."), milestone-based ("We are at . . ."), artifact-based ("I need a . . ."), and so on.

We launched the Standards and Guidelines materials and the SDPC library with a diverse communication strategy, targeting the right message to the right audience at the right time. We included a feedback process to ensure the evolution and applicability of the standards over time. This continuous improvement process includes (a) an inbound email account for receiving comments, suggestions, and ideas; (b) a review and update process including roles and milestones; and (c) timely updates to the SDPC and targeted printings of the Standards and Guidelines manual.

As a way to capture our lessons learned from each project, we adopted the After Action Review (AAR) process. Every project conducts an AAR upon completion, and we seek improvements to, and innovations beyond, our existing processes and templates. We institutionalize these improvements within our evolving Standards and Guidelines and the SDPC library. We solved our lessons learned dilemma (best practices were dutifully archived but rarely referenced) by incorporating each discovery within our Standards and Guidelines.

Standards and Guidelines, in and of themselves, are a means to an end. We found that simply publishing and communicating them is not enough to meaningfully impact our culture. To that end, we instituted a Quality Management Group (QMG) staffed with a small, diverse team of experts. The QMG both *enables* our projects with consulting services and education, and *ensures* our projects are in compliance with published Standards and Guidelines. With five inspection milestones, the QMG demonstrates our organizational commitment to best practices and continuous improvement.

1.20 A CONSULTANT'S VIEW OF PROJECT MANAGEMENT AND BEST PRACTICES

When companies begin their quest for excellence in project management, emphasis is usually placed upon external benchmarking. While benchmarking studies have merit, the information obtained may have limited or no real value for the company conducting the benchmarking even though they started out with the greatest of intentions. Perhaps the better choice is to hire consultants that can bring to the table expertise from a variety of companies, both large and small. The remainder of this section has been provided by Sandra Kumorowski, Marketing and Operations Consultant for Enakta.

Project Management in the Small Business Environment Generally, in the small business environment, small business owners are more inclined to use project management practices—sometimes not intentionally and mostly through common sense—than anywhere else. I have seen quite sophisticated project estimating, budgeting, and scheduling practices in the small business environment as small business owners consider it the number one skill to master because their livelihood relies on that skill. Scope and objectives are very well defined and expectations set at the beginning of the project. That can, however, vary by industry. Let me start with two examples from construction and health care.

Project Management in Construction

My husband has owned his painting (residential and commercial painting) business for over a decade now. Over so many years, there was not a single project on which he had lost money. His precise estimating skills, communication style, and managerial experience enable him to create quite seamless and timely project experiences. When he works with a general contractor, where the success of the construction project lies in the precise alignment of all subcontracting services, accurate project scheduling is critical. When he works with direct clients, precision and time are even more important as clients' lives can be directly impacted by the painting project.

The number one success factor in his business is a very intensive and continuous *training of his employees*. When he hires someone new, the employee goes through an intensive training. All employees are continuously updated and trained on new innovations and techniques in painting.

The number two success factor is *managed team autonomy*. Each team has a team leader who is fully responsible for the project performance and is required to update my husband everyday on project progress. My husband progressed to this managed team autonomy system from the previous system where he personally almost fully controlled each project (which took up too much of his time and was not necessarily more profitable) and found it very effective. The new system was met with very positive response from the employees who felt more empowered and actually started performing better, especially in terms of time. Now, they finish projects in shorter periods of time keeping up the same if not better quality standards.

My husband's natural project management skills are well-refined and he has never used any formal project management concepts or project management software. He has developed his own project management system that has been working well for him. However, there are many project management challenges in the construction industry. Therefore, I believe there is a great opportunity for these small business owners to be introduced to and educated about project management methodology from which they could benefit to a great extent. When surveyed, most of them considered it too time-consuming. But this opinion comes mainly from limited knowledge about project management and often limited computer skills.

Project Management in Health Care/Dental Category

Based on my 7-year experience in a periodontal specialty practice, I have to admit that the project and patient management practices were quite advanced. The practice was run by one doctor and his incredibly time-efficient and advanced surgical skills allowed for a large number of patients to be seen every day. In one day, the practice could see up to 60 patients divided between the doctor, the hygienist, and the postoperative assistant. That required quite sophisticated scheduling and time-monitoring practices. We used dental practice software Eagelsoft, which, although somewhat limited in reports capability, provided effective tracking and monitoring tools. On average, a patient would see us four to six times a year. To avoid cancellations and disruptions in the schedule, we attempted to schedule all appointments in advance for each patient and established a strict cancellation policy—something quite unusual in the dental category.

Each patient was like a miniproject and without the software and methodical scheduling, we would not be able to manage them successfully. The practice, thanks to its meticulous approach to operations management and scheduling in particular, has enjoyed continuous growth.

Project Management in Higher Education

Project work could be quite challenging for many college students. In higher education, most of the time, projects represent 50 percent of the final grade, but there is not enough attention paid to project management directions.

I did a survey in my class and, out of 44 students, only 2 were familiar with project management. About 70 percent of students perceived project work as a big challenge mainly because of the project management and team management issues.

So I have taken a proactive approach and dedicated about 10 percent of the total semester class time to project management practices and methods to make their project work more effective and successful.

Project Management in the Marketing and Advertising Category

Working in the marketing and advertising category for years now enabled me to study and analyze how project management was understood in the agency environment. Based on my experience, first of all, there is not enough awareness about project management practices and proven benefits. People do not believe in it and, most of the time, project management is perceived as a deterrent to creativity. But it is important to state that the way client accounts and projects are managed is quite consistent with a project management methodology. The problem, however, is that account planners do not reach out to the project management discipline to improve upon processes and make projects perform better. There is no formal relationship between project management and account management in marketing/advertising category.

When I introduced project management to my company of advertising executives as a way to improve our business, it took me some time to make them believe in it.

The story of how I did it follows.

Project Management Initiative for a Marketing Strategy Consultancy

About the Company My company was a small (22 employees) but high-profile successful marketing strategy consultancy. Founded in 2003 by an advertising executive, it focuses on delivery of research insights and related actionable strategy based on the analysis of online conversations. It caters to Fortune 500 companies and earned its reputation on coming up with the "big idea" in each project and helping the brands in building their long-term brand equity.

Need for a Project Management Initiative Due to the project-based nature of the company, I was approached by the top management to evaluate current project management practices and establish a more formal project management system. As the company grew, the need for a consistent system for our project planning and execution became very apparent.

Our projects ranged from two-week turnaround projects to eight-week research studies and there was a need to create a project portfolio management system to better control our future planning.

In addition, some more serious human issues began to surface. Most of the problems were in the initiation and planning phases where communication got lost mainly due to too many people being involved. Scopes were sometimes out of control and the team would find out at the end of the project that they were working on something different than what the client wanted. Main message (scope/objectives) would get lost.

Later in the project, the scoping mistakes would exponentially amplify into sometimes barely controllable situations where team members would have to stay late into the night and on weekends trying to bring the project to the satisfactory completion stage. The stress levels were sometimes very high.

Another issue that kept disturbing our projects was team dynamics. Each team consisted of two members—a senior strategist who was the project lead and a junior strategist. The role of the senior strategist on a project was defined as 80 percent strategy development and 20 percent data gathering and analysis. The junior strategist had to do 80 percent data gathering and was contributing only 20 percent to the strategy development. That led to some strong feelings of unfairness and some team members refused to work together. That needed to be changed.

There were seven main areas where the use of project management as an instrument for continuous improvement was apparent:

1. Client relationship management (CRM)
2. Team health/dynamics
3. Project quality/creativity
4. Project costing/budgeting
5. Project scope/timing
6. Project portfolio management
7. Company culture

The needed change can be seen in Figure 1–16.

Project Management Initiative Chronology

1. Need Recognition: Need for project management recognition based on scope creeps, unhappy team, lack of project control, and so on.
2. Initial Buy-In Meeting with Top Management: Establish immediate and relevant need
3. Interviews: Thorough interviews with each employee including top management (I created a questionnaire for each employee and then met with each of them individually to discuss it.)
4. Research and Analysis: Research of project management best practices, current processes analysis, interview analysis
5. Project Management Best Practices Research Period
6. Second Buy-In Meeting with Top Management: To reconfirm the need for project management, to review cost, timeline, and objectives of the project management initiative, to prioritize areas of focus for project management initiative
7. Project Management Initiative Goal Definition
8. Official Announcement: To all employees
9. Goal and Tactics Implementation: and testing
10. Monthly Progress Reports

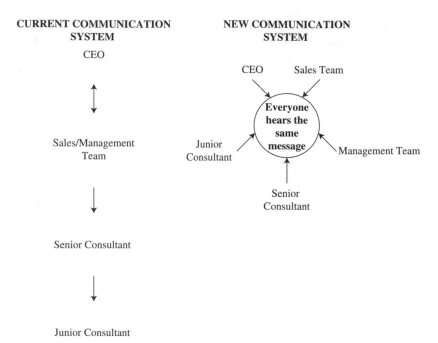

FIGURE 1–16. The before and after communication system.

Example of Project Management Initiative Initial Goals Since our company was new at project management, the changes took place in stages. Below (see Table 1–10) are some of the goals that were reached and actions taken as soon as the project management initiative was implemented.

Goal 1 CRM: To build long-term client relationship, increase word of mouth and reputation, increase sales
Action: List of all client meetings to be scheduled.
Goal 2 Team Health/Dynamics: To establish effective team collaboration and work load balance to achieve the best quality product
Action: 1. Schedule lecture on new project management processes, effective teamwork and leadership for the whole team FRIDAY. 2. Schedule meeting on leadership with project leaders.
Goal 3 Project Planning: Scope management/meeting effectiveness
Action: 1. Finalize and use internal kick-off meeting agenda template. 2. Finalize and use internal postmortem meeting agenda template.
Goal 4 PM Project Monitoring: To track projects and the success of the project management project
Action: 1. Develop ALL projects tracking format (white board). 2. Report on project management goals presented monthly.

TABLE 1–10. PROJECT MANAGEMENT INITIATIVE OVERVIEW

Priority Rank	Area	Action	Outcome/Benefit	Measure
1.	**Client Relationship Mgmt** REASON: RETURNING CLIENT/SALES	1. **Establish personal contact w/client** (add travel cost to contract), on longer, complex, opportunistic projects 2. **Improve planning phase** 3. **Scope/contract template** 4. **Post mortem template** 5. Establish meetings with standardized agendas & let everyone work on scope statement	• Long-term client relationships • Gaining credibility • More projects • Growing company Reputation	• Client survey • Client retention/return rate • Recommendation rate • Number of n-w/old client projects per month/year
2.	**Team Health/Dynamics** REASON: RELATIONSHIPS	1. **Assign 3 people in each project** (3rd for balance/control) 2. **40/60 workload distribution** (instead of 20/80 - Sr. Consultants/Jr. Consultants) 3. **New PM responsibilities** 4. Redefinition of current roles	• Better team collaboration • Less stress and issues • Jr. Consultants feel more valued • Sr. Consultants empowered with more responsibility	• Postmortem evaluation results (Jr. Consultant openly evaluates Sr. Consultant and vice versa?) • WESS (Weighed Employee Satisfaction Survey) ???
3.	**Creativity** REASON: LIMITED BY 2,4,5	1. **Standardize project planning** processes through project structure standardization 2. **Stick to well-defined roles** (everyone knows what to do)	• More time to create • Less time to worry	• Big idea in every project • Actionable insights in every project
4.	**Project Costing/Budgeting** REASON: CONTROL	1. Devise & implement best customized **project costing system (MS project)** 2. Cost estimates at the beginning of the project (MS project)	• Cost control throughout the project • Ability to price projects to avoid losses	• Earned value method • Cost variance (CV=EV-AC), *see Appendix* • CPI (Cost Performance Index)
5.	**Project Scope/Timing** REASON: CONTROL	1. Implement customized project **timing/planning system (MS project)** 2. Focus on planning phase to avoid **scope/time creep**	• Scope/time control throughout the project	• Earned value method • Schedule variance (SV=EV-PV), *see Appendix* • SPI (Schedule Perf. Index)
6.	**Project Portfolio Mgmt** REASON: BIG PICTURE, STRATEGY	1. Devise & implement all projects **tracking tool (MS project)** 2. **Analyze projects** based on cost, price, time, time-sensitivity, opportunity	• Effective project portfolio Management to make strategic business decisions • Decreased uncertainty & better future planning	• Profit margin • Effective planning & scheduling of future projects
7.	**Culture**	1. **Apply transparency** through open communication & active communication throughout each project	• Less fear & resentment • Innovation/creativity • Learning culture • Transparency	• Post-mortem evaluation results generalized & Learned from: best practices library

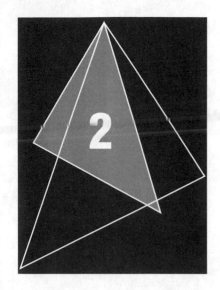

From Best Practice to Migraine Headache

2.0 INTRODUCTION

For almost 30 years, project management resided in relatively few industries such as aerospace, defense, and heavy construction. These industries were project-driven and implemented project management mainly to placate customer requests. Project management was considered as something nice to have but not a necessity. As a result, best practices in project management were never really considered as important.

Within the last two decades, project management has evolved into a management process that is mandatory for the long-term survival of the firm. Project management is now a necessity rather than a luxury. Project management permeates all aspects of a business. Companies are now managing their business by projects. Project management has become a competitive weapon. The knowledge learned from project management is treated as intellectual property and PMOs have been established as the guardians of the project management intellectual property, reporting to the senior levels of management and being given the task of capturing best practices in project management.

As with any new project management activity, benefits are accompanied by disadvantages and potential problems. Some of the problems are small and easy to correct while others are colossal migraine headaches and keep executives awake at night. The majority of the migraine headaches emanate from either a poor understanding of the benefits of project management or having expectations that are set too high. Other potential problems occur when an activity really is not a best practice and detrimental results occur.

2.1 GOOD INTENTIONS BECOMING MIGRAINES

Sometimes, the best intentions can turn into migraine headaches. As an example, one company quickly realized the importance of project management and made it a career path position. This was certainly the right thing to do. Internally, people believed that the

FIGURE 2–1. Risk growth.

company considered project management as a strategic competency and professionalism in project management evolved. Externally, their customers were quite pleased seeing project management as a career path discipline and the business improved.

These good intentions soon turned into problems. To show their support for excellence in project management, the project managers were provided with 14 percent salary increases whereas project team members and line managers received 3–4 percent. Within two years after implementing a project management career path, everyone was trying to become a project manager and climb the project management career path ladder of success, including critical line managers with specialized expertise. Everyone thought that "the grass was greener" in the project manager's yard than in his or her yard. Line managers with critical skills were threatening to resign from the company if they were not given the chance to become project managers. The company eventually corrected the problem by telling everyone that every career path ladder in the company had the same career path opportunities for advancement. The large differential in salary increases disappeared and was replaced by a more equitable plan. However, the damage was done. Team members and line managers felt that the project managers exploited them, and the working relationship suffered. Executives were now faced with the headache of trying to repair the damage.

Figure 2–1 illustrates why many other headaches occur. As project management grows and evolves into a competitive weapon, pressure is placed upon the organization to implement best practices, many of which necessitate the implementation of costly internal control systems for the management of resources, costs, schedules, and quality. The project management systems must be able to handle several projects running concurrently. Likewise, obtaining customer satisfaction is also regarded as a best practice and can come at a price. As the importance of both increases, so do the risks and the headaches. Maintaining parity between customer satisfaction and internal controls is not easy. Spending too much time and money on customer satisfaction could lead to financial disaster on a given project. Spending too much time on internal controls could lead to noncompetitiveness.

2.2 ENTERPRISE PROJECT MANAGEMENT METHODOLOGY MIGRAINE

As the importance of project management became apparent, companies recognized the need to develop project management methodologies. Good methodologies are best practices and can lead to sole-source contracting based upon the ability of the methodology to continuously deliver quality results and the faith that the customer has in the methodology. Unfortunately, marketing, manufacturing, information systems, R&D, and engineering may have their own methodology for project management. In one company, this suboptimization was acceptable to management as long as these individual functional areas did not have to work together continuously. Each methodology had its own terminology, life-cycle phases, and gate review processes.

When customers began demanding complete solutions to their business needs rather than products from various functional units, the need to minimize the number of methodologies became apparent. Complete solutions required that several functional units work together. This was regarded by senior management as a necessity, and senior management believed that this would eventually turn into a best practice as well as lead to the discovery of other best practices.

One company had three strategic business units (SBUs), which, because of changing customer demands, now were required to work together because of specific customer solution requirements. Senior management instructed one of the SBUs to take the lead role in condensing all of their functional processes into one enterprise project management (EPM) methodology. After some degree of success became apparent, senior management tried unsuccessfully to get the other two SBUs to implement this EPM methodology that was believed to be a best practice. The arguments provided were "We don't need it," "It doesn't apply to us," and "It wasn't invented here." Reluctantly, the president of the company made it clear to his staff that there was now no choice. Everyone would use the same methodology. The president is now facing the same challenge with globalization of acceptance of the methodology. Now cultural issues become important.

2.3 CUSTOMER SATISFACTION MIGRAINE

Companies have traditionally viewed each customer as a one-time opportunity, and after this customer's needs were met, emphasis was placed upon finding other customers. This is acceptable as long as there exists a large potential customer base. Today, project-driven organizations, namely those that survive on the income from a continuous stream of customer-funded projects, are implementing the "engagement project management" approach. With engagement project management, each potential new customer is approached in a way that is similar to an engagement in marriage where the contractor is soliciting a long-term relationship with the customer rather than a one-shot opportunity. With this approach, contractors are selling not only deliverables and complete solutions but also a willingness to make their EPM methodology compatible with the customer's methodology. To maintain customer satisfaction and hopefully a long-term relationship, customers are requested to provide input on how the contractor's EPM methodology can be extended into their organization. The last life-cycle phase in the EPM methodology used by ABB (Asea, Brown, and Boveri) is called "customer satisfaction management" and is specifically designed to solicit feedback from the customer for long-term customer satisfaction.

This best practice of implementing engagement project management is a powerful best practice because it allows the company to capitalize on its view of project management, namely that project management has evolved into a strategic competency for the firm leading to a sustained competitive advantage. While this approach has merit, it opened a Pandora's box. Customers were now expecting to have a say in the design of the contractor's EPM methodology. One automotive supplier decided to solicit input from one of the Big Three in Detroit when developing its EPM approach. Although this created goodwill and customer satisfaction with one client, it created a severe problem with other clients that had different requirements and different views of project management. How much freedom should a client be given in making recommendations for changes to a contractor's EPM system? Is it a good idea to run the risk of opening Pandora's box for the benefit of customer satisfaction? How much say should a customer have in how a contractor manages projects? What happens if this allows customers to begin telling contractors how to do their job?

2.4 MIGRAINE RESULTING FROM RESPONDING TO CHANGING CUSTOMER REQUIREMENTS

When project management becomes a competitive weapon and eventually leads to a strategic competitive advantage, changes resulting from customer requests must be done quickly. The EPM system must have a process for configuration management for the control of changes. The change control process portion of the EPM system must maintain flexibility. But what happens when customer requirements change to such a degree that corresponding changes to the EPM system must be made and these changes could lead to detrimental results rather than best practices?

One automotive tier 1 supplier spent years developing an EPM system that was highly regarded by the customers for the development of new products or components. The EPM system was viewed by both the customers and the company as a best practice. But this was about to change. Customers were now trying to save money by working with fewer suppliers. Certain suppliers would be selected to become "solution providers" responsible for major sections or chunks of the car rather than individual components. Several tier 1 suppliers acquired other companies through mergers and acquisitions in order to become component suppliers. The entire EPM system had to be changed and, in many cases, cultural shock occurred. Some of the acquired companies had strong project management cultures and their own best practices, even stronger than the acquirer, while others were clueless about project management. And to make matters even worse, all of these companies were multinational and globalization issues would take center stage. We now had competing best practices.

After years of struggling, success was now at hand for many component suppliers. The mergers and acquisitions were successful and new common sets of best practices were implemented. But, once again, customer requirements were about to change. Customers were now contemplating returning to component procurement rather than "solution provider" procurement, believing that costs would be lowered. Should this occur across the industry, colossal migraines will appear due to massive restructuring, divestitures, cultural changes, and major changes to the EPM systems. How do contractors convince customers that their actions may be detrimental to the entire industry? Furthermore, some companies that were previously financially successful as chunk or section manufacturers might no longer have the same degree of success as component manufacturers.

2.5 REPORTING LEVEL OF PMO MIGRAINE

Companies have established a PMO as the guardian of project management intellectual property. Included in the responsibilities of a PMO are strategic planning for project management, development of and enhancement to the EPM, maintenance of project management templates, forms and guidelines, portfolio management of projects, mentorship of inexperienced project managers, a hot line for project problem solving, and maintaining a project management best practices library. The PMO becomes the guardian of all of the project management best practices.

While the creation of a PMO is regarded as a best practice for most companies, it places a great deal of intellectual property in the hands of a few, and information is power. And with all of this intellectual property in the hands of three or four people in the PMO, the person to whom the PMO reports could possibly become more powerful than his or her counterparts. What is unfortunate is that the PMO must report to the executive levels of management and there appears to be severe infighting at the executive levels for control of the PMO.

To allay the fears of one executive becoming more powerful than another, companies have created multiple PMOs which are supposedly networked together and sharing information freely. Hewlett-Packard has multiple PMOs all networked together. Exel Corporation has PMOs in the United States, Europe, Asia, Mexico, and Brazil all networked together. Star Alliance has a membership of 24 airlines, each with a PMO and all networked together with a PMO in Germany as the lead. These PMOs are successful because information and project management intellectual property are shared freely.

Allowing multiple PMOs to exist may seem like the right thing to do to appease each executive, but in some cases it has created the headaches of project management intellectual property that is no longer centralized. And to make matters worse, what happens if every executive, including multinational executives, each demand their own PMO? This might eventually be viewed as an overmanagement expense, and unless the company can see a return on investment on each PMO, the concept of the PMO might disappear, thus destroying an important best practice because of internal politics.

2.6 CASH FLOW DILEMMA MIGRAINE

For many companies that survive on competitive bidding, the cost of preparing a bid can range from a few thousand dollars to hundreds of thousands. In most cases, project management may not appear until after the contract is awarded. The results can be catastrophic if benefit realization at the end of the project does not match the vision or profit margin expected during proposal preparation or at project initiation. When companies develop an EPM system and the system works well, most companies believe that they can now take on more work. They begin bidding on every possible contract believing that with the EPM system they can accomplish more work in less time and with fewer resources without any sacrifice of quality.

In the summer of 2002, a large, multinational company set up a project management training program in Europe for 50 multinational project managers. The executive vice president spoke for the first 10 minutes of the class and said, "The company is now going to begin turning away work." The project managers were upset over hearing this

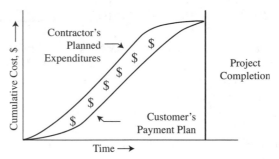

FIGURE 2–2. Spending curve.

and needed an explanation. The executive vice president put Figure 2–2 on the screen and made it clear that the company would no longer accept projects where profit margins would eventually be less than 4–6 percent because they were financing the projects for their customers. The company was functioning as a banker for its clients. Benefit realization was not being achieved. To reduce the costs of competitive bidding, the company was responding to proposal requests using estimating databases rather than time-phased labor. The cash flow issue was not being identified until after go-ahead.

While project financing has become an acceptable practice, it does squeeze profits in already highly competitive markets. To maintain profit margins, companies are often forced to disregard what was told to the customer in the proposal and to assign project resources according to the customer's payment plan rather than the original project schedule provided in the proposal. While this may lead to short-term profitability, it often results in elongated schedules, potential lawsuits, and customer dissatisfaction. The balance between customer satisfaction, long-term client relationships, and profitability is creating a huge headache. The best practice of creating a world-class EPM system can lead to detrimental results if profitability cannot be maintained.

2.7 SCOPE CHANGE DILEMMA MIGRAINE

For companies that require successful competitive bidding for survival, the pot of gold is often the amount of scope changes that occur after go-ahead. The original contract may be underbid in the hope that lucrative customer or contractor-generated scope changes will occur. For maximization of profit, a best practices scope change control process must be part of the EPM system.

Over the years, project managers have been encouraged by their superiors to seek out any and all value-added scope changes to be funded by the customers. But these scope changes are now playing havoc with capacity-planning activities and the assigning of critical resources needed for the scope changes and other projects. As companies mature in project management, the EPM systems become Web based. All individual project schedules are rolled up into a master schedule such that senior management can get a realistic picture of resources committed for the next 90 or 180 days. This allows a company to determine how much additional work it can undertake without overtaxing the existing labor base. And if a resource bottleneck is detected, it should be relatively clear how many additional resources should be hired and in which functional groups.

As capacity planning converts from an art to a science, the problems with obtaining qualified resources for unplanned scope changes grow. Maximization of profits on a particular project may not be in the best interest of the company, especially if the resources can be used more effectively elsewhere in the organization. Organizations today are understaffed, believing that it is better to have more work than people rather than more people than work. Executives must find a way to balance the need for added project resources, scope changes, portfolio selection of projects, and the strain on the working relationship between project and line managers. How do executives now convince project managers that scope changes are unnecessary and to forget profit maximization?

2.8 OUTSOURCE OR NOT MIGRAINE

One of the responsibilities of a PMO is debriefing the project team at the completion of the project. This includes capturing lessons learned, identifying opportunities for improving the EPM system, and updating the estimating database. As the estimating database improves, companies realize that they can outsource some project work at a significantly lower cost than performing the same work internally.

While this function can become an important best practice and can save the company some money, there may be detrimental results. A bank received significant negative publicity in local newspapers when it was discovered that the information systems division would be downsized in conjunction with cost-effective outsourcing. Another organization also outsourced its information systems work to such an extent that it had to begin providing its suppliers and contractors with company-proprietary data. Headaches occur when executives must balance short-term profitability with the long-term health of the corporation and community stakeholder needs and expectations.

Best practices are designed to benefit both the company and the workers. When the implementation of best practices leads to loss of employment, the relative importance of best practices can diminish in the eyes of the employees.

2.9 MIGRAINE OF DETERMINING WHEN TO CANCEL A PROJECT

Virtually every EPM system is based upon life-cycle phases. Each life-cycle phase terminates with an end-of-phase gate review meeting designed to function as a go/no-go decision point for proceeding to the next phase. Very few projects seem to be terminated at the early gate review meetings. One reason for this is that project managers do not necessarily provide all of the critical information necessary to make a viable decision. Project managers provide information in forecast reports on the estimated cost at completion and time at completion. What are missing are the expected benefits at completion and this value may be more important than time and cost. While it is understandable that this value may be difficult to obtain during early life-cycle phases, every attempt should be made to present reasonable benefits-at-completion estimates.

If a project comes in late or is over budget, the expected benefits may still be achievable. Likewise, if a project is under budget or ahead of schedule, there may be no reason to believe that the vision at the project's initiation will be met at completion. Intel has initiated a concept called map days where the team maps out its performance to date. This can be expanded to include possible benefits at completion.

While good project management methodologies are best practices and provide valuable information for the management of projects, the system must also be capable of providing the necessary information to senior management for critical decision-making. All too often, EPM systems are developed for the benefit of project managers alone rather than for the best interest of the entire company.

2.10 MIGRAINE OF PROVIDING PROJECT AWARDS

Perhaps the biggest headache facing senior management is the establishment of an equitable project award/recognition system that is part of the Wage and Salary Administration Program. Companies have recognized that project management is a team effort and that rewarding project teams may be more beneficial than rewarding individuals. The headache is how to do it effectively.

There are several questions that need to be addressed:

- Who determines the magnitude of each person's contribution to the project's success?
- Should the amount of time spent on the project impact the size of the award?
- Who determines the size of the award?
- Will the award system impact future estimating, especially if the awards are based upon underruns in cost?
- Will the size of the awards impact future personnel selection for projects?
- Will employees migrate to project managers that have a previous history of success where large awards are provided?
- Will people migrate away from high-risk projects where rewards may not be forthcoming?
- Will employees avoid assignments to long-term projects?
- Can union employees participate in the project award system?

Providing m onetary and nonmonetary recognition is a best practice as long as it is accomplished in an equitable manner. Failure to do so can destroy even the best EPM systems as well as a corporate culture that has taken years to develop.

2.11 MIGRAINE FROM HAVING WRONG CULTURE IN PLACE _____

Creating the right corporate culture for project management is not easy. However, when a strong corporate culture is in place and it actively supports project management such that other best practices are forthcoming, the culture is very difficult to duplicate in other companies. Some corporate cultures lack cooperation among the players and support well-protected silos. Other cultures are based upon mistrust while yet others foster an atmosphere where it is acceptable to persistently withhold information from management.

A telecommunications company funded more than 20 new product development projects which all had to be completed within a specific quarter to appease Wall Street and provide cash flow to support the dividend. Management persistently overreacted to bad news and information flow to senior management became filtered. The project management methodology was used sparingly for fear that management would recognize early on the seriousness of problems with some of the projects.

Not hearing any bad news, senior management became convinced that the projects were progressing as planned. When it was discovered that more than one project was in serious trouble, management conducted intensive project reviews on all projects. In one day, eight project managers were either relieved of their responsibilities or fired. But the damage was done and the problem was really the culture that had been created. Beheading the bearer of bad news can destroy potentially good project management systems and lower morale.

In another telecommunications company, senior management encouraged creativity and provided the workforce with the freedom to be creative. The workforce was heavily loaded with technical employees with advanced degrees. Employees were expected to spend up to 20 percent of their time coming up with ideas for new products. Unfortunately, this time was being charged back to whatever projects the employees were working on at the time, thus making the cost and schedule portion of the EPM system ineffective.

While management appeared to have good intentions, the results were not what management expected. New products were being developed but the payback period was getting longer and longer while operating costs were increasing. Budgets established during the portfolio selection of the projects process were meaningless. To make matters worse, the technical community defined project success as exceeding specifications rather than meeting them. Management on the other hand defined success as commercialization of a product. And given the fact that as many as 50–60 new ideas and projects must be undertaken to have one commercially acceptable success, the cost of new product development was bleeding the company of cash and project management was initially blamed as the culprit. Even the best EPM systems are unable to detect when the work has been completed other than by looking at money consumed and time spent.

It may take years to build up a good culture for project management, but it can be destroyed rapidly through the personal whims of management. A company undertook two high-risk R&D projects. A time frame of 12 months was established for each in hopes of making a technology breakthrough and, even if it could happen, the product would have a shelf life of about 1 year before obsolescence would occur.

Each project had a project sponsor assigned from the executive levels. At the first gate review meeting, both project managers recommended that their projects be terminated. The executive sponsors, in order to save face, ordered the projects to continue to the next gate review rather than terminate the projects while the losses were small. The executives forced the projects to continue on to fruition. The technical breakthroughs occurred 6 months late and virtually no sales occurred with either product. There was only one way the executive sponsors could save face—promote both project managers for successfully developing two new products and then blame marketing and sales for their inability to find customers.

Pulling the plug on projects is never easy. People often view bad news as a personal failure, a sign of weakness, and a blemish on their career path. There must be an understanding that exposing a failure is not a sign of weakness. The lesson is clear: Any executive who always makes the right decision is certainly not making enough decisions and any company where all of the projects are being completed successfully is not working on enough projects and not accepting reasonable risk.

2.12 SOURCES OF SMALLER MIGRAINES

Not all project management headaches lead to migraines. The following list identifies some of the smaller headaches that occurred in various companies but do not necessarily lead to major migraines:

- *Maintaining Original Constraints:* As the project team began working on the project, work began to expand. Some people believed that within every project there was a larger project just waiting to be recognized. Having multiple project sponsors all of whom had their own agendas for the project created this problem.
- *Revisions to Original Mission Statement:* At the gate review meetings, project redirection occurred as management rethought its original mission statement. While these types of changes were inevitable, the magnitude of redirections had a devastating effect on the EPM system, portfolio management efforts, and capacity planning.
- *Lack of Metrics:* An IT organization maintained a staff of over 500 employees. At any given time, senior management was unable to establish metrics on whether or not the IT group was overstaffed, understaffed, or just right. Prioritization of resources was being done poorly and resource management became reactive rather than proactive.
- *More Metrics:* In another example, the IT management team, to help identify whether or not projects were being delivered on schedule, recently implemented an IT balanced scorecard for projects. After the first six months of metric gathering, the conclusion was that 85 percent of all projects were delivered on time. From executive management's perspective, this appeared to be misleading, but there was no way to accurately determine whether or not this number was accurate. For example, one executive personally knew that none of his top 5 projects and all 10 of an IT manager's projects were behind schedule. Executive management believed the true challenge would be determining appropriate metrics for measuring a project's schedule, quality, and budget data.
- *Portfolio Management of Projects:* When reviewing project portfolios or individual projects, all of the plans were at different levels of detail and accuracy. For example, some plans included only milestones with key dates while other plans

had too much detail. The key issue became "what is the correct balance of information that should be included in a plan and how can all plans provide a consistent level of accuracy across all projects?" Even the term *accuracy* was not consistent across the organization.

- *Prioritization of Projects and Resources:* In one company, there were no mechanisms in place to prioritize projects throughout the organization, and this further complicated resource assignment issues in the organization. For example, the CIO had his top 5 projects, one executive had his top 10 projects, and an IT manager had his top 10 projects. Besides having to share project managers and project resources across all of these projects, there was no objective way to determine that the CIO's #3 project was more/less important than an executive's #6 project or an IT manager's #1 project. Therefore, when competing interests developed, subjective decisions were made and it was challenging to determine whether or not the right decision had been made.

- *Shared Accountability for Success and Failure:* The organization's projects traditionally were characterized as single-resource, single-process, and single-platform projects. Now, almost every project was cross team, cross platform, and cross process. This new model had not only increased the complexity and risk for many projects but also required increased accountability by the project team for the success/failure of the project. Unfortunately, the organization's culture and people still embraced the old model. For example, if one team was successful on its part of a project and another was not, the attitude would be "I am glad I was not the one who caused the project to fail" and "Even though the project failed, I succeeded because I did my part." While there was some merit to this, overall, the culture needed to be changed to support an environment where "If the project succeeds, we all succeed" and vice versa.

- *Measuring Project Results:* Many of the projects that were completed were approved based on process improvements and enhanced efficiency. However, after a process improvement project was completed, there were no programs in place to determine whether or not the improvements were achieved. In fact, because the company was experiencing double-digit growth annually, the executive team questioned whether or not approved process improvements were truly scalable in the long term.

- *Integrating Multiple Methodologies:* Application development teams had adopted the software development methodology (SDM) and agile methodology for software development. Both of these methodologies had excellent approaches for delivering software components that met quality, budget, and schedule objectives. The challenge the organization faced was whether or not components from both of these methodologies could be adapted to projects that were not software development related and, if so, how can this be accomplished? This debate had elevated to upper management for resolution and upper management had been reluctant to make a decision one way or the other. This difference in views on how projects should be managed, regardless of whether or not the project was software development related or not, had led to several different groups lobbying for others to join their efforts to support SDM and Agile for all projects. Overall, the lobbying efforts were not adding value to the organization and were wasted effort by key resources.

- *Organizational Communications:* Although there was a lot of communication about projects throughout the organization, many shortcomings existed with the existing process. For example, one executive stated that when he had his monthly status meeting with his direct reports, he was amazed when a manager was not aware of another manager's project, especially if the project was getting ready to migrate into production. The existing process led many managers to react to projects instead of proactive planning for projects. Additionally, the existing communication process did not facilitate knowledge sharing and coordination across projects or throughout the organization. Instead, the existing communication process facilitated individual silos of communication.
- *Meaning of Words:* A project was initiated from the staff level. The SOW contained numerous open-ended phrases with vague language such as "Develop a world-class control platform with exceptional ergonomics and visual appeal." The project manager and his team interpreted this SOW using their own creativity. There were mostly engineers on the team with no marketing members and the solution ended up being technically strong but a sales/marketing disaster. Months were lost in time to market.
- *Problem with Success:* A project was approved with a team charter that loosely defined the boundaries of the project. During the course of the project, some early successes were realized and word quickly spread throughout the organization. As the project moved forward, certain department managers began "sliding" issues into the project scope using their own interpretation of the SOW, hoping to advance their own agendas with this talented group. The project eventually bogged down and the team became demoralized. Senior management disbanded the group. After this, management had real trouble getting people to participate on project teams.
- *Authority Challenges:* A new cross-functional project team was assembled involving technical experts from numerous departments. The project manager was a consultant from an outside contractor. During the course of this large project, resource conflicts with production schedules began to arise. Inevitably, the line managers began to draw resources away from the project. The consultant promptly reported pending delays due to this action and the staff reiterated the consultant's concerns and the need for the organization to support the project. The struggles continued through the entire length of the project, creating stressful situations for team members as they tried to balance their workloads. The project finished late with significant cost overruns and indirectly caused a great deal of animosity among many of the participants.
- *Open-Ended Deliverables:* A project was launched to redesign and deploy the engineering change management system. The team received strong support throughout its duration. At a project closure meeting with the executive staff, the project manager presented the team's interpretation of the deliverables. Much to his surprise, the staff determined that the deliverables were not complete. In the end, this particular team worked on "spider webs" spawning off of their original SOW for over three years (the original closing presentation took place after nine months). The team was frustrated working on a project that never seemed to have an end and the staff grew

impatient with a team they felt was "milking" a job. The project management process at the company came under fire, threatening future efforts and staff support.

- *Cost Overruns:* Soon after a major product renovation project was commissioned, the project manager reported that the cost of completion was grossly understated. Unfortunately, the marketing department, in anticipation of a timely completion, had already gone to the marketplace with a promotion "blitz" and customer expectations were high in anticipation of the product's release. The senior staff was faced with a decision to have a runaway cost issue to complete the project on time or endure loss of face and sales in the marketplace to delay the project's completion.

Despite all of these headaches, project management does work and works well. But is project management falling short of expectations? Some people argue "yes" because project management is not some magical charm that can produce deliverables under all circumstances. Others argue that project management works well and nothing is wrong except that expectations of the executives are overinflated. Project management can succeed or fail, but the intent, commitment, and understanding at the executive levels must be there.

2.13 TEN UGLIES OF PROJECTS[1]

Introduction

Project management methodologies, classes, and books are adequate at explaining the mechanics of running projects and the tools used to do so. Understanding these mechanics is essential, but it is experience that distinguishes successful project managers. More specifically, it is the sum of all of the negative experiences that project managers have in their careers that teaches them what not to do. As Vernon Law explains, "Experience is a hard teacher because she gives the test first, the lesson afterwards."

In my many years of project management experience, I have come across several areas that consistently cause projects to experience difficulties. I call these the "uglies" of projects since these are the things that make projects turn ugly. These are also usually the things that, once recognized, are hard to fix easily.

This section will discuss the 10 project uglies and propose some resolutions. There are definitely other uglies out there, but these 10 are the ones that seem to be the most common and have the biggest impact based on my experience.

The Ten Uglies

Below are the 10 uglies with a description of each and some symptoms that indicate that these uglies may be happening.

1. Lack of Maintained Documentation: Oftentimes when projects are in a crunch, the first thing that gets eliminated is documentation. Sometimes documentation is not

---

1. This section was provided by Kerry R. Wills, PMP®, Director of Portfolio Management, Infrastructure Solutions Division. The Hartford. ©2005 by Kerry R. Wills. Reproduced by permission of Kerry R. Wills.

created even when projects do have the time. When documentation is createdproperly, as projects continue to progress, it is a rarity to see the documentation maintained.

Symptoms

- Requirement documents that do not match what was produced
- Technical documents that can not be used to maintain the technology because they are outdated
- No documentation on what decisions were made and why they were made
- No audit trail of changes made

This is a problem since documentation provides the stewardship of the project. By this I mean that future projects and the people maintaining the project once it has been completed need the documentation to understand *what* was created, *why* it was created, and *how* it was created. Otherwise, they wind up falling into the same traps that happened before—in this case "he who ignores history in documentation is doomed to repeat it."

2. Pile Phenomenon: "What is that under the rug?" is a question often asked toward the end of a project. The mainstream work always gets the primary focus on a project but it is those tangential things that get forgotten or pushed off until "later," at which point there are several piles (swept under the rug) that need to be handled. I call this the "pile phenomenon" because team members think of it as a phenomenon that all this "extra" work has suddenly appeared at the end.

Symptoms

- Any work that gets identified as "we will do this later" but is not on a plan somewhere
- Growing logs (issues, defects, etc.)
- Documentation assumed to be done at the end

There is no "later" accounted for in most project plans, and therefore these items either get dropped or there is a mad rush at the end to finish the work.

3. No Quality at Source: Project team members do not always take on the mantra of "quality at the source." There is sometimes a mentality that "someone else will find the mistakes" rather than a mentality of ownership of quality. Project managers do not always have the ability to review all work, so they must rely on their team members. Therefore, the team members must have the onus to ensure that whatever they put their name on represents their best work.

Symptoms

- Handing off work with errors before reviewing it
- Developing code without testing it
- Not caring about the presentation of work

There are several studies that show that quality issues not found at the source have an exponential cost when found later in the project.

4. Wrong People on the Job. Project roles require the right match of skills and responsibilities. Sometimes a person's skill set does not fit well with the role that he or she has been given. I also find that work ethic is just as important as skills.

Symptoms

- Team members being shown the same things repeatedly
- Consistent missing of dates
- Consistent poor quality

As project managers, all we have are our resources. Not having the right fit for team members will result in working harder than necessary and impacts everyone else on the team who has to pick up the slack. There is also a motivational issue here: When team members are in the wrong roles, they may not feel challenged or feel that they are working to their potential. This has the impact of those persons not giving their best effort, not embodying a solid work ethic when they normally would, feeling underutilized, and so on.

5. Not Involve the Right People: The people who know how to make the project successful are the team members working on the project. Not involving the right team members at the right time can set the project up for failure before it begins.

Symptoms

- Having to make changes to work already completed
- Constant scope changes from the customer
- Lack of team buy-in to estimates
- Lack of ownership of decisions

Not involving the right people up front in a project always results in changes to work. Not involving team members in decisions and estimates causes them to feel like they have no control over their work or the outcomes of the project.

6. Not Having Proper Sponsorship: Projects need internal and customer executive sponsorship to be successful. Sponsors act as tiebreakers and eliminate organizational politics/roadblocks that are holding up the project.

Symptoms

- Inadequate support from different areas of the organization and from customer stakeholders
- Issues taking very long before being resolved
- Decisions not being made efficiently

Not having proper sponsorship can result in projects "spinning their wheels." Also, when a change effort is involved, not having proper sponsorship can keep impacted employees from buying in to a project (i.e., not cascading the messages from the top down to the "masses").

7. No Rigor around Process: Almost every company uses a methodology for implementing projects. The success of these methodologies depends on the amount of rigor used on the project. Often, processes are not adhered to and projects run awry.

Symptoms
- Incomplete/nonexistent deliverables
- Inconsistencies within the project
- Lack of understanding of the project's big picture
- Lack of repeatable processes ("reinventing the wheel" unnecessarily)

Processes are only as valuable as the rigidity placed on them. In some companies, there are too many project management methodologies used. Some are necessary due to the varying nature of work, but basic project management practices and principles (and even tools, i.e., using Project vs. Excel) could easily be standardized but are not. When one manager has to transition to another, this creates an extra layer of complexity, because a common language is not being used between the two people (it is like trying to interpret someone else's code when they have not followed the standards you have been using).

8. No Community Plan: Project managers spend a significant amount of time on planning, estimating, and scheduling activities. If these results are not shared with team members, then they do not know what they are working toward and cannot manage their own schedules. This includes the communication of goals and items that are a big picture for the team.

Symptoms
- Lack of knowledge about what is due and when it is due
- Missed dates
- Lack of ownership of deliverables
- Deliverables get forgotten

Not having a community plan will result in not having an informed community. Having a shared plan and goals helps to build a cohesiveness and a greater understanding of how the work of each individual fits overall.

9. Not Plan for Rework: Estimation techniques often focus on the time that it takes to create units of work. What usually gets left out is the time spent on rework. By this I mean work that was done incorrectly and needs to be revisited as opposed to scope management. When rework is required, it either takes the place of other work which now comes in late or is pushed off until later (see ugly 2).

Symptoms

- Missed dates
- Poor quality

Never assume that anything is done right the first time.

10. Dates Are Just Numbers: Schedule is a major driver of project success. I am amazed at the number of people who think of dates as "suggestions" rather than deadlines. Because of interdependencies on projects, a missed date early on could ripple through the schedule for the remainder of the project.

Symptoms

- Consistently missed dates
- Items left open for long periods of time
- Incomplete/nonexistent deliverables
- Lack of a sense of urgency on the project team

Without structure around the management of dates, success requires a lot more effort. One other issue here is that of communication—these dates need to be communicated clearly and people must agree that this is their target. Also, they must understand what is on the critical path and what has slack, so if they slip on a critical path item, they know there is an impact on the project or on another project within the same program.

Possible Remedies Upon analyzing the uglies I observed that they are all interrelated. For example, not having rigor around processes (#7) can result in not having a shared plan (#8), which can result in people not caring about dates (#10), and so on. (See Figure 2–3.)

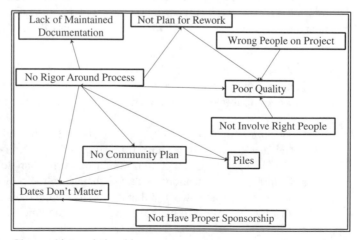

FIGURE 2–3. Observed interrelationships.

I also realized that a few remedies could mitigate these uglies. The trick here is to proactively solve them rather than react to them since by the time you realize that there is an ugly, YOUR PROJECT IS ALREADY UGLY.

Proactive Management   Proactive management means spending the appropriate amount of time up front to minimize the number of "fires" that need to get put out later. Proactive management includes the following actions:

- Creation of a detailed plan.
- Always looking at the plan to see what is coming up and preparing for it:
  - Thinking about the upcoming work and running down any issues that may be coming. I think of the team as running a marathon and it is my job to "clear the road" in front of them so they can keep on running.
  - Setting up logistics. Something as trivial as not having a conference room booked in advance can cause a schedule delay.
  - Lining up the appropriate people to be ready when the work comes their way.
  - Know people's vacation schedules.
- Constant replanning as information becomes more available.
- Understanding what is going on with the project. I see so many project managers in the "ivory tower" mode where they find out about things about THEIR project for the first time on a status report. By this time, as much as a week has gone by before the project manager is aware of issues.

There will always be unexpected issues that arise, but proactive management can help to mitigate those things that are controllable. This can be treated as an investment of time, in that you will spend far more time (and money) reacting to problems than you will focusing on ensuring that the process be followed properly. This is difficult for some project managers because it requires the ability to always look ahead of the current state of the project rather than just focusing on the problem of the day. A key element of proactive management is having the ability to make decisions efficiently.

"Do It While You Do It"   Now that you are not reacting to fires, you can focus team members on maintaining their work as they go. This means staying focused on all aspects of the current work and thinking of implications. Characteristics of this include:

- Documenting as work is being done and not at the end. I am sure that this will get the knee-jerk "we don't have time" reaction but I really believe (and have proved) that documenting as you go takes far less time than doing it at the end.
- Thinking of implications as things change on the project. For example, if a document changes, the owner of that document should think about any other deliverables that may be impacted by the change and communicate it to the appropriate person.
- Check all work before passing it on to others.
- Use the process/plan as a guideline for what work has to be done. I have heard this referred to as "living the plan."

The result of this technique will be an even distribution of work across the project and minimal spikes at the end. Rather than the notorious "death march," the worst case could be considered an "uncomfortable marathon."

Empower the Team   Project managers must realize that project structures resemble an inverse pyramid where the project manager works FOR the team. It is the team members who do the work on the project, so the project manager's primary role is to support them and address obstacles that may keep them from completing their work. This includes:

- Involving team members in project planning so they can not say that they were just given a deadline by management.
- Ask team members how things are doing and then act on their concerns. Asking for feedback and then doing nothing about it is worse than not asking at all because it suggests an expectation that concerns will be addressed.
- Celebrate the successes of the team with the team members.
- Be honest with the team members.

I am a big fan of W. Edwards Deming, who revolutionized the manufacturing industry. His 14 points of management revolve around empowerment of the team and apply very much to projects. Excerpts are noted below with my opinion of how they relate to project management:

Deming Point	Observation
8. Drive out fear, so that everyone may work effectively for the company.	This means that the "iron fist" technique of project management is not such a great idea. People will be averse to giving their opinions and doing a quality job.
10. Eliminate slogans, exhortations, and targets for the workforce asking for zero defects and new levels of productivity. Such exhortations only create adversarial relationships, as the bulk of the causes of low quality and low productivity belong to the system and thus lie beyond the power of the workforce.	I take this to mean that project managers should not just throw out targets, but rather involve the team members in decisions. It also means that project managers should look at the process for failure and not the team members.
12. Remove barriers that stand between the hourly worker and his (or her) right to pride of workmanship.	This is my marathon metaphor—where project managers need to remove obstacles and let the team members do their work.
13. Institute a vigorous program of education and self-improvement.	Allow the team members to constantly build their skill sets.

Empowering the team will enable the project manager to share information with the team members and will also enable the team members to feel like they have control over their own work. The result is that each team member becomes accountable for the project.

Results of the Remedies   The results of applying these remedies to the uglies are shown below. I call my vision of the new way of doing things the "attractive state" since it attracts people to success.

Attractive State Characteristics

Ugly Number	Ugly Name	Ugly State Characteristics	Proactive Management	Do It While You Do It	Empower
1	Maintained documentation	• No idea what decisions were made • Do not know why decisions were made • Cannot rely on accuracy of documents • Cannot use on future projects	• Updated documentation • Documentation will be planned for • Anyone can understand decisions	• Done during the project • No extra work at the end of the project	• Team members will own documentation
2	Piles	• Put off until later • May never get done	• Manageable work • If piles do exist, they will be scheduled in the plan	• Will be worked on as people go, so they should never grow out of control	• Minimized because people will take ownership of work
3	Quality at the source	• No ownership of work • Poor quality • Expensive fixes	• Better quality because you have spent appropriate time upfront	• Quality will be focused on as people do their work rather than assumed at a later time	• Quality will be upheld as people take ownership of work
4	People fit	• Bad project fit	• The ability to recognize resource issues and resolve them before they seriously impact the project • Proper resource fit from the start	• Manage work so resource issues are identified early	• Other team members may take on work for failing colleagues
5	People involvement	• Changes after work has been done • No ownership of work • No accountability for results	• Involving the right people up front to avoid rework later	• Involve people during work rather than have them react to it later	• Involve people during work rather than have them react to it later. • Empowered team members take ownership for work
6	Sponsorship	• Cannot resolve problems • Caught up in organizational politics	• Engaging stakeholders early will enable their support when really needed	• Rapid and effective decisions as needed	• May be improved due to better understanding of issues

#				
7	Process rigor	• No rigor • Poor quality • Inconsistent work	• Proper rigor is the essence of proactive management • Repeatable processes • Looking ahead will ensure proper attention to process	• Rigor will be followed as team members follow the process • Ensures that process steps are not missed	• Ownership of work will enable better rigor around process
8	Community plan	• No idea what is due and when • Team members do not take accountability for work—the plan is for the project manager	• Have the ability to share a plan and goals with the team	• Everyone is working to the same plan and knows where they are going	• Everyone is informed—shared goals • People can manage their own work
9	Rework	• Not planned for • Trade-off between doing other work or fixing issues	• Anticipating areas where may be rework or scope creep and working with key stakeholders early to address those • Planned for	• Rework will be accounted for as the team members work • By staying on top of the project, you will be aware of the magnitude of rework and can replan as needed	• Should be minimized due to motivation and ownership of work
10	Dates	• Dates do not matter • No accountability • Missing deliverables	• Dates (and impacts of missing them) clearly communicated	• Will matter and items will be closed when they are due	• Team members take ownership of dates

Conclusion   Focusing on proactive management, keeping up with work, and
                                        empowering your teams are key to running a successful project. There
is nothing in this section that has not been written of or spoken of hundreds of times before.
Nothing should sound new to a project manager. And yet, we keep seeing the uglies over
and over again. That leads me to a conclusion that it is the application of these concepts
that is the challenge. I find that after I read a good paper or attend a management course,
I have great enthusiasm to try out the new techniques but at the first signs of trouble I
revert back to my comfort zone. Therefore, I propose that there is a fourth remedy for the
uglies—being conscious. This is nothing more than being aware of what is going on and
how you are managing your project.

I come to work every morning a little earlier than the rest of the team so I can have
my quiet time and think about what work needs to be done (not just for that day but in the
upcoming days). I also give myself reminders that trigger my "step-back-and-think" mode.
An excellent series that goes into this technique are the "emotional intelligence" books by
Daniel Goleman.

There will always be uglies on your projects, but if you are conscious of them, then
you can identify them when they are happening and you may be able to prevent them from
throwing your projects into chaos. Best of luck.

References
W. E. Deming, *Out of the Crisis: Quality, Productivity and Competitive Position,* Cambridge University
    Press, Cambridge, 1982, 1986.
D. Goleman, *Working with Emotional Intelligence,* Bantam Books, New York, 1998.

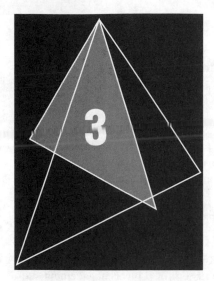

Journey to Excellence

3.0 INTRODUCTION

Every company has its own forces, or driving forces, as we discussed in Chapter 1, that force the company to embark upon a journey for excellence in project management. Some companies complete the journey in two or three years, while others may require a decade or more. In this chapter, we will discuss the approaches taken by a variety of companies. Each company took a different path, but they all achieved some degree of excellence in project management.

Some companies embark on the journey at the request of their own workers whereas other companies are forced into it by the actions of competitors and customers. In any event, there are driving forces that propagate the quest to excel in project management.

The driving forces for excellence, as discussed previously, include:

- Capital projects
- Customer expectations
- Competitiveness
- Executive understanding
- New product development
- Efficiency and effectiveness

Even the smallest manufacturing organization can conceivably spend millions of dollars each year on capital projects. Without good estimating, good cost control, and good schedule control, capital projects can strap the organization's cash flow, force the organization to lay off workers because the capital equipment either was not available or was not installed properly, and irritate customers with late shipment of goods. In non-project-driven organizations and manufacturing firms, capital projects are driving forces for maturity.

Customers' expectations can be another driving force. Today, customers expect contractors not only to deliver a quality product or quality services but also to manage this activity using sound project management practices. This includes effective periodic reporting of status, timely reporting of status, and overall effective customer communications. It should be no surprise that low bidders may not be awarded contracts because of poor project management practices on previous projects undertaken for the client.

The third common driving force behind project management is competitiveness. Companies such as Nortel and Hewlett-Packard view project management as a competitive weapon. Project-driven companies that survive on contracts (i.e., income) from external companies market their project management skills through virtually every proposal sent out of house. The difference between winning and losing a contract could very well be based upon a firm's previous project management history of project management successes and failures.

The most common form of competitiveness is when two or more companies are competing for the same work. Contracts have been awarded based upon previous project management performance, assuming that all other factors are equal. It is also not uncommon today for companies to do single-source procurement because of the value placed upon the contractor's ability to perform. A subset of this type of competitiveness is when a firm discovers that outsourcing is cheaper than insourcing because of the maturity of their contractor's project management systems. This can easily result in layoffs at the customer's facility, disgruntled employees, and poor morale. This creates an environment of internal competition and can prevent an organization from successfully implementing and maturing in project management.

A fourth driving force toward excellence is executive buy-in. Visible and participative executive support can reduce the impact of many obstacles. Typical obstacles that can be overcome through executive support include:

- Line managers who do not support the project
- Employees who do not support the project
- Employees who believe that project management is just a fad
- Employees who do not understand how the business will benefit
- Employees who do not understand customers' expectations
- Employees who do not understand the executives' decision

Another driving force behind project management is new product development. The development of a new product can take months or years and may well be the main source of the company's income for years to come. The new product development process encompasses the time it takes to develop, commercialize, and introduce new products to the market. By applying the principles of project management to new product development, a company can produce more products in a shorter period of time at lower cost than usual with a potentially high level of quality and still satisfy the needs of the customer.

In certain industries, new product development is a necessity for survival because it can generate a large income stream for years to come. Virtually all companies are involved in one way or another in new product development, but the greatest impact may very well be with the aerospace and defense

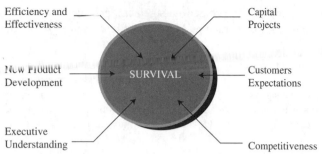

FIGURE 3–1. Components of survival.

contractors. For them, new product development and customer satisfaction can lead to multiyear contracts, perhaps for as long as 20 or more years. With product enhancements, the duration can extend even further.

Customers will pay only reasonable prices for new products. Therefore, any methodology for new product development must be integrated with an effective cost management and control system. Aerospace and defense contractors have become experts in earned value measurement systems. The cost overruns we often hear about on new government product development projects are attributed not necessarily to ineffective project management or improper cost control but more to scope changes and enhancements.

Improvement in the overall efficiency and effectiveness of the company is sometimes difficult, if not impossible. It often requires change in the corporate culture, and culture changes are always painful. The speed at which such changes accelerate the implementation of project management often depends on the size of the organization. The larger the organization, the slower the change.

Obviously, the most powerful force behind project management excellence is survival. It could be argued that all of the other forces are tangential to survival (see Figure 3–1). In some industries, such as aerospace and defense, poor project management can quickly lead to going out of business. Smaller companies, however, certainly are not immune.

Sometimes, there are additional driving forces:

- Increase in project size mandated by the necessity to grow
- Customers demanding faster implementation
- Customers demanding project management expertise for some degree of assurance of success completion
- Globalization of the organization mandated by the need to grow
- Consistency in execution in order to be treated as a partner rather than as a contractor

3.1 THE LIGHT AT THE END OF THE TUNNEL

Most people seem to believe that the light at the end of the tunnel is the creation of an enterprise project management methodology that is readily accepted across the entire organization and supports the need for survival of the firm. Actually, the goal should be to achieve excellence in project management, and the methodology is the driver for this. According to a spokesperson at AT&T, excellence can be defined as:

> A consistent Project Management Methodology applied to all projects across the organization, continued recognition by our customers, and high customer satisfaction. Also our project management excellence is a key selling factor for our sales teams. This results in repeat business by our customers. In addition there is internal acknowledgement that project management is value-added and a must have.

While there may be some merit to this belief that excellence begins with the creation of a methodology, there are other elements that must be considered, as shown in Figure 3–2. Beginning at the top of the triangle, senior management must have a clear vision of how project management will benefit the organization. The two most common visions are for the implementation of project management to provide the company with a sustained competitive advantage and for project management to be viewed internally as a strategic competency.

Once the vision is realized, the next step is to create a mission statement, accompanied by long- and short-term objectives that clearly articulate the necessity for project management. As an example, look at Figure 3–3. In this example, a company may wish to be recognized by its clients as a solution provider rather than as a supplier of products or services. Therefore, the mission might be to develop a customer-supported enterprise project management methodology that provides a continuous stream of successful solutions for the customers whereby the customers treat the contractor as a strategic partner rather than as just another supplier. The necessity for the enterprise project management methodology may appear in the wording of both the vision statement and the mission statement.

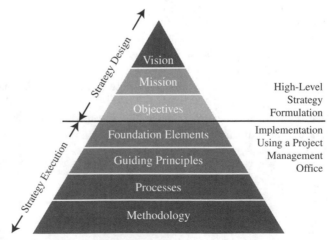

FIGURE 3–2.   Enterprise project management.

FIGURE 3–3. Identifying the mission.

FIGURE 3–4. Identifying the metrics.

Mission statements can be broken down into near- and long-term objectives. For example, as seen in Figure 3–4, the objectives might begin with the establishment of metrics from which we can identify the CSFs and the KPIs. The CSFs focus on customer satisfaction metrics within the product, service, or solution. The KPIs are internal measurements of success in the use of the methodology. The CSFs and KPIs are the drivers for project management to become a strategic competency and a competitive advantage. Notice also in Figure 3–4 that the CSFs and KPIs can be based upon best practices.

The top three levels of the triangle in Figure 3–2 represent the design of the project management strategy. The bottom four levels involve the execution of the strategy beginning with the foundation elements. The foundation elements are the long- and short-term factors that must be considered perhaps even before beginning with the

TABLE 3–1.   FOUNDATION ELEMENTS

Long Term	Short Term
Mission	Primary and secondary processes
Results	Methodology
Logistics	Globalization rollout
Structure	Business case development
Accountability	Tools
Direction	Infrastructure
Trust	
Teamwork	
Culture	

development of an enterprise project management methodology (Table 3–1). While it may be argumentative as to which factors are most important, companies seem to have accelerated to excellence in project management when cultural issues are addressed first.

To achieve excellence in project management, one must first understand the driving forces that mandate the need for excellence. Once the forces are identified, it is essential to be able to identify the potential problems and barriers that can prevent successful implementation of project management. Throughout this process, executive involvement is essential. In the following sections, these points will be discussed.

3.2 MANAGING ASSUMPTIONS

Whenever we discuss the journey to excellence, people expect to see a chronology of events as to how the company matured in project management. While this is certainly important, there are other activities that happen that can accelerate the maturity process. One such factor is an understanding of the assumptions that were made and a willingness to track the assumptions throughout the project. If the assumptions were wrong or have changed, then perhaps the direction of the project should change or be canceled.

Planning begins with an understanding of the assumptions. Quite often, the assumptions are made by marketing and sales personnel and then approved by senior management as part of the project selection and approval process. The expectations for the final results are based upon the assumptions made.

Why is it that, more often than not, the final results of a project do not satisfy senior management's expectations? At the beginning of a project, it is impossible to ensure that the benefits expected by senior management will be realized at project completion. While project length is a critical factor, the real culprit is changing assumptions.

Assumptions must be documented at project initiation using the project charter as a possible means. Throughout the project, the project manager must revalidate and challenge the assumptions. Changing assumptions may mandate that the project be terminated or redirected toward a different set of objectives. The journey to excellence must necessitate a way to revalidate assumptions. The longer the project, the greater the chance that the assumptions will change.

A project management plan is based upon the assumptions described in the project charter. But there are additional assumptions made by the team that are inputs to the project management plan.[1] One of the primary reasons why companies use a project charter is that project managers were most often brought on board well after the project selection process and approval process were completed. As a result, project managers were needed to know what assumptions were considered.

3.3 MANAGING ASSUMPTIONS IN CONSERVATION PROJECTS—WWF INTERNATIONAL[2]

In 2005, in collaboration with other conservation organisations,[3] the World Wide Fund for Nature (WWF), agreed upon and began to roll out a set of *Standards for Conservation Project and Programme Management ("the Programme Standards").*[4] These Standards are rooted in a long history of project and programme planning and management within WWF, across other conservation organizations, and in other disciplines. The Programme Standards are designed to help project managers and staff describe what they intend to conserve, identify their key assumptions, develop effective strategies, measure their success, and then adapt, share, and learn over time.

Adaptive Management and Challenges in Conservation Projects

Although there exists significant research and documentation on project management in the private sector, the principles of which apply equally to the non-profit sector, conservation projects face additional challenges. Beyond the usual processes of project execution and control, conservation projects must operate amid significant uncertainty and complex systems influenced by biological, political, social, economic, and cultural factors.

---

1. See *A Guide to the Project Management Body of Knowledge®,* 4th ed., Project Management Institute, Newtown Square, PA, 2008, p. 79.
2. Any reproduction in full or in part of this article must mention the title and credit WWF as the copyright owner. © text 2009 WWF–World Wide Fund For Nature (also known as World Wildlife Fund). All rights reserved. Material was provided by William Reidhead, MSc, Manager, Programme Management Capacity Building, Conservation Strategy and Performance Unit, WWF International.
3. The WWF Programme Standards are closely based on the Open Standards for the Practice of Conservation developed by the Conservation Measures Partnership, a partnership of 11 conservation organizations working together to seek better ways to design, manage, and measure the impacts of their conservation actions (www.conservationmeasures.org).
4. For more information on the WWF Programme Standards, please visit www.panda.org/standards.

In defining the project context, conservationists must consider uncertainty on the status of biodiversity, on the functioning of ecological systems, and on how humans bring about changes to the ecological systems and are in turn affected by them. Similarly, when designing interventions aimed at improving the status of biodiversity, conservation projects face challenges in selecting amongst a number of untested strategies, in knowing which will be the most effective, and in measuring and communicating the impact of these strategies. All of this takes place in the context of limited human and financial resources, information, and political capital, and increased calls for transparency and impact from the donors and governments supporting the projects.

As a result, the WWF Programme Standards follow an experimental approach to managing conservation projects, integrating project definition, design, management, and monitoring to systematically test assumptions in order to adapt and learn. The adaptive management process requires that project teams explicitly identify the assumptions under which they are operating and systematically test each assumption to see if it holds in their project context. This provides a method for making more informed decisions about strategies, testing the effectiveness of strategies used, and learning and adapting to improve strategies.

Described below are two tools that are recommended best practice within the WWF Programme Standards and are key for determining and managing project assumptions.

Conceptual Models

A *conceptual model* (variously known as a "problem tree" or "map of the problematic") is a diagram representing a set of assumed causal relationships between factors that are believed to impact one or more of the biodiversity targets (species or habitats) that the project aims to conserve. A good conceptual model should explicitly link biodiversity targets to the direct threats impacting on them and the indirect threats and opportunities influencing the direct threats. It will also highlight the assumptions that have been made about causal relationships and will advise paths along which strategic activities can be used to positively influence these relationships. In summary, a conceptual model portrays the present situation at the project site and provides the basis for determining where project teams can intervene with strategic activities. Note that each arrow connecting two boxes in Figure 3–5 indicates causality and represents an assumption that can be tested.

Results Chains

Conservation project teams implement strategies that they believe will contribute to conserving the biodiversity in their site, but may not formally state their assumptions about exactly how the strategy will lead to threat reduction and conservation of biodiversity. In fact, it is likely that they have many implicit assumptions— assumptions which may even differ across team members and project partners—about how their strategies will contribute to achieving conservation. If these assumptions are not made explicit, however, they cannot be tested nor can their validity be determined over time.

A *results chain* is a tool that clarifies these assumptions, a diagram that maps out a series of causal statements that link factors in an "if . . . then" fashion. Results chains help teams to specify and model their theories of change. In some organisations, results chains are also termed "logic models" or "solutions trees." The results chains are built from the conceptual model, and as shown in Figure 3–6, are composed of a strategy (a group of activities), desired outcomes, and the ultimate impact that these results will have on the biodiversity

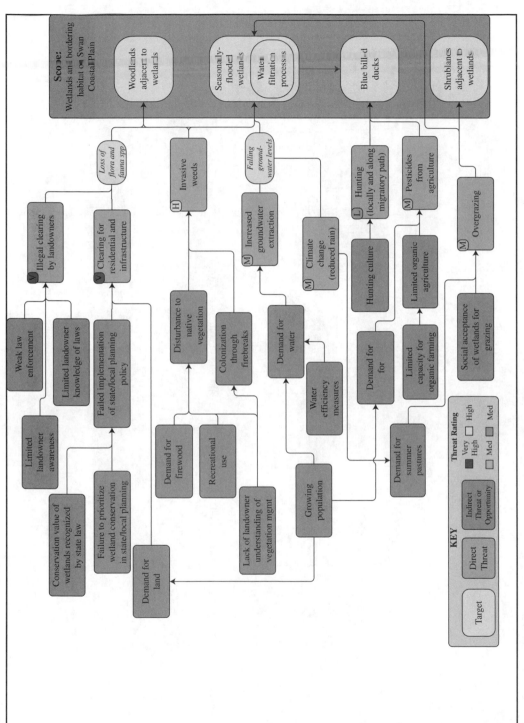

FIGURE 3–5. Example (simplified) of a conceptual model for Swan Coastal Plain in southwest Australia.

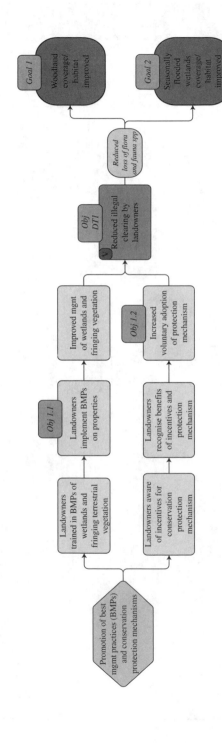

FIGURE 3-6. Example (simplified) of a results chain for Swan Coastal Plain in southwest Australia.

target. A goal is a formal statement of a desired impact on a biodiversity target and an objective is a formal statement of a desired outcome, frequently the reduction of a threat.

In this manner, a well-constructed results chain will provide a project with a set of strategic activities to be executed on the ground, as well as goals and objectives, in short, a conservation action plan. Results chains also provide the basis for financial/operational plans, as well as for formulating indicators and monitoring and control plans. In addition to elucidating assumptions and developing plans for project execution, conceptual models and results chains are both useful tools for monitoring and control during project implementation, for assessing impact, and for diagnosing any bottlenecks that may arise.

It is worth nothing that the above two tools fall within the WWF Programmes Standards planning steps known as Define and Design. These steps also include other tools for assessing the viability of biodiversity targets, for ranking threats to biodiversity, for analysis of stakeholders, for assessing risks, etc. Further steps include best practices for Implementation, for Analysing results and Adapting plans, and for Sharing.

3.4 PROJECT GOVERNANCE

Most companies begin the journey to excellence with the development of a project management methodology. The purpose of the methodology is to provide not only a road map of how to proceed but also the project manager with the necessary and timely information for decision-making. Decision-making requires some form of governance and too often this is discovered late in the journey toward excellence.

A methodology is a series of processes, activities, and tools that are part of a specific discipline, such as project management, and designed to accomplish a specific objective. When the products, services, or customers have similar requirements and do not require significant customization, companies develop methodologies to provide some degree of consistency in the way that projects are managed. These types of methodologies are often based upon rigid policies and procedures.

As companies become reasonably mature in project management, the policies and procedures are replaced by forms, guidelines, templates, and checklists. This provides the project manager more flexibility in how to apply the methodology to satisfy a specific customer's requirements. This leads to a more informal application of the project management methodology.

Today, we refer to this informal project management approach as a framework. A framework is a basic conceptual structure that is used to address an issue, such as a project. It includes a set of assumptions, concepts, values, and processes that provide the project manager with a means for viewing what is needed to satisfy a customer's requirements. A framework is a skeleton support structure for building the project's deliverables.

Frameworks work well as long as the project's requirements do not impose severe pressure upon the project manager. Unfortunately, in today's chaotic environment, this pressure appears to be increasing because:

- Customers are demanding low-volume, high-quality products with some degree of customization.

- Project life cycles and new product development times are being compressed.
- Enterprise environmental factors are having a greater impact on project execution.
- Customers and stakeholders want to be more actively involved in the execution of projects.
- Companies are developing strategic partnerships with suppliers, and each supplier can be at a different level of project management maturity.
- Global competition has forced companies to accept projects from customers that are all at a different level of project management maturity.

These pressures tend to slow down the decision-making processes at a time when stakeholders want the processes to be accelerated. This slowdown is the result of:

- The project manager being expected to make decisions in areas where he or she has limited knowledge
- The project manager hesitating to accept full accountability and ownership for the projects
- Excessive layers of management being superimposed on top of the project management organization
- Risk management is being pushed up to higher levels in the organization hierarchy
- The project manager demonstrating questionable leadership ability

These problems can be resolved using effective project governance. Project governance is actually a framework by which decisions are made. Governance relates to decisions that define expectations, accountability, responsibility, the granting of power, or verifying performance. Governance relates to consistent management, cohesive policies and processes, and decision-making rights for a given area of responsibility. Governance enables efficient and effective decision-making to take place.

Every project can have different governance even if each project uses the same enterprise project management methodology. The governance function can operate as a separate process or as part of project management leadership. Governance is designed not to replace project decision-making but to prevent undesirable decisions from being made.

Historically, governance was provided by the project sponsor. Today, governance is a committee. The membership of the committee can change from project to project and industry to industry. The membership may also vary based upon the number of stakeholders and whether the project is for an internal or external client.

3.5 SEVEN FALLACIES THAT DELAY PROJECT MANAGEMENT MATURITY _____

All too often, companies embark upon a journey to implement project management only to discover that the path they thought was clear and straightforward is actually filled with obstacles and fallacies. Without sufficient understanding of the looming roadblocks and how to overcome them, an organization may never reach a high level of project management maturity. Their competitors, on the other hand, may require only a few years to implement an organization-wide strategy that predictably and consistently delivers successful projects.

One key obstacle to project management maturity is that implementation activities are often spearheaded by people in positions of authority within an organization. These people often have a poor understanding of project management, yet are unwilling to attend training programs, even short ones, to capture a basic understanding of what is required to successfully bring project management implementation to maturity. A second key obstacle is that these same people often make implementation decisions based upon personal interests or hidden agendas. Both obstacles cause project management implementation to suffer.

The fallacies affecting the maturity of a project management implementation do not necessarily prevent project management from occurring. Instead, these mistaken beliefs elongate the implementation time frame and create significant frustration in the project management ranks. The seven most common fallacies are explained below.

Fallacy 1: Our ultimate goal is to implement project management. Wrong goal! The ultimate goal must be the progressive development of project management systems and processes that consistently and predictably result in a continuous stream of successful projects. A successful implementation occurs in the shortest amount of time and causes no disruption to the existing work flow. Anyone can purchase a software package and implement project management piecemeal. But effective project management systems and processes do not necessarily result. And successfully completing one or two projects does not mean that only successfully managed projects will continue.

Additionally, purchasing the greatest project management software in the world cannot and will not replace the necessity of people having to work together in a project management environment. Project management software is not:

- A panacea or quick fix to project management issues
- An alternative for the human side of project management
- A replacement for the knowledge, skills, and experiences needed to manage projects
- A substitute for human decision-making
- A replacement for management attention when needed

The right goal is essential to achieving project management maturity in the shortest time possible.

Fallacy 2: We need to establish a mandatory number of forms, templates, guidelines, and checklists by a certain point in time. Wrong criteria! Project management maturity can be evaluated only by establishing time-based levels of maturity and by using assessment instruments for measurement. While it is true that forms, guidelines, templates, and checklists are necessities, maximizing their number or putting them in place does not equal project management maturity. Many project management practitioners—me included—believe that project management maturity can be accelerated if the focus is on the development of an organization-wide project management methodology that everyone buys into and supports.

Methodologies should be designed to streamline the way the organization handles projects. For example, when a project is completed, the team should be debriefed to capture lessons learned and best practices. The debriefing session often uncovers ways to minimize or combine processes and improve efficiency and effectiveness without increasing costs.

Fallacy 3: We need to purchase project management software to accelerate the maturity process. Wrong approach! Purchasing software just for the sake of having project management software is a bad idea. Too often, decision-makers purchase project management software based upon the *bells and whistles* that are packaged with it, believing that a larger project management software package can accelerate maturity. Perhaps a $200,000 software package is beneficial for a company building nuclear power plants, but what percentage of projects require elaborate features? Project managers in my seminars readily admit that they use less than 20 percent of the capability of their project management software. They seem to view the software as a scheduling tool rather than as a tool to proactively manage projects.

Consider the following example that might represent an average year in a midsize organization:

- Number of meetings per project: 60
- Number of people attending each meeting: 10
- Duration of each meeting: 1.5 hours
- Cost of one fully loaded man-hour: $125
- Number of projects per year: 20

Using this information, the organization spends an average of $2.25 million (U.S.) for people to attend team meetings in one year! Now, what if we could purchase a software package that reduced the number of project meetings by 10 percent? We could save the organization $225,000 each year!

The goal of software selection must be the benefits to the project and the organization, such as cost reductions through efficiency, effectiveness, standardization, and consistency. A $500 software package can, more often than not, reduce project costs just as effectively as a $200,000 package. What is unfortunate is that the people who order the software focus more on the number of packaged features than on how much money using the software will save.

Fallacy 4: We need to implement project management in small steps with a small breakthrough project that everyone can track. Wrong method! This works if time is not a constraint. The best bet is to use a large project as the breakthrough project. A successfully managed large project implies that the same processes can work on small projects, whereas the reverse is not necessarily true.

On small breakthrough projects, some people will always argue against the implementation of project management and find numerous examples why it will not work. Using a large project generally comes with less resistance, especially if project execution proceeds smoothly.

There are risks with using a large project as the breakthrough project. If the project gets into trouble or fails because of poorly implemented project management, significant damage to the company can occur. There is a valid argument for starting with small projects, but the author's preference is larger projects.

Fallacy 5: We need to track and broadcast the results of the breakthrough project. Wrong course of action! Expounding a project's success benefits only that project rather than the entire company. Illuminating how project management caused a project to succeed benefits the entire organization. People then understand that project management can be used on a multitude of projects.

Fallacy 6: We need executive support. Almost true! We need *visible* executive support. People can easily differentiate between genuine support and lip service. Executives must *walk the talk.* They must hold meetings to demonstrate their support of project management and attend various project team meetings. They must maintain an open-door policy for problems that occur during project management implementation.

Fallacy 7: We need a project management course so our workers can become PMPs. Once again, almost true! What we really need is lifelong education in project management. Becoming a PMP is just the starting point. There is life beyond the *PMBOK® Guide.* Continuous organization-wide project management education is the fastest way to accelerate maturity in project management.

Needless to say, significantly more fallacies than discussed here are out there, waiting to block your project management implementation and delay its maturity. What is critical is that your organization implements project management through a well-thought-out plan that receives organization-wide buy-in and support. Fallacies create unnecessary delays. Identifying and overcoming faulty thinking can help fast-track your organization's project management maturity.

3.6 MOTOROLA

"Motorola has been using project management for well over 20 years, but institutionalized within the last nine years," according to a spokesperson at Motorola.[5] The forces that drove the company to recognize the need to become successful in project management were "increasing complexity of projects coupled with quality problems, schedule and cost overruns, which drove senior management to seek an alternative management solution to what previously existed.

A chronology of what Motorola did to get where it is today as well as some of the problems encountered are as follows:

- 1995: Hire a director of project management
- 1996: First hire project managers—formal role definition and shift in responsibilities for scheduling and ship acceptance
- 1998: Formal change control instituted—driven by project managers
- 1998: Stage–gates rolled out and deployed across all projects
- 2000: Deployment of time-tracking tool

---

5. H. Kerzner, *Best Practices in Project Management: Achieving Global Excellence*, Wiley, Hoboken, 2006, p. 88.

- 2001: Deployment of a more formal resource tracking
- 2002: Improved resource planning and tracking
- 2004: Project cost accounting

Initially, program management was viewed as an overhead activity, with engineering managers reluctant to give up program control and status communication. It was only through senior management commitment to formal project management practices that a PMO was created and roles and responsibilities shifted. Full engineering management acceptance did not occur until after several years of project management demonstrating the value of structured program management practices which resulted in consistent on-time product delivery. These include formal, integrated, and complete project scheduling, providing independent cross-functional project oversight, communicating unbiased program status, coordinating cross-functional issue resolution, and the identification and management of program risks. Later, project management responsibilities increased to include other key areas such as customer communications, scope control and change management, cost containment, and resource planning.

Executive support was provided through sponsorship of the development of the program management function. The reporting structure of the function has been carefully kept within an appropriate area of the organization, ensuring independence from undue influences from other functional areas so that objective and independent reporting and support would be provided.

3.7 TEXAS INSTRUMENTS[6]

A critical question facing companies is whether the methodology should be developed prior to establishing a project management culture. Companies often make the fatal mistake of believing that the development of a project management methodology is the solution to their ailments. While this may be true in some circumstances, the excellent companies realize that people execute methodologies and that the best practices in project management might be achieved quicker if the focus is on the people rather than the tools.

Texas Instruments recognized the importance of focusing on people as a way to accelerate project success. Texas Instruments developed a success pyramid for managing global projects. The success pyramid is shown in Figure 3–7. A spokesperson at Texas Instruments describes the development and use of the success pyramid for managing global projects at Texas Instruments:

By the late 1990s, the business organization for sensors and controls had migrated from localized teams to global teams. I was responsible for managing 5–6 project managers who were in turn managing global teams for NPD (new product development). These teams typically consisted of 6–12 members from North America, Europe, and Asia. Although we were operating in a global business environment, there were many new and unique

6. H. Kerzner, *Advanced Project Management: Best Practices in Implementation*, Wiley, Hoboken, NJ, 2004, pp. 46–48.

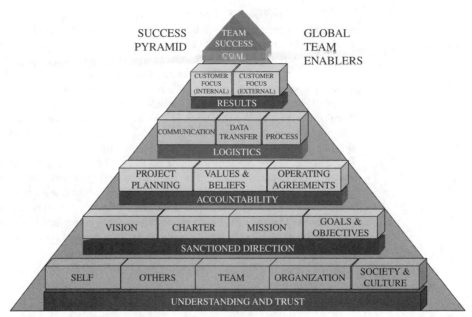

FIGURE 3–7. Texas Instruments success pyramid.

difficulties that the teams faced. We developed the success pyramid to help these project managers in this task.

Although the message in the pyramid is quite simple, the use of this tool can be very powerful. It is based on the principle of building a pyramid from the bottom to the top. The bottom layer of building blocks is the *foundation* and is called "understanding and trust." The message here is that for a global team to function well, there must be a common bond. The team members must have trust in one another, and it is up to the project manager to make sure that this bond is established. Within the building blocks at this level, we provided additional details and examples to help the project managers. It is common that some team members may not have ever met prior to the beginning of a project, so this task of building trust is definitely a challenge.

The second level is called "sanctioned direction." This level includes the team charter and mission as well as the formal goals and objectives. Since these are virtual teams that often have little direct face time, the message at this level is for the project manager to secure the approval and support from all the regional managers involved in the project. This step is crucial in avoiding conflicts of priorities from team members at distant locations.

The third level of the pyramid is called "accountability." This level emphasizes the importance of including the values and beliefs from all team members. On global teams, there can be quite a lot of variation in this area. By allowing a voice from all team members, not only can project planning be more complete but also everyone can directly buy into the plan. Project managers using a method of distributed leadership in this phase usually do very well. The secret is to get people to transition from attitude of obligation to a willingness of accepting responsibility.

The next level, called "logistics," is where the team lives for the duration of the project and conducts the day-to-day work. This level includes all of the daily, weekly, and monthly communications and is based on an agreement of the type of development process that will be followed. At Texas Instruments, we have a formal process for NPD projects, and this is usually used for this type of project. The power of the pyramid is that this level of detailed work can go very smoothly, provided there is a solid foundation below it.

Following the execution of the lower levels in the pyramid, we can expect to get good "results," as shown in the fifth level. This is driven in the two areas of internal and external customers. Internal customers may include management or may include business center sites that have financial ownership of the overall project.

Finally, the top level of the pyramid shows the overall goal and is labeled "team success." Our experience has shown that a global team that is successful on a one- to two-year project is often elevated to a higher level of confidence and capability. This success breeds added enthusiasm and positions the team members for bigger and more challenging assignments. The ability of managers to tap into this higher level of capability provides competitive advantage and leverages our ability to achieve success.

At Texas Instruments, the emphasis on culture is a best practice. It is unfortunate that more companies do not realize this.

3.8 EDS

When a company can recognize the driving forces for excellence in project management and understands that project management potentially could be needed for the survival of the firm, good things can happen quickly for the betterment of both the company and its clients. Doug Bolzman, Consultant Architect, PMP®, ITIL Service Manager at EDS, describes the forces affecting project management success at EDS and some of the problems they faced and overcame:

The significant emo-ptional events that we have experienced in client environments are:

- Loss of market share
- Not knowing the baseline timing or budgets for projects, thus not knowing if they are doing good or poorly
- Not having the ability for speed to market
- Understanding that there is much bureaucracy in the organization due to no design of a single project management capability

Doug Bolzman discusses three problems that EDS faced:

Problem 1: Management not knowing or understanding the relevance of full-time, professionally trained and certified project management staff. This problem generated several business symptoms that were removed once the root cause of the problem was eliminated. To remove this problem, management needs to analytically understand all of the roles and responsibilities performed by project management, the deliverables produced, and the time required. Once they understand it is a significant effort, they can start to budget and plan for the role separate from the work at hand.

Onc manager was convinced that the engineering team was not working at the capacity they should until it was demonstrated that the amount of project management work they were required to perform was over and above their engineering responsibilities. Since the work was distributed to every engineer, their overall output was reduced. The leader of the engineering team demonstrated the roles and the time commitments that were project management related for the executive to assign the role to a full-time project manager. The output of the team was restored to expected levels.

Problem 2: Everyone is overworked, and there is no time to implement a project management discipline. Since this is a common problem, the way to work around this situation is to generate a tactical Project Management Governance Board to determine the standards, approaches, and templates that will be considered "best practice" from previous projects and leveraged to future projects. To not place additional scope or risk to existing projects, they are "grandfathered" from the new standards. As a project charter and team are generated, they are trained and mentored in the new discipline. The Governance Board meets when needed to approve new project management structures and measure conformance by project managers.

Problem 3: Project managers were working at a higher level of maturity than the organization can benefit. Project managers often use all the tools and templates at their disposal to manage a project but are incoherent of the client's level of business maturity. For such cases, the saying goes, "That manager is using 30 pounds of project management to manage a 10-pound project." If the client is in an unstable, ever-changing environment, the project manager spends most of the time formally administering change management, adjusting all of the appropriate costing tools, and does not further the project. To remedy this situation, guidelines must be given to the project managers to balance the level of project management maturity to the client's business environment. This is done while working with the client to demonstrate how the maturity of the business environment costs additional time, money, and resources.

With regard to the role of executives during project management implementation, Doug Bolzman commented:

Many executives take a mild "management commitment" role during the implementation since project management and framework implementations are foundational capabilities and are not recognized as market facing, revenue generating, or exciting! Usually the executive approves a low budget plan where the majority of resources are absorbed from the organization and [there] will be a motivational speaker at the kickoff meeting. I attended a kickoff meeting last week where the sponsor told the team that he did not expect the effort to be successful or change the culture. The team wondered if the sponsor was providing motivation by instituting a challenge.

Executives understand business language and do not tolerate or listen to project management techno talk. Executives are results oriented, and if the project teams can simply translate the environment into business terms and create a business cast for incremental improvements to provide business value, the executive will be receptive to assist. If the executives are expected to generate the strategy or plan for improvement, or define project management's role within the organization, the implementation will fail. The majority of implementation success comes from the immediate business leaders and the project managers themselves who are tired of the status quo and want to implement improvements.

As for the chronology of events at EDS, Doug Bolzman continues:

When planning for the implementation of a framework, such as the implementation of a project, change, or release management (all using project management disciplines), we

TABLE 3–2. ORGANIZATIONAL MILESTONES

Milestone	Activity	Value
1. Establish governance board structure	Development of all participants' roles and responsibilities for implementing the improvements	• Implementing of a working "best practice" governance structure • All roles integrated and approved
2. Governance assignments	The sponsors (executives) name who will play each role, assigning accountability and authorization	• Executives establish priority through assignments • Everyone is trained in their role; expectations are set
3. Generation of attributes	The attributes describe the requirements, standards, capabilities, and metrics that will be used to define the improvements and measure the results	• Improvement is measurable, not emotional • Team can demonstrate value in business terms
4. Generation of improvement plan	Incremental plans for improvement are generated based on a maturity model or business improvement objectives	• The environment is improved based upon the speed the organization can afford • The plans can be adjusted based on business changes
5. Implementation	Each implementation is a release of the environment, is measured, and demonstrates business value	• Incremental improvements realized • Business invests incrementally, based on need

have learned that the organization needs to successfully progress through a series of organizational milestones. An example is shown in Table 3–2. The chronology is similar to a person who decides they need to lose weight. The person first makes the determination due to an event, such as clothes getting tight, peoples' observations, or health issues. Then the person realizes a cultural change is required, and if they are not willing to change behavior, they will not lose weight. The person has to encounter the "significant emotional event" for them to justify the discomfort of the change of behavior, such as exercising, not eating at night, or changing food types.

For clients, a basic approach is defined and reviewed. Many times the client attempts short cuts but then realizes that every step provides a foundational value for the larger journey. For one client, this approach was implemented seven years ago, is still in place, has generated nine major releases of their change management environment, and has weathered five major corporate reorganizations.

Figure 3–8 reflects how the client needs to transform from a functional to a matrix directorate to establish a common framework and how the programs are then measured for how they conform to the framework.

3.9 EXEL CORPORATION[7]

One of the characteristics of a well-managed company is not only early recognition that excellence in project management is needed but also developing a structured approach to achieve some degree of maturity and excellence in project management in a reasonable time

---

7. Material on Exel Corporation was graciously provided by Julia Caruso, PMP®, Project Manager, Regional PM Group, Americas; Todd Daily, PMP®, Business Process Manager, DHL Global Forwarding, International Supply Chain; Stan Krawczyk, PMP®, Director, Regional PM Group, Americas; and David Guay, Global VP for Effective Operations.

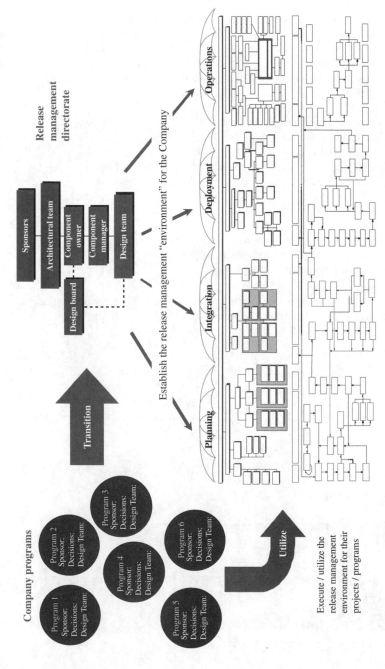

FIGURE 3–8. The transformation.

frame. Any structured approach to implementing project management must be done first and foremost for the benefit of improving the business. This occurred at Exel. According to Stan Krawczyk, Director, Regional Project Management Group, Americas for Exel Corporation:

> Beginning in 1997, Exel set out on a journey to understand the business needs of the organization with regard to projects and how projects were managed. This journey has now spanned nearly twelve years and has evolved into developing project management (PM) into a global core competency. The following list is a summary of major PM changes that have occurred in recent years that indicate a direction leading toward maturity and excellence in PM:

> 1. Operations to project-centric mindset
> 2. Defined PM processes and tools
> 3. Defined PM methodology
> 4. Senior management directive for project management and support
> 5. Project management value and maturity awareness and tracking
> 6. Global PM rollout
> 7. Global training curriculum
> 8. Centralized PM center of excellence (Global PMO)
> 9. Strong PM marketing and communication efforts
> 10. Establishment of a project manager career path
> 11. Multilingual tools and training

> The four phases listed below outline this journey toward PM growth and maturity. See Figure 3–9. Also listed below are the barriers that had to be overcome.

FIGURE 3–9. Exel project management journey.

Phase I—Business Needs Analysis (1997)

Situation

- Exel is operationally focused
- Poor overall project execution
- Inconsistent PM approach
- Rely heavily on individual heroics

Barriers

- Perception of the organization by management as an operating company versus a project-centric organization
- No PM expertise demand from customers
- Undefined PM processes, roles, and responsibilities
- Organizational perception of PM as non-value added

Critical Success Factors and Accomplishments

- Senior management directive for development of single, global, business delivery process
- Investment in outside consultancy to facilitate change program over nine-month period
- Project management begins to be perceived as enabler to change
- Informal PMO established as internal thought leadership team
- Assessment and tracking of PM value and maturity

Phase II—Process and Methodology (1997–1998)

Situation

- Exel establishes formal business delivery process (BDP)
- Three phases (strategy, business delivery, supporting infrastructure)
- Nine steps from pursuit through implementation to closing
- Identified three new focus areas (strategic alignment, business delivery, supporting infrastructure)
- No singular PM methodology established

Barriers

- Multiple organizational "agendas"
- Ad hoc PM approach by department
- Buy-in across business units
- No available proof source for PM
- No reference point for industry

Critical Success Factors and Accomplishments

- Project management process must fit overall BDP
- Simple is better to enable rapid engagement

- Previous project failures utilized to reinforce need for consistent processes across business units/divisions
- Methodology standardized
- Executive sponsorship obtained

Phase III—Tools, Training, and Organizational Structure (1998–2000)
Situation

- Defined PM process with basic tools
- Small PMO thought leadership focus
- Division-based PM teams forming
- Limited engagement in organization

Barriers

- Identification of internal PM proponents
- Shift to matrix team structure
- Reluctance to invest in PM development
- Quantification of PM value

Critical Success Factors and Accomplishments

- Project management defined as a high priority within the organization
- Acceptance of project-centric organization
- Appeal to broad audience
- Initial participants highly supportive
- Project management tools and training must connect with all stakeholders and BDP
- Global training curriculum
- Develop KPIs to illustrate progress/value
- Divisional teams responsible for tactical application and implementation
- Centralized PM center of excellence (EPM group)
- Avoidance of mandated tools that may not be globally acceptable/appropriate
- Ensure training of tools prior to release and utilization
- Determine appropriate level of technology that supports processes
- Emphasis on professional certification in PM

Phase IV—Visibility, Collaboration, and Globalization (2000–2009)
Situation

- Well-established business delivery processes
- Global merger/organization
- Geographically and culturally diverse organization
- Varying degrees of PM maturity

- Multi-industry service offerings
- More complex customer expectations
- Track record of reliability and predictability

Barriers

- Geography and time zones
- Culture and languages
- Multicultural project teams
- Global PM visibility
- Geographically dispersed project teams
- Project resource constraints (internal and customer)
- Variations in cultural issues

Critical Success Factors and Accomplishments

- Global PMO group established and linked to divisional and regional PMOs
- Global PMO focused solely on strategic development, training, tools, processes, and knowledge management
- Strong, global communication and broad marketing efforts
- Centralized knowledge base (e.g., Corporate Internet)
- Collaboration among all divisions/departments via enterprise management software (PlanView) and collaboration software (ProjectPlace, Sharepoint)
- Global collaboration on PM tool development and deployment
- Consistent, multilingual PM tools with standard "look-n-feel"
- Globally consistent PM training curriculum

For growth and maturity in project management to occur in a reasonable time frame, there must exist not only a structured approach, such as the Exel journey shown above, but also visible executive support. This requires that senior management visibly spearhead the journey and define their expectations at journey's end. This was initiated at Exel by Bruce Edwards, Global CEO. Mr. Edwards provided Exel with a clear definition of excellence in project management. According to Mr. Edwards this included, in no specific order:

1. Project-centric organization
2. Project-based culture
3. Strong organizational and leadership support for project management
4. Matrix team structure
5. Focus on project management skill development and education
6. Emphasis on project management skill track
7. Globally consistent project management training curriculum
8. Globally consistent project management processes and tools
9. Template-based tools versus procedures
10. Multilingual tools and training
11. Acknowledgment and support of advanced certification in project management (PMPCAPM)

12. Internal PMP and CAPM support programs for associates
13. Strong risk management
14. Project management knowledge sharing
15. Organizational visibility to portfolio of projects and status via enterprise software (PlanView)

One of the reasons behind Exel's success was that Exel viewed its journey to growth and maturity in project management as a strategic planning effort. Todd Daily, PMP®, Business Process Manager, International Supply Chain division, explains the strategic planning process for project management excellence at Exel:

Exel's Global PMO, with guidance from Harold Kerzner's paper *"Strategic Planning for Project Management and the Project Office" [Project Management Journal,* 2004], has put together a strategic planning matrix for each of the critical activities identified in the paper. Using these identified activities and a few Exel-specific categories, the group conducted a gap analysis of project management (PM) strengths and weaknesses that can be used for planning purposes. See example below for strength/weakness areas:
    The critical activities are:

1. Project management information system (PMIS)
2. Project failure information system (PFIS)
3. Dissemination of information
4. Mentorship
5. Development of standards and templates
6. Project management benchmarks
7. Business case development
8. Customized training in project management
9. Stakeholder and relationship management
10. Continuous improvement
11. Capacity planning/resource management
12. Reporting and organizational structure
13. Internal/functional projects
14. Project management maturity

An extract from the strategy-planning matrix is illustrated in Table 3–3. According to Todd Daily, there were significant driving forces that provided indications that excellence in project management was needed:

1. *Increasing project size.* As projects become more complex and involved, the level of control and management is essential to minimize risk and improve quality.
2. *Customer demand for faster implementations.* Exel provides full-service supply chain solutions for our customers that often lead to competing with smaller, niche players. This demands a rapid implementation approach. Project management enables better control, more efficient implementations, and higher quality deliverables.
3. *Customer demand for project management expertise and proficiency.* Today, more and more customers, like Exel, are realizing the importance of project management

TABLE 3–3. EXEL PROJECT MANAGEMENT: STRATEGIC PLANNING MATRIX

Critical Activities	Description	Kerzner's Guidelines	Exel's Current Status	Rank	Gaps/Suggestions	Risks, Challenges, Next Steps
Project management information system	Project and tools for capturing intellectual information relating to value measurement and risk management	• Company PM intranet • Project websites and databases • PMIS/enterprise systems	• Corporate intranet • PlanView • PMO Global DB	○	• More PM-focused systems • Systems to support global metrics	• Global, sector needs vary • Funds, desire to implement true global system
Performance failure information system	Method for gathering project performance information to enhance future project successes	• PM databases • Lessons-learned activities • Postmortem analysis	• Lessons-learned formalized • FEMA (Auto sector) • Operational reviews • Newsletter	○	• Develop lessons-learned and risk DB	• Some information considered confidential
Dissemination of information	Communication of critical project management information such as KPIs, CSF, data, news, and training	• KPIs • Line management support • Senior management support • Methodology	• Training • Process guides • Online PM overview	○	• Lack of global time/cost information	• Continue to promote PM value • Need to work with ops groups to get right people trained
Mentorship	Project office assumes mentorship role for inexperienced project managers seeking advice and guidance in the practice of project management	• CSF • Time • Cost • Quality • Scope • PM guidance and mentorship of functional PMs	• Provide guidance to other PMOs • Development of formal mentorship program • New project review tool	○ ○	• Regional acceptance of mentorship and project review processes	• Expand mentorship efforts globally
Development of standards and templates	Design and implementation of project management standards and tools that foster teamwork by creating a common "PM" language	• Dotted-line reporting structure • Customized templates, forms, and checklists • Limited formalized policies/procedures • Simple and dynamic • Managed and maintained by project office	• Developed and own global PM tools and templates • Global governance policy to ensure consistency	○	• Develop 3rd tier of PM tools and templates • Institute routine template audits	• Desires to tweak tools in different business units and locations • Language consistency may become an issue

in their businesses. It is strategically advantageous for Exel to possess project management excellence when being considered as a supply chain solution provider for customers.

5. *Globalization of the organization.* Over the past five years, Exel has completed a significant global merger and a number of strategic acquisitions that have integrated a number of different approaches with regard to project management. As the new organization evolves and broadens its service offerings, a consistent, go-to-market project management approach is essential. Geographical and cultural diversity is a key characteristic of Exel's business. With excellence in project management Exel can establish globally diverse project teams, working in any region, with one common project management methodology and a common understanding of all tools. It also enables the delivery of a consistent message to our global customers.

6. *Varying degrees of project management maturity.* Exel is structured in a sector based (or divisional) format by industry type—for example, automotive, technology, consumer products, health care, retail, and chemical. Based on assessments of each sector, it was apparent there were varying levels of project management maturity by sector. Assessments of each sector enabled Exel to develop an approach to balance project management awareness, knowledge, and expertise.

7. *Senior management mandate for consistent project management processes.*

8. *Shared success stories among and across sectors (lessons learned).*

9. *Management by objectives (MBOs) tied to project management training and advancement.*

10. *Business development and account teams using project management as a selling tool to clients.*

In the fall of 2003, Exel embarked on a global change management program designed to identify, develop, and instill consistent best practice processes and appropriate behaviors for all customer-related activities throughout every part of the organization. This program was entitled the Exel Way.

Project management has become an essential enabler for Exel in the achievement of its strategic initiatives. The capability to successfully lead, manage, and deliver projects or new operational solutions in a timely and cost-effective manner while adhering to specifications of our customers is core to Exel's delivery approach.

Based on these principles, a project management module team was established, along with six other modules ranging from value proposition through on-going customer relationship management. The project management module focused on global growth and maturity of the discipline. The mission was to further develop and embed Exel's leading project management methodology (DePICT®) across the organization, by leveraging organizational and industry best practices that enable continuous, effective delivery of customer solutions.

By design, each module of the Exel Way program was assigned a sponsor from the executive ranks of the organization and a team of module cosponsors representing all industry sectors and geographic regions. This ensured executive buy-in and participation and served as a means to overcome existing barriers and hurdles in the development and implementation of new practices and procedures.

Bruce Edwards, Global CEO, sponsored the project management module, with a module team comprised of key project management stakeholders from the Americas, Europe, and Asia

In early 2004, a comprehensive worldwide survey was conducted among 400 executives, stakeholders, and project management practitioners. The survey results helped establish a baseline on project management maturity within the organization and provide a framework on which to build and enhance new and existing processes. Follow-up interviews were conducted with approximately 100 respondents consisting of key business leaders and practitioners to obtain additional information and clarification.

Based on results obtained in the survey, six guiding principles were identified and used for developing our solution:

1. *Utilize Best Practices:* Apply a comprehensive project management methodology throughout Exel's global businesses that leverages organizational and industry best practices and enables continuous, effective delivery of customer solutions.
2. *Drive Risk Mitigation:* The discipline of recognizing and managing risks before they evolve into bigger problems is at the core of how project management brings value.
3. *Strive for Consistency*: A consistent process to enable reliable execution for our customers and is why project management is part of the Exel Way.
4. *Support the Business Strategy*: Align resources and effort with project priorities as determined by each business unit.
5. *Keep It Simple:* Harness knowledge capital gathered across the regions to facilitate the ease of implementation and avoid "reinventing the wheel."
6. *Assure a Cultural Fit:* A scalable and flexible model that can be culturally adopted across all regions and maintain a suitable level of rigor for repeated success.

Further, as a result of these guiding principles and outcomes from the survey, four key workstreams were created within the project management module to address key subjects:

1. Awareness and knowledge sharing
2. Methodology, tools, and reporting
3. Training and career development
4. Environment and structure

Following 18 months of effort, the project management module efforts concluded with a global rollout of initiatives, processes, and tools and an enhanced project management structure for sustaining project management across all areas of Exel.

Key Outcomes and Accomplishments

Some of the key initiatives delivered as a result of the Exel Way project management effort included:

1. A global project management organizational structure compiled of three regional project management offices (Americas, Europe/Middle East and Africa, and Asia-Pacific)
2. An enhanced global project management toolkit and consistent global templates
3. Knowledge-sharing tools and practices, including online databases and a project management portal (currently in use on the corporate intranet)
4. Enhanced training curriculum that added an advanced project management training course, project management value proposition training for sales and business development teams, online introduction to Exel's project management methodology, IT project delivery training, and enhanced support and sponsorship for advanced certification of candidates through PMI® and Prince2 methodologies

5. Project manager recruiting guidelines and an on-boarding program to better source and retain key project management resources
6. Redefined and formalized project management role profiles to better reflect regional and cultural differences
7. Comprehensive sales support collateral and communications materials
8. A four-volume project management guidebook which includes an executive summary of Exel's project management methodology, an Exel project manager's "Practitioner's Guide," project management tools and templates, and a project management training and career development guide

To date, Exel has almost 100 certified Project Management Professionals (PMP®).

Having completed the key development and design phases of the Exel Way project management module, rollout to the global organization continued through 2006.

In 2007, Exel realized that additional efforts were needed to continue progress in meeting their customers' needs. This resulted in the development of the First Choice program designed to ensure that Exel was the first choice of their customers. This program consisted of activities in six areas:

Process improvement
Performance management
Employee engagement
Customer management
Quality assurance
Project management

The intent of First Choice was to identify critical customer service touchpoints and to imbed key disciplines within the organization to be able to meet or exceed customer expectations on a consistent basis. The inclusion of project management in the First Choice program was recognition that project management was still deemed critical to business success. The continuous improvement in Exel's global project management processes and tools was defined as the prime target of this program. As David Guay, Exel's Global VP for Effective Operations states:

> Project management is a key component of the First Choice initiative. It was clear to the organization that for Exel to deliver consistent services and solutions to our customers, that a standard effective project management methodology was essential to delivering improved customer satisfaction. Implementation of new business is a critical time in a customer relationship. Trained project managers using a consistent global methodology not only provides a standard approach to global customers, but also provides a common language for the management team to effectively understand and manage projects across wide geographies and cultures. Project Management is one of six building blocks that comprise the foundation to drive customer satisfaction in Exel.

3.10 HEWLETT-PACKARD

Since 1992, Hewlett-Packard's management made the decision to focus on developing maturity and excellence in project management. A new group of dedicated project resources was formed within the Services organization and given the charter to become professional project management "experts." Hewlett-Packard established an aggressive project management training program as well as an informal "mentor" program where senior project managers would provide guidance and direction for the newly assigned people. In addition to the existing internal training courses, new project management courses were developed. When necessary, these courses were supplemented with external programs that provided comprehensive education on all aspects of project management. Efforts to achieve industry-recognized certification in project management became a critical initiative for the group.

Hewlett-Packard recognized that demonstrating superior project management skills could expand its business. In large, complex solution implementations, project management was viewed as a differentiator in the sales process. Satisfied customers were becoming loyal customers. The net result was additional support and product business for Hewlett-Packard. Hewlett-Packard recognized also that its customer's either did not have or did not want to tie up their own resources, and Hewlett-Packard was able to educate customers in the value of professional project management. Simply stated, if Hewlett-Packard has the skills, then why not let Hewlett-Packard manage the project?

According to Jim Hansler (PMP®), Project Manager at Hewlett-Packard, the following benefits were obtained:

> First, we are meeting the implementation needs of our customers at a lower cost than they can achieve. Second, we are able to provide our customers a consistent means of implementing and delivering a project through the use of a common set of tools, processes, and project methodologies. Third, we are leveraging additional sales using project management. Our customers now say, "Let HP do it!"

Hewlett-Packard recognized early on that it was no longer in the business of selling only products, but more in the business of providing "solutions" to its customers. HP sells solutions to its customers whereby HP takes on all of these responsibilities and many more. In the end, the customer is provided with a complete, up-and-running solution without the customer having to commit significant company resources. To do this successfully and on a repetitive basis, HP must also sell its outstanding project management capabilities. In other words, customers expect HP to have superior project management capability to deliver solutions. This is one of the requirements when customers' expectations are the driving force.

Mike Rigodanzo, Former Senior Vice President, HP Services Operations and Information Technology, stated that:

> In the services industry, how we deliver is as important as what we deliver. Customers expect to maximize their return on IT investments from our collective knowledge and experience when we deliver best-in-class solutions.
>
> The collective knowledge and experience of HP Services is easily accessible in HP Global Method. This integrated set of methodologies is a first step in enabling HPS to optimize our efficiency in delivering value to our customers. The next step is to know what is available and learn how and when to apply it when delivering to your customers.
>
> HP Global Method is the first step toward a set of best-in-class methodologies to increase the credibility as a trusted partner, reflecting the collective knowledge and expertise of HP Services. This also improves our cost structures by customizing pre-defined proven approaches, using existing checklists to ensure all the bases are covered and share experiences and learning to improve Global Method.

Hewlett-Packard clearly identifies its project management capabilities in its proposals. The following material is an example of what typically is included in HP proposals.

HP Services' Commitment to Project Management

Why HP Services Project Management HP Services considers strong project management a key ingredient to providing successful solutions to our customers. Our project managers are seasoned professionals with broad and deep experience in solutions, as well as managing projects. Our rigorous business processes make sure you are satisfied. A program roadmap provides an overall architecture of the project lifecycle while senior HP Services management conduct regular progress reviews to ensure quality. Our world-class project management methodology combines industry best practices with HP's experience to help keep everything on track. Our knowledge management program enables project managers and technology consultants to put our experience around the globe to work for you.

PM Processes and Methodology HP Services uses rigorous processes to manage our programs. The Program Roadmap provides an overall architecture for the project lifecycle. It includes the Solution and Opportunity Approval and Review (SOAR) process that approves new business as well as conducts implementation progress reviews to ensure quality and resolve problems quickly.

HP Services' project management methodology uses industry best practices with the added value of our experience implemented through web-based technology to allow quick updates and access throughout the world. It has over 20,000 web pages of information available to support our project teams. The methodology includes extensive knowledge management databases such as lessons learned and project experience from prior engagements that our project managers can use to help in managing their projects.

3.11 DTE ENERGY

Several maturity models are available in the marketplace to assist companies in their quest for growth, excellence, and success in project management. Table 1–1 from Chapter 1 is one such model. The purpose of the model is simply to provide some sort of structure to the maturity process. Tim Menke, PMP, Senior Continuous Improvement Expert, DO Performance Management, describes the growth process at DTE Energy:

> DTE Energy embraces project management to achieve timely and cost effective delivery of high quality products and services for our diversified utility customers. Our application of project management to engineering and construction efforts dates back decades. Expanding project management to areas other than major projects has been increasing in recent years. For example, our Information Technology organization started a formal drive to increase project management maturity in 2000. Individual mastery of project management skills was encouraged through training, practical experience, and formal certification. Job codes and a career progression model were created. Simultaneously, a cross-functional group of IT leaders, representing all areas of IT, was established to manage the portfolio of IT

projects. Over time, other departments have migrated to a similar model including the use of formal job codes and progression paths in an effort to increase control of their projects.

A focus on project management continues in our company. For example, we recently created a new department—Major Enterprise Projects—headed by a Senior Vice President, reporting directly to the CEO. This department was created to oversee major projects and their interactions. An example of one such project is the potential build of a Nuclear Power Generator.

Our motivation to increase project management maturity was previously internally driven (Year 2000 IT Remediation, Merger with MichCon, Enterprise Resource Planning System Implementation, CMMI Certification, etc.). However, the recent economic shock-wave being felt both nationally and internationally has resonated loudly in Southeast Michigan. We are using Lean Six Sigma continuous improvement to drive waste out of our processes and further streamline our operations. Project management is crucial to bringing our continuous improvement projects to fruition quickly and efficiently. We have revised our Continuous Improvement Project Management methodology to include a feedback mechanism to allow for placing on hold or canceling projects without losing sight of them in the future(Project Identification and Project Selection Phases). We have provided for scalability in our approach by recognizing a range of project types and corresponding requirements. All of these efforts have been championed by senior levels of DTE Energy leadership.

3.12 COMAU[8]

In the second edition of my text, *Strategic Planning for Project Management Using a Project Management Maturity Model*, I stated that the path to maturity can be accelerated with (1) the implementation of a PMO early on in the process; (2) having the PMO report directly to the executive levels of the company; and (3) having visible executive-level support for project management. Companies that accomplish all three of these items seem to outperform their competitors with regard to project management. Such was the case with Comau.

Background about Comau

Comau provides full services for automotive and aerospace product engineering, production systems, and maintenance services. Comau has operations in 19 countries. The Comau business portfolio is composed of body and welding, powertrain, engineering, services, and robotics products. The business centers are supported by the central functions (Contract & Project Management; Finance, Administration and Control; Human Resources; Legal & Corporate Affairs; Compliance Officer), all operating all over the world.

Description of the Problem

During the 1980s, Comau's success began to flourish. As with most successful companies, Comau recognized the opportunities that could

8. Material on Comau was provided by Mauro Fenzi, Comau Vice President, and Alexandre Sörensen Ghisolfi, Corporate PMO Manager.

come from acquisitions. In the 1990s, Comau started performing global acquisitions. By 1999, Comau's headcount had increased by 30 percent to approximately 20,000 employees. The problems with acquisitions soon became apparent because each acquisition could be at the different level of maturity in project management, and no corporate standards existed for project management. Up until a few years ago, project management in Comau, generally speaking, was executed in a very fragmented way and there was a lack of culture, methodologies, processes, and guidelines for project management in place. In 2007, an urgent, efficient, and global approach needed to be implemented. The goal was simple: the global integration of project management knowledge across the entire company such that a competitive advantage would occur.

The Solution—"From a Cluster to an Effective Network . . . " In 2007, Comau decided to reinforce the project management culture by establishing the Contract & Project Management corporate function. As with most companies that understand project management and recognize the need for executive leadership in project management, the new organization was headed up by a vice-president of contract and project management. The main guidelines for the organization's mission statement included:

1. Company organizational development and the implementation of global organizational policies related to project management
2. The reinforcement of enterprise project management systems and the project management office structure
3. Creation of the Comau Project Management Academy—the continuous development of knowledge and both hard and soft Skills

Comau correctly recognized the importance of the PMO in achieving the mission. Unlike less mature companies, Comau viewed itself as a solution provider, satisfying the business needs of its global customers. The PMO was viewed as the mechanism by which internal business solutions would be provided.

Comau approached the three main guidelines as follows:

● *Company Development and the Implementation of the Organizational Policies.* During 2007, a global project management policy was developed combined with an intensive training program based on the *PMBOK® Guide*. Project management benchmarks were established to measure the maturity level and an action plan was created to continue improvement of the maturity process. The global policy was a project management policy that described the job that all the projects teams must perform and it had to be applied globally. The policy was directly connected to the PMI® *PMBOK® Guide* best practices. One important point to mention was that the company decided to combine both contract management and project management into a single unit. Comau believed that this was an important innovation, and that positive results would be achieved, as discussed below.
● *The Reinforcement of the Enterprise Project Management and the Project Management Office Structure.* Today the Comau portfolio included in the PMO is

a multimillion-euro revenue group of global automotive and aerospace projects, conducted in more than 30 countries. The Comau Project Management Office, referred to as the Corporate PMO Office, coordinates the efforts of four regional PMOs: PMO North America, PMO South America, PMO Europe, and PMO Asia. This is shown in Figure 3–10. Comau's global organization is shown in Figure 3–11. The project management team, which employs international experts in project, program, and portfolio management, is called the project management family; it is composed of project managers, program managers, planners, and team members. The whole company is focused on providing its customers with high-quality products, projects, and services. The missions for the Comau PMO are shown in Figure 3–12.

To support the Comau missions, the PMO prepared a high-level roadmap for 2007–2009, which included the following:

- 2007:
  - Perform benchmarking and maturity baselining
  - Define the concept of the PMO
  - Develop operational policies for project management
  - Develop project management training programs
- 2008:
  - Establish a corporate PMO
  - Establish regional PMOs
  - Implement innovation actions according to the maturity assessment
  - Establish the Project Management Academy
- 2009:
  - Manage ongoing activities for projects, programs, and the portfolio of projects
  - Perform external benchmarking on project management maturity
  - Expand the activities of the Project Management Academy
  - Manage selected strategic and special projects

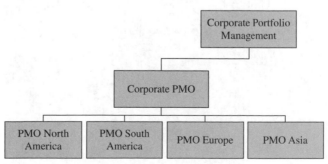

FIGURE 3–10. Comau PMO network.

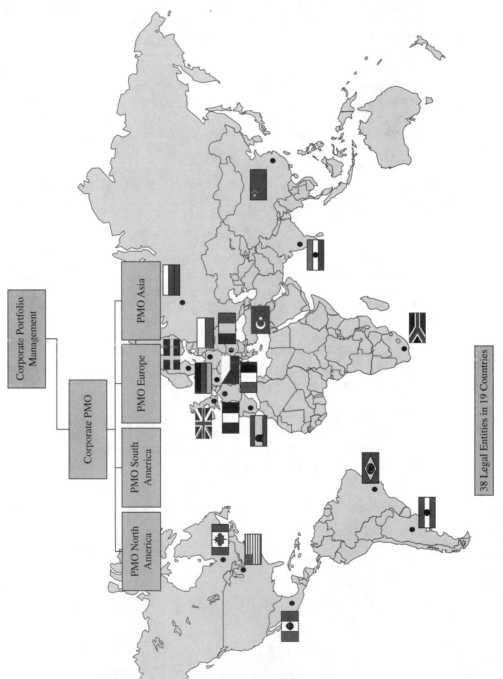

FIGURE 3-11. Comau worldwide.

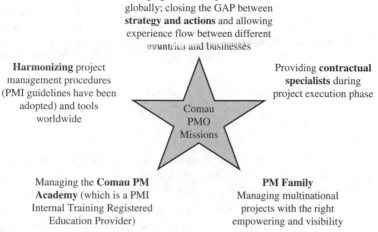

Managing Comau Projects Portfolio globally; closing the GAP between **strategy and actions** and allowing experience flow between different countries and businesses

Harmonizing project management procedures (PMI guidelines have been adopted) and tools worldwide

Providing **contractual specialists** during project execution phase

Comau PMO Missions

Managing the **Comau PM Academy** (which is a PMI Internal Training Registered Education Provider)

PM Family Managing multinational projects with the right empowering and visibility

FIGURE 3–12. Comau PMO missions.

As stated previously, Comau viewed the PMO as the primary mechanism for providing internal business solutions. Some of the benefits achieved by Comau included:

1. Recognition by its customers as the number one integrator of complex work, thus adding value to the value/supply chain.
2. Development of a high standard international class-compliant project management culture/approach.
3. Better support for the sales team resulting in increasing project success through proactive project planning and risk reduction strategies.
4. Development of a culture capable of synchronizing language with Comau customers, reducing misunderstandings in the projects definitions and executions, and supporting trustful communications through the projects.
5. Development of one of the best workload optimization techniques capable of reducing costs for its customers.
6. Develop the technical language when working to standardize a global approach, e.g., WBS Powertrain Italy and France. In this case, making it possible to exchange parts between products and project teams in different countries, achieving better planning, execution, cost control, workload planning and leveling, risk management, communication, and quality.
7. Identification and the management of out-of-project-scope situations, resulting in better benefits for customers, Comau, and providers.
8. Optimize processes and the reporting systems, avoiding waste of time and making more time available for managing critical issues.
9. Contribute to the company integration of work, sharing information, visions, and strategies, including start-ups of strategic projects.
10. Create a strong team of high-skilled managers and technicians capable of supporting difficult projects and high-pressure situations.

The PMO team has become the change agents inside each of the organizational business units. The result has been several "quick wins" solutions. Comau has been able to get better control of its indirect cost as well as providing value-added opportunities for both Comau and its global customer base. All the managers are now delegating authority to a greater extent than in the past and the Vice President and Country CEOs are functioning as strong sponsors.

The regional PMOs report to the Corporate PMO and also to the CEO of the regional area. The CEOs have demonstrated a sincere desire to function as executive-level sponsors and make the PMO evolution happen.

The Comau Project Management Academy—A Common Base of Knowledge and Skills

In 2008, the Comau Academy became a PMI® REP®—Registered Education Provider. Under the Corporate PMO, the Academy provides a training curriculum to ensure that the technical workforce shares a common base of knowledge and skills. Comau also strongly encourages project managers to obtain PMP® certification.

Results—Benefits   Good project management practices provide meaningful results. Some results are tangible, such as return on investment (ROI), cash flow, and profitability. Other results also exist at Comau but are not as easy to measure, such as the results of company integration efforts, more effective leadership, coaching, mentoring, and so on. Additional benefits that Comau found included:

● Growth in the PM Global organizational policies applications, maturity, and growth.
● The project management family has demonstrated the viability of applying global work processes, and now, other functional groups are following the steps and guidelines organizing similar actions.
● The Comau Project, Program and Leadership Academy is becoming mature and project management is leading the business.

Innovations

When companies become good in project management, the next logical step is the capturing of lessons learned and converting some of the lessons learned into best practices or innovations. Comau now documents all best practices as shown in Figure 3–13. Some of these innovations have provided important "quick wins" for the organization and the PMO. The innovations include:

● *Consistency in the Report to the Customer Process:* We now have better customer relationships. Our customers are recognizing the importance of Professional Project Management at Comau and many have expressed their satisfaction when working with our process and project management team.
● *The Project Management Family (Project Managers, Planners, and Controllers):* A real perception of integration and teamwork has been developed. People are now sharing best practices and trying to solve common problems together. In this family we started joining project and program managers, planners and controllers. The concept of "professional families" is growing.

COMAU

PROJECT MANAGEMENT

BOOK OF BEST PRACTICES

First Edition – March 2008

Developed by Comau Corporate Project Management Office

Contract and Project Management

FIGURE 3–13. The Comau Book of Best Practices.

- *Standardization:* The Project Management integrated approach is now being applied to all of the company's business units and countries. All business units now share the same PM goals, guidelines, training and reporting systems, and have made it possible to synchronize the different departments, achieve better quality in the communications, and define and measure the key performance indicators.

- *PMP Certified Growth:* From January 2008 to December 2008, about 20 [of] our employees have become PMP Certified; this represents approx 15% of the total number of Project Managers, and more are expected to be certified.

- *Organizational Common Language—The Comau Project Management Book of Best Practices:* A collection of project management articles and organization policies, guidelines and templates have been developed, it was important to disseminate the message not only in Italy but globally.

- *Comau Project Management Family On-Line—Global Virtual Community:* Using a collaboration tool, for the first time in the Company, the Comau project management global family has started sharing knowledge between projects, products and innovation ideas and now other groups are starting using it.

- *Portfolio—Program and Projects Reporting—Dashboard and Traffic Lights:* We have adopted [a] professional approach to the Portfolio Management System. A simple and common approach was applied to all countries and business units. A systematic project review is already in place where the company's Board of Directors can analyze the project status, define priorities, etc. It is a monthly review in each business unit.

- *Straight Connection between Contract Management and Project Management and Execution:* This has brought consistency and avoided risks and cost overruns.

- *Development of a Dedicated Human Resources Structure, Dedicated to the Project Management Family Development:* An important action bringing corporate attention to project management and empowering the application of Project Management.

- *The Project, Program and Leadership Academy (described in the previous pages).*

- *Company Development: High Support to PMP Certification, from the Board of Directors*

- *Company Development and Motivation:* For example, the creation of the "Comau Project Management Award," where Comau is receiving many professional and real examples of the application of our project management global policy. It has an interesting action to create motivation, innovation and also discipline when performing project management. In 2009, the winners are going to be awarded in a special event.

- *Application of Global Policies, Project Management Procedures:* For the first time, all different teams in the globe are applying the same practices. Many audits have also been performed without resistances and generating positive improvement actions.

- *Continuous PM Meetings, Round Tables, and Feedback Sessions to Share Strategy and Actions to All Levels of Management:* Many "informal meetings" have taken place with the goal of sharing the information and the strategy for the next actions. It has also generated the required environment to make gradual and positive changes in the company culture. The "buy-in" feeling has also been developed during these meetings.

- *Monthly Communication Reporting the News and Actions to All Levels of the Company:* Using the intranet, consistent monthly feedbacks for all the company are available, showing that something was happening with real and positive results: It creates a chain of positive reactions.

- *Strategic Planning:* The PMO has contributed to the perception that Comau is a Global Corporation, provided important strategic decisions and assisted with project start-ups.

Some Lessons Learned   The true success of a project is not only whether it has made a profit, but whether knowledge has been obtained that can benefit the entire corporation. Comau has prepared a list of lesson learned which were obtained through their endeavors in project management.

- The ability to share best practices and organizational standards is even more important when working globally. It allows the exchange of information, work products, workload sharing, resource leveling, etc.
- People participation is critical. It is a difficult task getting all the people to understanding the future benefits of project management and the importance of effective participation; it is related to the different knowledge, experiences and power/authority people have. The ability of getting people to participate is critical for change, speed and buy-in.
- Developing "quick-wins" is an important motivator.
- Making people feel secure about the future is a buy-in motivator. Today the importance of project management as a business solution for a global company is a strong motivator for the PM Family.
- Increasing effective communication, transparency and use of the "open door" philosophy is a strong motivator.
- Developing the Project Management and Program Management approach is important but not enough; the whole company must understand the business model of management by projects.
- Education is the basis for development. Our people must learn from somebody else's mistakes, not their own! Development of both "hard skills" and "soft skills" is essential to create the foundation for change and PM application.
- Education in effective leadership is becoming a key issue on a daily basis.
- Efforts should be applied with consistency. In the beginning it is not so much a question about quality, but a question about culture and discipline; better quality will arrive in the next few rounds.
- All of the actions have demonstrated to Comau that the organization must apply consistent efforts in project management best practices to achieve better business results and improve the way of doing things. From sales, contract management, production and after sales, all the company is passing through an educational program and also acting to support the application. It really represents a big change for the company.
- The PMO is demonstrating value through the application of the principles of integration, simplicity, results orientation, flexibility, discipline and continuous maturity improvement.
- The Corporate PMO is also providing important value to the business through the integration, empowering opportunities and performing strategic projects. When performing strategic projects the perception of value and the return of investment are directly and positively impacted. It is a natural consequence since this strategic PMO has a wide vision of the company entities, customers and products.

- The project management approach has demonstrated that other functional groups must be equally receptive to the acceptance of the project management, reducing barriers and generating better products.

3.13 VISTEON[9]

Background

Visteon was formed in 1997 (see Figure 3–14) as an auto parts supplier to Ford Motor Company. Composed of former business units of Ford Motor Company, Visteon originally had six divisions, 23 strategic business units, and 82,000 employees worldwide. Visteon became an independent, publicly traded company in 2000 with a culture strongly focused on the core competencies of product design and manufacturing, and the concepts and practices of project and commercial management were not widely used across the company.

Project Management Evolution As Visteon grew into an independent tier 1 automotive supplier (see Figure 3–15), the organizational culture changed. Customer business groups were created, a strong central global program management office (GPMO) was formed, and a global product development process—called Visteon Product Development Process (VPDP)—was created. This process was based on the fundamentals of traditional project management. In practice, the process was a sequential list of deliverables and forms required to deliver products. A primary focus was placed institutionalizing the process and deliverables were mandatory. An executive gate review process was implemented to ensure process execution.

Visteon's GPMO launched a suite of project management tools (see Figure 3–16) to support the new project management organization. The initial tool set used Microsoft Project and Excel to track the master list of projects. As the usage expanded, GPMO launched PM Plus and Primavera—Enterprise Project Portfolio Management. Concurrently, Visteon launched a business operating system, called VBOS. This business model included the roles and responsibilities, decision rights, and metrics for each business unit. VBOS contained more than 400 functional business processes.

The Project—The Team, The Process, The Tool

In 2005, Visteon reached a new structural agreement with Ford that reshaped Visteon into a leaner, more competitive company (See Figure 3–17). The corporate culture focused on creating, designing, and manufacturing leading automotive solutions driven by applying "automotive intellect." Three strong product groups emerged: climate, interiors, and electronics. Product

---

9. Material on Visteon was provided by Darlene Taylor, PMP, Visteon Senior Manager, Global Program Management Office, Global Information Technology—Product Lifecycle Management.

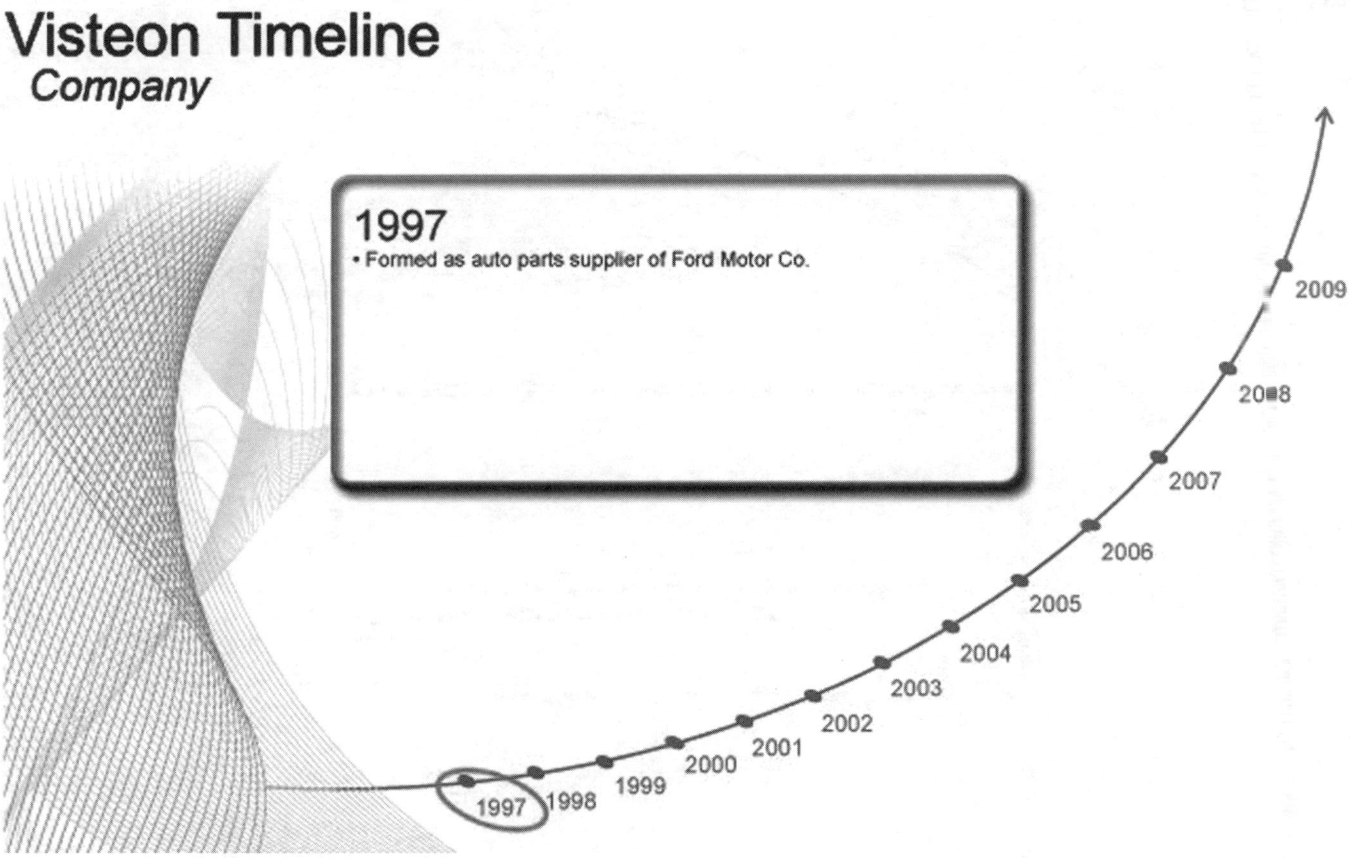

Visteon Timeline
Company

1997
- Formed as auto parts supplier of Ford Motor Co.

2009
2008
2007
2006
2005
2004
2003
2002
2001
2000
1999
1998
1997

FIGURE 3–14. The formation of Visteon in 1997.

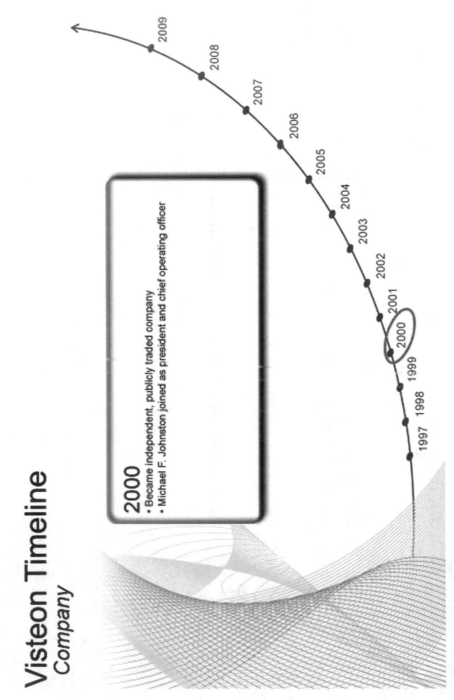

FIGURE 3–15. Visteon in 2000.

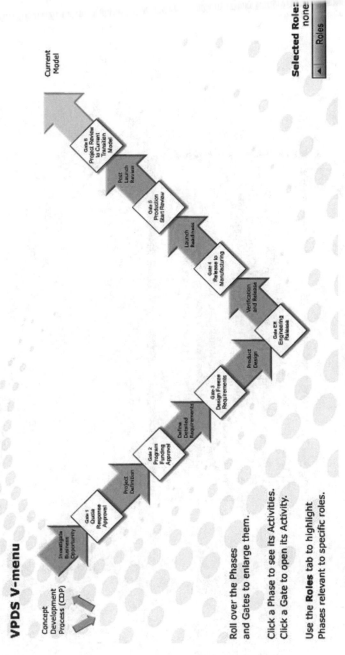

FIGURE 3–16. VPDS menu.

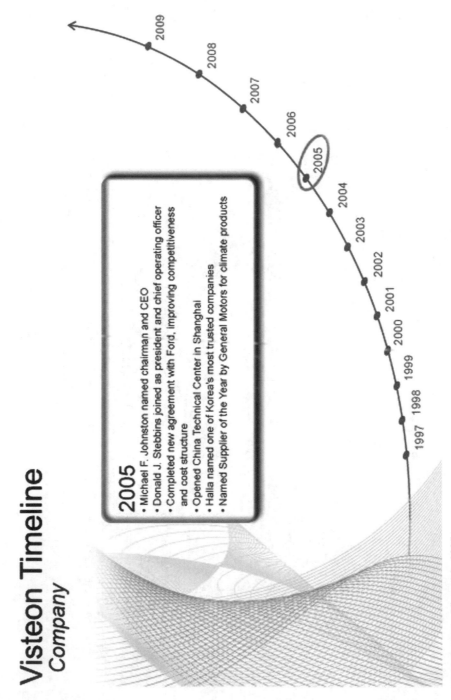

Visteon Timeline
Company

2005
- Michael F. Johnston named chairman and CEO
- Donald J. Stebbins joined as president and chief operating officer
- Completed new agreement with Ford, improving competitiveness and cost structure
- Opened China Technical Center in Shanghai
- Halla named one of Korea's most trusted companies
- Named Supplier of the Year by General Motors for climate products

1997 1998 1999 2000 2001 2002 2003 2004 2005 2006 2007 2008 2009

FIGURE 3–17. Visteon in 2005.

group teams were supported by a small, central GPMO responsible for project management process and tool oversight. A cross-product group program management council was added to ensure GPMO roadmap alignment. The executive review process evolved into a streamlined, high-level review focused on the project team's ability to manage risk and project business case. "The success of Visteon's global program management office is dependent on our ability to support the business with flexible tools and processes," said Darlene Taylor, Senior Manager, Visteon information technology and global program management office. (See Figure 3–18.) "The processes must efficiently balance project management using a lean global staff, while delivering value-added and highly complex projects; and the tools must provide rapid access to accurate information to support decision making—all delivered in a cost-effective, highly-reliable manner that can rapidly respond to changing business needs."

Today, VPDS is an optimized, build-based process that enables project leaders to match and align deliverables and outputs with customer product development cycles. See Figure 3–19. Since most customers structure their product development cycle around build events—either prototype or production—project leaders can match the number of VPDS phases with customer phases. VPDS deliverables are reviewed for applicability and value-add, and outputs are aligned using the custom project management tool that simplifies document storage, eliminates forms, and facilitates collaborative teamwork. VBOS is now streamlined to contain the critical few global cross-product group processes required to successfully operate the business.

Visteon's project management tools evolved from a large-scale, highly complex Primavera-based system to the customizable project management tool, enterProj. Visteon uses enterProj as its companywide online project management collaboration tool. It maintains project timing, executes VPDS and Visteon's IT solutions delivery process (ISDP) and serves as the central repository for project data. Within enterProj, teams can communicate real-time information related to all project management activities.

enterProj has the capability to manage:

- Timing
- Risks, issues, and opportunities
- Change management
- Financial business case—an interface with eFIN financial reporting solution
- Suppliers—interfaces with supplier management tool and supports a "request for quotation" solution
- Parts—interfaces with product life-cycle management (PLM) and supports a bill of materials solution
- Plant and launch readiness
- Facilities and tooling
- Document management
- Lessons learned

"By implementing a build-based VPDS and enterProj, project managers have a simplified solution to easily share real-time information with global team members," said

Global I.T.
Enterprise Solutions

Team

The Project Manager partners with our customers to lead and manage the project and its deliverables with support from a cross-functional team.

This team is responsible from the Pre-Pursuit through Post Launch phases.

Project Teams are supported by the Program Management Council dedicated to delivering Operational Excellence through the continuous improvement of Visteon's Program Management processes and tools.

Process

VPDS is a build-based process that enables Project Leaders to match and align VPDS deliverables and outputs with the customer product development cycle.

Outputs are aligned to the enterProj project management tool to enable collaboration and team member communication and simplify document storage.

Users can select deliverables in several languages; Spanish, French, German, Portuguese and Chinese.

Software

enterProj is Visteon's enterprise wide online project management collaboration tool. It is used to maintain Project Timing, execute VPDS and serves as the central repository for key project data.

With enterProj, teams have the ability to globally communicate real time information related to all activities associated with managing Forward and Current Model projects: Timing, Risks, Issues, Opportunities, Plant and Launch Readiness Information, End Item and Component Parts, Facilities and Tooling, Suppliers, Document Management, TGR/TGW (lessons learned), Program and operational changes, VQF management, Customer Relationship management, and eFIN project financial management.

FIGURE 3–18. Visteon global IT solution.

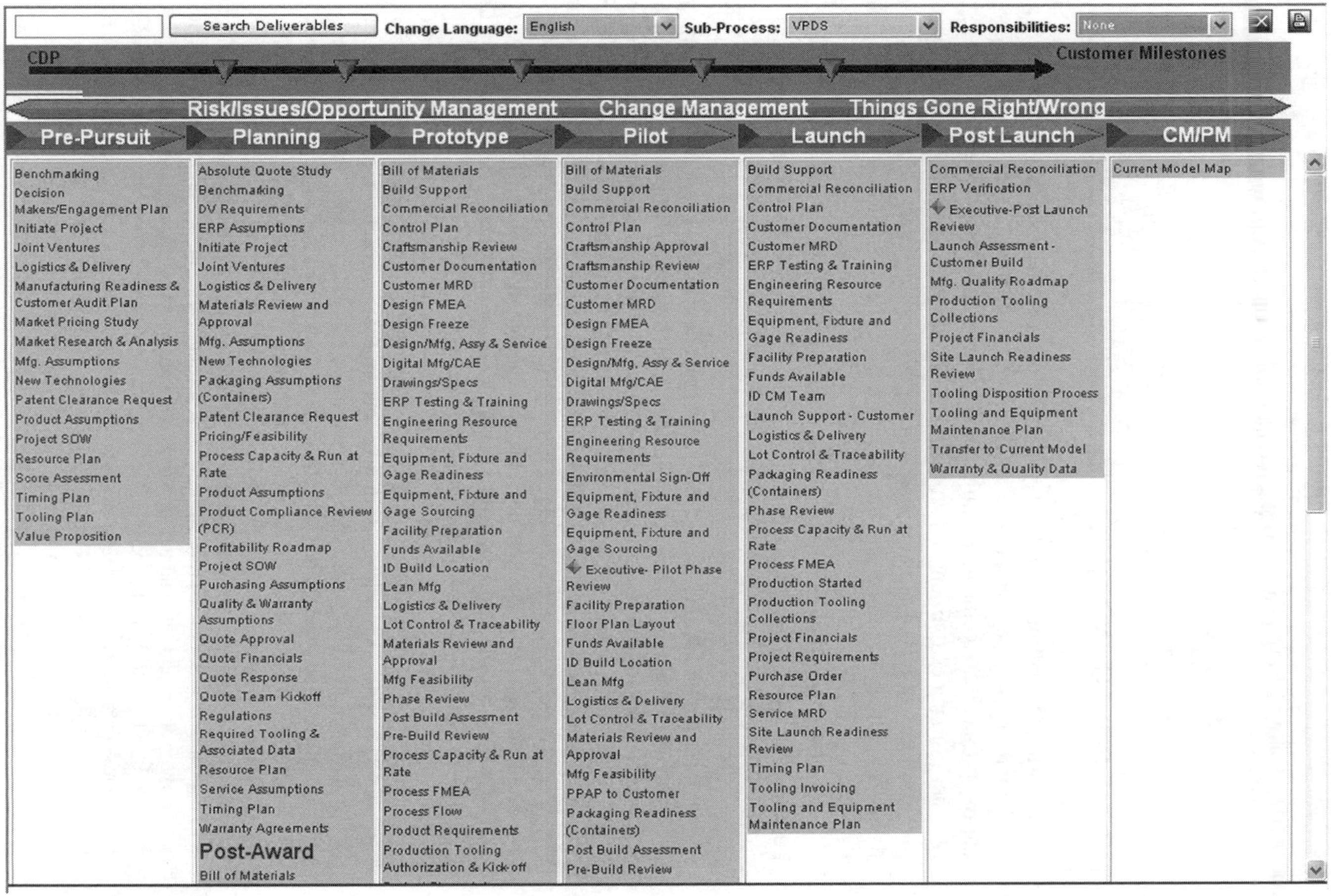

CDP

Customer Milestones

RiskIssues/Opportunity Management		Change Management		Things Gone Right/Wrong		
Pre-Pursuit	**Planning**	**Prototype**	**Pilot**	**Launch**	**Post Launch**	**CM/PM**

Pre-Pursuit	Planning	Prototype	Pilot	Launch	Post Launch	CM/PM
Benchmarking	Absolute Quote Study	Bill of Materials	Bill of Materials	Build Support	Commercial Reconciliation	Current Model Map
Decision Makers/Engagement Plan	Benchmarking	Build Support	Build Support	Commercial Reconciliation	ERP Verification	
Initiate Project	DV Requirements	Commercial Reconciliation	Commercial Reconciliation	Control Plan	◆ Executive-Post Launch Review	
Joint Ventures	ERP Assumptions	Control Plan	Control Plan	Customer Documentation	Launch Assessment-	
Logistics & Delivery	Initiate Project	Craftsmanship Review	Craftsmanship Approval	Customer MRD	Customer Build	
Manufacturing Readiness &	Joint Ventures	Customer Documentation	Craftsmanship Review	ERP Testing & Training	Mfg. Quality Roadmap	
Customer Audit Plan	Logistics & Delivery	Customer MRD	Customer Documentation	Engineering Resource	Production Tooling	
Market Pricing Study	Materials Review and	Design FMEA	Customer MRD	Requirements	Collections	
Market Research & Analysis	Approval	Design Freeze	Design FMEA	Equipment, Fixture and	Project Financials	
Mfg. Assumptions	Mfg. Assumptions	Design/Mfg, Assy & Service	Design Freeze	Gage Readiness	Site Launch Readiness	
New Technologies	New Technologies	Digital Mfg/CAE	Design/Mfg, Assy & Service	Facility Preparation	Review	
Patent Clearance Request	Packaging Assumptions	Drawings/Specs	Digital Mfg/CAE	Funds Available	Tooling Disposition Process	
Product Assumptions	(Containers)	ERP Testing & Training	Drawings/Specs	ID CM Team	Tooling and Equipment	
Project SOW	Patent Clearance Request	Engineering Resource	ERP Testing & Training	Launch Support - Customer	Maintenance Plan	
Resource Plan	Pricing/Feasibility	Requirements	Engineering Resource	Logistics & Delivery	Transfer to Current Model	
Score Assessment	Process Capacity & Run at	Equipment, Fixture and	Requirements	Lot Control & Traceability	Warranty & Quality Data	
Timing Plan	Rate	Gage Readiness	Environmental Sign-Off	Packaging Readiness		
Tooling Plan	Product Assumptions	Equipment, Fixture and	Equipment, Fixture and	(Containers)		
Value Proposition	Product Compliance Review	Gage Sourcing	Gage Readiness	Phase Review		
	(PCR)	Facility Preparation	Equipment, Fixture and	Process Capacity & Run at		
	Profitability Roadmap	Funds Available	Gage Sourcing	Rate		
	Project SOW	ID Build Location	◆ Executive- Pilot Phase	Process FMEA		
	Purchasing Assumptions	Lean Mfg	Review	Production Started		
	Quality & Warranty	Logistics & Delivery	Facility Preparation	Production Tooling		
	Assumptions	Lot Control & Traceability	Floor Plan Layout	Collections		
	Quote Approval	Materials Review and	Funds Available	Project Financials		
	Quote Financials	Approval	ID Build Location	Project Requirements		
	Quote Response	Mfg Feasibility	Lean Mfg	Purchase Order		
	Quote Team Kickoff	Phase Review	Logistics & Delivery	Resource Plan		
	Regulations	Post Build Assessment	Lot Control & Traceability	Service MRD		
	Required Tooling &	Pre-Build Review	Materials Review and	Site Launch Readiness		
	Associated Data	Process Capacity & Run at	Approval	Review		
	Resource Plan	Rate	Mfg Feasibility	Timing Plan		
	Service Assumptions	Process FMEA	PPAP to Customer	Tooling Financials		
	Timing Plan	Process Flow	Packaging Readiness	Tooling and Equipment		
	Warranty Agreements	Product Requirements	(Containers)	Maintenance Plan		
	Post-Award	Production Tooling	Post Build Assessment			
	Bill of Materials	Authorization & Kickoff	Pre-Build Review			

FIGURE 3–19. Process map.

Todd Smith, Senior Manager, climate product group project management. "It's effectively eliminated the need to populate forms and create 'evidence books,' making all project information globally available on-demand."

Visteon takes pride in its project management methodology as evidenced by its commitment to exceeding customer expectations.

3.14 CONVERGENT COMPUTING

In the previous sections, we discussed some of the problems that companies face when implementing project management. Some companies believe that education can accelerate the project management learning curve because it is better to learn from the mistakes of other companies than one's own mistakes. While education does not always eliminate the possibility of mistakes, it does accelerate the learning process, especially when executives also participate in and support the learning process. Such was the case at Convergent Computing (CCO).

Anne Walker, Enterprise Project Manager with Convergent Computing, describes her company's experience with project management as well as some of the problems that had to be overcome during implementation:

> CCO has been utilizing project management methodologies over 15 years. I have been with CCO going on 9 years and have used project management ongoing since the day I started. I was encouraged to participate in a UC-Berkeley extension project management program that consisted of six courses that we took at night after work. Two of CCO's business owners had already graduated from the program and felt the coursework and results were impressive. Other co-workers were receiving the same encouragement and many enrolled and had support from our managers and the organization. Others in our company have gone through the program on an on-going basis and it truly enhances our abilities to deliver successful projects for our clients.
>
> CCO manages a wide variety of projects and in general each involves our consultants partnering with our clients to best understand their objectives for the project and then delivering them successfully. The project many times will involve employees from different levels within the organization and, in some cases, consultants from other consulting firms. Many times our clients look to our project managers to help train their employees so that they gain experience in managing projects and can then project manage on future projects.
>
> CCO is a medium-sized, privately owned company that partners with our clients who look to CCO for professional and knowledgeable IT consulting services. Staying current on technology and having advanced knowledge of those technologies mean CCO needs to have technical expertise. Our professional services department initially included high-end consultants and engineers who did not have any project management experience or knowledge. Our expertise in IT is what our clients come to us for, so that is a priority throughout the organization. It is our "bread and butter," so to speak, so this made it difficult when consultants were asked to understand project management methodology as well. CCO consultants and engineers tend to be passionate when it comes to IT and have labs

at home and enjoy tackling new technology. We look for this skill when we hire and many were not accustomed to identifying the scope of a project, attending decision-making meetings, or having to take a structured approach to testing the technology being implemented. Technical engineers tend to want to get in and fix the problem or take a structured approach to testing the technology being implemented. Technical engineers tend to want to get in and fix the problem or implement the new technology. If a client wanted additional tasks added—and they involved technology—most technical people want to dig in and enjoy taking on the additional challenge. What was missing was someone asking about the scope of the project and whether this was within the scope, would change our deadlines, would impact the budget, and so on. Getting our engineers to understand and work through the objectives of a project as defined by the client was not an easy task.

Many of the consultants and engineers didn't understand why they should have to learn and deal with project management and this made it difficult to get them "on board." Many of them did not want to take the additional time to learn project management methodologies and felt they had enough on their hands just staying knowledgeable with the ever-changing technology. When project managers were assigned to projects and began scheduling kick off meetings or asking for a regular status update, consultants felt they were burdened with additional tasks or asked to attend meetings they didn't see as necessary.

As we added technical writers and project managers to our professional services organization, our sales team had to learn how to help clients understand why a project manager on an IT consulting project was necessary. Many times, clients would agree with our approach in a proposal but would want to cut the project management resource or hours allocated to project management out of the project or process.

I believe CCO has come a long way in "winning over" the engineers and consultants in regards to project management. Some key factors in this are:

- Senior management support.
- Successful project outcomes when project methodology was incorporated into the project.
- The benefits of project management could not be ignored. Multiple engineer projects were running more smoothly, communication improved, people on the team and the client knew where the project was, and our teams were able to respond to situations proactively rather than putting out a "fire" after the fact. As more and more successful projects and happy clients resulted, our professional services engineers and consultants began to expect kickoff meetings and status reports as it made them more successful in their client engagements.
- CCO's own project management expertise and knowledge grew. This reinforced the importance of project management.

Two of the owners of CCO graduated from the UC-Berkeley extension program and encouraged others to participate in the program as well. CCO paid tuition fees for employees who participated in the program and the employees gave up one or two nights a week for two years to complete the program. Executives also worked closely with our sales staff to help them better justify project managers for our consulting services. In delivering our IT consulting services, when a project manager was involved, we inevitably had better success and those projects were delivered on time. We continue to have clients who are very satisfied with what CCO has delivered, especially when we are able to have project management involved in the projects.

3.15 AVALON POWER AND LIGHT

Avalon Power and Light (a disguised case) is a mountain states utility company that, for decades, had functioned as a regional monopoly. All of this changed in 1995 with the beginning of deregulation in public utilities. Emphasis was now being placed on cost cutting and competitiveness.

The Information Systems Division of Avalon was always regarded as a "thorn in the side" of the company. The employees acted as prima donnas and refused to accept any of the principles of project management. Cost-cutting efforts at Avalon brought to the surface the problem that the majority of the work in the Information Systems Division could be outsourced at a significantly cheaper price than performing the work internally. Management believed project management could make the division more competitive, but would employees now be willing to accept the project management approach?

According to a spokesperson for Avalon Power and Light:

> Two prior attempts to implement a standard application-development methodology had failed. Although our new director of information systems aggressively supported this third effort by mandating the use of a standard methodology and standard tools, significant obstacles were still present.
>
> The learning curve for the project management methodology was high, resulting in a tendency of the leads to impose their own interpretations on methodology tasks rather than learning the documented explanations. This resulted in an inconsistent interpretation of the methodology, which in turn produced inconsistencies when we tried to use previous estimates in estimating new projects.
>
> The necessity to update project plans in a timely manner was still not universally accepted. Inconsistency in reporting actual hours and finish dates resulted in inaccurate availabilities. Resources were not actually available when indicated on the departmental plan.
>
> Many team leads had not embraced the philosophy behind project management and did not really subscribe to its benefits. They were going through the motions, producing the correct deliverables, but managing their projects intuitively in parallel to the project plan rather than using the project plan to run their projects.
>
> Information systems management did not ask questions that required use of project management in reporting project status. Standard project management metrics were ignored in project status reports in favor of subjective assessments.

The Information Systems Division realized that its existence could very well be based upon how well and how fast it would be able to develop a mature project management system. By 1997, the sense of urgency for maturity in project management had permeated the entire Information Systems Division. When asked what benefits were achieved, the spokesperson remarked:

> The perception of structure and the ability to document proposals using techniques recognized outside of our organization has allowed Information Systems to successfully compete against external organizations for application development projects.

Better resource management through elimination of the practice of "hoarding" preferred resources until another project needs staffing has allowed Information Systems to actually do more work with less people.

We are currently defining requirements for a follow-on project to the original project management implementation project. This project will address the lessons learned from our first two years. Training in project management concepts (as opposed to tools training) will be added to the existing curriculum. Increased emphasis will be placed on why it is necessary to accurately record time and task status. An attempt will be made to extend the use of project management to non-application-development areas, such as network communications and technical support. The applicability of our existing methodology to client-server development and Internet application development will be tested. We will also explore additional efficiencies such as direct input of task status by individual team members.

We now offer project management services as an option in our service-level agreements with our corporate "customers." One success story involved a project to implement a new corporate identity in which various components across the corporation were brought together. The project was able to cross department boundaries and maintain an aggressive schedule. The process of defining tasks and estimating their durations resulted in a better understanding of the requirements of the project. This in turn provided accurate estimates that drove significant decisions regarding the scope of the project in light of severe budget pressures. Project decisions tended to be based on sound business alternatives rather than raw intuition.

3.16 ROADWAY

In the spring of 1992, Roadway Express realized that its support systems (specifically information systems) had to be upgraded in order for Roadway Express to be well positioned for the twenty-first century. Mike Wickham, then President of Roadway Express, was a strong believer in continuous change. This was a necessity for his firm, because the rapid changes in technology mandated that reengineering efforts be an ongoing process. Several of the projects to be undertaken required a significantly larger number of resources than past projects had needed. Stronger interfacing between functional departments would also be required.

At the working levels of Roadway Express, knowledge of the principles and tools of project management was minimal at best in 1992. However, at the executive levels, knowledge of project management was excellent. This would prove to be highly beneficial. Roadway Express recognized the need to use project management on a two-year project that had executive visibility and support and that was deemed strategically critical to the company. Although the project required a full-time project manager, the company chose to appoint a line manager who was instructed to manage his line and the project at the same time for two years. The company did not use project management continuously, and the understanding of project management was extremely weak.

After three months, the line manager resigned his appointment as a project manager, citing too much stress and being unable to manage his line effectively while performing project duties. A second line manager was appointed on a part-time basis and, as with his predecessor, he found it necessary to resign as project manager.

The company then assigned a third line manager but this time released her from all line responsibility while managing the project. The project team and selected company personnel were provided with project management training. The president of the company realized the dangers of quick implementation, especially on a project of this magnitude, but was willing to accept the risk.

After three months, the project manager complained that some of her team members were very unhappy with the pressures of project management and were threatening to resign from the company if necessary simply to get away from project management. But when asked about the project status, the project manager stated that the project had met every deliverable and milestone thus far. It was quickly apparent to the president, Mike Wickham, and other officers of the company that project management was functioning as expected. The emphasis now was how to "stroke" the disgruntled employees and convince them of the importance of their work and how much the company appreciated their efforts.

To quell the fears of the employees, the president assumed the role of the project sponsor and made it quite apparent that project management was here to stay at Roadway Express. The president brought in training programs on project management and appeared at each training program.

The reinforcement by the president and his visible support permeated all levels of the company. By June of 1993, less than eight months after the first official use of project management, Roadway Express had climbed further along the ladder to maturity in project management than most other companies accomplish in two to three years due to the visible support of senior management.

Senior management quickly realized that project management and information systems management could be effectively integrated into a single methodology. Mike Wickham correctly recognized that the quicker he could convince his line managers to support the project management methodology, the quicker they would achieve maturity. According to Mike Wickham, President of Roadway Express at that time (and later Chairman of the Board):

> Project management, no matter how sophisticated or how well trained, cannot function effectively unless all management is committed to a successful project outcome. Before we put our current process in place, we actively involved all those line managers who thought it was their job to figure out all of the reasons a system would never work! Now, the steering committee says, "This is the project. Get behind it and see that it works." It is a much more efficient use of resources when everyone is focused on the same goal.

3.17 DEFCON CORPORATION

A defense contractor that wishes to remain nameless (we call it Defcon Corporation) had survived for almost 20 years on fixed-price, lump-sum government contracts. A characteristic of a fixed-price contract is that the customer does not audit your books, costs, or perhaps even your project management system. As a result, the company managed its projects rather loosely between 1967 and 1987. As long as deliveries were on time, the capabilities of the project management system were never questioned.

By 1987, the government-subcontracting environment had changed. There were several reasons for this:

- The DoD was undergoing restructuring.
- There were cutbacks in DoD spending, and the cutbacks were predicted to get worse.
- The DoD was giving out more and more cost-reimbursable contracts.
- The DoD was pressuring contractors to restructure from a traditional to a product-oriented organizational form.
- The DoD was pressuring contractors to reduce costs, especially the overhead rates.
- The DoD was demanding higher quality products.
- The DoD was now requiring in its proposals that companies demonstrate higher quality project management practices.

Simply to survive, Defcon had to bid on cost-reimbursable contracts. Internally, this mandated two critical changes. First, the organization had to go to more formal rather than informal project management. Second, the organization had to learn how to use and report earned-value measurement. In order to be looked upon favorably by the government for the award of a cost-reimbursable contract, a company must have its earned-value cost control/reporting system validated by the government.

A manager within one such company that was struggling made the following comments on how "survival" had forced the organization to climb the ladder to maturity over the past 10 years:

> Formal project management began with the award of the first major government program. There was a requirement to report costs by contract line item and to report variances at specific contract levels. A validated system was obtained to give us the flexibility to submit proposals on government programs where cost schedule reporting was a requirement of the RFP (request for proposal).
>
> We have previous experience in PERT (program evaluation and review technique) networking, work breakdown structures, and program office organizations. Management was also used to working in a structured format because of our customer's requirements for program reviews. After system validation in 1987, it took six months to a year to properly train and develop the skills needed by cost account managers and work package supervisors. As you move along in a program, there is the need to retrain and review project management requirements with the entire organization.
>
> We visited other companies and divisions of our own company that had prior experience in project management. We sent people to seminars and classes held by experts in the field. We conducted internal training classes and wrote policies and procedures to assist employees with the process of project management. Later we purchased canned reporting packages to reduce the cost of internal programming of systems.
>
> We established dedicated teams to a contract/program. We have program office organizations for large programs to follow through and coordinate information to internal

management and our customer. We adjusted our systems and reports to meet both our internal and external customers' needs.

Implementation of integrated systems will provide data on a timelier basis. These data will allow management to react quickly to solve the problem and minimize the cost impact.

Project management has allowed us to better understand costs and variances by contract/program. It provides us with timely data and makes tracking of schedule issues, budget issues, and earned values more manageable. Project management has given us visibility into the programs that is useful in implementing cost reductions and process improvements. Having a validated system allows us to remain competitive for bidding on those programs that require formal cost schedule control systems.

3.18 KOMBS ENGINEERING

The company described above was very fortunate to have identified the crises and taken the time to react properly. Some companies are not so fortunate. Although the next two companies appear to be outdated, there are valuable lessons that can be learned about what not to do when embarking on the path to maturity. Consider the Michigan-based Kombs Engineering (name of the company is disguised at company's request).

In June 1993, Kombs Engineering had grown to a company with $25 million in sales. The business base consisted of two contracts with the Department of Energy (DoE), one for $15 million and one for $8 million. The remaining $2 million consisted of a variety of smaller jobs for $15,000 to $50,000 each.

The larger contract with the DoE was a five-year contract for $15 million per year. The contract was awarded in 1988 and was up for renewal in 1993. The DoE had made it clear that, although it was very pleased with the technical performance of Kombs, the follow-on contract must go through competitive bidding by law. Marketing intelligence indicated that the DoE intended to spend $10 million per year for five years on the follow-on contract with a tentative award date of October 1993. On June 21, 1993, the solicitation for proposal was received at Kombs. The technical requirements of the proposal request were not considered to be a problem for Kombs. There was no question in anyone's mind that on technical merit alone Kombs would win the contract. The more serious problem was that the DoE required a separate section in the proposal on how Kombs would manage the $10 million/year project as well as a complete description of how the project management system at Kombs functioned.

When Kombs won the original bid in 1988, there had been no project management requirement. All projects at Kombs were accomplished through the traditional organizational structure. Only line managers acted as project leaders.

In July 1993, Kombs hired a consultant to train the entire organization in project management. The consultant also worked closely with the proposal team in responding to the DoE project management requirements. The proposal was submitted to the DoE during the second week of August. In September 1993, the DoE provided Kombs with a list of questions concerning its proposal. More than 95 percent of the questions involved project management. Kombs responded to all questions.

In October 1993, Kombs received notification that it would not be granted the contract. During a postaward conference, the DoE stated that it had no "faith" in the Kombs project management system. Kombs Engineering is no longer in business.

Kombs Engineering is an excellent case study to give students in project management classes. It shows what happens when a subcontractor does not recognize how smart the customer has become in project management. Had Kombs been in close contact with its customers, the company would have had five years rather than one month to develop a mature project management system.

3.19 WILLIAMS MACHINE TOOL COMPANY

The strength of a culture can not only prevent a firm from recognizing that a change is necessary but also block the implementation of the change even after need for it is finally realized. Such was the situation at Williams Machine Tool Company (another disguised case).

For 75 years, the Williams Machine Tool Company had provided quality products to its clients, becoming the third largest U.S.-based machine tool company by 1980. The company was highly profitable and had an extremely low employee turnover rate. Pay and benefits were excellent.

Between 1970 and 1980, the company's profits soared to record levels. The company's success was due to one product line of standard manufacturing machine tools. Williams spent most of its time and effort looking for ways to improve its "bread and butter" product line rather than to develop new products. The product line was so successful that other companies were willing to modify their production lines around these machine tools, rather than asking Williams for major modifications to the machine tools.

By 1980, Williams Company was extremely complacent, expecting this phenomenal success with one product line to continue for 20 to 25 more years. The recession of 1979–1983 forced management to realign its thinking. Cutbacks in production had decreased the demand for the standard machine tools. More and more customers were asking either for major modifications to the standard machine tools or for a completely new product design.

The marketplace was changing and senior management recognized that a new strategic focus was necessary. However, attempts to convince lower-level management and the workforce, especially engineering, of this need were meeting strong resistance. The company's employees, many of them with over 20 years of employment at Williams Company, refused to recognize this change, believing that the glory days of yore would return at the end of the recession.

In 1986, the company was sold to Crock Engineering. Crock had an experienced machine tool division of its own and understood the machine tool business. Williams Company was allowed to operate as a separate entity from 1985 to 1986. By 1986, red ink had appeared on the Williams Company balance sheet. Crock replaced all of the Williams senior managers with its own personnel. Crock then announced to all employees that Williams would become a specialty machine tool manufacturer and the "good old days" would never return. Customer demand for specialty products had increased threefold in just the last 12 months alone. Crock made it clear that employees who would not support this new direction would be replaced.

The new senior management at Williams Company recognized that 85 years of traditional management had come to an end for a company now committed to specialty products. The company culture was about to change, spearheaded by project management, concurrent engineering, and total quality management.

Senior management's commitment to project management was apparent by the time and money spent in educating the employees. Unfortunately, the seasoned 20+ year veterans still would not support the new culture. Recognizing the problems, management provided continuous and visible support for project management in addition to hiring a project management consultant to work with the people. The consultant worked with Williams from 1986 to 1991.

From 1986 to 1991, the Williams Division of Crock Engineering experienced losses in 24 consecutive quarters. The quarter ending March 31, 1992, was the first profitable quarter in over six years. Much of the credit was given to the performance and maturity of the project management system. In May 1992, the Williams Division was sold. More than 80 percent of the employees lost their jobs when the company was relocated over 1500 miles away.

Williams Machine Tool Company did not realize until too late that the business base had changed from production-driven to project-driven. Living in the past is acceptable only if you want to be a historian. But for businesses to survive, especially in a highly competitive environment, they must look ahead and recognize that change is inevitable.

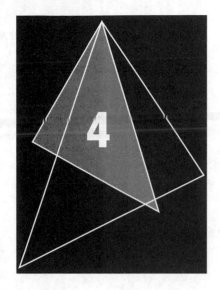

Project Management Methodologies

4.0 INTRODUCTION

In Chapter 1 we described the life-cycle phases for achieving maturity in project management. The fourth phase was the growth phase, which included the following:

- Establish life-cycle phases.
- Develop a project management methodology.
- Base the methodology upon effective planning.
- Minimize scope changes and scope creep.
- Select the appropriate software to support the methodology.

The importance of a good methodology cannot be understated. Not only will it improve your performance during project execution, but it will also allow for better customer relations and customer confidence. Good methodologies can also lead to sole-source or single-source procurement contracts.

Creating a workable methodology for project management is no easy task. One of the biggest mistakes made is developing a different methodology for each type of project. Another is failing to integrate the project management methodology and project management tools into a single process, if possible. When companies develop project management methodologies and tools in tandem, two benefits emerge. First, the work is accomplished with fewer scope changes. Second, the processes are designed to create minimal disturbance to ongoing business operations.

This chapter discusses the components of a project management methodology and some of the most widely used project management tools. Detailed examples of methodologies at work are also included.

4.1 EXCELLENCE DEFINED

Excellence in project management is often regarded as a continuous stream of successfully managed projects. Without a project management methodology, repetitive successfully completed projects may be difficult to achieve.

Today, everyone seems to agree somewhat on the necessity for a project management methodology. However, there is still disagreement on the definition of excellence in project management, the same way that companies have different definitions for project success. In this section, we will discuss some of the different definitions of excellence in project management.

Some definitions of excellence can be quite simple and achieve the same purpose as complex definitions. According to a spokesperson from Motorola[1]:

Excellence in project management can be defined as:

- Strict adherence to scheduling practices
- Regular senior management oversight
- Formal requirements change control
- Formal issue and risk tracking
- Formal resource tracking
- Formal cost tracking

A spokesperson from AT&T defined excellence at AT&T as:

Excellence [in project management] is defined as a consistent Project Management Methodology applied to all projects across the organization, continued recognition by our customers, and high customer satisfaction. Also our project management excellence is a key selling factor for our sales teams. This results in repeat business from our customers. In addition there is internal acknowledgement that project management is value-added and a must have.

According to Colin Spence, Project Manager/Partner at Convergent Computing (CCO):

While CCO does not have a formal definition of "excellence in project management," our project managers or resources tasked with the job of acting as a project manager on a project should seek to achieve the following goals:

- To ensure on-budget, on-time, on-scope delivery of professional services as defined in the proposal or conceptual plan
- To ensure that deliverables are of the highest quality and meet the goals of the project
- To provide strategic oversight of the services delivery process to ensure that all CCO resources are working effectively and efficiently on the project

---

1. H. Kerzner, *Best Practices in Project Management: Achieving Global Excellence*, Wiley, Hoboken, NJ, 2006, p. 136.

● To oversee CCO resources to ensure that each is fulfilling the requirements of their role in the client team
● To facilitate communications within the client team and with the client
● To monitor any yellow or red flags in projects and drive toward resolution and escalate if needed
● To develop professional services feedback and performance measurement

Some of the skills that are expected of a CCO project manager include:

● Strong communications and management skills
● Three to five years of project management experience
● Strong organizational skills
● Experience with delivering technology solutions related to the specific project
● Experience with Microsoft Project software
● Knowledge of CCO processes and policies
● Ability to oversee and coordinate multiple resources and activities

Doug Bolzman, Consultant Architect, PMP®, ITIL Service Manager at EDS, discusses his view of excellence in project management:

Excellence is rated, not by managing the pieces, but by understanding how the pieces fit together, support each other's dependencies, and provide value to the business. If project management only does what it is asked to do, such as manage 300 individual projects in the next quarter, it is providing a low value-added function that basically is the "pack mule" that is needed, but only does what it is asked—and no more. Figures 4–1 and 4–2 demonstrate that if mapping project management to a company's overall release management framework, each project is managed independently with the characteristics shown

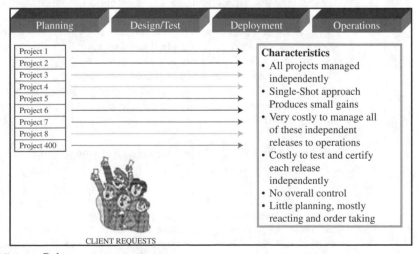

| Planning | Design/Test | Deployment | Operations |

Project 1
Project 2
Project 3
Project 4
Project 5
Project 6
Project 7
Project 8
Project 400

Characteristics
• All projects managed independently
• Single-Shot approach Produces small gains
• Very costly to manage all of these independent releases to operations
• Costly to test and certify each release independently
• No overall control
• Little planning, mostly reacting and order taking

CLIENT REQUESTS

FIGURE 4–1. Release management stages.

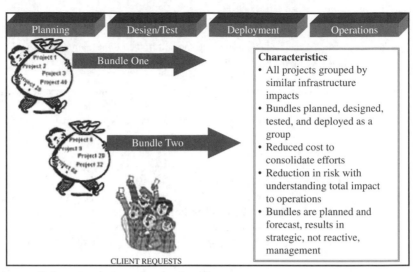

FIGURE 4–2. Release management stages: bundling requests.

Using the same release framework and the same client requests, project management disciplines can understand the nature of the requirements and provide a valuable service to bundle the same types of requests (projects) to generate a forecast of the work, which will assist the company in balancing its financials, expectations, and resources. This function can be done within the PMO.

At DTE Energy, excellence in project management is defined using the project management methodology. According to Jason Schulist, Manager—Continuous Improvement, Operating Systems Strategy Group:

> DTE Energy leverages a four-gate/nine-step project management model for continuous improvement. (See Figures 4–3 and 4–4.) In gate1 the project lead clearly defines the metric that will measure the success in the project. In gate 2 of the project, after the ideal state has been defined, the project lead gains approval from the champion confirming not only the appropriate metrics but also the target.
>
> This target for success is maintained throughout the project and the project does not close (gate 4) unless the target is achieved. If the project does not achieve the target through the committed actions, the project reverts to gate 2 where the ideal state is revisited and further actions are determined.

Allan Dutch, PMP®, Senior Project Manager, Software Engineering, Methods and Staffing at DTE Energy, believes that:

> A DTE Energy information technology (IT) project exhibiting success and excellence in project management is one where the project manager directs her or his team much like a conductor directs her or his orchestra. Business value is demonstrable and recognized by the business unit while interactions with support organizations and infrastructure requirements are coordinated smoothly and according to plan.

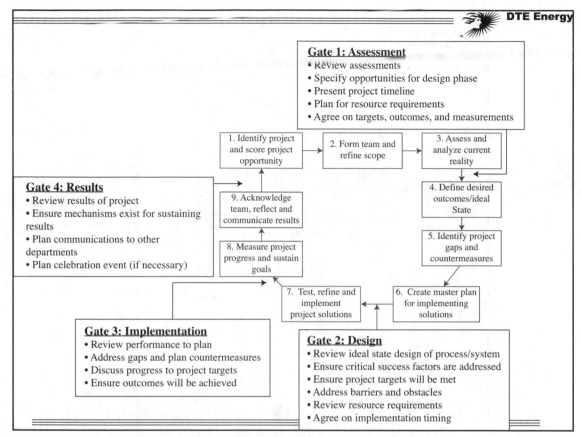

FIGURE 4–3. Four-gate/nine-step project management model.

4.2 RECOGNIZING THE NEED FOR METHODOLOGY DEVELOPMENT

Simply having a project management methodology and following it do not lead to success and excellence in project management. The need for improvements in the system may be critical. External factors can have a strong influence on the success or failure of a company's project management methodology. Change is a given in the current business climate, and there is no sign that the future will be any different. The rapid changes in technology that have driven changes in project management over the past two decades are not likely to subside. Another trend, the increasing sophistication of consumers and clients, is likely to continue, not go away. Cost and quality control have become virtually the same issue in many industries. Other external factors include rapid mergers and acquisitions and real-time communications.

Project management methodologies are organic processes and need to change as the organization changes in response to the ever-evolving business climate. Such changes, however, require that managers on all levels be committed to the changes and develop a

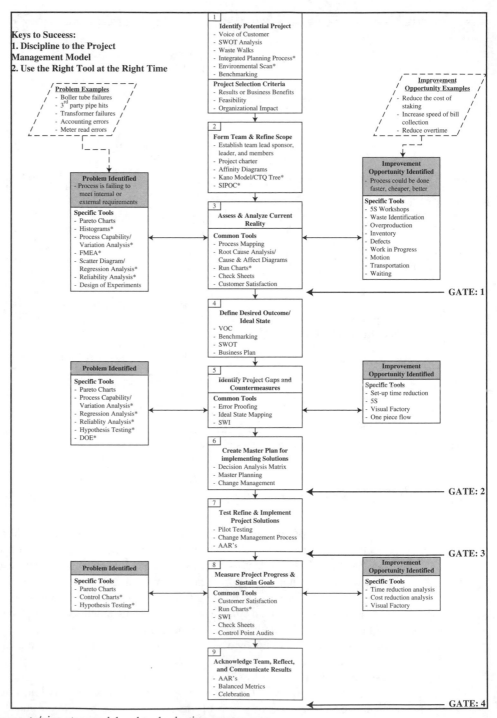

FIGURE 4–4. Four-gate/nine-step model and tool selection process map.

vision that calls for the development of project management systems along with the rest of the organization's other business systems.

Today, companies are managing their business by projects. This is true for both non project-driven and project-driven organizations. Virtually all activities in an organization can be treated as some sort of project. Therefore, it is only fitting that well-managed companies regard a project management methodology as a way to manage the entire business rather than just projects. Business processes and project management processes will be merged together as the project manager is viewed as the manager of part of a business rather than just the manager of a project.

Developing a standard project management methodology is not for every company. For companies with small or short-term projects, such formal systems may not be cost effective or appropriate. However, for companies with large or ongoing projects, developing a workable project management system is mandatory.

For example, a company that manufactures home fixtures had several project development protocols in place. When they decided to begin using project management systematically, the complexity of the company's current methods became apparent. The company had multiple system development methodologies based on the type of project. This became awkward for employees who had to struggle with a different methodology for each project. The company then opted to create a general, all-purpose methodology for all projects. The new methodology had flexibility built into it. According to one spokesman for the company:

> Our project management approach, by design, is not linked to a specific systems development methodology. Because we believe that it is better to use a (standard) systems development methodology than to decide which one to use, we have begun development of a guideline systems development methodology specific for our organization. We have now developed prerequisites for project success. These include:
>
> ● A well-patterned methodology
> ● A clear set of objectives
> ● Well-understood expectations
> ● Thorough problem definition

During the late 1980s, merger mania hit the banking community. With the lowering of costs due to economies of scale and the resulting increased competitiveness, the banking community recognized the importance of using project management for mergers and acquisitions. The quicker the combined cultures became one, the less the impact on the corporation's bottom line.

The need for a good methodology became apparent, according to a spokesperson at one bank:

> The intent of this methodology is to make the process of managing projects more effective: from proposal to prioritization to approval through implementation. This methodology is not tailored to specific types or classifications of projects, such as system development efforts or hardware installations. Instead, it is a commonsense approach to assist in prioritizing and implementing successful efforts of any jurisdiction.

In 1996, the information services (IS) division of one bank formed an IS reengineering team to focus on developing and deploying processes and tools associated with project management and system development. The mission of the IS reengineering team was to improve performance of IS projects, resulting in increased productivity, cycle time, quality, and satisfaction of the projects' customers.

According to a spokesperson at the bank, the process began as follows:

> Information from both current and previous methodologies used by the bank was reviewed, and the best practices of all these previous efforts were incorporated into this document. Regardless of the source, project methodology phases are somewhat standard fare. All projects follow the same steps, with the complexity, size, and type of project dictating to what extent the methodology must be followed. What this methodology emphasizes are project controls and the tie of deliverables and controls to accomplishing the goals.

To determine the weaknesses associated with past project management methodologies, the IS reengineering team conducted various focus groups. These focus groups concluded that there was:

- Lack of management commitment
- Lack of a feedback mechanism for project managers to determine the updates and revisions needed to the methodology
- Lack of adaptable methodologies for the organization
- Lack of training curriculum for project managers on the methodology
- Lack of focus on consistent and periodic communication on the methodology deployment progress
- Lack of focus on the project management tools and techniques

Based on this feedback, the IS reengineering team successfully developed and deployed a project management and system development methodology. Beginning June 1996 through December 1996, the target audience of 300 project managers became aware and applied a project management methodology and standard tool (MS Project).

The bank did an outstanding job of creating a methodology that reflects guidelines rather than policies and provides procedures that can easily be adapted on any project in the bank. Below the selected components of the project management methodology are discussed.

Organizing

With any project, you need to define what needs to be accomplished and decide how the project is going to achieve those objectives. Each project begins with an idea, vision, or business opportunity, a starting point that must be tied to the organization's business objectives. The project charter is the foundation of the project and forms the contract with the parties involved. It includes a statement of business needs, an agreement of what the project is committed to deliver, an identification of project dependencies, the roles and responsibilities of the team members involved, and the standards for how project budget and project management should be approached. The project charter defines the boundaries of the project.

Planning
Once the project boundaries are defined, sufficient information must be gathered to support the goals and objectives and to limit risk and minimize issues. This component of project management should generate sufficient information to clearly establish the deliverables that need to be completed, define the specific tasks that will ensure completion of these deliverables, and outline the proper level of resources. Each deliverable affects whether or not each phase of the project will meet its goals, budget, quality, and schedule. For simplicity sake, some projects take a four-phase approach:

- *Proposal:* Project initiation and definition.
- *Planning:* Project planning and requirements definition.
- *Development:* Requirement development, testing, and training.
- *Implementation:* Rollout of developed requirements for daily operation.

Each phase contains review points to help ensure that project expectations and quality deliverables are achieved. It is important to identify the reviewers for the project as early as possible to ensure the proper balance of involvement from subject matter experts and management.

Managing
Throughout the project, management and control of the process must be maintained. This is the opportunity for the project manager and team to evaluate the project, assess project performance, and control the development of the deliverables. During the project, the following areas should be managed and controlled:

- Evaluate daily progress of project tasks and deliverables by measuring budget, quality, and cycle time.
- Adjust day-to-day project assignments and deliverables in reaction to immediate variances, issues, and problems.
- Proactively resolve project issues and changes to control scope creep.
- Aim for client satisfaction.
- Set up periodic and structured reviews of the deliverables.
- Establish a centralized project control file.

Two essential mechanisms for successfully managing projects are solid status-reporting procedures and issues and change management procedures. Status reporting is necessary for keeping the project on course and in good health. The status report should include the following:

- Major accomplishment to date
- Planned accomplishments for the next period
- Project progress summary:
  - Percent of effort hours consumed
  - Percent of budget costs consumed
  - Percent of project schedule consumed

- Project cost summary (budget versus actual)
- Project issues and concerns
- Impact to project quality
- Management action items

Issues-and-change management protects project momentum while providing flexibility. Project issues are matters that require decisions to be made by the project manager, project team, or management. Management of project issues needs to be defined and properly communicated to the project team to ensure the appropriate level of issue tracking and monitoring. This same principle relates to change management because inevitably the scope of a project will be subject to some type of change. Any change management on the project that impacts the cost, schedule, deliverables, and dependent projects is reported to management. Reporting of issue and change management should be summarized in the status report denoting the number of open and closed items of each. This assists management in evaluating the project health.

Simply having a project management methodology and using it does not lead to maturity and excellence in project management. There must exist a "need" for improving the system toward maturity. Project management systems can change as the organization changes. However, management must be committed to the change and have the vision to let project management systems evolve with the organization.

4.3 ENTERPRISE PROJECT MANAGEMENT METHODOLOGIES

Most companies today seem to recognize the need for one or more project management methodologies but either create the wrong methodologies or misuse the methodologies that have been created. Many times, companies rush into the development or purchasing of a methodology without any understanding of the need for one other than the fact that their competitors have a methodology. Jason Charvat states[2]:

Using project management methodologies is a business strategy allowing companies to maximize the project's value to the organization. The methodologies must evolve and be "tweaked" to accommodate a company's changing focus or direction. It is almost a mind-set, a way that reshapes entire organizational processes: sales and marketing, product design, planning, deployment, recruitment, finance, and operations support. It presents a radical cultural shift for many organizations. As industries and companies change, so must their methodologies. If not, they're losing the point.

Methodologies are a set of forms, guidelines, templates, and checklists that can be applied to a specific project or situation. It may not be possible to create a single enterprisewide methodology that can be applied to each and every project. Some companies have been successful doing this, but there are still many companies that successfully

---

2. J. Charvat, *Project Management Methodologies*, Wiley, Hoboken, NJ, 2003, p. 2.

maintain more than one methodology. Unless the project manager is capable of tailoring the enterprise project management methodology to his or her needs, more that one methodology may be necessary.

There are several reasons why good intentions often go astray. At the executive levels, methodologies can fail if the executives have a poor understanding of what a methodology is and believe that a methodology is:

- A quick fix
- A silver bullet
- A temporary solution
- A cookbook approach for project success[3]

At the working levels, methodologies can also fail if they:

- Are abstract and high level
- Contain insufficient narratives to support these methodologies
- Are not functional or do not address crucial areas
- Ignore the industry standards and best practices
- Look impressive but lack real integration into the business
- Use nonstandard project conventions and terminology
- Compete for similar resources without addressing this problem
- Don't have any performance metrics
- Take too long to complete because of bureaucracy and administration[4]

Deciding on the type of methodology is not an easy task. There are many factors to consider, such as:

- The overall company strategy—how competitive are we as a company?
- The size of the project team and/or scope to be managed
- The priority of the project
- How critical the project is to the company
- How flexible the methodology and its components are[5]

Project management methodologies are created around the project management maturity level of the company and the corporate culture. If the company is reasonably mature in project management and has a culture that fosters cooperation, effective communications, teamwork, and trust, then a highly flexible methodology can be created based upon guidelines, forms, checklists, and templates. Project managers can pick and choose the parts of the methodology that are appropriate for a particular client. Organizations that do not possess either of these two characteristics rely heavily upon methodologies constructed with rigid policies and procedures, thus creating significant paperwork requirements with

---

3. Ibid., p. 4.
4. Ibid., p. 5.
5. Ibid., p. 66.

accompanying cost increases and removing the flexibility that the project manager needs for adapting the methodology to the needs of a specific client.

Jason Charvat describes these two types as light methodologies and heavy methodologies.[6]

Light Methodologies   Ever-increasing technological complexities, project delays, and changing client requirements brought about a small revolution in the world of development methodologies. A totally new breed of methodology—which is agile and adaptive and involves the client every part of the way—is starting to emerge. Many of the heavyweight methodologists were resistant to the introduction of these "lightweight" or "agile" methodologies (Fowler, 2001[7]). These methodologies use an informal communication style. Unlike heavyweight methodologies, lightweight projects have only a few rules, practices, and documents. Projects are designed and built on face-to-face discussions, meetings, and the flow of information to the clients. The immediate difference of using light methodologies is that they are much less documentation oriented, usually emphasizing a smaller amount of documentation for the project.

Heavy Methodologies   The traditional project management methodologies (i.e., SDLC approach) are considered bureaucratic or "predictive" in nature and have resulted in many unsuccessful projects. These heavy methodologies are becoming less popular. These methodologies are so laborious that the whole pace of design, development, and deployment slows down—and nothing gets done. Project managers tend to predict every milestone because they want to foresee every technical detail (i.e., software code or engineering detail). This leads managers to start demanding many types of specifications, plans, reports, checkpoints, and schedules. Heavy methodologies attempt to plan a large part of a project in great detail over a long span of time. This works well until things start changing, and the project managers inherently try to resist change.

Enterprise project management methodologies can enhance the project planning process as well as providing some degree of standardization and consistency. Companies have come to the realization that enterprise project management methodologies work best if the methodology is based upon templates rather than rigid policies and procedures. The International Institute for Learning has created a Unified Project Management Methodology (UPMM™) with templates categorized according to the *PMBOK® Guide* Areas of Knowledge[8]:

Communication:

Project Charter
Project Procedures Document

---

6. Ibid., pp. 102–104.
7. M. Fowler, *The New Methodology, Thought Works*, 2001. Available: www.martinfowler.com/articles.
8. Unified Project Management Methodology (UPMM™) is registered, copyrighted and owned by International Institute for Learning, Inc., © 2009; reproduced by permission.

Project Change Requests Log
Project Status Report
PM Quality Assurance Report
Procurement Management Summary
Project Issues Log
Project Management Plan
Project Pe.rformance Report

Cost:

Project Schedule
Risk Response Plan and Register
Work Breakdown Structure (WBS)
Work Package
Cost Estimates Document
Project Budget
Project Budget Checklist

Human Resources:

Project Charter
Work Breakdown Structure (WBS)
Communications Management Plan
Project Organization Chart
Project Team Directory
Responsibility Assignment Matrix (RAM)
Project Management Plan
Project Procedures Document
Kick-off Meeting Checklist
Project Team Performance Assessment
Project Manager Performance Assessment

Integration:

Project Procedures Overview
Project Proposal
Communications Management Plan
Procurement Plan
Project Budget
Project Procedures Document
Project Schedule
Responsibility Assignment Matrix (RAM)
Risk Response Plan and Register
Scope Statement
Work Breakdown Structure (WBS)
Project Management Plan

Project Change Requests Log
Project Issues Log
Project Management Plan Changes Log
Project Performance Report
Lessons Learned Document
Project Performance Feedback
Product Acceptance Document
Project Charter
Closing Process Assessment Checklist
Project Archives Report

Procurement:

Project Charter
Scope Statement
Work Breakdown Structure (WBS)
Procurement Plan
Procurement Planning Checklist
Procurement Statement of Work (SOW)
Request for Proposal Document Outline
Project Change Requests Log
Contract Formation Checklist
Procurement Management Summary

Quality:

Project Charter
Project Procedures Overview
Work Quality Plan
Project Management Plan
Work Breakdown Structure (WBS)
PM Quality Assurance Report
Lessons Learned Document
Project Performance Feedback
Project Team Performance Assessment
PM Process Improvement Document

Risk:

Procurement Plan
Project Charter
Project Procedures Document
Work Breakdown Structure (WBS)
Risk Response Plan and Register

Scope:

Project Scope Statement
Work Breakdown Structure (WBS)
Work Package
Project Charter

Time:

Activity Duration Estimating Worksheet
Cost Estimates Document
Risk Response Plan and Register Medium
Work Breakdown Structure (WBS)
Work Package
Project Schedule
Project Schedule Review Checklist

4.4 BENEFITS OF A STANDARD METHODOLOGY

For companies that understand the importance of a standard methodology, the benefits are numerous. These benefits can be classified as both short- and long-term benefits. Short-term benefits were described by one company as:

- Decreased cycle time and lower costs
- Realistic plans with greater possibilities of meeting time frames
- Better communications as to "what" is expected from groups and "when"
- Feedback: lessons learned

These short-term benefits focus on KPIs or, simply stated, the execution of project management. Long-term benefits seem to focus more upon critical success factors (CSFs) and customer satisfaction. Long-term benefits of development and execution of a world-class methodology include:

- Faster "time to market" through better scope control
- Lower overall program risk
- Better risk management, which leads to better decision-making
- Greater customer satisfaction and trust, which lead to increased business and expanded responsibilities for the tier 1 suppliers
- Emphasis on customer satisfaction and value-added rather than internal competition between functional groups
- Customer treating the contractor as a "partner" rather than as a commodity
- Contractor assisting the customer during strategic planning activities

Perhaps the largest benefit of a world-class methodology is the acceptance and recognition by your customers. If one of your critically important customers develops its own methodology, that customer could "force" you to accept it and use it in order to remain a supplier. But if you can show that your methodology is superior or equal to the customer's, your methodology will be accepted, and an atmosphere of trust will prevail.

One contractor recently found that its customer had so much faith in and respect for its methodology that the contractor was invited to participate in the customer's strategic planning activities. The contractor found itself treated as a partner rather than as a commodity or just another supplier. This resulted in sole-source procurement contracts for the contractor.

Developing a standard methodology that encompasses the majority of a company's projects and is accepted by the entire organization is a difficult undertaking. The hardest part might very well be making sure that the methodology supports both the corporate culture and the goals and objectives set forth by management. Methodologies that require changes to a corporate culture may not be well accepted by the organization. Nonsupportive cultures can destroy even seemingly good project management methodologies.

During the 1980s and 1990s, several consulting companies developed their own project management methodologies, most frequently for information systems projects, and then pressured their clients into purchasing the methodology rather than helping their clients develop a methodology more suited to the clients' needs. Although there may have been some successes, there appeared to be significantly more failures than successes. A hospital purchased a $130,000 project management methodology with the belief that this would be the solution to its information system needs. Unfortunately, senior management made the purchasing decision without consulting the workers who would be using the system. In the end, the package was never used.

Another company purchased a similar package, discovering too late that the package was inflexible and the organization, specifically the corporate culture, would need to change to use the project management methodology effectively. The vendor later admitted that the best results would occur if no changes were made to the methodology.

These types of methodologies are extremely rigid and based on policies and procedures. The ability to custom design the methodology to specific projects and cultures was nonexistent, and eventually these methodologies fell by the wayside—but after the vendors made significant profits. Good methodologies must be flexible.

4.5 CRITICAL COMPONENTS

It is almost impossible to become a world-class company with regard to project management without having a world-class methodology. Years ago, perhaps only a few companies really had world-class methodologies. Today, because of the need for survival and stiffening competition, there are numerous companies with good methodologies.

The characteristics of a world-class methodology include:

- Maximum of six life-cycle phases
- Life-cycle phases overlap
- End-of-phase gate reviews
- Integration with other processes
- Continuous improvement (i.e., hear the voice of the customer)
- Customer oriented (interface with customer's methodology)
- Companywide acceptance
- Use of templates (level 3 WBS)
- Critical path scheduling (level 3 WBS)
- Simplistic, standard bar chart reporting (standard software)
- Minimization of paperwork

Generally speaking, each life-cycle phase of a project management methodology requires paperwork, control points, and perhaps special administrative requirements. Having too few life-cycle phases is an invitation for disaster, and having too many life-cycle phases may drive up administrative and control costs. Most companies prefer a maximum of six life-cycle phases.

Historically, life-cycle phases were sequential in nature. However, because of the necessity for schedule compression, life-cycle phases today will overlap. The amount of overlap will be dependent upon the magnitude of the risks the project manager will take. The more the overlap, the greater the risk. Mistakes made during overlapping activities are usually more costly to correct than mistakes during sequential activities. Overlapping life-cycle phases requires excellent up-front planning.

End-of-phase gate reviews are critical for control purposes and verification of interim milestones. With overlapping life-cycle phases, there are still gate reviews at the end of each phase, but they are supported by intermediate reviews during the life-cycle phases.

World-class project management methodologies are integrated with other management processes such as change management, risk management, total quality management, and concurrent engineering. This produces a synergistic effect which minimizes paperwork, minimizes the total number of resources committed to the project, and allows the organization to perform capacity planning to determine the maximum workload that the organization can endure.

World-class methodologies are continuously enhanced through KPI reviews, lessons-learned updates, benchmarking, and customer recommendations. The methodology itself could become the channel of communication between the customer and contractor. Effective methodologies foster customer trust and minimize customer interference in the project.

Project management methodologies must be easy for workers to use as well as covering most of the situations that can arise on a project. Perhaps the best way is to have the methodology placed in a manual that is user friendly.

Excellent methodologies try to make it easier to plan and schedule projects. This is accomplished by using templates for the top three levels of the WBS. Simply stated, using

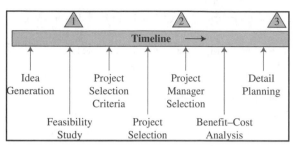

FIGURE 4–5. When to prepare the charter.

WBS level 3 templates, standardized reporting with standardized terminology exists. The differences between projects will appear at the lower levels (i.e., levels 4–6) of the WBS. This also leads to a minimization of paperwork.

Today, companies seem to be promoting the use of the project charter concept as a component of a methodology, but not all companies create the project charter at the same point in the project life cycle, as shown in Figure 4–5. The three triangles in Figure 4–5 show possible locations where the charter can be prepared:

- In the first triangle, the charter is prepared immediately after the feasibility study is completed. At this point, the charter contains the results of the feasibility study as well as documentation of any assumptions and constraints that were considered. The charter is then revisited and updated once this project is selected.
- In the second triangle, which seems to be the preferred method, the charter is prepared after the project is selected and the project manager has been assigned. The charter includes the authority granted to the project manager, but for this project only.
- In the third method, the charter is prepared after detail planning is completed. The charter contains the detailed plan. Management will not sign the charter until after detail planning is approved by senior management. Then, and only then, does the company officially sanction the project. Once management signs the charter, the charter becomes a legal agreement between the project manager and all involved line managers as to what deliverables will be met and when.

4.6 FROM MYTH TO REALITY

Sooner or later, all companies recognize the need for a project management methodology, perhaps even an enterprisewide project management methodology (EPM). While recognition is easy, the actual development and ultimately implementation can be difficult if the company bases its decisions on myths rather than reality. In Table 4–1 are some of the characteristics of well-developed methodologies, at least in the author's opinion. The second column illustrates the myth, whereas the third column identifies what mature companies seem to do.

TABLE 4–1. DESIGN CHARACTERISTICS OF A GOOD METHODOLOGY

Characteristic	Myth	Reality
Description of the Methodology	Excessive details are needed, perhaps hundreds of pages with illustrations, graphs, and tables	Description is a high-level framework, short and easy to read and understand; perhaps 25 pages or less
Readability	Excessive details and complex illustrations	Narrative descriptions supported mainly by flowcharts
Applicability	Need multiple methodologies based upon the types of programs, projects, deliverables,and functional areas	One high-level methodology can be used for both projects and programs
Alignment	Methodology should be aligned to the deliverables of the project or programs	Methodology should be aligned to the business and applicable to both projects and programs
Adaptability	Methodology must be based upon rigid policies and procedures	Methodology should have flexibility so that it can be adapted to all programs and projects and also be readily adaptable to necessary changes
PMBOK® Guide mapping	Methodologies have unique characteristics and are based upon rigid policies and procedures	The basic structure of the methodology should be mapped against the *PMBOK® Guide*
Role delineation	The methodology should clearly delineate the roles of all players; project managers, line managers, executives, and customers	Because of the need for flexibility and adaptability, only the role of the project manager needs to be delineated, specifically integration responsibility
Lessons learned	Methodology usage ends when the deliverables are met	Methodologies should have a final step to capture best practices and lessons learned

A capital equipment producer for the electronics industry has developed a methodology that follows most of these reality characteristics. For example, the opening paragraph in the Teradyne's methodology states[9]:

> One purpose of this document is to present a high level description of the Program/Project Manager's function within the product development cycle. A Program Manager is responsible for multiple related product aligned projects culminating in a major product deliverable where a Project Manager will focus on a single major product level deliverable. The purpose of the Program/Project Manager is to: "Partner with design leads in directing and integrating cross-functional project teams to ensure stakeholders objectives are met and deliver products in accordance with the approved project plan".... The value proposition for program management could be stated: A Program/Project Manager (PM) provides the organized, industry-aligned disciplined framework to integrate functional engineering and operational groups in driving product-based projects or programs to completion on-time and on-budget, while meeting market demands/needs/requirements. Another purpose of

---

9. The remaining quotes in this section were provided by Dr. Steve Lyons, Applications Project Manager at Teradyne, a capital equipment manufacturer in the electronics industry and adapted from their methodology for managing projects and programs. The material is reproduced with permission of Teradyne.

this document will be to present a high level description of the work instruction framework that Program Management uses in managing product development projects/programs. . . . Given the unique integrative role of the PM, this document will also attempt to link tie points with functional groups such as: marketing, engineering and operations.

The document also goes on to describe the role of the project and program managers, as well as the fact that the methodology applies to both programs and projects. In addition, the methodology also identifies the fact that the entire methodology may not be required on certain projects. This fact alone provides considerable flexibility in its use and illustrates the company's faith in the ability of their project/program managers.

This document will pertain to all Programs and Projects managed by the Division Program Management Office (PMO). A Project Manager will focus on a single major product level deliverable where a Program Manager is responsible for multiple related product aligned projects culminating in a major product deliverable. Hereafter in this document the title Project Manager and Program Manager will be referenced using the acronym, PM, with the meaning taken from the context of the passage. *There will be variation in the level of effort that project or program managers will expend for any one project or program depending on multiple factors such as: project/program budget size, technical or schedule complexity.*

The PMO has established the criteria to determine when a Project should be considered a Program. On the other end of the spectrum, it has also established criteria to define which project should use the "PM Lite" process, which does not require use of the PM tools & techniques identified in this document. Both definitions can be found in the "PM Lite" process document available on the PMO web page. This document will assume a situation where there is a need for a PM and the full range of standardized project tools/techniques available would be in use.

As mentioned previously, good methodologies focus on flowcharting that is easy to understand. Documents with complex charts and exhibits are not only difficult to follow but also make it difficult for the employees to follow and use. Complex or poorly constructed flowcharts can severely limit the project manager's flexibility. An excellent example of an easy-to-follow flowchart can be shown from the statement below taken from the Teradyne's methodology:

> Figure 4–6 shows the alignment between product development and program management and the fact that heaviest PM involvement is in Path II, Phases 1 through 4. The company's functional groups and the program managers operate in a matrix, composite organization (*PMBOK® Guide* 2004, Figure 2.12). The program managers are responsible for the scope, schedule, and cost of a project but do not directly manage human resources. The program manager(s) generally work closely with a design lead(s) (the "two in a box" concept) in controlling the project's 'triple-constraints' of scope, schedule, and cost. Note that path II, phase 1 can be divided into optional A and B segments. This would typically be recommended for large platform projects. Phase 2 is also divided into segments A & B on all projects.

The two circles in Figure 4–6 represent Dr. Deming's cycle of continuous improvement. This shows that quality is embedded in the process flow and the company's commitment to quality and the total quality management processes.

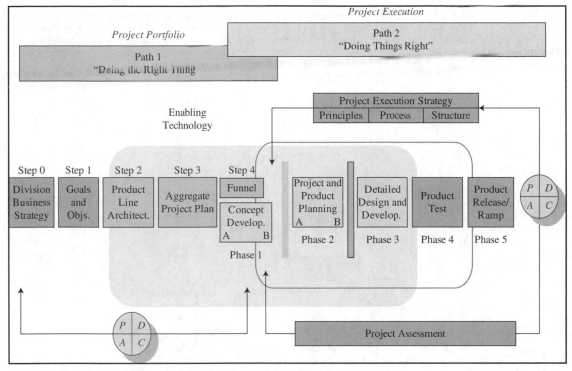

FIGURE 4–6. Alignment of product development and program management.

Another characteristic of many companies is the alignment of their methodology to the *PMBOK® Guide* processes. According to Teradyne's methodology:

> Table 4–2 shows the mapping of the phase deliverables to *PMBOK® Guide* processes. Phase 2 (planning) may require the use of all 9 process knowledge areas and is generally considered to be the phase requiring the largest time commitment from the PM because of the amount of actual "pick and shovel" work. For example, during Phase 2 the PM creates the detailed schedule, budget, resource plan, and risk management plans. Phases 3 and 4 require continued PM effort because in Phase 2 the work is planned but in Phases 3 and 4 you work the plan. In reality no project ever goes according to plan so the PM role of monitoring, adjusting, and re-planning (within the scope, schedule, and cost constraints) is challenging. PM involvement with Phase 5 is minimal at this time although end of project assessments do prove invaluable in a capturing the "lessons learned."

Finally, methodologies undergo changes and enhancements. Just like with configuration management and scope change control, the revisions are tracked for traceability. This is shown below in Table 4–3 which has been adapted from Teradyne's methodology.

TABLE 4–2.   MAPPING OF PHASE DELIVERABLES TO *PMBOK® GUIDE* PROCESSES

Description	*PMBOK® Guide* Process Groups	*PMBOK® Guide* Process Knowledge Areas
Concept development	Initiation Planning	Integration Scope Cost Human resources
Project and product planning	Planning Monitoring and control	Integration Scope Time Cost Quality Human resources Communications Risk Procurement
Detail design and development	Execute Monitor and control	Integration Scope Time Cost Quality Human resources Communications Risk
Product test and verification	Execute Monitor and control	Integration Scope Time Cost Quality Human resources Communications Risk
Product release and ramp	Closing	Integration Procurement

TABLE 4–3.   HISTORY OF REVISIONS

Date	Page(s)	Section	Comments	Rev.
3/15/2009	6–8	2.13	Expansion of Roles and Responsibilities	1.0

4.7 PROJECT MANAGEMENT FUNCTIONS

Over the years, companies have come to accept the *PMBOK® Guide* as the gospel of project management. But we must remember that the *PMBOK® Guide* is still just a guide and not necessarily the actual body of knowledge. Creating a perfect body of knowledge would have to be based upon the size, nature, and complexity of a firm's projects, the type of industry in which they compete, and the percentage of work within the firm that must utilize project management. Some project management educators argue that there are three ways to manage projects: the right way, the wrong way, and the *PMBOK® Guide* way. Although companies create project management methodologies based upon the *PMBOK® Guide*, the methodologies rarely follow it exactly. A large percentage of the areas of knowledge may be used, but the focus could be more heavily aligned to the domain areas rather than the areas of knowledge. Processes may be more important than knowledge areas. As an example, consider the following comments from Carl Manello, PMP, Solutions Lead—Program and Project Management, Slalom Consulting:

> For all that project managers are responsible for, depending on their role (e.g., small or large projects, business focused programs, corporate mergers/acquisitions), there are several models for representing what each needs to know to become an expert in the practice of project management. One model, the *PMBOK® Guide*, breaks this tacit knowledge into nine basic knowledge areas. A model that I developed has further dissected the knowledge areas into twelve project management functions (PM*f*). There are scores of other models crowding the internet. Whether nine or twelve areas, the models help us to define what we as project managers do.
>
> In my experience, all project managers should know about and be experienced in the project management functions defined in my model. The basic PM*f* (see Figure 4–7) represents the functions as I have defined them for a project/program manager. The PM*f* are robust and change as the project management focus changes (i.e., project managers' functions are different from those of an Enterprise Program Office).[10] If we leverage the PM*f* model, these functions can be established as *standard practices* for project managers, but may have to be customized or adapted to a specific project or client. In my dealings with corporations—both large and small—I specifically avoid the term *best practice*, which has always been puzzling to me. If each company determines their own "best practices," then the term itself is diluted and loses its meaning. As an industry, project management cannot have hundreds of *best* practices. While it is certainly possible that the practices adopted within one enterprise may be *best*, it has been my experience that these are more often the "preferred practices" of the implementing organization or key leader at the time of implementation.
>
> If we release ourselves from a focus on preferred, best or super-stupendous as a qualifier, the project management functional areas can be seen as key practice areas that project managers should be able to rally around. Each company must choose which functional areas are most important to them, define the right level or maturity to strive for, put plans in place to achieve mastery, reassess and continue on the never-ending path of evolving capabilities. Each organization will implement project management to the best of their ability and whether or not that is a true level of "best"—for which there is no globally defined standard—it will be the level most appropriate for that time and place.

10. For more information on the PM*f*, readers are referred to C. Manello's forthcoming book, *Value Project Management*.

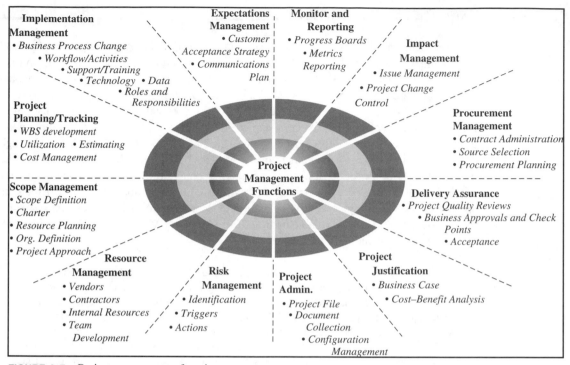

FIGURE 4–7. Project management functions.

Unfortunately, in my discussions with corporations, there appears to be an equality in the mind of many senior managers that the implementation of standard practices can only be realized with the implementation of project management tools (e.g., Clarity, PlanView, MS/Project). It has been my observation that companies that try to short-cut the establishment of process first and instead chase a tool to solve their project management challenges typically do not succeed with the implementation of any good practices, at least in a reasonable period of time. While I'm certain that this is not the case for all organizations—otherwise the tool vendors would be out of business—it should be clear that process cannot be ignored or substituted with a tool.

I have worked with a variety of industries. Within each industry I have seen companies that have spent millions of dollars on the acquisition, customization, implementation, education, and administration of project management tools. Each has had less than adequate results. A common reason for poor performance can be attributed to:

● Believing that the implementation of project management software is the same as the implementation of project management

There is certainly a need for software in project management. But the critical question is as follows: Should we select a project management software tool and then design our project management methodology around that tool, or should we design our methodology first and then select the appropriate tool to fit our methodology? The latter is certainly the better choice. Software is simply a tool. Projects are managed by people, not tools. People manage tools; tools do not manage people. A proper understanding of software and its capability is essential at the onset of project management implementation but should not be viewed as a replacement for good processes. Carl Manello states:

> As a consultant, I have had the opportunity [to] work with many companies and to observe multiple examples of tool implementations to support project management: both successes and failures. The life-cycle phases of the successful project management tool implementation projects are usually the same, but I do not think one can easily apply a repeatable time frame to a chronology to this effort. For example, the national implementation of a well known software package as the tool of record for the financial replacement program for a major airline manufacturer—spanning several years and including over 800 full and part time resources—was not the same as the implementation within the engineering division of a major cell phone manufacturer. Each implementation is different. What is extensible, however, is the process by which an organization approaches the attainment of standard practices. One should begin by assessing where the organization is at in its capabilities (leveraging the PM*f*), determine where the definition, construction and implementation of process will support growth, select supporting toolsets, analyze, customize, and implement. It is crucial that *process* precede the *tools*.

4.8 REPLACING METHODOLOGIES WITH FRAMEWORKS

While project management methodologies seem to be in favor today, there is a growing trend toward replacing methodologies with framework models. Methodologies do have disadvantages and companies that see one and only one methodology, namely their own, may have a difficult time recognizing the pitfalls. But management consultants that have had the exposure to a variety of methodologies in various industries often have a much better understanding of the limitations. Carl Manello, PMP, Solutions Lead—Program and Project Management, Slalom Consulting, states that:

> In the past twenty years I have seen a fair share of detailed methodologies. There are of course great success stories around methodologies, especially when hordes of highly trained consultants are brought in to run enormous enterprise-wide initiatives. However, for the small garden variety project or those instances when training an entire project team on how to use the methodology is unrealistic, methodologies can cause as many problems as they solve.
>
> While working at a large manufacturing company in the late 80's I had the privilege of working with a team of folks from the Technology Center and the Enterprise Program Office (of which I was a member). We leveraged a vendor framework, adopted it to our needs and put it in place for both IT and R&D projects. In my first stint as a Solution Lead for Program and Project Management with a different consultancy, I had the opportunity to take the approach of *frameworks* to other clients. Now, at Slalom, I am continuing to advocate frameworks over detailed methodologies. Slalom Consulting does not have a proprietary PM methodology. Instead we leverage frameworks.
>
> The difference is that methodologies tend to be defined as a series of inflexible steps that must be completed, usually in sequential order. "Frameworks" on the other hand, outline steps that may be followed. The difference is that the steps of a framework may be implemented differently per project, may be passed over, or may be implemented at

varying degrees of detail. At one of my prior employers, I created a PM life-cycle frame-work and gave it an internal marketing tag of "Flexibly Applying a Rigor Process." The marketing phrase was developed not only to help sell the concept of a new process but also to help illustrate the key difference between a framework and a methodological tome. The framework's process was designed and built with a significant amount of detail (e.g., project steps, deliverables, metrics, approval processes). However the implementation rigor varied depending on the type of project, size, criticality, etc. Large projects had to adhere to the PM life-cycle with exacting detail, while small projects had the flexibility to omit specific portions of the approach. Not all projects are created equally.

At one of my former clients, there is an exhaustive methodology defined and in place, which employs scores of templates, numerous processes and multiple governance teams that each own different pieces of the process. The challenge of this implementation of *methodology* is its inflexibility. Project managers spend too much time trying to navigate the process and are distracted or prevented from focusing on their PM activities. They spend more time doing project management process related work than adding value to their projects as project managers. This in turn forces the client to hire more people to do "project management" activities (having some project managers to project manage and some to do methodology compliance), thereby increasing the time spent on project management. This extra time spent cannot be shown to have a direct impact on the project or the financial benefits, and in fact erodes whatever initial benefits the programs claimed to be able to realize. Similarly, there is insufficient training, coaching, documentation or other support available to help project managers navigate their way through the rigorous process. Not only do project managers spend an inordinate amount of time complying with what they understand, they must also chase around the organization looking for the right governance office to provide them help on what to do. If the current enterprise-wide system replacement program should not do well, the project management methodology itself (which should have been enabling success) will certainly be one of the key elements in the downfall.

4.9 LIFE-CYCLE PHASES

Determining the best number of life-cycle phases can be difficult when developing a project management methodology. As an example, let's consider IT. During the 1980s, with the explosion in software, many IT consulting companies came on the scene with the development of IT methodologies using systems development life-cycle (SDLC) phases. The consultants promise their client phenomenal results if the client purchases the package along with the accompanying training and consulting efforts. Then, after spending hundreds of thousands of dollars, the client reads the fine print that states that the methodology must be used as is, and no customization of the methodology will take place. In other words, the client must change their company to fit the methodology rather than vice versa. Most of the IT consultancies that adopted this approach no longer exist.

For an individual company, agreeing on the number of life-cycle phases may be difficult at first. But when an agreement is finally reached, all employees should live by the same phases. However, for today's IT consulting companies, the concept of one-package-fits-all will not work. Whatever methodology they create must have flexibility in it so that client customization is possible. In doing so, it may be better to focus on processes rather than phases, or possibly a framework approach that combines the best features of each. Carl Manello, PMP, Solutions Lead—Program and Project Management, Slalom Consulting, states that:

> While a project management life-cycle parallels an SDLC, they are separate and distinct (especially because many projects have nothing to do with IT or "systems"). With some special client-specific exceptions, the standard phases are:
>
> 1. Project Conceptualization—where the project begins to take form (usually an informal unstructured process)
> 2. Project Initiation—an approved initiative begins its journey and begins to complete the rigor of the life-cycle
> 3. Analysis & Design—begin to decompose project scope into the meat of what is to be accomplished
> 4. Development/Test /Implement—standard SDLC phases, but tailored to meet non-IT projects as needed
> 5. Close Down/Turn Over—ending the "project" and initiation of the "ongoing maintenance" (as needed)
> 6. Benefits Realization Tracking—the woefully neglected phase where we prove the realization of benefits claimed back in conceptualization
>
> During one of my corporate lives, I helped define a project management life cycle that built upon these six core phases (see Figure 4–8). As a framework, phases should be malleable to the needs of the situation. Our Project Management Office was part of the IT Governance function, which was primarily the financial arm for IT. The equivalent of the IT CFO wanted to ensure we allowed sufficiently detailed processes to enable the activation (or administrative start-up) of a project. Activation represented setting up the right time tracking buckets, financial buckets and establishing service level agreements. Similarly, we added the Quality Assurance and Testing phase because there was a newly formed organization that focused on these functions and IT wanted to ensure their prominence in the minds of project teams. Lastly, to help erase the years of short-comings at implementation, I created a separate phase for implementation. This late phase detailed the roles, responsibilities, deliverables and associated check-points that would enable us to manage a project past the "go live" date. As discussed above, each initiative flowed through this framework in a "flexible" way, leveraging the parts that made sense given the nature of the project.

4.10 AT&T

AT&T maintains a project management methodology. As with any good methodology that must be custom designed to fit the needs of a large company, the methodology may follow the *PMBOK® Guide* but also includes company-specific policies, procedures, and guidelines. According to a spokesperson at AT&T:

> The AT&T Project Management Methodology is a simple, scalable, and standardized methodology and can be applied enterprise-wide. (See Figure 4–9.) It is based upon the industry standard *A Guide to the Project Management Body of Knowledge* (*PMBOK® Guide*), produced by the Project Management Institute (PMI®), as well as AT&T specific

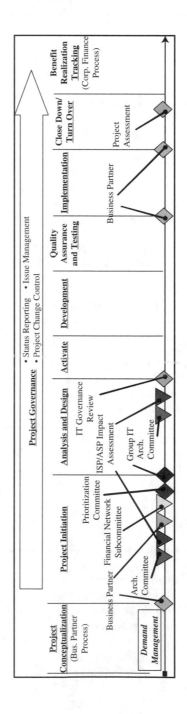

FIGURE 4–8. Extended PM framework.

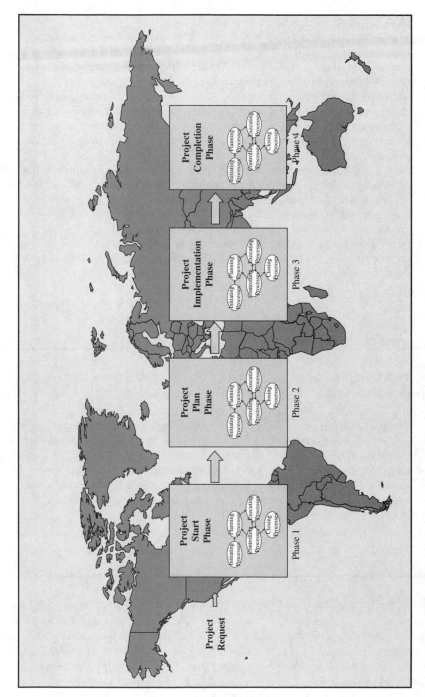

FIGURE 4–9. AT&T worldwide project management methodology.

processes and procedures. Because a project may be determined by many factors: strategic importance, size, scope, schedule, cost and duration, as well as many others, this methodology is scalable to accommodate all types of projects.

The Project Management Methodology utilizes a four-phase project life cycle. These four phases of a project, plus the inputs, activities and deliverables key to the phases comprise the methodology. Throughout each of the four distinct *project phases*, the five iterative *process groups* of Initiating, Planning, Executing, Monitoring and Controlling and Closing will be used. Each of the five processes is applied within each project phase. Many times changes occur within the life cycle of a project, and process groups must be repeated.

Project Start: Recognizes that a new project is being considered. During this phase, basic information is gathered, evaluated, and based upon the information a go/no go decision is made.

Project Plan Phase: Establishing the project's approach, and planning how to achieve the desired results and baselines for the project in terms of scope, schedule, and cost.

Project Implementation Phase: Implementing the Project Plan to produce the agreed upon deliverables, monitoring the project progress, and ensuring that deliverables meet expectations.

Project Completion Phase: Completing the project. Ensuring that the project was delivered as expected and ensuring that there is final/formal acceptance in order to close out the project.

4.11 CHURCHILL DOWNS, INCORPORATED

Churchill Downs, Incorporated has created a project management methodology that clearly reflects its organization. According to Chuck Millhollan, Director of Program Management:

> While we based our methodology on professional standards, we developed a graphic (and used terminology) understood by our industry to help with understanding and acceptance. For example, we have a structured investment request, approval and prioritization process. (See Figure 4–10). We used the analogy of bringing the thoroughbred into the paddock prior to race and then into the starting gate. The project, or race, is not run until the thoroughbred has entered the starting gate (approved business case and project prioritization).

4.12 INDRA

As mentioned previously in Chapter 3, the quest for excellence in project management is almost always accompanied by the development of a project management methodology. Such was the case at Indra. Indra defines excellence in project management as follows: "Excellence in project management is achieved by being able of repeatedly reaching the project targets, creating business opportunities, and improving the management process itself when managing the assigned projects." Enrique Sevilla Molina, PMP, Corporate PMO Director, discusses the journey to excellence:

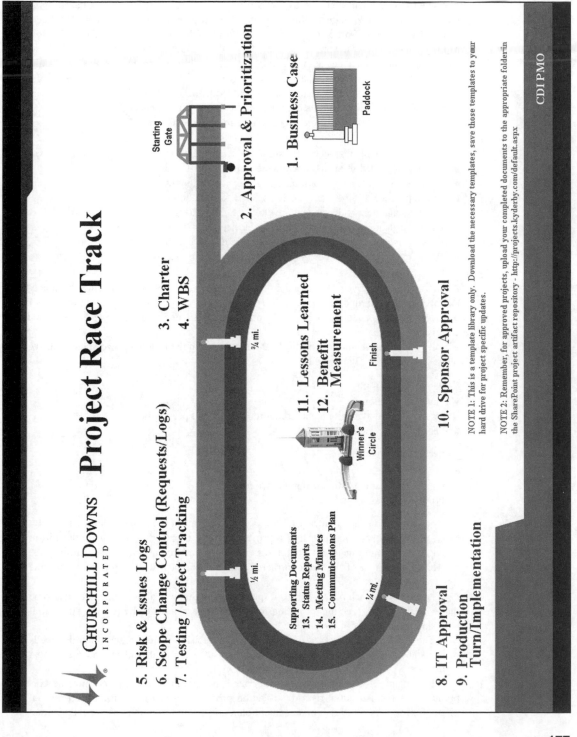

FIGURE 4-10. The Churchill Downs, Incorporated methodology.

A project management methodology was formally defined in the mid 90's based upon the experience gained in our major international contracts. The main problems we faced were related to the definition of the scope, the limits of the methodology, and the adoption of the correct strategy to spread this knowledge throughout the company. To solve these issues, our management chose to hire an external consulting company to act as a dynamic factor that boosted and drove the cultural change.

Yes, the process was carefully sponsored from the beginning by senior executives and closely followed-up until its complete deployment in all areas of the company.

The major milestones of the process have roughly been:

- Project management strategy decision.............mid 90's
- Methodology definition and documentation......mid-late 90's
- Tools definition and preparation ....late 90's
- Training process start.....................................2000
- Risk management at department level.................2002
- PMP certification training start 2004
- Risk management process defined at corporate level...2007
- Program and portfolio management processes definition start...2008

A PM methodology was developed in the early 90's and formalized during that decade. It has been eventually updated to cope with the company and the industry evolution. It is being used as a framework to develop and maintain the PMIS, and to train the PMs throughout the company.

It is based on the project lifecycle and structured in the following two stages and six phases as shown in Figure 4–11:

Precontractual stage
- Phase 1. Initiation
- Phase 2. Concept development, creation of offers and proposals
- Phase 3. Offer negotiation

Contractual stage
- Phase 4. Project planning
- Phase 5. Execution, Monitoring and Control
- Phase 6. Closure

The pre-contractual and contractual stages are both part of the project and its lifecycle. Most problems that appear during a project's lifespan originate during its definition and in the negotiation of its objectives, contents and scope with the customers. A proper management of the pre-contractual stage is the best way to prevent problems later on.

At the end of each phase there is a specific result that will allow a key decision to be made, focusing and directing the actions of the next phase and thereby reducing the initial risks and uncertainties of the project.

The decision on the stages and phases was a decision mainly based upon the needs of our standard cycle of a project conception and development, and based on the most significant types of projects we were involved with.

Risk management processes are integrated into the methodology and into the corporate PM tools. An initial risk identification process is performed during the proposal phase, followed by a full risk management plan during the planning phase of the contract

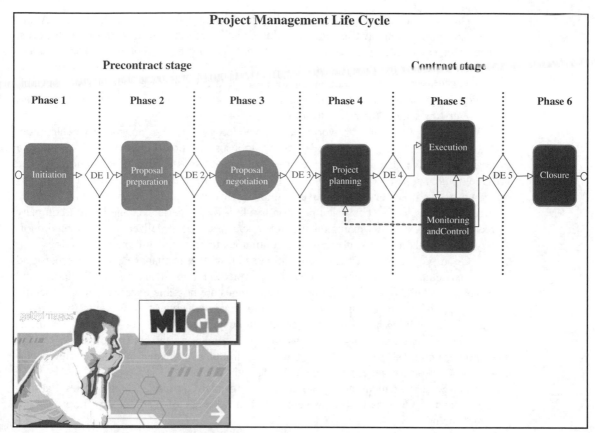

FIGURE 4–11. Project management life cycle.

stage, and the subsequent monitoring processes during the execution phase of the project. QA and change control processes are considered main support processes in the methodology.

4.13 IMPLEMENTING METHODOLOGY

The physical existence of a methodology does not convert itself into a world-class methodology. Methodologies are nothing more than pieces of paper. What converts a standard methodology into a world-class methodology is the culture of the organization and the way the methodology is implemented.

The existence of a world-class methodology does not by itself constitute excellence in project management. The corporatewide acceptance and use of it do lead to excellence. It is through excellence in execution that an average methodology becomes a world-class methodology.

One company developed an outstanding methodology for project management. About one-third of the company used the methodology and recognized its true long-term benefits. The other two-thirds of the company would not support the methodology. The president eventually restructured the organization and mandated the use of the methodology.

The importance of execution cannot be underestimated. One characteristic of companies with world-class project management methodologies is that they have world-class managers throughout their organization.

Rapid development of a world-class methodology mandates an executive champion, not merely an executive sponsor. Executive sponsors are predominantly on an as-needed basis. Executive champions, on the other hand, are hands-on executives who drive the development and implementation of the methodology from the top down. Most companies recognize the need for the executive champion. However, many companies fail to recognize that the executive champion position is a life-long experience. One Detroit company reassigned its executive champion after a few successes were realized using the methodology. As a result, no one was promoting continuous improvement to the methodology.

Good project management methodologies allow you to manage your customers and their expectations. If customers believe in your methodology, then they usually understand it when you tell them that no further scope changes are possible once you enter a specific life-cycle phase. One automotive subcontractor carried the concept of trust to its extreme. The contractor invited the customers to attend the contractor's end-of-phase review meetings. This fostered extreme trust between the customer and the contractor. However, the customer was asked to leave during the last 15 minutes of the end-of-phase review meetings when project finances were being discussed.

Project management methodologies are an "organic" process, which implies that they are subject to changes and improvements. Typical areas of methodology improvement might include:

- Improved interfacing with suppliers
- Improved interfacing with customers
- Better explanation of subprocesses
- Clearer definition of milestones
- Clearer role delineation of senior management
- Recognition of need for additional templates
- Recognition of need for additional metrics
- Template development for steering committee involvement
- Enhancement of the project management guidebook
- Ways to educate customers on how the methodology works
- Ways of shortening baseline review meetings

4.14 IMPLEMENTATION BLUNDERS

Even though companies recognize the driving forces which indicate a need for project management improvement, the actual decision to make an investment to do it may not happen until some crisis occurs or a significant amount of red ink appears on the company's balance sheet. Recognizing a need is a lot easier than doing something about it because doing it requires time and money. Too often, executives procrastinate giving the go-ahead in hopes that a miracle will occur and project management improvements will not be necessary. And while they procrastinate, the situation often deteriorates further. Consider the following comments from Carl Manello, PMP, Solutions Lead—Program and Project Management, Slalom Consulting:

> I find that the greatest motivation for my clients to invest in improvements is how they view the impact of project management on their initiatives. When their track record at driving large scale business initiatives is less than stellar (lacking sufficient process, methods, tools or skills), they begin to understand the need to invest. Unrealized project implementations, blown budgets and poor quality all speak loudly and capture the attention of senior executive management. The challenge is instead to arrest executive attention before millions of dollars are squandered.
>
> At first, many corporations are unlikely to want to invest in improving PM infrastructure like the PM*f*. "There are *real* projects with hard-core benefits to be realized instead." However, after these same organizations begin to struggle, understand their weaknesses and the need for improvement in basic project management, they begin to focus on those improvements.

Delayed investment in project management capabilities is just one of many blunders. Another common blunder, which can occur in even the best companies, is the failure to treat project management as a profession. In some companies, project management is a part-time activity to be accomplished in addition to one's primary role. The career path opportunities come from the primary role, not through project management. In other companies, project management may be regarded merely as a specialized skill in the use of scheduling tools. Carl Manello continues:

> While the PMI has done a super job, especially in the last 10 years, advocating project management as a specialized skill that should be left to the professionals, I find that many companies still believe project management is a skill, not a profession. Whether in marketing or engineering organizations, someone is often randomly assigned to be the project manager, regardless of their training, demonstrated skill level or capabilities as a project manager. This lack of attention to project management as a profession may be one of the contributing factors to projects around the world which continue to perform poorly. Too many projects do not have qualified experienced project managers at the helm.

4.15 OVERCOMING DEVELOPMENT AND IMPLEMENTATION BARRIERS _____

Making the decision that the company needs a project management methodology is a lot easier than actually doing it. There are several barriers and problems that surface well after the design and implementation team begins their quest. Typical problem areas include:

- Should we develop our own methodology or benchmark best practices from other companies and try to use their methodology in our company?
- Can we get the entire organization to agree upon a singular methodology for all types of projects or must we have multiple methodologies?
- If we develop multiple methodologies, how easy or difficult will it be for continuous improvement efforts to take place?
- How should we handle a situation where only part of the company sees a benefit in using this methodology and the rest of the company wants to do its own thing?
- How do we convince the employees that project management is a strategic competency and the project management methodology is a process to support this strategic competency?
- For multinational companies, how do we get all worldwide organizations to use the same methodology? Must it be intranet based?

These are typical questions that plague companies during the methodology development process. These challenges can be overcome, and with great success, as illustrated by the companies identified in the following sections.

4.16 PROJECT MANAGEMENT TOOLS

Project management methodologies require software support systems. As little as five years ago, many of the companies described in this book had virtually no or limited project management capabilities. How did these companies implement project management so fast? The answer came with the explosion of personal computer-based software for project planning, estimating, scheduling, and control. These were critical for methodology development.

Until the late 1980s, the project management tools in use were software packages designed for project scheduling only. The most prominent were:

- Program evaluation and review technique (PERT)
- Arrow diagramming method (ADM)
- Precedence diagramming method (PDM)

These three networking and scheduling techniques provided project managers with computer capabilities that far surpassed the bar charts and milestone charts that had been in use. The three software programs proved invaluable at the time:

- They formed the basis for all planning and predicting and provided management with the ability to plan for the best possible use of resources to achieve a given goal within schedule and budget constraints.
- They provided visibility and enabled management to control one-of-a-kind programs.

● They helped management handle the uncertainties involved in programs by answering such questions as how time delays influence project completion, where slack exists among elements, and which elements are crucial to meeting the completion date. This feature gave managers a means for evaluating alternatives.
● They provided a basis for obtaining the necessary facts for decision-making.
● They utilized a so-called time network analysis as the basic method of determining manpower, material, and capital. requirements as well as providing a means for checking progress.
● They provided the basic structure for reporting information.

Unfortunately, scheduling techniques cannot replace planning. And scheduling techniques are only as good as the quality of the information that goes into the plan. Criticisms of the three scheduling techniques in the 1980s included the following:

● Time, labor, and intensive effort were required to use them.
● The ability of upper-level management to contribute to decision-making may have been reduced.
● Functional ownership of the estimates was reduced.
● Historical data for estimating time and cost were lost.
● The assumption of uninvited resources was inappropriate.
● The amount of detail required made full use of the scheduling tools inappropriate.

Advancements in the memory capabilities of mainframe computer systems during the 1990s eventually made it possible to overcome many of the deficiencies in the three scheduling techniques being used in project management in the 1970s and 1980s. There emerged an abundance of mainframe software that combined scheduling techniques with both planning and estimating capabilities. Estimating then could include historical databases, which were stored in the mainframe memory files. Computer programs also proved useful in resource allocation. The lessons learned from previous projects could also be stored in historical files. This improved future planning as well as estimating processes.

The drawback was that mainframe project management software packages were very expensive and user unfriendly. The mainframe packages were deemed more appropriate for large projects in aerospace, defense, and large construction. For small and medium-sized companies, the benefits did not warrant the investment.

The effective use of project management software of any kind requires that project teams and managers first understand the principles of project management. All too often, an organization purchases a mainframe package without training its employees in how to use it in the context of project management.

For example, in 1986, a large, nationally recognized hospital purchased a $130,000 mainframe software package. The employees in the hospital's information systems department were told to use the package for planning and reporting the status of all projects. Less than 10 percent of the organization's employees were given any training in project management. Training people in the use of software without first training them in project management principles proved disastrous. The morale of the organization hit an all-time low point and eventually no one even used the expensive software.

Generally speaking, mainframe software packages are more difficult to implement and use than smaller personal computer–based packages. The reason? Mainframe packages require that everyone use the same package, often in the same way. A postmortem study conducted at the hospital identified the following common difficulties during the implementation of its mainframe package:

- Upper-level managers sometimes did not like the reality of the results.
- Upper-level managers did not use the packages for planning, budgeting, and decision-making.
- Day-to-day project planners sometimes did not use the packages for their own projects.
- Some upper-level managers sometimes did not demonstrate support and commitment to training.
- Clear, concise reports were lacking.
- Mainframe packages did not always provide for immediate turnaround of information.
- The hospital had no project management standards in place prior to the implementation of the new software.
- Implementation highlighted the inexperience of some middle managers in project planning and in applying organizational skills.
- Neither the business environment nor the organization's structure supported the hospital project management/planning needs.
- Sufficient/extensive resources (staff, equipment, etc.) were required.
- The business entity did not determine the extent of and appropriate use of the systems within the organization.
- Some employees viewed the system as a substitute for the extensive interpersonal skills required of the project manager.
- Software implementation did not succeed because the hospital's employees did not have sufficient training in project management principles.

Today, project managers have a large array of personal computer–based software available for planning, scheduling, and controlling projects. Packages such as Microsoft Project have almost the same capabilities as mainframe packages. Microsoft Project can import data from other programs for planning and estimating and then facilitate the difficult tasks of tracking and controlling multiple projects.

The simplicity of personal computer–based packages and their user friendliness have been especially valuable in small and medium-sized companies. The packages are so affordable that even the smallest of companies can master project management and adopt a goal of reaching project management excellence.

Clearly, even the most sophisticated software package can never be a substitute for competent project leadership. By themselves, such packages cannot identify or correct task-related problems. But they can be terrific tools for the project manager to use in tracking the many interrelated variables and tasks that come into play in contemporary project management. Specific examples of such capabilities include the following:

- Project data summary; expenditure, timing, and activity data
- Project management and business graphics capabilities
- Data management and reporting capabilities
- Critical-path analyses
- Customized as well as standardized reporting formats
- Multiproject tracking
- Subnetworking
- Impact analysis
- Early-warning systems
- Online analyses of recovering alternatives
- Graphical presentations of cost, time, and activity data
- Resource planning and analyses
- Cost and variance analyses
- Multiple calendars
- Resource leveling

Figure 4–12 shows that right now approximately 95 percent of the project management software focus on planning, scheduling, and controlling projects. In the future, we can expect more software to be created for the initiation of a project and the closure of a project.

Perhaps the biggest problems today with methodologies are that companies are not taking advantage of the full capabilities of the tools they are purchasing. A perfect example of this is with the earned-value measurement (EVM) system. Companies appear reluctant to use EVM probably for fear of the financial reality that the EVM numbers will show.

The intent behind EVM is to make sure that the person leading the project is functioning as a project manager rather than a project monitor. Project monitors simply record numbers and report them to a higher authority for decision-making. The project manager, on the other hand, measures the variances from the baselines, develops contingency plans,

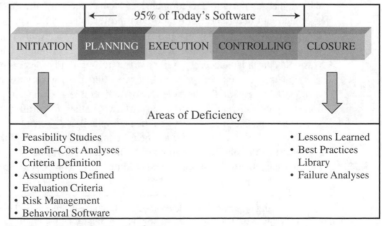

FIGURE 4–12. Project management software.

gets the plans approved, implements the plans, and measures the new variances to see if the improvements worked. Without full use of an EVM system, it may be difficult to fully manage a project.

Enrique Sevilla Molina, PMP, Corporate PMO Director at Indra, states:

> Earned valued techniques are used during project execution. Earned value measurement is performed regularly by the project manager making use of the corporate tool.
>
> Variances are analyzed monthly at a work package level and also at the project level.
>
> If corrective action is necessary, it is initiated by the project manager who analyses a set of project indicators (risk status, overall cost, schedule and margin deviations, project cash flow, . . .), and it is followed by the expert judgment by the area controller who also considers the analysis of the information provided by the project team. If it is the case, a specific Executive report can be requested from the PM for providing more detailed information on deviations by work package and the status of the program. Once the overall picture of the project is clear, a decision is taken on the need for corrective actions.

Perhaps EVM is the best approach for determining progress, status, and eventually forecasts of where we will end up. The critical component of EVM is earned value, which is the amount of work that has physically been completed and measured in either hours or dollars. All too often, companies accept EVM as a way of life without understanding the complexities in using EVM. Memorizing the 12–15 equations necessary to use EVM is easy. The hard part is capturing the data that go into the equations. Carl Manello, PMP, Solutions Lead—Program and Project Management, Slalom Consulting, states that:

> Earned Value (EV) is not for the beginner. Conceptually a straightforward measure of work done/remaining to be done, the effort needed to collect the necessary data and to make the calculations is no simple undertaking. Many of my clients have done such a poor job getting a handle on their time capture, and don't have sufficient enough project plans, that capturing information at required levels for EV is almost impossible. After working with numerous Fortune 250 companies, I have found only one that tracked SPI/CPI.[11] The overhead required to collect data on that project, perform quality assurance and maintain the information accurately in a project management tool required significant overhead (the program employed more than a dozen people on the program whose sole accountability was data entry, data accuracy and reporting through the project tool).
>
> For the less mature projects, I have often used a variation of earned value. This method is based on the completion of deliverables. Using simple spreadsheets, I am able to track deliverables (assigned different values for draft or final), milestones and key activities. The spreadsheet generates graphs that represent how much work has been delivered, the cost associated with the scheduled time spent, and the earned value actuals compared to the plan (see Figures 4–13 through 4–15).Each deliverable is set against a planned schedule for completion and credit is only realized with any delivered value when

---

11. SPI is the schedule performance index and CPI is the cost performance index. SPI and CPI identify the trends or direction in which cost and schedule are heading. During executive briefings on project status, it is often the case that SPI and CPI are the first two numbers discussed.

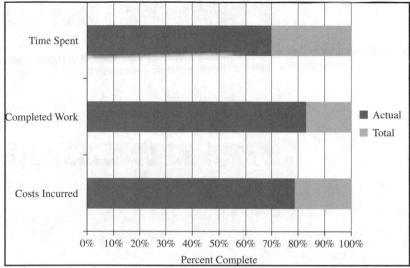

FIGURE 4–13. Actual versus total costs.

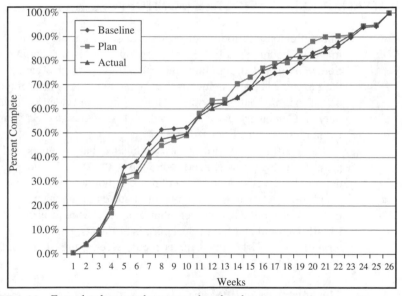

FIGURE 4–14. Earned-value actuals compared to the plan.

it is complete (i.e., something 90% complete gets a big fat zero in this model. There are many circumstances where this over-simplistic approach is not appropriate and a partially completed deliverable *is* of value. In this implementation, however, the choice is all or none). While this tool also requires some overhead to create the initial list based on the WBS, track completion and update status to reflect progress, the overhead is in the order of hours instead of multiple full-time equivalents.

Status			
Deliverables	Schedule	Budget	Overall
10%	12%	9%	
○	○	○	○
Key:	○ On Target	◉ At Risk	● In Danger

Key Indicators			
Scope	Delivery	Cost	Staffing
⇧	⇔	⇧	⇔
Key:	⇔ Holding Steady	⇧ Increasing	⇩ Decreasing

FIGURE 4–15. Status and indicator directions.

4.17 SATYAM: PROJECT PROCESS MONITORING

In the previous section we discussed some of the tools that are used with project management methodologies. EVM and scheduling tools are common to most methodologies. Some companies design tools that are unique to their organization and then capitalize on their expected benefits. Such is the situation with Satyam Computer Services. The remainder of the information in this section has been provided by Krishna Gali, Quality Group, Satyam Computer Services Limited, and Anu Khendry, Satyam Learning World, Satyam Computer Services Limited.

Satyam is a software services company that conducts its business through software projects. These projects need to comply with critical product and process parameters as defined by the customers. Project process monitoring (PPM) is an assurance activity that enables project managers, their teams, and senior management to understand these critical parameters and the related compliance levels of the project. PPM is a mandatory activity for all fixed-bid and high-impact/critical projects, which are defined based on certain criteria of size, effort, and customer focus. (See Figures 4–16 and 4–17, which contain two parts of PPM.)

The objectivise of PPM are to:
- Provide early alerts on potential project problems with suggestions for corrective and preventive measures
- Identify and accelerate process improvements in projects
- Identify best practices and share them across the company

The unique features of the PPM process include:
- Assessment on 12 key project focus areas with predefined parameters

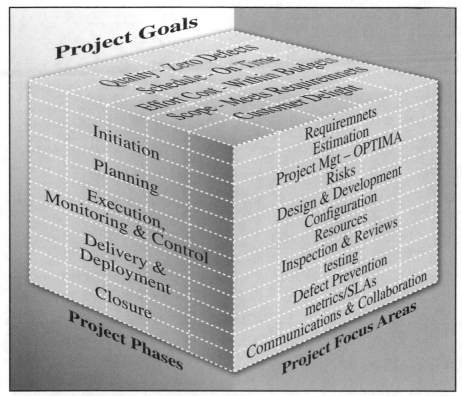

FIGURE 4–16. Project process monitoring.

- Quantitative rating of the project on a scale of 100
- Automation and integration with other project management tools at Satyam.
- Process assessment aligned to CMMI and International Organization for Standardization (ISO) 9001

The PPM process includes:

- Self assessment by the project manager on all focus areas and parameters that are applicable to the project.
- Review and validation by quality assurance on project compliance. This is done by assessing project artifacts against applicable product/process requirements. At the end of this step the project has a compliance rating on a scale of 100 and a red/amber/green status based on this rating.
- Reports of project ratings to senior management.
- The deficiencies observed during PPM are followed up in the subsequent month.

The PPM process has enabled identification of right project issues at the right time for immediate resolution. PPM reports depict real-time data on project performance. Process automation has also helped in conducting aging analysis on project issues and their tracking toward closure.

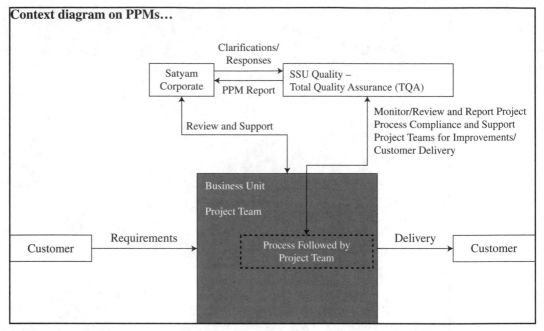

FIGURE 4–17.   Project process monitoring.

The PPM reviews highlight the impact of noncompliance on the three basic project parameters: quality, cost, and time. The success of this process is evident and it has become a key input supporting all business performance reviews at the corporate level with quantitative project issue reporting, month after month.

Customer One   In the life cycle of a software project in Satyam, software work products are created, reviewed, and tested in multiple phases, such as unit testing, integration testing, and system testing. The focus during these phases is on the detection of defects and fulfillment of customer requirements.

The final phase of testing prior to delivery is user acceptance testing (UAT). During this phase, project teams review, inspect, and test the final deliverable from the customer perspective. Any oversight at this time can lead to defect leakage to the customer, resulting in rework and customer dissatisfaction. This not only impacts efficient and effective project delivery but also impacts the overall relationship with the customer. The Customer One (C1) process in Satyam ensures that the final deliverables are as per the expectations of the customer and will lead to customer delight. (See Figure 4–18.)

The objectives of the C1 process are:

● External validation of the final project deliverables from the perspective of the customer

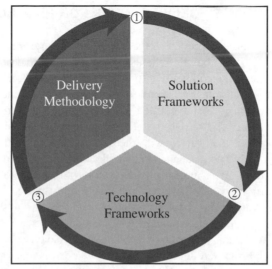

FIGURE 4–18. C1: assessment components driving contractual commitments.

- Go/no-go decision on the release of project deliverables to the customer

The focus of C1 assessment is to validate that the project deliverables not only meet the customer's explicit specifications and requirements but also conform to the implied requirements and intended business value. C1 thus helps to prevent any unpleasant surprises for the customer and subsequent escalations on the quality of delivery.

Projects are selected for C1 assessment based on project criticality to the customer, pricing model, and project type. The assessment is conducted by a team of experts with relevant competencies/skills. The C1 assessment team is supported with detailed guidelines on multiple focus areas, ranging from requirements, technology/tools and technical architecture, delivery methodology, testing, and predeployment activities.

The C1 assessments for any projects are broadly categorized into two components:

1. Assessment of solution effectiveness against SOW/requirements
2. Assessment of project deliverables against business requirements and specifications

The assessment involves:
- Performing pseudo customer role
- Bringing customers' view in the assessment of solution/product/services
- Focusing on contractual commitments (SOW)
- Unearthing potential risks

The C1 assessment process is automated and integrated with other project management tools in Satyam. The success of this process has been seen through a significant increase in customer appreciations on quality of deliverables.

4.18 SATYAM: CUSTOMER DELIGHT INDEX FOR PROJECTS _____

What's more important to the average company: maintaining alliances with existing clients or seeking out new clients? We could argue that both are equally important. However, because of the competitive forces in the marketplace, the edge goes to customer management and customer relations efforts. There is a tendency today to add in a life-cycle phase after contract closure entitled "customer satisfaction management" or "customer relations management." The purpose of this phase is to review how well satisfied the customer was throughout the project, with the end results, and with the flow of information provided by your project management methodology. In this phase, you are basically asking the customer, "What improvements would you like us to make prior to the next project we perform for you?" The customer may request more involvement in scope changes, a different format for status reporting or even better ways to track the status of their project through your project management methodology.

While this approach of having a life-cycle phase called customer satisfaction management has merit, it does have a downside risk that customer satisfaction is looked at only at the end of the project. Good companies track customer satisfaction throughout the project, not merely at the end. Satyam has found a rather interesting way to do this. The remaining information has been provided by Hirdesh Singhal and Anu Khendry, School of Program and Project Management, Satyam Computer Services Limited:

Satyam is a software services company that manages projects that have been outsourced to it by its customers. Customer satisfaction or delight plays a critical role in getting repeat business from customers and this is important for the sustenance of Satyam. Also, retaining a customer costs significantly less than acquiring a new customer. The customer delight index (CDI) framework has been developed in Satyam to continuously measure and improve customer delight.

The CDI is measured for each project in Satyam. (See Figure 4–19.) CDI is essentially the project manager's perspective on what he or she perceives about customer perception for the services provided depending on the communications and interactions with the stakeholders at the customer place. It helps senior management or investors of projects to proactively take actions and support the projects if required. For project managers, it helps in sending right signals proactively to the management that they need their support and intervention.

The goals of CDI:

- Assess the health of projects and take proactive steps to avoid customer escalations.
- Track project critical issues that directly affect customer delight to closure.
- Identify and record reasons for customer delight.

The CDI value is entered based on the perspective of the project manager on "how the customer is feeling about the project." It is arrived at based on customer feedback and interactions. Project managers need to enter the CDI, project issues, and/or reasons for delight on an agreed time frequency. Program managers review the status of the CDI on the same frequency.

FIGURE 4–19. Customer delight index process.

There are four types of status for CDI:

If the CDI status is *delighted,* the project is expected to have one or more reasons for customer delight. Project managers need to record the reason(s) and save the same.

If a project manager fails to enter the CDI within the specified period of one week, then the status is recorded as "Data Not Entered" (represented by "gray" color) for the week.

If the CDI status is *dissatisfied*, the project is expected to have a minimum of one open issue that reflects the reason for customer dissatisfaction. Poject managers need to record the issue and save the issue.

After having updated the issues/reasons for delight for the project, the appropriate CDI rating is selected and the CDI status is saved (Figure 4–21).

Project Issues   Project issues need to be mapped against the predefined set of categories and subcategories. When the CDI status is other than delighted, project managers need to specify the project issues using the categories and subcategories.

⬤	Not Entered
☼	Dissatisfied
△	Satisfied
⬛	Delighted

FIGURE 4–20. Summary categories of satisfaction.

Parameter\ Interpretation	Delighted	Satisfied	Dissatisfied
Scope	Fulfill scope with higher business value solution/ service	Fulfill scope with in acceptable levels of solution/service performance	Scope not fulfilled
Quality	Zero-defect products/ services delivered	Products/services delivered with in customer acceptable level of defects	Products/services delivered with defects beyond customer acceptable level of defects
Schedule	Ahead of schedule/ turnaround time	On time/mutually acceptable delay	Delayed

CDI rating for the project can be selected based on a combination of the above factors. CDI rating shall be the lowest of the interpretations of all the above parameters put together. For example, if the interpretation for Scope for a given project is "Delighted" while that of "Quality" and "Schedule" is "Satisfied", the overall CDI rating for the project shall be "Satisfied".

FIGURE 4–21. Guidelines for CDI rating selection.

Category	Sub category	Category	Sub category
1. Customer Management	Customer Communication	4. Human Resources	Competency
	Scope		Attrition
	Customer Style		Productivity
	Customer Escalations		Motivation
	Customer Sign-off		Attitude
	Customer Expectation		Teaming
			Resource Availability
2. Project Management	Project Schedule		Domain Knowledge
	Estimation		
	Collaboration	5. Infrastructure and Services	Equipment
	Communication		Office Space
	Process Compliance		Communication Facilities
	Project Ownership		Network Connectivity and Bandwidth
	Risks		Travel and Boarding
	Project Tracking and Control		Immigration
	Project Planning		
		6. Financial	Project Contract
3. Software Engineering	Requirements		PO/SOW
	Design		Invoicing
	Coding		Project Cost
	Testing		Project Profitability
	Delivery		
	Deployment		
	Technology		
	Software Quality		
	Configuration Management		
	Environment (Development, Testing,...)		

FIGURE 4–22. Detailed categoris of satisfaction.

It is mandatory for a project to have a minimum of one open issue that is the likely cause for customer dissatisfaction when the CDI status of the project is dissatisfied. Projects where the CDI status is *satisfied* or delighted can also have issues that either have a potential to impact customer delight or are internal issues that need to be tracked to closure.

A project that has the CDI status as dissatisfied will not be allowed to move to satisfied or delighted without all the open issues being closed (Figure 4–22).

Project Reasons   In cases where the CDI status is delighted and/or satisfied, project managers need to enter the project reasons that have helped them delight the customers. In the same screen as Customer Delight, the project manager needs to enter the project reasons against the options that are listed. Based on the selection, the reason field will be enabled. Find below options available to enter the project the reason for delighted status.

4.19 GENERAL MOTORS POWERTRAIN GROUP

For companies with small or short-term projects, project management methodologies may not be cost effective or appropriate. For companies with large projects, however, a workable methodology is mandatory. General Motors Powertrain Group is another example of a large company achieving excellence in project management. The company's business is based primarily on internal projects, although some contract projects are taken on for external customers. The size of the group's projects ranges from $100 million to $1.5 billion. Based in Pontiac, Michigan, the GM Powertrain Group developed and implemented a four-phase project management methodology that has become the core process for its business. The company decided to go to project management in order to get its products out to the market faster. According to Michael Mutchler, former Vice President and Group Executive:

> The primary expectation I have from a product-focused organization is effective execution. This comprehends disciplined and effective product program development, implementation, and day-to-day operations. Product teams were formed to create an environment in which leaders could gain a better understanding of market and customer needs, to foster systems thinking and cross-functional, interdependent behavior, and to enable all employees to understand their role in executing GM Powertrain strategies and delivering outstanding products. This organizational strategy is aimed at enabling a large organization to be responsive and to deliver quality products that customers want and can afford.

The program management process at GM Powertrain is based upon common templates, checklists, and systems. The following lists several elements that were common across all GM Powertrain programs during the 1990s:

- Charter and contract
- Program team organizational structure with defined roles and responsibilities
- Program plans, timing schedules, and logic networks
- Program-level and part-level tracking systems
- Four-phase product development process
- Change management process

Two critical elements of the GM Powertrain methodology are the program charter and program contract. The program charter defines the scope of the program with measurable objectives, including:

- Business purpose
- Strategic objective
- Results sought from the program
- Engineering and capital budget
- Program timing

The program contract specifies how the program will fulfill the charter. The contract becomes a shared understanding of what the program team will deliver and what the GM Powertrain staff will provide to the team in terms of resources, support, and so on.

Although the information here on GM Powertrain may appear somewhat dated, it does show that GM was several years ahead of most companies in the development of an enterprise project management methodology. GM has made significant changes to its methodology since then. What GM accomplished more than a decade ago many companies are just beginning to develop. Today, GM uses the above-mentioned methodology for new product development and has a second methodology for software projects.

4.20 ERICSSON TELECOM AB

General Motors Corporation and the bank were examples of project management methodologies that were internal to the organization (i.e., internal customers). For Ericsson Telecom AB, the problem is more complicated. The majority of Ericsson's projects are for external customers, and Ericsson has divisions all over the world. Can a methodology be developed to satisfy these worldwide constraints?

In 1989, Ericsson Telecom AB developed a project management methodology called PROPS.[12] Though it was initially intended for use at Business Area Public Telecommunications for technical development projects, it has been applied and appreciated throughout Ericsson worldwide, in all kinds of projects. In the author's opinion, PROPS is one of the most successful methodologies in the world.

New users and new fields of applications have increased the demands on PROPS. Users provide lessons-learned feedback so that their shared experiences can be used to update PROPS. In 1994, a second generation of PROPS was developed, including applications for small projects, concurrent engineering projects, and cross-functional projects and featuring improvements intended to increase quality on projects.

PROPS is generic in nature and can be used in all types of organizations, which strengthens Ericsson's ability to run projects successfully throughout the world. PROPS can be used on all types of projects, including product development, organizational development, construction, marketing, single projects, large and small projects, and cross-functional projects.

PROPS focuses on business, which means devoting all operative activities to customer satisfaction and securing profitability through effective use of company resources. PROPS uses a tollgate concept and project sponsorship to ensure that projects are initiated and procured in a business-oriented manner and that the benefits for the customer as well as for Ericsson are considered.

The PROPS model is extremely generic, which adds flexibility to its application to each project. The four cornerstones of the generic project model are:

- Tollgates
- The project model
- The work models
- Milestones

---

12. The definition of the acronym PROPS is in Swedish. For simplicity sake, it is referred to as PROPS throughout this book.

Tollgates are superordinate decision points in a project at which formal decisions are made concerning the aims and execution of the project, according to a concept held in common throughout the company. In PROPS, five tollgates constitute the backbone of the model. The function and position of the tollgates are standardized for all types of projects. Thus the use of PROPS will ensure that the corporate tollgate model for Ericsson is implemented and applied.

The project sponsor makes the tollgate decision and takes the overall business responsibility for the entire project and its outcome. A tollgate decision must be well prepared. The tollgate decision procedure includes assessment and preparation of an executive summary, which provides the project sponsor with a basis for the decision. The project and its outcome must be evaluated from different aspects: the project's status, its use of resources, and the expected benefit to the customer and to Ericsson. At the five tollgates, the following decisions are made:

- Decision on start of project feasibility study
- Decision on execution of the project
- Decision on continued execution, confirmation of the project or revision of limits, implementation of design
- Decision on making use of the final project results, handover to customer, limited introduction on the market
- Decision on project conclusion

The project model describes which project management activities to perform and which project documents to prepare from the initiation of a prestudy to the project's conclusion. The project sponsor orders the project and makes the tollgate decisions while most of the other activities described in the project model are the responsibility of the project manager. The project model is divided into four phases: prestudy, feasibility study, execution, and conclusion phases.

The purpose of the prestudy phase is to assess feasibility from technical and commercial viewpoints based on the expressed and unexpressed requirements and needs of external and internal customers. During the prestudy phase a set of alternative solutions is formulated. A rough estimate is made of the time schedule and amount of work needed for the project's various implementation alternatives.

The purpose of the feasibility study phase is to form a good basis for the future project and prepare for the successful execution of the project. During the feasibility study, different realization alternatives and their potential consequences are analyzed, as well as their potential capacity to fulfill requirements. The project goals and strategies are defined, project plans are prepared, and the risks involved are assessed. Contract negotiations are initiated, and the project organization is defined at the comprehensive level.

The purpose of the execution phase is to execute the project as planned with respect to time, costs, and characteristics in order to attain the project goals and meet the customer's requirements. Technical work is executed by the line organization according to the processes and working methods that have been decided on. Project work is actively controlled; that is, the project's progress is continuously checked and the necessary action taken to keep the project on track.

The purpose of the conclusion phase is to break up the project organization, to compile a record of the experiences gained, and to see to it that all outstanding matters are taken care of. During the conclusion phase, the resources placed at the project's disposal are phased out, and measures are suggested for improving the project model, the work models, and the processes.

Besides describing the activities that will be performed to arrive at a specific result, the work model also includes definitions of the milestones. However, to get a complete description of the work in a specific project, one or more work models should be defined and linked to the general project model. A work model combined with the general project model is a PROPS application. If there are no suitable work models described for a project, it is the project manager's responsibility to define activities and milestones so that the project plan can be followed and the project actively controlled.

A milestone is an intermediate objective that defines an important, measurable event in the project and represents a result that must be achieved at that point. Milestones link the work models to the project model. Clearly defined milestones are essential for monitoring progress, especially in large and/or long-term projects. Besides providing a way of structuring the time schedule, milestones will give early warning of potential delays. Milestones also help to make the project's progress visible to the project members and the project sponsor. Before each milestone is reached, a milestone review is performed within the project in order to check the results achieved against the milestone criteria. The project manager is responsible for the milestone review.

Ericsson's worldwide success can be partially attributed to the acceptance and use of the PROPS model. Ericsson has shown that success can be achieved with even the simplest of models and without the development of rigid policies and procedures.

4.21 ROCKWELL AUTOMATION: QUEST FOR A COMMON PROCESS[13]

Rockwell Automation was formed by bringing two major automation companies together in the late 1980s. These two companies, Allen-Bradley and Reliance Electric, were the foundation of what is now Rockwell Automation. Over the years, Rockwell Automation has continued to acquire leading automation suppliers as a growth strategy and also as a way to bring new advanced automation technologies into the company. In 2005, as Rockwell Automation was planning the rollout of a new SAP business system, we recognized the need for a new "common" product development process that would be defined based on company best practices combined with what was considered the industry's best practices for product development. This effort resulted in a "common product development" (CPD) process that was defined in a way to allow for enterprisewide adoption. This is shown in Figure 4–23. This means that 16 different product businesses ranging from high-volume

---

13. This section on Rockwell Automation was provided by James C. Brown, PgMP, PMP, OPM3 AC, MPM, CIPM, CSP, CSSMBB, Director, A&S Enterprise Program Management Office; Karen Wojala, Manager, Business Planning; and Matt Stibora, Lean Enterprise Manager.

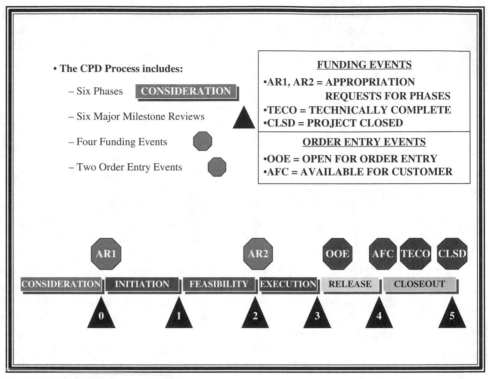

FIGURE 4–23. CPD process: basic concepts.

component suppliers to complex continuous process control systems solution suppliers all use the same high-level process framework for their new product developments.

The resulting process is made up of six phases with a stage-gate review after each phase. The six phases are:

Consideration

● To develop a high-level business case and project proposal to justify AR1 funding for the execution of initiation and feasibility phase activities.

Initiation

● To refine the high-level business case document (BCD) created in the consideration phase into a set of customer requirements sufficient for the project team to create solution concepts, product requirement, and functional requirement documents in the feasibility phase. (See Figure 4–24.)

Feasibility

● To evaluate solution concepts to address the customer requirements from the initiation phase.
● To define the product requirements and functional requirements.

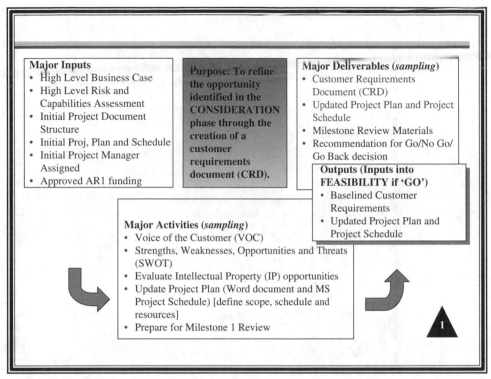

Major Inputs
- High Level Business Case
- High Level Risk and Capabilities Assessment
- Initial Project Document Structure
- Initial Proj, Plan and Schedule
- Initial Project Manager Assigned
- Approved AR1 funding

Purpose: To refine the opportunity identified in the CONSIDERATION phase through the creation of a customer requirements document (CRD).

Major Deliverables (*sampling*)
- Customer Requirements Document (CRD)
- Updated Project Plan and Project Schedule
- Milestone Review Materials
- Recommendation for Go/No Go/ Go Back decision

Outputs (Inputs into FEASIBILITY if 'GO')
- Baselined Customer Requirements
- Updated Project Plan and Project Schedule

Major Activities (*sampling*)
- Voice of the Customer (VOC)
- Strengths, Weaknesses, Opportunities and Threats (SWOT)
- Evaluate Intellectual Property (IP) opportunities
- Update Project Plan (Word document and MS Project Schedule) [define scope, schedule and resources]
- Prepare for Milestone 1 Review

1

FIGURE 4–24. Initiation phase summary.

- To complete all project planning and scheduling to update the project plan for all activities and resources required to complete the execution, release, and closeout phases of the project.
- To develop a BCD that justifies the investment required to execute the project plan for the solution concept chosen. (See Figure 4–25.) Execution
- To develop the product or service according to the baselined functional requirement specification (FRS) from the feasibility phase; performing the necessary reviews; making approved requirement and/or design changes as the project progresses.

Release
- To finalize all test, certification, and other product verification documentation.
- To build and validate pilot production.
- To open for order entry and execute the commercial launch.

Close
- To position the product for transition to continuation/sustaining engineering; documentation cleanup, postmortems, lessons learned, record retention, and complete all financial transactions.

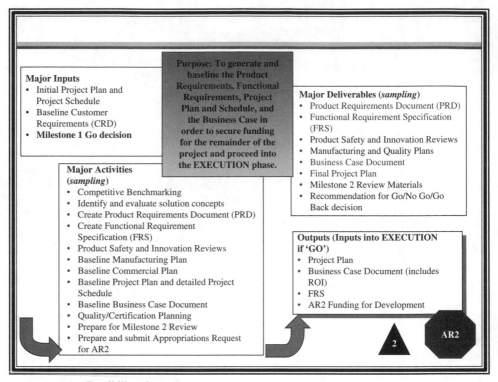

Major Inputs
- Initial Project Plan and Project Schedule
- Baseline Customer Requirements (CRD)
- **Milestone 1 Go decision**

Purpose: To generate and baseline the Product Requirements, Functional Requirements, Project Plan and Schedule, and the Business Case in order to secure funding for the remainder of the project and proceed into the EXECUTION phase.

Major Deliverables (*sampling*)
- Product Requirements Document (PRD)
- Functional Requirement Specification (FRS)
- Product Safety and Innovation Reviews
- Manufacturing and Quality Plans
- Business Case Document
- Final Project Plan
- Milestone 2 Review Materials
- Recommendation for Go/No Go/Go Back decision

Major Activities
(*sampling*)
- Competitive Benchmarking
- Identify and evaluate solution concepts
- Create Product Requirements Document (PRD)
- Create Functional Requirement Specification (FRS)
- Product Safety and Innovation Reviews
- Baseline Manufacturing Plan
- Baseline Commercial Plan
- Baseline Project Plan and detailed Project Schedule
- Baseline Business Case Document
- Quality/Certification Planning
- Prepare for Milestone 2 Review
- Prepare and submit Appropriations Request for AR2

Outputs (Inputs into EXECUTION if 'GO')
- Project Plan
- Business Case Document (includes ROI)
- FRS
- AR2 Funding for Development

2   AR2

FIGURE 4–25.   Feasibility phase summary.

Our goal was to achieve a rapid, repeatable framework that consistently results in high quality output. A major focus of the team that produced this new process was to drive the product businesses to be more disciplined in how innovation was embraced when deciding what projects proposals were funded and which ones were not. There were too many examples of projects receiving management support and funding without meeting a set of minimum criteria that would result in a higher probability of commercial success. Investment proposals were not always based on an ideation process that was driven by our customers.

We found examples of funded projects enjoying support and funding without any customer-driven commercial basis. The justification for these projects was based on new interesting technologies, investing in product family coverage for the sake of coverage without real market demand, providing niche solutions with limited potential driven by a single customer, etc.

To solve this problem the team's original focus was on two aspects of what was defined as best practice. There are numerous theories that attempt to describe the best way to capture customer needs and use them as the basis for creating effective new product concepts. Our goal was to understand the customer problems before we produced solution concepts and product solutions. We accomplished this goal by breaking apart an existing tool called the Marketing Requirements Document (MRD) and process used by Product Management into two tools.

We wanted the product owners to understand the market and target customer's problems before they considered solutions. By breaking the MRD into two deliverables, the first (Customer Requirements Document) focused on the market need and customer problems and the second (Product Requirements Document) focused on the solution concepts and product requirements, and locating these tools and activities in separate phases divided by a management stage gate review, we forced our product managers to break away from the "continuously evolve the product death march" that we were on.

Of course, accepted practice and company culture is hard to break so governance is critical in driving change. This simple step is the beginning of what will be a significant improvement in the new product development practices of this company.

The driving force behind management's commitment to implement this new process and to drive the cultural change was the vision of a common consistent methodology for new product development across the enterprise. This consistency was prioritized from the top (direct management involvement in the stage-gate reviews) down, in order to realize benefit as soon as possible.

All too often, businesses were forced to deny funding for strategic projects due to the never ending incremental product improvements that just kept coming. By forcing business management to approve each project's passage from one phase to the next, we pushed the visibility of every project, every resource and every dollar up to the decision makers who wrestled with trying to find dollars to fund the real game changers. This visibility made it easier for the business owners to kill projects with questionable returns, or to delay a project in order to free up critical resources. Once management began to see the returns from these decisions in the form of product introductions that really moved the needle, we began to focus on the fine tuning of the process and methodologies employed. (See Figure 4–26.)

The stage-gate review is the most important event of a project. Previously, under the old way of executing a project, these reviews were informal and haphazard. Teams were able to continue spending and even overspending without any real fear of cancellation. This new process ensures that every dependent organization is represented at the appropriate review and given the chance to agree or disagree with the project manager that all deliverables are available. The intent is to have the go/no-go decision made by both the primary organization responsible for the deliverables during the previous phase and the primary organization responsible for the deliverables in the subsequent phase. Both these organizations are required at each stage-gate review. If done correctly, we will be able to avoid surprises during the later phases by ensuring transparency during the earlier phases.

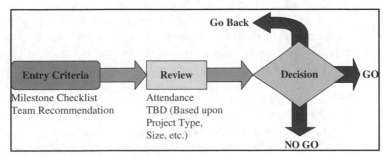

FIGURE 4–26. The decision-making process.

Once a project manager is assigned and a project team is formed, the importance of well-defined deliverables that are easy to locate and use became evident. The 12 months following the original launch of the new process was spent continuously improving the phase definitions, procedure documents, deliverable templates, and governance policies. There is a fine line between rigor and burden; the trick is to push this line hard to ensure rigorous implementation without slowing the progress of the project team down.

At the end of the feasibility phase, as the project enters the execution phase with requested funding secured, the project plan becomes the bible. The project plan drives all activities through the execution phase, release phase, and finally project close. Any issue that the project team is faced with that requires a change in course must be recognized in an updated project plan. At the conclusion of the project, the plan must represent what actually occurred.

Prior to any new product being released for customer shipment, all impacted stakeholders must agree that the product is ready before giving the final approval.

There is a dimension of introducing an end-to-end process that has been assumed but must be mentioned. The company is built from many related but very different product businesses. Each business segment was at a different maturity level relative to all aspects of product development, even the existence of a formal project management organization.

Project managers are instrumental in the execution of a product development process. If consistency, transparency, and risk mitigation are important to a business, and they are to Rockwell Automation, then a formal well-recognized and managed project management entity is paramount.

Rockwell Automation is pursuing the discipline of project management at all levels in its organization.

4.22 SHERWIN-WILLIAMS

There are several ways that a company can develop a methodology for project management. Outsourcing the development process to another company can be beneficial. Some companies have template methodologies that can be used as a basis for developing their own methodology. This can be beneficial if the template methodology has enough flexibility to be adaptable to their organization. The downside is that this approach may have the disadvantage that the end result may not fit the needs of the organization or the company's culture. Hiring outside consultants may improve the situation a little, but the end result may still be the same unfavorable result as well as being more costly. This approach may require keeping contractors on the payroll for a long time such that they can fully understand the company's culture and the way it does business.

Benchmarking may be effective, but by the time the benchmarking is completed, the company could have begun the development of its own methodology. Another downside risk of benchmarking is that the company may not be able to get all of the needed information or the supporting information to make the methodology work.

Companies that develop their own methodology internally seem to have greater success, especially if they incorporate their own best practices and lessons learned from other activities. This occurred in companies such as General Motors, Lear, Johnson Controls, Texas Instruments, Exel, Sherwin-Williams, and many other organizations

The information below was provided by Sherwin-Williams.

Company Background   The Sherwin-Williams Company engages in the development, manufacture, distribution, and sale of paints, coatings, and related products to professional, industrial, commercial, and retail customers in North and South America, the United Kingdom, Europe, China, and India. It operates in four segments: paint stores, consumer, Latin America, and global finishes. The paint stores segment sells paint, coatings, and related products to end-use customers. This segment markets and sells Sherwin-Williams branded architectural paints and coatings, industrial and marine products, and original equipment manufacturer product finishes and related items. As of December 31, 2008, it operated 3346 paint stores. The consumer segment engages in the development, manufacture, and distribution of paints, coatings, and related products to third-party customers and the paint stores segment. The Latin American and global finishes segments develop, license, manufacture, distribute, and sell architectural paint and coatings, industrial and marine products, automotive finishes and refinish products, and original equipment manufacturer coatings and related products. These segments also license certain technology and trade names as well as distribute Sherwin-Williams branded products through a network of 541 company-operated branches, direct sales staff, and outside sales representatives to retailers, dealers, jobbers, licensees, and various third-party distributors. The company was founded in 1866 and is headquartered in Cleveland, Ohio.

The Corporate Information Technology (IT) Department for The Sherwin-Williams Company provides shared services support for the three operating divisions, described above, that make up the organization.

Case Study Background   During the summer of 2002, the Corporate IT Department engaged in activities surrounding the conversion of international, interstate, intrastate, and local telecommunications services from the company's present voice telecommunications carrier to a new carrier. Project management disciplines and best practices, using a structured project management methodology, were utilized on this project, ultimately leading to a successful project outcome.

The project was implemented using a phased approach consisting of the major phases as described below. The phases were established to include many of the principles stated in the *PMBOK® Guide* and also included many of the best practices that had been developed previously at The Sherwin-Williams Company. The phases could overlap, if necessary, allowing for a gradual evolvement from one phase to the next. The overlapping also allowed the company to accelerate schedules, if need be, but possibly at an additional risk. Project reviews were held at the end of each phase to determine the feasibility of moving forward into the next phase, to make "go/no-go" decisions, to evaluate existing and future risks, and to determine if course corrections are needed.

- *Initiate*: The first phase is the initiate phase where the project team is formed, a project kickoff meeting is held, needs and requirements are identified, and roles and responsibilities are defined.

- *Planning*: The planning phase is the next phase and is regarded by most project managers as the most important phase. Most of the project's effort is expended in the planning phase, and it is believed that the appropriate time and effort invested in this phase ensure the development of a solid foundation for the project. Management wholeheartedly supports the efforts put forth in this phase because this is where many of the best practices have occurred. Also, a solid foundation in this phase allows for remaining phases of the project to be accomplished more efficiently, giving senior management a higher degree of confidence in the ability of project managers to produce the desired deliverables and meet customer expectations.

A series of meetings are typically held throughout this phase to identify at the lowest level the project needs, requirements, expectations, processes, and activities/steps for the processes. The results of these meetings are several deliverables, including a needs and requirements document, a project plan, a risk management plan, an issue log, and an action item list. Additional documents maintained include quality management and change management plans. Together these documents provide management with an overview of the entire project and the effort involved to accomplish the goal of transitioning services by the target date established by management.

- *Execution*: The third phase in implementation is execution. This phase is evolved into gradually once the majority of planning has been completed. All activities outlined in the processes during the planning phase come to fruition at this time as actual communication line orders begin to take place as well as the installation of equipment where necessary. Services begin to be transitioned by the division/segment and implementation moves forward aggressively for this project due to a stringent timeframe. It is of vital importance that activities in this phase be monitored closely in order to facilitate the proactive identification of issues that may negatively impact the timeline, cost, quality, or resources of the project.

To facilitate monitoring and control of the project, weekly status meetings were held with the vendor and the project team, as well as short internal daily meetings to review activities planned for each day. Ad hoc meetings also occurred as necessary.

- *Closure:* The final phase of the project is closure. In this phase, there is typically a closure meeting to identify any remaining open issues and to determine the level of client satisfaction. This phase also included any "clean-up" from the project, administrative closeout, the communication of postimplementation support procedures, and a review of lessons learned.

Best practices that worked notably well for The Sherwin-Williams Company included the establishment of success criteria, consisting of project objectives and a needs/requirements analysis, regular communications both within the project team and with stakeholders, dedicated resources, defined roles and responsibilities, knowledge transfer between cross-functional teams, teamwork, the development of a fun, synergistic working environment, and a review of lessons learned.

One of the best practices in project management is that maturity and excellence in project management can occur quickly when senior management not only actively supports project management but also articulates to the organization their vision of where they expect project management to be in the future. This vision can motivate the organization to excel, and best practices improvements to a project management methodology seem to occur at a rapid rate. Such was the case at The Sherwin-Williams Company. Tom Lucas, Chief Information Officer at The Sherwin-Williams Company, comments on his vision for The Sherwin-Williams Company:

> The future of project management at The Sherwin-Williams Company includes the integration of project management disciplines and best practices, combined with portfolio management techniques, to deliver high value project results on a consistent basis. The Sherwin-Williams Company anticipates that the use of a PMO will not only instill the best practices of project management as core competencies, but also aid in the growth of the organization's project management maturity.
>
> One goal has been to unify the goals and objectives of individual departments by applying a universal yet flexible project management framework in pursuit of better across-the-board results. We have made significant strides in this regard. The Sherwin-Williams Company desires to learn from past successes, as well as mistakes, make processes more efficient, and develop people's skills and talents to work more effectively through the establishment of standardized procedures within the company. Above all, we must demonstrate real business value in using professional project management.
>
> While project management professionals may reside in multiple operating units, so as to be as close as possible to our internal clients, our intent is to have a core group of project management professionals that would be the standards setting body and provide for best practices identification and sharing.
>
> We have all managed projects at one time or another, but few of us are capable of being Project Managers. Herein lies one of the biggest impediments to implementing professional project management. We can have the best-trained project managers, we can have all the right process in place, we can use all the right words, yet the PMO will either fail or be only a shell of what it can be. Staff and management have a hard time appreciating the power, and improved results, of a professionally managed project. Until staff and management become involved themselves, until they feel it, until they personally see the results, the distinction between managing a project and project management is just semantics.
>
> The difference between managing projects and professional project management is like the difference between getting across the lake in a rowboat versus a racing boat. Both will get you across the lake but the rowboat is a long and painful process. But how do people know until you give them a ride?
>
> The 2002 telecom case study was just such a ride. While the focus of the case study discussion was to articulate the mechanics of the PMO process, the real story is the direct per share profitability improvement resulting from this successful initiative. In addition, there was legitimate concern from the business on the potential impact this change may have on our internal clients and external customers should something go wrong during the transition. The professional project management that was used gave everyone the cautious optimism to proceed and the results made the staff and management "believers" in the process.
>
> Project by project, success by success, a cultural transition is in process. As we demonstrate improved business results because of professional project management

we are able to offer services to a wider audience and are able to take on projects outside of IT where the PMO got its start.

By staying focused on business results, by staying close to our clients so we understand their needs well, and by constantly challenging ourselves to improve our underlying processes our PMO services are maturing more and more every day. It becomes a fun ride for everybody.

4.23 PEROT SYSTEMS: MATURING YOUR METHODOLOGY[14] _____

This section will discuss how Perot Systems Corporation advanced its project management methodology to meet the demands of an expanding global enterprise and increased its ability to replicate project delivery excellence for its clients. The material in the remainder of this section was provided by Michele A. Caputo, PMP, Enterprise PMO Leader and P³MM Program Manager; Linda L. Wilson, PMP, P³MM Program Team; David Brodrick, PMP, EPMO Governance Leader; and Darin A. Hart, MBA, PMP, EPMO Methodology Leader.

There were several driving forces which indicated the need to become excellent in project management:

- The need to increase the ability to replicate excellence across all projects
- Standardized project delivery processes support consistency in execution, with reliable, predictable results
- A methodology that provides early warning through quantitative metrics and routine monitoring and reporting
- What is not measured cannot be managed or controlled

Body   As a 20-year-old company, Perot Systems Corporation would be considered young by many standards. Yet this IT technology services company, headquartered in Plano, Texas, valued project management even in its early years and grew the capability along with the expansion of its business. The company's Project Management Methodology (PMM) consisted of guidelines supported by tools and templates. Internal training expanded use of the PMM and the library of artifacts that allowed project managers to share ideas and best practices. Governance resided within the project management offices (PMOs) of business units or departments and grew considerably with the rollout of a Project Management Quality Program (PMQP).

With a solid framework in place, gaps still existed and challenges grew as the company's annual revenues from operations in 23 countries increased to over $2 billion:

---

14. Perot Systems is a worldwide provider of information technology services and business solutions. Through its flexible and collaborative approach, Perot Systems integrates expertise from across the company to deliver custom solutions that enable clients to accelerate growth, streamline operations, and create new levels of customer value. Headquartered in Plano, Texas, Perot Systems reported 2008 revenue of $2.8 billion. The company has more than 23,000 associates located in the Americas, Europe, Middle East, and Asia Pacific. Additional information on Perot Systems is available at http://www.perotsystems.com/.

- Tools and templates were insufficient for larger or more complex projects or programs spanning business units and continents.
- Governance varied among business units making centralized reporting difficult and hampering visibility into early warning signs of underperforming projects.
- Inconsistencies in quality and standards across the business units negatively affected the teams' ability to work most efficiently and in some cases impacted client satisfaction.
- Lack of mature project management and PMO skills, processes, and tools contributed to a higher risk of challenged projects.

To address these needs and better position the company for expansion, the chief operating officer (COO) sponsored the launch of the P³MM (for Project, Program, and Portfolio Management). The COO challenged the P³MM team to start with the best components of the PMM and then search the company for best practices already contributing to operational success. Through collaboration with representatives from all business units and delivery teams across Perot Systems, the P³MM program was undertaken with the expectation of the following benefits:

- Appropriate scaling of project management and PMO operations for each new client through increased involvement in the sales cycle
- Increased capability to monitor and control the performance of the enterprisewide portfolio through better governance, including standardized, quantitative performance measures providing early warning, a centralized project status reporting application, and processes for proactive intervention and remediation
- Reduced administrative burden and increased effectiveness of the project managers and PMOs through automated and streamlined project management activities and tools
- Increased knowledge and skills of project managers through more flexible and cost-effective global virtual classroom training

P³MM Highlights

The expanded methodology closely ties to the PMI PMBOK© knowledge areas and incorporates templates, checklists, software tools, custom applications, standard operating procedures (SOPs), metrics, and customized training for project management but also extends the methodology to incorporate program and portfolio management. A Project Management Information System (PMIS) supports and automates critical project management processes. In addition, the P³MM incorporates the associated "people" programs that guide the selection, assignment, certification, and management of project manager and PMO resources.

For all projects across the enterprise, the P³MM establishes a holistic set of key performance and compliance metrics by building them into the standardized processes, tools, and governance. To ensure consistency in the reporting of performance results, the methodology relies on seven metrics selected for cost, schedule, and quality performance, including industry best practices such as earned-value computations and key customer satisfaction metrics. The resulting measurements of project delivery provide a balanced "scorecard" for leadership.

An important concept within the P³MM is scalability. The methodology's tools and templates are not intended to be applied uniformly across all types and all sizes of projects or engagements. Project managers and PMOs apply the P³MM more rigorously depending on the size, complexity, and risk of a project or program. The online Web-based "packaging" uses a metadata-driven navigational structure to allow associates to locate the appropriate materials required for each level and phase of a project or material specific to a knowledge area. The project duration, project type (applications development, infrastructure, etc.), team size, and geography also drive decisions about appropriate application of the P³MM components.

Scaling occurs on many levels:

- People-level scalability through (1) PMO organizational design materials and a toolkit for determining the staffing levels and roles based on the portfolio of projects and engagement type and (2) scaling the project manager certification and skill requirements based on the project levels in the portfolio
- Process-level scalability that allows the project manager to adjust the degree of structure and rigor of process to fit the size, complexity, and risk of the project
- Template-level scalability that gives the project manager the flexibility to customize and streamline templates as appropriate
- PMIS tool-level scalability with PMO toolkits scaled with alternate versions of tools that scale based on volume of data, size of project, or size of portfolio

Governance: Managing Success from the Top Down
Rollout of this extensive program required a team committed to practicing its own methodology. The P³MM team established a two-year program plan with senior project managers and subject matter specialists from across the corporation who participated in developing the new methodology, assisting business units with adoption of the P³MM, and establishing governing processes. While governance is executed primarily at the client account PMO level, a cross-business unit(BU) governing group sets enterprise standards and reviews performance reporting to ensure consistency in application and to provide support as needed through a committed group of coaches and mentors.

P³MM governance includes quality checkpoints and oversight processes such as:

- Certifications of the project plans during the sales cycle and project planning phase
- Certification of the PMO organization on all new deals
- Project performance and compliance reporting
- Project health assessments
- Intervention and remediation support
- Training and tool oversight
- Guidelines for governance organization structure, including roles and responsibilities at all levels of the organization from the project/program level to the client account, industry segment, business unit, and enterprise level

In addition to internal governance of project delivery performance, project quality, and methodology compliance, the P³MM addresses the standards for project portfolio governance and portfolio management on client engagements in which Perot Systems has oversight responsibility for the portfolio of projects. As an IT technology services company, Perot Systems regards portfolio governance as critical to maximizing a client's project investments because governance *specifies the decision rights* and provides an *accountability framework* to drive better IT usage behaviors and ensure IT outcomes are aligned with the client's business strategies.

The P³MM provides processes and an associated toolkit to assess and develop a more mature IT governance model that addresses the following questions:

- What decisions must be made to ensure effective management and use of IT?
- Who will make those decisions?
- What process will be used to make those decisions?
- How will the decisions be communicated?
- How will the outcomes of those decisions be monitored?

Tools included in the P³MM aid Perot Systems in assessing and designing a best practice IT governance model for a client:

- An IT governance maturity model and associated IT governance maturity assessment tool
- An IT governance model selector
- An IT governance participant framework
- An IT governance value discipline review
- Prioritization of IT governance capabilities
- Assessment of portfolio management strategic business drivers

The P³MM defines eight IT governance capabilities. These capabilities are a collection of practices that help optimize IT-enabled investments, ensure service delivery, and provide measures against which to judge when situations require corrective intervention. These practices ensure that IT is aligned with the client's business strategies, enables their business practices, uses IT resources responsibly, and manages risks appropriately.

The eight IT governance capabilities emphasized in the P³MM are:

- IT governance framework
- IT strategic planning process
- IT enterprise architecture process
- Governing IT organization and resource management
- Governing IT investment and project portfolio management
- Communicating governance
- Governing IT risks
- Governing IT regulatory compliance

Perot Systems has found that a successful IT governance implementation is defined by the following characteristics:

- Managers in leadership positions can accurately describe the IT governance framework, processes, and decision rights structure
- Effective use of communication about governance issues and outcomes
- Senior leaders directly involved with IT governance (both from the business and IT sides)
- Clear business objectives for IT investment
- Clearly stated and highly differentiated business strategies
- Fewer renegade projects and more formally approved exceptions
- Fewer changes to IT governance from year to year

The IT governance capability maturity model is part of the P³MM toolkit that PMO leaders can use to assess the maturity level of the IT governance capabilities practiced on an account. The PMO leader can combine the results of a capability maturity assessment with knowledge of the client's primary business strategies to design PMO operational processes and deliverables appropriate to the client's current-state maturity of IT governance and establish a path toward an optimal IT governance model that maximizes the effectiveness of the portfolio management processes for the client.

The Project Manager: Managing Success from the Bottom Up

Success of a methodology as robust as the P³MM depends on the individual project manager's commitment to performing well within the new environment. The P³MM's Project Workbook tool was designed to help each project manager self-manage the P³MM requirements and implement the checks and balances required for a project's success. The macro-driven Project Workbook file is developed around the feasibility, initiation, planning, execution, and shutdown phases of the project life cycle. Rather than providing a rote checklist, the Project Workbook delves into the quality of the content by leading the project manager through a series of questions organized by PMI PMBOK© knowledge area.

During feasibility, for example, the Workbook challenges the project manager to understand and integrate the requirements of the contract or statement of work and align to the cost elements as they are defined. In addition, the questions prompt the project manager to identify all necessary roles and responsibilities to properly begin human resource management.

Throughout initiation and planning, the Workbook focuses on activities such as mapping requirements to deliverables and acceptance criteria, obtaining customer acceptance of and commitment to the plan, and ensuring the project manager has the organizational support required for the level of project. P³MM tools for these phases are referenced in the Workbook, and the questions guide the project manager in the application of those tools and of the new processes, such as the requirement for management plans, project controls, or waivers for P³MM components deemed not appropriate for the project.

Workbook questions during execution prove the most challenging, as they deal with the collection of cost and schedule metrics and cover such vital processes as change request management and risk mitigation, as well as ensuring critical client signoffs are obtained.

At shutdown, the Workbook calculates across all categories to provide scoring that guides the project manager and the PMO in assessing overall project success and identifying growth opportunities through lessons learned and project performance data that project managers can leverage for future projects.

Ensuring Success   The P³MM team began its effort to change the project management culture with solid executive support and a group of highly skilled subject matter specialists. To address any resistance to change, the team employed techniques from Perot Systems' organizational change management (OCM) methodology, beginning with a change readiness assessment and identification of champions within each organization across the globe. The team engaged stakeholders at all levels during planning and implementation, provided feedback opportunities, and shared ownership and accountability. The team also assured success among business units and delivery teams by providing training and support to build a solid foundation for long-term adoption and sustainability.

Perot Systems' P³MM is successfully meeting the COO's challenge, and results from the first year of implementation show impressive improvements in project delivery performance. As the program matures across the corporation, Perot Systems expects the full impact of the P³MM to be reflected in improved profitability, in consistently high client satisfaction, and in the company's ability to maintain a culture of operational excellence.

4.24 ANTARES MANAGEMENT SOLUTIONS[15]

Some companies have found that some of the readily available methodologies that can be purchased or leased have enough flexibility to satisfy their needs. This is particularly true in the IT area. The following information was provided by John Frohlich, Director, Information Systems Development at Antares, and Dan Halicki, Planning Coordinator at Antares.

Introduction   Industry analysts believe that a quarter of all IT projects are delivered on time, and fewer are within budget. In response to this problem, various project management approaches have emerged with the primary objective of ensuring project success. The business challenge is to find the proper balance between rigorous methodology requirements and the realities of containing administrative project overhead. More is not always better in the world of project management.

In 1989, the systems development team at Antares Management Solutions adopted selected aspects of the Navigator Systems Series methodology to manage its IT projects. In 2005, Antares IT improved upon this methodology by introducing a project management methodology that blended the best of the Navigator Systems Series with the Project Management Institute's (PMI) standards for project management. In 2007, Antares integrated a project management office (PMO) into its project management methodology. The PMO is responsible for the governance and oversight of the overall project management

---

process of Antares. During the past 20 years, the use of Navigator, PMI concepts, and now a PMO have contributed to overall project success and helped Antares to grow and prosper in the competitive world of IT and IT outsourcing.

What, then, are the best practices that Antares has followed to bring the right amount of discipline to its project management approach? The following discusses the techniques and deliverables that constitute the standard project management elements of initiation, planning, execution, controlling and monitoring, and closing. Additionally, it is important to look beyond the elements of the methodology itself to include the support system that makes it all work, such as executive management buy-in, training, and follow-up activity.

Make Use of Project Management Concepts and Terminology Throughout the Enterprise, Not Just in Information Systems

Antares uses a strategic planning process that requires collaboration between client management and IT planning to effectively align business vision with technical solutions. To achieve the desired goals, related project proposals are drafted using the standardized project charter format. Consequently, the client management team has a clear idea of how a project is to be defined and structured from the very beginning. Customer commitment, project expectations and criteria for success are indicated up front.

Provide Ongoing Project Management Training Throughout the Enterprise

The term "project management" can mean different things to different people. An instructor-led class explains how Antares applies its improved "project management methodology" to ensure that all project participants, including the customer community, have a common understanding of the process, especially stakeholder participation and joint ownership. With a duration of 24 hours over three days, the class provides instruction and a hands-on workshop on the major components of the project management process, for example, project initiation, planning, execution, monitoring and control, and project closure.

Structure Every Project in a Consistent Manner, Including Scope, Responsibilities, Risks, and High-Level Milestones

Central to the project management process is the creation of the project charter, which documents and formalizes the agreement of all concerned parties, including business sponsors, project manager, and project team members.

The charter spells out what will be accomplished and when, who is responsible for getting it done, known risks and the course of action to mitigate the risk, quality assurance measures, and the major milestones to be met.

Communicate and Update Project Plans and Status on an Ongoing Basis Using Online Tools When Appropriate

Once an approved project charter is in place, the project plan becomes the next focus of attention. The plan and the regular status reports that document progress versus the plan are the primary communication vehicles for keeping management and staff informed. Since 2005, Antares has utilized an online real-time project status center to keep project planning, issues, change control, and project information current and readily available.

Maintain an Official Project Issues List, Including Who Is Responsible, What Are the Potential Impacts, and How the Issue Was Resolved

As problems arise during the execution of a project, it is vital to maintain an accurate issues log to properly document the problems and what is being done about them. In project meetings, the issues log serves as a meaningful tool to prevent unproductively rehashing old information and assures concerned areas that their problems have not been forgotten. Along with the charter and the regular progress reports, the issues log can be accessed via the online project status center and critical issues highlighted in the project's status reporting dashboard.

Use a Formal Change Control Process, Including an Executive Steering Committee to Resolve Major Changes

As issues arise that could require additional resources, the nemesis known as "scope creep" can begin to surface and impede a smooth-running project. A formal change control process is essential to help the project manager with the predicament of pleasing the customer or going over budget or missing milestones. The Antares change control process (request and assessment) can be accessed via the online project status center and critical change control highlighted in the project's status reporting dashboard.

Conclude Every Major Project with an Open Presentation of the Results to Share Knowledge Gained, Demonstrate New Technology, and Gain Official Closure

Many organizations rush through the project conclusion phase in order to get on with the next assignment. However, taking the time to fully document accomplishments, outstanding issues, deferred deliverables, and lessons learned is a worthwhile exercise that gives proper credit for successful completions, reduces misunderstandings over omissions, and facilitates transitions to follow-up projects. Postimplementation review is conducted one to three months after project closure to check whether benefits, as defined in the business case, have been achieved and identify opportunities for future improvement.

Periodically Audit and/ or Benchmark the Project Management Process and Selected Projects to Determine How Well the Methodology Is Working and to Identify Opportunities for Improvement

With the establishment of project management standards and the sizeable investment in training, it makes sense to step back occasionally and assess if the overall objectives are being met. Where projects ran into difficulty, would a different approach have helped? Where projects went well, what aspects of the project execution ought to be shared by other projects? These are the kinds of questions the audit exercise tries to address. Benchmark efforts are also valuable to assess progress over time.

Adapt to Meet the Business Needs of the Client

Antares utilizes different software development methodologies to fulfill business requirements. Methodologies may include development models (sequential, incremental, iterative) used in conjunction with one or more techniques (prototyping, object oriented). Regardless of the model or technique used in the development effort, delivery of a successful project is dependent on the project management framework of common processes, for example, initiation, planning, execution, monitor and control, and closure. This framework provides common vocabulary for project managers and allows all stakeholders to understand the project phases more clearly.

Conclusion

One of the major reasons projects lose control is because project managers operate reactively rather than anticipating possible stumbling blocks. When the stumbling block finally appears, they are hesitant to make effective decisions, which often results in situations that can kill a project. Antares's experience has been that the aforementioned best practices help foster a more proactive environment in partnership with the customer, resulting in successful project delivery more consistently. Moreover, the outlined approach strikes a balance that requires minimum project overhead while obtaining maximum benefit.

4.25 HOLCIM

Holcim is one of the world's leading suppliers of cement and aggregates (crushed stone, gravel, and sand) as well as further activities such as ready-mix concrete and asphalt-including services. Holcim Group holds majority and minority interests in more than 70 countries on all continents. The material in this section has been provided by Roberto Nores, Holcim Group Support Ltd., Corporate Human Resources, Global Learning—Project Management

The Project Management Approach

Standard Methodology for All Projects Based on PMI Standards
The cement industry is a capital-intensive business. Executing projects within expected cost and time targets is crucial for the long-term success of such companies. In the late 1990s, Holcim decided to introduce a standard project management approach (PMA) for developing and executing all types of projects within the business.

This methodology consists of 5 phases and 25 steps covering the whole project management life cycle. It tackles most of the 9 knowledge areas of PMI's *PMBOK® Guide*, and it is supported by a series of project templates. Special emphasis is put on the definition, closing, and evaluation phases to enable knowledge sharing within the Holcim Group.

A Cost-Effective Roll-Out Methodology to Ensure Sustainability

The PMA roll-out across Holcim was based on three principles:

- Active management involvement: Projects require a regular involvement of management throughout their life. Project clients (owners, or sponsors) participate in workshops regularly to define, steer, and close the project.
- On-the-job training: Training happens through application of the methodology on actual projects with the support of internally trained facilitators. Training and coaching for the five phases last on average three days per team.
- Local trainers available all over the world: In each country and company where Holcim has operations, there is a team of local trainers who provide support and training in the PMA to local employees. It is based on train-the-trainer principles and is therefore cost effective and sustainable since expertise and support are available locally in each country.

Further Project Management Standards for CAPEX Projects

In order to maintain and improve its business results, over the last decade Holcim has been investing on average 50 percent of its cash flow from operations in so-called CAPEX projects—projects to maintain, improve, and expand Holcim's production facilities. As a further step in standardization, Holcim decided to expand PMA for application on CAPEX projects, incorporating the most important and crucial tasks and deliverables. Four new methodologies were developed and are being implemented. They consist of the same five phases of PMA, but the number of steps has been extended to between 50 and 200; each one is supported by tools and guidelines. The new methodologies are:

- ProMap: Methodology for CAPEX projects in the cement business. It is divided in two submethodologies, ProMap for small projects (investments below 5 million Swiss francs) and ProMap for medium and large projects (investments above 5 million CHF).
- AggPro: Methodology for CAPEX projects in the aggregates business.
- RMXPro: Methodology for CAPEX projects in the ready-mixed concrete business.
- AsphPro: Methodology for CAPEX projects in the asphalt business.

Since most project managers know PMA, they can apply the CAPEX-specific methodologies with limited additional training.

Project Management Training in Holcim

Throughout the years, Holcim has established a series of project management training programs to increase the level of competency of project managers. Currently Holcim works with a three-level project management training concept as follows:

Local Training Programs
PMA: This training aims at ensuring that project managers and teams understand the standard PMA and the basic project management principles behind it. It lasts three days, split into three workshops.

ProMap: This training aims at ensuring CAPEX project managers understand the additional steps and tools of this approach. It lasts two days.

Regional Training Programs PM Seminar: This program is aimed at helping project managers deal with the most important and complex projects in each country. It develops both hard project management skills (scheduling, earned-value management, etc.) as well as soft competencies (leadership, stakeholder engagement, etc). It lasts five days and has several pre- and postseminar activities to maximize the impact of the new skills on their projects. It facilitates the creation of networks of project managers in geographical regions (Asia, Europe, Latin America, etc.).

Global Training CAPEX Project Management Forum: It is an exchange platform where project managers and key stakeholders of largest CAPEX projects get together to exchange experiences and discuss best practices. It lasts 3.5 days and is held at a construction site of the largest and most complex projects of Holcim.

Project Portfolio Management

As a consequence of its efforts in project management, Holcim realized the need to establish clear links between its yearly, high-level planning (business planning and budgeting processes) and the daily, operational execution of projects. An average Holcim Group company runs between 100 and 300 projects each year. There was a need to optimize project prioritization, cash flows, resource allocation, and portfolio monitoring. Consequently, a project portfolio management methodology was established that supports top management in each country to manage its annual project portfolio.

Since investments in off-the-shelf systems for its project portfolio were very high in the early 2000s, Holcim decided to develop in-house software to support the most basic processes of its project portfolio. So far this software is in place in a few operations, and it has good acceptance from employees.

Results Achieved

Back in 1999 the Holcim Executive Committee made PMA a worldwide standard for its operations. Ten years after its introduction, PMA has become the common language for project management within the Group. The majority of the 5000+ project managers around the world make use of this approach, adapting it to the nature of each project.

The application of these methodologies helped to:

- Reduce time to execute projects (e.g., reducing by 30 percent the time needed to erect greenfield ready-mixed concrete (RMX) plants)
- Reduce cost overruns as well as costs for investments (e.g., less usage of contingency costs)
- Improve capability to execute projects (e.g., more projects executed with the same personnel)
- Improve collaboration and team work (e.g., interplant projects for continuous improvement)

These efforts in project management helped the Holcim Group to become one of the leaders in the construction materials sector.

4.26 WESTFIELD GROUP

Developing a project management methodology may not be as complicated as one believes. There are activities that a company can do to make the process of methodology development easier. The information in this section has been provided by Janet Kungl, PMP, Program Manager, Westfield Group.

There are four essential building blocks that, if present, allow for the development of an effective project management methodology in a reasonably short period of time:

- Recognition and support at the executive level of the need for project management expertise
- Establishment of a project-focused organization with a vision of how the project management discipline will be integrated in the organization
- Leveraging the *PMBOK® Guide* for methodology development
- Commitment to developing and/or hiring skilled project managers

Westfield Insurance (westfieldinsurance.com), an insurance, banking and related financial services group of businesses headquartered in Westfield Center, Ohio, recognized in the late 1990s that the needs of the business had changed with regard to information technology and information systems projects. In order to meet those needs, senior management recognized that project management capabilities were required as well as an organization that was designed to "partner" with the business and specialize in providing high-quality business and technology solutions. Westfield reorganized its information systems department to include a delivery unit focused on project delivery via the use of "virtual teams" comprised of technology and business team members.

In order to build the project management capability, one of Westfield's project managers was assigned the project of developing the project management methodology. The approach was to use the *PMBOK® Guide* as a starting point and customize the processes to reflect Westfield's culture and organizational structure. A key component of the ongoing support for the methodology was having the other project managers assigned to work on developing the processes.

The methodology starts with an explanation of the project phases (see Figure 4–27). Within each of these phases, the activities to be performed are defined and cross-referenced according to the nine knowledge areas of the *PMBOK® Guide*. An example of this is shown in Figure 4–28. Each of the activities in Figure 4–28 can then be exploded into a flowchart showing the detailed activities needed for accomplishment of the deliverables for this

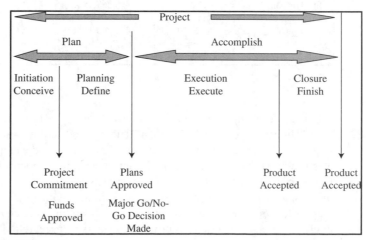

FIGURE 4–27. Life-cycle phases.

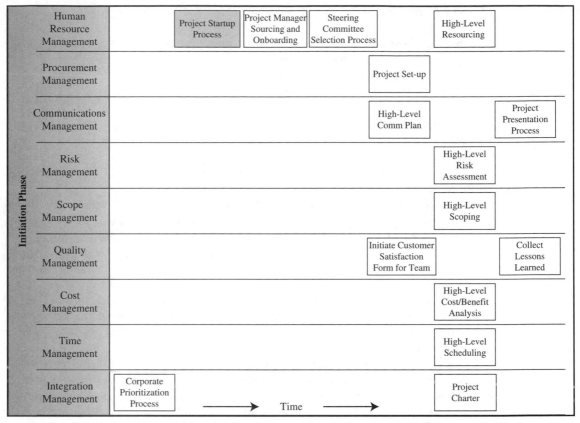

FIGURE 4–28. Initiation phase activities.

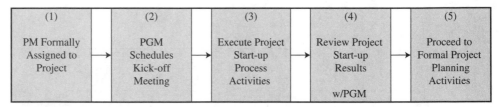

FIGURE 4–29. Project startup process.

activity. For example, the "Project Startup Process" activity in Figure 4–28 is cross-listed under Human Resource Management. The detailed activities are shown by the flowchart in Figure 4–29 Artifact templates and examples were also developed and are accessible in an electronic standard project folder. A continuous improvement process for the methodology was also implemented. Process improvements include the development of project success metrics, formalized project reviews, and defining the integration points with other processes [e.g., software development life cycle (SDLC), architecture governance, and portfolio planning]. Over time, experienced project managers were recruited from other organizations to

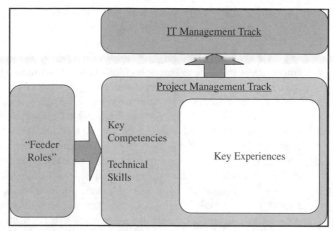

FIGURE 4–30.   The project manager experience path.

supplement internally developed project managers. To support their professional development, a project manager forum was established in 2006. This biweekly meeting allows the project managers to share best practices and lessons learned with their peers. A project management experience path was documented in 2007 to identify key experiences and competencies that influence a project manager's career trajectory and define the path into, through, and out of the project management career track (see Figure 4–30).

The project management function gained acceptance by successfully delivering projects. Increased support from all levels of the organization is evident as more business customers request partnerships with teams led by project managers. Putting the right building blocks in place created the foundation necessary for moving ahead. Currently, Westfield is experiencing increased success in delivery and improved overall results with key business initiatives. These results validate the value of project management. Building toward excellence in project management continues at Westfield Insurance.

4.27 EXEL[16]

Exel was placed in a position of having to develop a methodology that would be global in nature and company intranet based. Julia Caruso, Project Manager, Regional Project Management Group, describes Exel's project management methodology:

> Exel has an established project management methodology (DePICT®)[17] which was introduced in 1997 based on a detailed assessment of the project management processes.

16. Material on Exel was graciously provided by Julia Caruso, PMP®, Project Manager, Regional PM Group, Americas; Todd Daily, PMP®, Business Process Manager, DHL Global Forwarding, International Supply Chain; and Stan Krawczyk, PMP®, Director, Regional PM Group, Americas.
17. DePICT® is a registered trademark of Exel, Inc.

This methodology is based on specific needs of the organization with strong influence from the Project Management Institute's *Project Management Body of Knowledge* (*PMBOK® Guide*)

The acronym DePICT® signifies the five phases of an Exel project: Define, Plan, Implement, Control, and Transition. The DePICT® methodology was developed to formalize a structured approach to all projects with globally consistent processes and tools.

The methodology is Intranet based to ensure all associates have access to the latest version of the toolkit and instructions. The Global PMO maintains the Project Management Portal which is the official site for information on the DePICT® methodology. (See Figure 4–31.) In addition, it provides an executive overview on project management, PMO contacts, the global PM training schedule, PM course descriptions, class registration instructions, additional information such as newsletters, presentations & references, and a Value Message Toolkit. The Value Message Toolkit contains a template for responding to RFP's on the topic of project management, as well as a library of presentation slides for anyone who gives presentations involving project management. These tools are to promote consistent messaging—both internally and externally—on DePICT® and project management at Exel.

The intranet site, or portal, utilizes Google search technology, which has the capability to search the full text of the portal pages as well as its downloadable material. By typing in key search words, the system will provide a link list by relevancy any web page or downloadable file that matches the search criteria.

In Exel's case, the methodology is stable. However, concepts, approaches, tools, and training information are reviewed on a continual basis, based on demands from the Exel project management community, customers, and new industry trends. This includes the ongoing enhancement of tools to support efficient and consistent business practices.

Exel receives customer feedback with respect to its project management methodology/approach through various modes. The most common is direct feedback on project team performance from customers to Exel account owners and customer lessons-learned exercises. In many cases, customers will provide feedback on Exel to various media outlets and industry publications. Today, most lessons-learned exercises are conducted by individual project teams and archived by each specific sector/department. Exel is evaluating methods of collecting, disseminating, and distributing key lessons learned across all sectors.

Recommended improvements are reviewed for applicability across the organization by the Global PMO on an annual basis, and a new version of DePICT® is released each year incorporating all of the approved changes. Interim changes, if critical, will be made available mid-release; however, Exel has improved its discipline of annual versioning in the past few years to reduce confusion with its users on what version they should be using, as well as streamline communications and training regarding the DePICT® updates.

Finally, a Project Management Americas mailbox is set up for all inquiries from project managers company-wide and is accessible by all members of the regional PMO. The regional PMO Americas also maintains an e-mail distribution list of those who have voiced an interest in receiving PM-related communications. This PM community numbers over 300 associates.

As companies begin the development of project management methodologies, emphasis seems to be on developing templates first. Exel Corporation has achieved success through its project management templates. According to Todd Daily, PMP®, Business Process Manager, International Supply Chain Division:

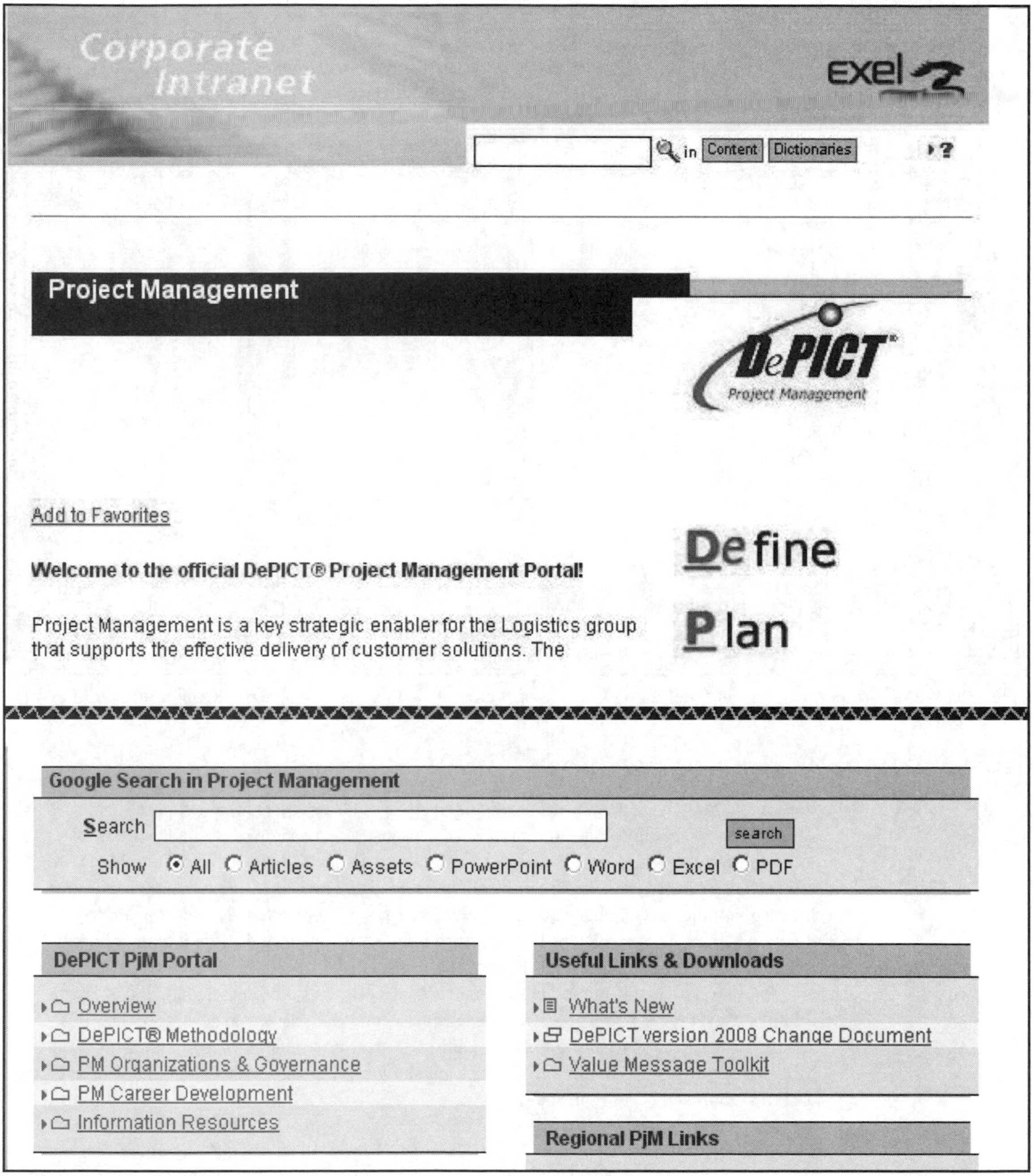

FIGURE 4–31. DePICT® project management intranet portal.

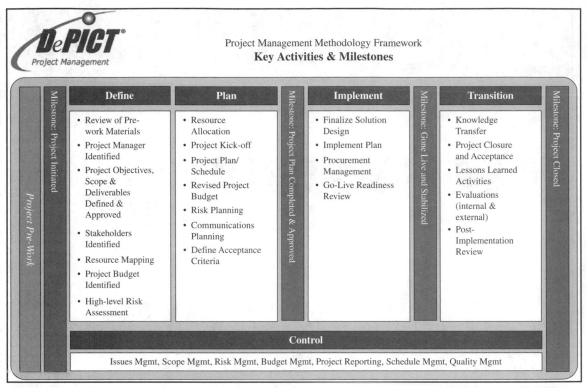

FIGURE 4–32.   Project management methodology: key activities and milestones (global). © Copyright 2005 DPWN.

Exel uses a project management toolkit comprised of 27 templates that support the phases of a project from initiation through closure. The use of the templates enables the project manager to satisfy the specific needs of his or her project without overburdening them with unnecessary paperwork. These templates have been created in Microsoft Excel Word, or Powerpoint and are embedded in a standardized file structure that is used for all project documentation. The consistent file structure allows managers and project team members to easily locate documents, regardless of the project.

The layout of the toolkit is designed to support Exel's project management methodology DePICT ®. Tools specific to each phase are categorized accordingly. DePICT® stands for the five phases of a project: Define, Plan, Implement, Control, and Transition. Milestone controls and major deliverables are shown in Figure 4–32. The global DePICT® toolkit is represented in Figure 4–33.

Additionally, a subset of tools has been identified as the bare-minimum requirement for each project. These are the Project Charter, Project Plan, Risk Assessment & Log, Issues Log, Scope Change Request Form & Log, Status Report, Lessons Learned, and the Project Closure & Acceptance Form. The use of templates increases understanding of the tool functions, provides a sufficient level of guidance for less experienced project managers, enables flexible/dynamic adaptation to all projects—pursuit or implementation, reduces time, and provides quicker adaptability for associates.

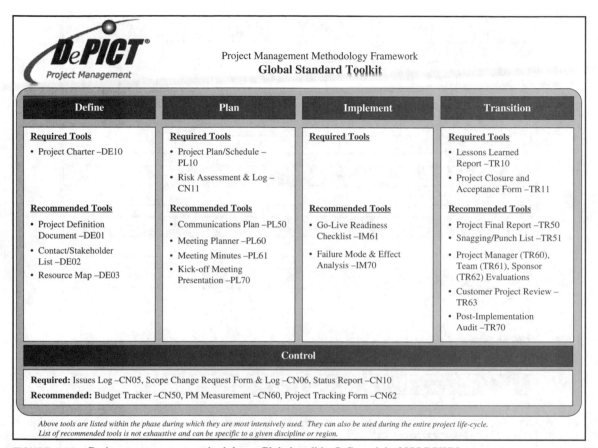

FIGURE 4–33. Project management methodology: Global toolkit. © Copyright 2005 DPWN.

Exel uses templates for all phases of projects. Typical templates appear in Figures 4–34 through 4–39. The two primary types of projects at Exel are pursuit projects, which involve the pursuit of new business via request-for-proposal (RFP) responses and solution design, and implementation projects, which are much larger, involving the implementation and delivery of the designed solution for our clients. Pursuit projects range in length from a couple of weeks to more than a month. Implementation projects can run from a couple of months to more than one year, depending on complexity. IT-specific projects tend to follow a traditional system development life-cycle format with links to general project management tools, such as a charter and issues list. Due to the fact that large implementation projects usually involve an IT component, an additional set of tools and templates focused on the IT aspect of projects has been developed and implemented globally. Exel also begun development of a third tier of templates that have been developed by the project management community. These templates are made available on the Intranet, and feedback is gathered on the usefulness of these tools. Should the community feel that there is broad applicability for these tools, consideration by the Global PMO will be given to inclusion in the official Global DePICT® toolkit.

PROJECT CHARTER

Project Name/ID #	
Project Manager	
Project Owner/Sponsor	
Customer Name	
Sector/Business Unit	
Version Date	

DePICT®
Project Management

Executive Summary: *(brief description of situation, need for project & project benefits)*

Project Objectives: *(SMART - Specific, Measurable, Agreed by all stakeholders, Realistic, Time-framed)*

Project Scope & Deliverables:

In Scope
•

Out of Scope
•

Management Approach: *(governance & key roles & responsibilities)*

Project Management Approach:

The project management approach will follow DHL/Exel's Project Management methodology, DePICT®, and the associated tools defined within that methodology.

Project Structure:

A core Project Team will be established drawn from DHL/Exel and <**CUSTOMER**> resources. This team will be accountable for the successful delivery of the Project objectives and will be led by a DHL/Exel Project Manager. Roles and responsibilities are to be further defined in the DE03 Resource Map.

Roles identified to date are as follows:

1. Project Sponsors: <**DHL/Exel & Customer names**>
2. Dedicated DHL/Exel Project Manager: <**Name**>
3. Dedicated Customer Project Manager: <**Name**>
4. Dedicated DHL/Exel IT Project Manager: <**Name**>
5. Workstream Leads: (as necessary depending on the complexity of the workstream; e.g. Operations, Systems, Human Resources, Real Estate, Transportation)

FIGURE 4–34. Project charter template.

ISSUES LOG

Project Name/ID#:		Customer Name:	
Project Manager:		Project Sponsor:	
Total # of Issues	0 Total	Version Date:	
# of Red Issues:	0 Red		

#	Priority	Status R/A/G	Function / Category	Issue Description/Impact	Owner	Origin Date	Due Date	Closed Date	O /C	Actions/Updates

FIGURE 4–35. Project issues log template.

Project Plan / Schedule

Project Name/ID#		Customer Name	
Project Manager		Project Start Date	10-Feb-09
Project Owner/Sponsor		Target End Date	
Sector/Bus Unit		Version	

OK	WBS#	Task Description	Resource	Start Date	Due Date	Duration	Actual Finish	% Complete	Predecessor
		Define							
		Assign Project Manager		10-Feb-09	10-Feb-09	1			
		Pre Kick-off planning		10-Feb-09	10-Feb-09	1			
		Complete Project Charter		10-Feb-09	10-Feb-09	1			
		Identify key stakeholders		10-Feb-09	10-Feb-09	1			
		Set up project management infrastructure		10-Feb-09	10-Feb-09	1			
		Review documentation		10-Feb-09	10-Feb-09	1			
		Plan							
		Develop project organisation and establish suppliers		10-Feb-09	10-Feb-09	1			
		Conduct kick-off meeting		10-Feb-09	10-Feb-09	1			

FIGURE 4–36. Project plan/schedule template.

4.28 CONVERGENT COMPUTING

It is extremely difficult, if not impossible, for information systems consulting organizations to survive in today's business environment without a successful project management methodology. Colin Spence, Project Manager/Partner at Convergent Computing (CCO), describes the methodology in his organization:

Region AMERICAS
Sub-Region Africa
Sector Auto
Country Code USA
Project Number UNIQUE NO.

Project Status Report
v20070919

DePICT® Project Management

Project Team & Key Dates			
Project Title	Enter Title Here		
Sponsor		Owner	
Project Manager		Account Manager	
Customer		Project Type	
Go-Live Date	1-Jan-2006	Report Month	April
Project Phase	Define (P1)		

Status Highlights					
OVERALL PROJECT STATUS	Amber			↔	
Sponsorship	Amber	↔	Budget	Amber	↔
Schedule	Amber	↔	Risk	Amber	↔
Scope	Amber	↔	Resource	Amber	↔

Project Description

Items causing a Non-Green Project Status

Milestone (Bold = Critical Path)	Planned Date	Revised Date	Actual Date	Status
(List milestones on critical path in **Bold**)				
(List "spotlight" milestones)				

Issues Management	Total Issues	Total Red Issues
		0

UNIQUE NO. 1 of 2 2/10/2009 12:05 PM

Region AMERICAS
Sub-Region Africa
Sector Auto
Country Code USA
Project Number UNIQUE NO.

Project Status Report
v20070919

DePICT® Project Management

Major Accomplishments This Period

Risk Management	Total Risks	Total Red Risks
	0	0

Plans For Next Period

Other Activities and Awareness items

Functional Group/Area Status			
Legal	Amber	Operations	Amber
Commercial Contract Management	Amber	Finance	Amber
Procurement	Amber	Solution Design	Amber
Internal IT	Amber	Health, Safety and Environmental	Amber
Human Resources	Amber	Customer IT	Amber
Real Estate	Amber	Other	Amber

Comments:

Project Budget (,000)					
Currency	USD	Original	Revised	Actual	Estimated At Completion
Capital		0	0	0	0
Expense		0	0	0	0
Total Spend		0	0	0	0

Comments:

UNIQUE NO. 2 of 2 2/10/2009 12:05 PM

FIGURE 4–37. Project status report template.

Risk Assessment

Project Name/ID#:	**[Project Name/ID#]**
Project Manager:	[Project Manager]
Project Owner/Sponsor:	[Project Owner/Sponsor]
Customer Name:	[Customer Name]
Origin Date & By Whom:	
Version Date:	
Probability Ratings:	3 = High = 75-100% occurrence 2 = Med = 40-75% occurrence 1 = Low = 0-40% occurrence
Impact Ratings:	4 = Critical = Failure to overall project 3 = High = Significant impact on cost or schedule 2 = Med = Moderate impact on cost or schedule 1 = Low = Minor impact on cost or schedule

Risk #:	1	Risk Owner:	
Probability:		Impact:	Severity:
Risk Name:			Function/Category:

Risk Description:

Impacts (where possible, quantify the impact):	Budget	Schedule	Customer

Probable Causes:

Preventative Steps & Actions:	Owner	Date Due	Date Completed

Contingency Plan:	Contingency Triggers:

Comments/Outcome:

FIGURE 4–38. Project risk assessment template.

Risk Log													
Project Name/ID#:	[Project Name/ID#]			Customer Name				[Customer Name]					
Project Manager:	[Project Manager]			Project Owner/ Sponsor:				[Project Owner/Sponsor]					
Total # of Risks	0 Total Open			Version Date:									
Total # of Red Risks	0 Red												

#	Function/ Category	Risk Name	Impact (Budget, Schedule, Customer)	Probability	Impact	Severity	Risk Response Plan		Completed	Trigger/ Due Date	Owner	Date Closed
							Preventive Action	Contingency Plan				
1	0	0		0	0						0	
2	0	0		0	0						0	
3	0	0		0	0						0	
4	0	0		0	0						0	
5	0	0		0	0						0	
6	0	0		0	0						0	
7	0	0		0	0						0	
8	0	0		0	0						0	
9	0	0		0	0						0	
10	0	0		0	0						0	
11	0	0		0	0						0	
12	0	0		0	0						0	
13	0	0		0	0						0	

FIGURE 4–39. Project risk log template.

A project manager is to be assigned to all new scoped projects and will work closely with the team members assigned to the project. A project manager's primary responsibility is to ensure that CCO is executing projects on time, on budget, and within scope to fulfill the obligations outlined in the proposal or scope of work. To this end, the project manager should provide oversight and strategic guidance through the services delivery process.

If the client does not approve funds for a project manager, the technical lead on the project will be expected to fulfill this role and receive assistance "behind the scenes" from a company project manager.

Project management involvement should normally start with approval of the proposal or scope of work (SOW) document(s) as these will be the primary documents that the project manager will use to guide the activities of the resources and to ensure the success of the project and satisfaction of the client. The objectives of the project, SOW, major activities, deliverables, and roles and responsibilities will be monitored as well as overall team effectiveness and progress.

During the project, all primary client deliverables (such as statements of work, consulting documents, project plans, budget reports, as-builts, etc.) should be run through the project manager to ensure our deliverables are consistent with project objectives and of the highest quality.

The project manager should be an escalation point for scope changes or any project flags and will keep the account manager involved in all scope change requests. The project manager will also provide updates on budget and timelines as needed and schedule regular customer satisfaction check point meetings.

A CCO project (using the methodology) is approached as having six distinct steps, as follows:

1. *Project Definition.* This step is designed to allow CCO to determine the scope of the project and sell the client on CCO's abilities. This is typically a "brainstorming" session, with the goal of gathering information rather than CCO recommending a solution.

2. *Internal Strategy.* This step allows CCO to validate the information gathered and craft a strategy for proposing our services. Time is allocated for the CCO team to strategize on the best solution to the client's needs, and a draft of the proposal or statement of work is created.

3. *Proposal.* This step allows CCO to ensure alignment with the client in the areas of budget, roles and responsibilities, goals, and objectives of the project and create the proposal. The draft SOW is delivered/presented to the client and then finalized based on client input. At this point the budget is solidified and the timeline discussed.

4. *Internal Readiness.* This step ensures that all of the internal administrative paperwork is done and that an internal work definition is created if the proposal or SOW does not provide enough information. This phase also ensures that all resources involved in the project understand the SOW to be performed, timeline, and budget.

5. *Project Services Delivery.* The details of this step will vary based on the size and nature of the project and may be split into subphases. Essentially, during this phase, the services are delivered as outlined in the proposal/SOW. There are typically internal team checkpoint meetings as well as customer satisfaction checkpoints with the customer. All deliverables, whether technology implementations, hardware configurations, training services, or documentation, are reviewed by the consultant in charge of the project and the project manager involved with the project.

6. *Sign-Off and Customer Satisfaction.* The final step involves getting the client sign-off indicating the project is complete in their eyes and gathering customer satisfaction information to help CCO improve its services and to set a tone that CCO is committed to high-quality delivery and highly satisfied clients.

There is also a business development step that is typically referred to as step 0 which is required for new clients, to "open the door" and allow CCO the opportunity to meet with the client, which is referred to as step 1.

Note that the activities, resources involved, and level of effort are commensurate with the complexity, strategic importance, and size of the project. So the duration and number of resources involved in each project will vary depending on who the customer is and the nature and complexity of the project.

Figure 4–40 is a flowchart and checklist that outlines the project life cycle created by Colin Spence. The project manager assigned to the project should use this list to ensure that no steps are missed, so check boxes are provided on the form.

This process of deciding upon the number of life-cycle phases involved many discussions and meetings between the managers. While the final structure seems to deemphasize the actual on-site work that needs to be done, it ensures that the project is well defined, that the right team of people is assembled, and that once the work is done, attention is paid to ensuring that the client is satisfied. CCO management had found through experience that too often resources were being assigned to tasks without a good enough definition of the needs of the client and the exact deliverables the client was expecting. Additionally, there

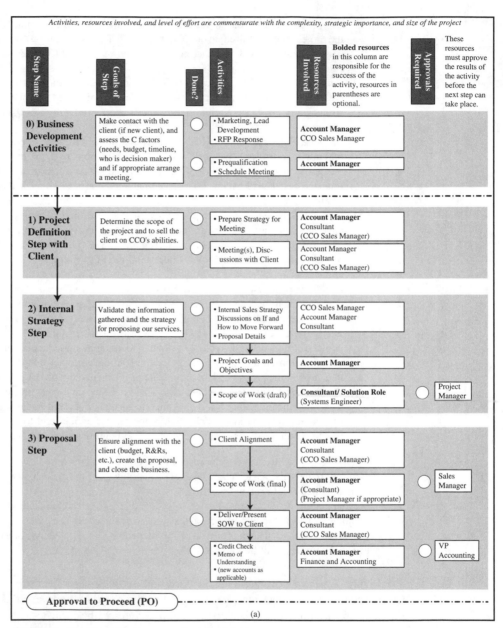

FIGURE 4–40. Project management methodology.

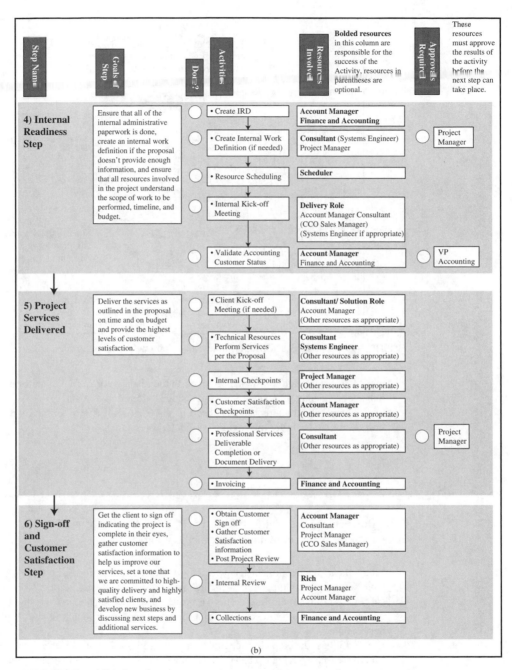

FIGURE 4–40. (*Continued*)

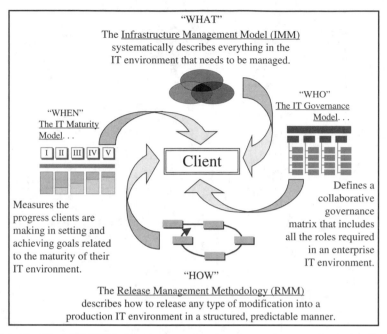

FIGURE 4–41. The EDS ITEM: "An integrated approach to an integrated solution."

was often not enough assistance provided to the resource or resources in terms of project management support and customer satisfaction suffered as a result.

This phase structure has now been in place close to five years and has drastically improved CCO's ability to provide top-notch professional services.

4.29 EDS

Doug Bolzman, Consultant Architect, PMP®, ITIL Service Manager at EDS, discusses the foundation for a project management methodology:

> Many clients have a project management methodology and, in addition, most companies have several methodologies that include project management disciplines. EDS has been successful in integrating the definitive project management method into the other methodologies to allow for leveraging of project management standards and tools without duplication.
>
> EDS has developed a client facing framework called Information Technology Enterprise Management (ITEM). This framework assists clients in mapping their strategic direction into feasible releases. ITEM is a preintegrated framework of three models and a methodology as illustrated in Figure 4–41.
>
> The release management methodology consists of four stages (planning, integration, deployment, and operations). Each stage consists of phases, activities, or tasks, as shown in Table 4–4.

TABLE 4–4. RELEASE METHODOLOGY STAGES

Stage	Description
Planning	The environment that is used to establish and manage the vision and strategic direction of the enterprise IT environment and proactively define the content and schedule of all IT releases. To provide a common means for the client and service providers to clearly and accurately plan the enterprise IT environment and manage all aspects of planning, estimating a release, and setting appropriate client expectations as to what each release will deliver.
Integration	The environment that is used to finalize the design of a planned infrastructure release and perform all of the required testing and client validation, preparing the release to be deployed to the user community. To provide a common means for the client, EDS, and service providers to clearly and accurately validate the accuracy, security, and content of each release and to finalize all development. To provide the client a clear and accurate portrait of the outcome of the release and to set proper expectations of the deployment and operations activities, costs, and schedules.
Deployment	The process that is used by the organization to implement new releases of the enterprise IT design (business, support, and technical components) to a target environment. To provide a common means for the client, EDS, and service providers to clearly and accurately schedule, deploy, and turn over to production the updated environment.
Operations	The production environment that is used to sustain and maintain the IT components and configurable items that are part of the enterprise IT environment. To provide a stable IT environment that is required by the IT users to support their business roles and responsibilities.

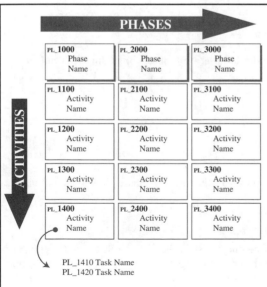

Description:
The sequential hierarchy of events should there be no need to deviate. This demonstrates the highest percentage of occurrence when the process is executed. Main intent for this chart is to expose the tasks subordinate to each activity.

Value:
- Each work element can be identified and mapped internal or external to the component
- All activities divided into tasks (justified)
- All work is represented once (even if executed many times)

Structure:
- All work elements have a unique identifier
  - Phase numbered by the thousands
  - Activity numbered by the thousands
  - Task numbered by the tens
- All work elements preceded by Component Name Identifier (PL_1000; PL_1100)
- Activity and Task Names kept relatively short; descriptive of work complete, not sentences
- All tasks make up the scope of the activity
- This is NOT a relationship model

FIGURE 4–42. Activity phase mapping.

In deciding upon the number of life-cycle phases we used the 9 × 9 rule, as shown in Figure 4–42. If there are more than 9 phases across a stage and more than 9 activities to each phase (81 total units of work), then the scope will become too large, and our numbering scheme will become unstable.

4.30 DTE ENERGY

Henry Campbell, Principal Analyst, Customer Service Project Management Office and Time Menke, PMP, Senior Continuous Improvement Expert, DO Performance Management at DTE Energy, discusses the growth of project management at DTE Energy:

> DTE Energy has created a common approach to ensure projects are implemented in a consistent manner. In the past, multiple project management methodologies were used throughout the enterprise. Several business units (specifically, the Customer Service (CS), Distribution Operations (DO), Gas Operations (GO), Operating Systems Strategy Group (OSSG), and Information Technology Services (ITS) groups) agreed that a "common" approach was needed.
>
> The "CI Project Portfolio Management Process" (Figure 4–43) was developed utilizing internal and external best practices. This approach was developed to provide a common framework for Continuous Improvement projects, however other types of projects (capital, ad-hoc, etc.) have also benefited from its structure. The process contains seven project management phases:
>
> I. Identification
>
> Potential projects are identified from a variety of sources including (but not limited to):
>
> • Corrective Action Requests
> • After Action Reviews
> • Benchmarking
> • Swarms
>
> Projects deemed "Just Do It (JDI)" move past the Selection phase, directly into Initiation.
>
> II. Selection
>
> Potential projects are then ranked according to established Project Selection Criteria, business cases are created if applicable, the project is registered with the appropriate organization and work is started on the project charter.
>
> III. Initiation

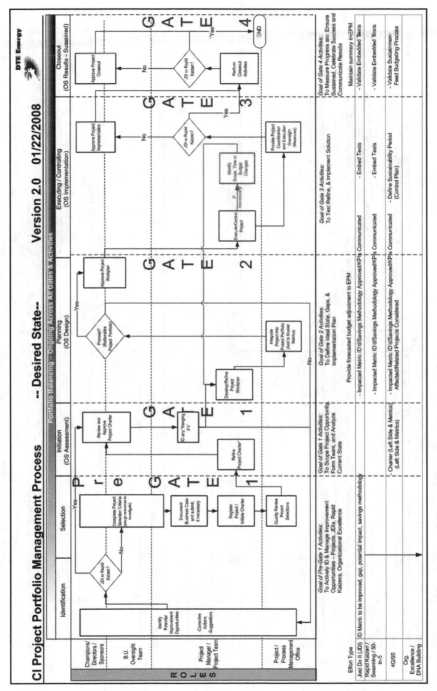

FIGURE 4-43. CI project portfolio management process.

PROJECT CHARTER: Version

Date:

Project Information and Gate Sign-Off

Gate One through Four Sign Off- Initial and Date at Review

	Assess		Design		Implement		Results	
	G1	Date	G2	Date	G3	Date	G4	Date
Project Name								
Team Lead								
Champion								
Process Owner								

Business Impact Analysis Reviewed- G1	Yes	No

Current State- G1

Problem Statement-G1

Future State- G2

Metrics	Baseline- G1	Target- G2	Current- G3	Final - G4	% Time

Gaps- G2

Team Members- G1	Role	
	SME	
	Core Team	

Four Milestone Dates- G1

Planned Date-	Target Date	Rev. 1	Rev. 2	Rev. 3
Gate One	Assessment			
Gate Two	Design			
Gate Three	Implementation			
Gate Four	Results			

Results Summary- G4

Cost and Resource Assessment Reviewed- G1	Yes	No

In Scope Out of Scope- G1

FIGURE 4-44. Project charter template.

The Initiation phase includes an assessment of the current state of business and why a change is needed. The Project Charter is refined and a decision to proceed is made. (Figure 4–44).

The Project Charter contains:

- Project Description and Purpose
- Team Members and Roles
- Timeline
- Case for Change
- Assumptions, Risks and Challenges
- Current State Identification
- Ideal State Identification
- Gaps Assessment (Current and Ideal States)
- Major Milestones
- Metrics (Baseline and Target)

IV. Planning

The Planning phase establishes the desired state of the change with an emphasis on process, technical and cultural (re)design. Project managers are given several tools to assist with the overall project planning. One of the tools is the "Project Plan Template" (Figure 4–45). The template contains a standard planning methodology for Customer Service Projects. It also provides project managers with standard lag and lead times for certain resource intensive tasks. Each project manager customizes the template respective to their initiative.

The Project Plan Template contains:

- Key Project Activities
- Project Plan Elements (Common Activities, Start and Finish Dates, and Assigned Resources)
- High Level Milestones

V/VI. Execution / Controlling

The actual process and procedural changes, technical coding, training, and necessary "course corrections" take place during the "Executing and Controlling" phases. If a project change is required that is related to scope, time, or financial resources, the project manager completes a "Change Request" (Figure 4–46) form. In addition, the project manager provides a bi-weekly status, used to populate the "Project Performance Board" (Figure 4–47).

The Project Performance Board tracks:

- Project Information
- Project Name and ID
- Project Manager Name
- Percent Complete
- Phase Completion (Gate Approvals)
- Issues

Activity ID	Activity Name	Performance % Complete	Start	Finish	Activity Owner
CSPMv3 Project Plan Template		0%	27-Jan-..	27-Jan-..	
CSPMv3.0 Benefits		0%	27-Jan-..	27-Jan-..	
CSPMv3 Gate 1 Initiation / Assessment		0%	27-Jan-..	27-Jan-..	
A3050	Review Project Management Process and Methodology	0%	27-Jan-..	27-Jan-..	
A1030	Identify Project Scope	0%	27-Jan-..	27-Jan-..	
A1040	Secure Project Resources / Form Team	0%	27-Jan-..	27-Jan-..	
A1050	Assess Current Processes	0%	27-Jan-..	27-Jan-..	
A1070	Milestone: Gate 1 Approval	0%		27-Jan-..	
A1005	Project Charter	0%	27-Jan-..	27-Jan-..	
CSPMv3 Gate 2 Planning - Design		0%	27-Jan-..	27-Jan-..	
A1080	Complete Change Management Form	0%	27-Jan-..	27-Jan-..	
A1090	Develop Project Workplan	0%	27-Jan-..	27-Jan-..	
A1170	Milestone: Gate 2 Approval	0%		27-Jan-..	
CSPMv3 Gate 2.2 Process		0%	27-Jan-..	27-Jan-..	
A1110	Define Desired State of Processes	0%	27-Jan-..	27-Jan-..	
A1120	Conduct Gap Assessment	0%	27-Jan-..	27-Jan-..	
A1130	Develop Countermeaures / Prespecifications / Embedded Tests	0%	27-Jan-..	27-Jan-..	
CSPMv3 Gate 2.1 Technology		0%	27-Jan-..	27-Jan-..	
A1140	Develop Technical Requirements	0%	27-Jan-..	27-Jan-..	
A1150	Review Technical Requirements with IT	0%	27-Jan-..	27-Jan-..	
A1160	Design (IT)	0%	27-Jan-..	27-Jan-..	
CSPMv3 Gate 3 Execution / Implementation		0%	27-Jan-..	27-Jan-..	
A1480	Conduct After Action Review	0%	27-Jan-..	27-Jan-..	
A1490	Milestone: Gate 3 Approval	0%		27-Jan-..	
CSPMv3 Gate 3.5 Training / Communications		0%	27-Jan-..	27-Jan-..	
A3060	Implement Communication Plan	0%	27-Jan-..	27-Jan-..	
CSPMv3 Gate 3.5.1 Develop Training and Communications Plan		0%	27-Jan-..	27-Jan-..	
A3180	Content	0%	27-Jan-..	27-Jan-..	
A3190	Media(s)	0%	27-Jan-..	27-Jan-..	
A3200	Schedule	0%	27-Jan-..	27-Jan-..	
A3210	Logistics	0%	27-Jan-..	27-Jan-..	
CSPMv3 Gate 3.5.2 Organization Rollout of Training		0%	27-Jan-..	27-Jan-..	
A3070	Revenue Management & Protection	0%	27-Jan-..	27-Jan-..	
A3080	Field Operations	0%	27-Jan-..	27-Jan-..	
A3090	Billing	0%	27-Jan-..	27-Jan-..	
A3100	Data Acquisition	0%	27-Jan-..	27-Jan-..	
A3110	Consumer Affairs	0%	27-Jan-..	27-Jan-..	
A3120	Gas Operations	0%	27-Jan-..	27-Jan-..	
A3130	Other Organizations	0%	27-Jan-..	27-Jan-..	
CSPMv3 Gate 3.5.2.1 Call Center		0%	27-Jan-..	27-Jan-..	
A3140	Notify the CCC Board of Project Scope / Hi Level Training Requirements	0%	27-Jan-..	27-Jan-..	
A3150	Arrange for Project SME's to provide floor support (if necessary)	0%	27-Jan-..	27-Jan-..	
A3160	Conduct Training (or communicate) for Non Customer Reps	0%	27-Jan-..	27-Jan-..	
A3170	Conduct Training for Customer Reps	0%	27-Jan-..	27-Jan-..	
CSPMv3 Gate 3.1 Process		0%	27-Jan-..	27-Jan-..	
A1180	Develop / Modify Processes	0%	27-Jan-..	27-Jan-..	
A1190	Develop/Modify Procedures (SWIs)	0%	27-Jan-..	27-Jan-..	
A1200	Develop Implementation Plan	0%	27-Jan-..	27-Jan-..	
CSPMv3 Gate 3.2 Technology		0%	27-Jan-..	27-Jan-..	
A1220	Development (IT)	0%	27-Jan-..	27-Jan-..	
A1230	System Testing (IT)	0%	27-Jan-..	27-Jan-..	
A1240	Conduct User Acceptance Testing (UAT	0%	27-Jan-..	27-Jan-..	
A1235	Develop UAT Test Plan (Scenarios)	0%	27-Jan-..	27-Jan-..	
A1241	Report any UAT defects	0%	27-Jan-..	27-Jan-..	
A1242	Re-test fixed defects	0%	27-Jan-..	27-Jan-..	
CSPMv3 Gate 3.4 Conduct Pilot (Placeholder, if necessary)		0%	27-Jan-..	27-Jan-..	
A1430	Design	0%	27-Jan-..	27-Jan-..	
A1440	Develop	0%	27-Jan-..	27-Jan-..	
A1450	Implement	0%	27-Jan-..	27-Jan-..	
A1470	Modify Processes (Implement pilot learnings)	0%	27-Jan-..	27-Jan-..	
CSPMv3 Gate 4 Closeout / Results		0%	27-Jan-..	27-Jan-..	
A1500	Develop Sustainability Plan (if necessary)	0%	27-Jan-..	27-Jan-..	
A1510	Conduct Project After Action Review	0%	27-Jan-..	27-Jan-..	
A1520	Monitor Implementation Results	0%	27-Jan-..	27-Jan-..	
A1540	Make adjustments to implementation (if necessary)	0%	27-Jan-..	27-Jan-..	
A1550	Acknowledge and Release Project Resources	0%	27-Jan-..	27-Jan-..	
A1560	Prepare Project Closeout Document	0%	27-Jan-..	27-Jan-..	
A1580	Milestone: Gate 4 Approval / Project Complete	0%		27-Jan-..	

FIGURE 4–45. Project plan template.

Customer Service Program Management   **Project Change Template**

| Request Date | <date> | Organization | |

| Project Manager | <pm name> | Project Number | <date> |
| Project Name | <project name> | Director/Sponsor | <date> |

| Priority | High ☐ Medium ☐ Low ☐ | Business Case Impacted? | Yes ☐ No ☐ |
| Re-Baseline Required? | Yes ☐ No ☐ | | |

Describe How this Change Impacts the Project

o   Include affected budget IT dollars

o   State as change amount and change percent, if possible.

For example: Project cost increased by x dollars which represents y percent of approved project cost.

Scope: ☐ _____

Time: ☐ _____

Cost: ☐ _____

Other: ☐ _____

Benefit of Change Request

Tracking Information—To be completed by Project Management Office

Additional Impact Analysis Required: Yes ☐ No ☐	Review Date:
	Reviewer:
Issues/Concerns with Change Request:	

Sponsor Approval – To be completed by Project Management Office

Approved ☐ Rejected ☐	Approval Date:
Reason:	
Attachments: *Please attach: (a) revised cost estimate; (b) proposed schedule, as needed.*	

APPROVALS

_____   _____
Project Sponsor   Director of CSPMO

_____
Project Manager

Version 2   Page 1 of 1

FIGURE 4–46.   Project change template.

Customer Service Project Management

Project Scorecard

	Project Name / Contact	Start Date / End Date	% Complete	Gate / Phase*	Description	Comments	Project Type	2009 Savings: Target	2009 Savings: Actual	Full Benefit Due Date	Assigned Resources
Customer Service Program Management											
Customer Billing											
Data Acquisition											
Customer Care Center											
Consumer Affairs											

Cancelled: Project was Cancelled
Deferred: The Project has been Postponed until
Planned: Charter in progress
Active: Project is currently in progress
Implemented: Project has completed Gate 3

***OSSG 4 Gate = PMI Phase**
Gate 1 Assessment = Initiation
Gate 2 Design = Planning
Gate 3 Implementation = Execution
Gate 4 Sustain = Closeout

FIGURE 4–47. Project performance board.

VII. Closeout

During the "Closeout" phase, the project manager qualifies the actual project deliverables against the plan. The project is formally closed when the desired targets are attained. The project manager completes the "Project Closeout Report" as part of the closeout process.
The Closeout Report contains:

- Project Summary
- Objectives Achieved
- Objectives Not Met
- Measures/Metrics
- Lessons Learned
- Further Actions

Every phase of the Common Project Management Process ends with a "Gate Approval" from the project sponsor. If approval is granted, the project formally moves to the next phase. Where appropriate, each business unit agrees to implement projects utilizing this common approach. While the process is the same across DTE Energy, the project management procedures are tailored to the needs of each specific department or organization.

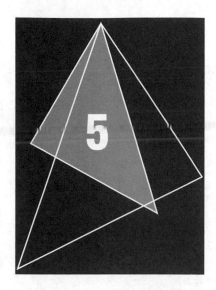

Integrated Processes

5.0 INTRODUCTION

Companies that have become extremely successful in project management have done so by performing strategic planning for project management. These companies are not happy with just matching the competition. Instead, they opt to exceed the performance of their competitors. To do this on a continuous basis requires processes and methodologies that promote continuous rather than sporadic success.

Figure 5–1 identifies the hexagon of excellence. The six components identified in the hexagon of excellence are the areas where the companies excellent in project management exceed their competitors. Each of these six areas is discussed in Chapters 5–10. We begin with *integrated processes*.

FIGURE 5–1. Six components of excellence. *Source:* Reprinted from H. Kerzner: *Excellence in Project Management,* Wiley, New York, 1998, p. 14.

5.1 UNDERSTANDING INTEGRATED MANAGEMENT PROCESSES _____

As we discussed in Chapter 1, several new management processes since 1985 (e.g., concurrent engineering) have supported the acceptance of project management. The most important complementary management processes and the years they were introduced are listed below:

- 1985: Total quality management (TQM)
- 1990: Concurrent engineering
- 1992: Employee empowerment and self-directed teams
- 1993: Reengineering
- 1994: Life-cycle costing
- 1995: Change management
- 1996: Risk management
- 1997–1998: Project offices and centers of excellence
- 1999: Colocated teams
- 2000: Multinational teams
- 2001: Maturity models
- 2002: Strategic planning for project management
- 2003: Intranet status reporting
- 2004: Capacity-planning models
- 2005: Six Sigma integration with project management
- 2006: Virtual project management teams
- 2007: Lean/agile project management
- 2008: Best practices libraries
- 2009: Project management methodologies
- 2010: Project management business process certification

The *integration* of project management with these other management processes is key to achieving sustainable excellence. Not every company uses every process all the time. Companies choose the processes that work the best for them. However, whichever processes are selected, they are combined and integrated into the project management methodology. Previously we stated that companies with world-class methodologies try to employ a single, standard methodology based upon integrated processes. This includes business processes as well as project management–related processes.

The ability to integrate processes is based on which processes the company decides to implement. For example, if a company implemented a stage gate model for project management, the company might find it an easy task to integrate new processes such as concurrent engineering. The only precondition would be that the new processes were not treated as independent functions but were designed from the onset to be part of a project management system already in place. The four-phase model used by the General Motors Powertrain Group and the PROPS model used at Ericsson Telecom AB readily allow for the assimilation of additional business and management processes.

Previously, we stated that project managers today are viewed as managing part of a business rather than just a project. Therefore, project managers must understand the business and the processes to support the business as well as the processes to support the project. Companies such as Visteon and Johnson Controls understand this quite well and have integrated business processes either into or with their project management methodology.

This chapter discusses each of the management processes listed and how the processes enhance project management. Then we look at how some of the integrated management processes have succeeded using actual case studies.

5.2 EVOLUTION OF COMPLEMENTARY PROJECT MANAGEMENT PROCESSES

Since 1985, several new management processes have evolved parallel to project management. Of these processes, TQM and concurrent engineering are the most relevant. Companies that reach excellence are the quickest to recognize the synergy among the many management options available today. Companies that reach maturity and excellence the quickest are those that recognize that certain processes feed on one another. As an example, consider the seven points listed below. Are these seven concepts part of a project management methodology?

- Teamwork
- Strategic integration
- Continuous improvement
- Respect for people
- Customer focus
- Management by fact
- Structured problem solving

These seven concepts are actually the basis of Sprint's TQM process. They could just as easily have been facets of a project management methodology.

During the 1990s, Kodak taught a course entitled Quality Leadership. The five principles of Kodak's quality leadership program included:

Customer focus	"We will focus on our customers, both internal and external, whose inputs drive the design of products and services. The quality of our products and services is determined solely by these customers."
Management leadership	"We will demonstrate, at all levels, visible leadership in managing by these principles."
Teamwork	"We will work together, combining our ideas and skills to improve the quality of our work. We will reinforce and reward quality improvement contributions."
Analytical approach	"We will use statistical methods to control and improve our processes. Data-based analyses will direct our decisions."
Continuous improvement	"We will actively pursue quality improvement through a continuous cycle that focuses on planning, implementing, and verifying of improvements in key processes."

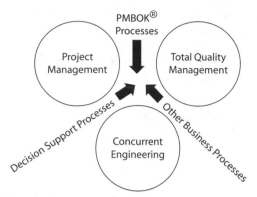

FIGURE 5–2. Totally uncoupled processes.

Had we just looked at the left-hand column we could argue that these are the principles of project management as well.

Figure 5–2 shows what happens when an organization does not integrate its processes. The result is totally uncoupled processes. Companies with separate methodologies for each process may end up with duplication of effort, possibly duplication of resources, and even duplication of facilities. Although there are several processes in Figure 5–2, we will focus on project management, TQM, and concurrent engineering only.

As companies begin recognizing the synergistic effects of putting several of these processes under a single methodology, the first two processes to become partially coupled are project management and TQM, as shown in Figure 5–3. As the benefits of synergy and integration become apparent, organizations choose to integrate all of these processes, as shown in Figure 5–4.

Excellent companies are able to recognize the need for new processes and integrate them quickly into existing management structures. During the early 1990s, integrating project management with TQM and concurrent engineering was emphasized. Since the middle 1990s, two other processes have become important in addition: risk management and change management. Neither of these processes is new; it's the emphasis that's new.

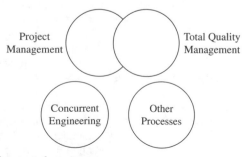

FIGURE 5–3. Partially integrated processes.

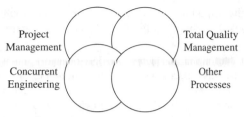

FIGURE 5–4. Totally integrated processes.

During the late 1990s, Steve Gregerson, formerly Vice President for Product Development at Metzeler Automotive Profile System, described the integrated processes in its methodology[1]:

> Our organization has developed a standard methodology based on global best practices within our organization and on customer requirements and expectations. This methodology also meets the requirements of ISO 9000. Our process incorporates seven gateways that require specific deliverables listed on a single sheet of paper. Some of these deliverables have a procedure and in many cases a defined format. These guidelines, checklists, forms, and procedures are the backbone of our project management structure and also serve to capture lessons learned for the next program. This methodology is incorporated into all aspects of our business systems, including risk management, concurrent engineering, advanced quality planning, feasibility analysis, design review process, and so on.

Clearly, Metzeler sees the integration and compatibility of project management systems and business systems. Another example of integrated processes is the methodology employed by Nortel. During the late 1990s, Bob Mansbridge, then Vice President, Supply Chain Management at Nortel Networks, believed[2]:

> Nortel Networks project management is integrated with the supply chain. Project management's role in managing projects is now well understood as a series of integrated processes within the total supply chain pipeline. Total quality management (TQM) in Nortel Networks is defined by pipeline metrics. These metrics have resulted from customer and external views of "best-in-class" achievements. These metrics are layered and provide connected indicators to both the executive and the working levels. The project manager's role is to work with all areas of the supply chain and to optimize the results to the benefit of the project at hand. With a standard process implemented globally, including the monthly review of pipeline metrics by project management and business units, the implementation of "best practices" becomes more controlled, measurable, and meaningful.

---

1. H. Kerzner, *Advanced Project Management: Best Practices on Implementation*, Wiley, Hoboken, NJ, 2000, p. 188.
2. Ibid.

The importance of integrating risk management is finally being recognized. According to Frank T. Anbari, Professor of Project Management, Drexel University:

> By definition, projects are risky endeavors. They aim to create new and unique products, services, and processes that did not exist in the past. Therefore, careful management of project risk is imperative to repeatable success. Quantitative methods play an important role in risk management. There is no substitute for profound knowledge of these tools.

Risk management has been a primary focus among health care organizations for decades, for obvious reasons, as well as financial institutions and the legal profession. Today, in organizations of all kinds, risk management keeps us from pushing our problems downstream in the hope of finding an easy solution later on or of the problem simply going away by itself. Change management as a complement to project management is used to control the adverse effects of scope creep: increased costs (sometimes double or triple the original budget) and delayed schedules. With change management processes in place as part of the overall project management system, changes in the scope of the original project can be treated as separate projects or subprojects so that the objectives of the original project are not lost.

Today, excellent companies integrate five main management processes (see Figure 5–5):

● Project management
● Total quality management
● Risk management
● Concurrent engineering
● Change management

Self-managed work teams, employee empowerment, reengineering, and life-cycle costing are also combined with project management in some companies. We briefly discuss these less widely used processes after we have discussed the more commonly used ones.

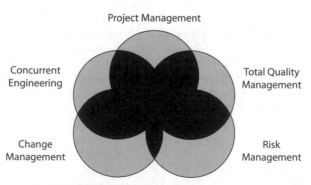

FIGURE 5–5. Integrated processes for twenty-first century.

5.3 ZURICH AMERICA INSURANCE COMPANY[3]

One of the benefits of having integrated processes is that it allows for more comprehensive and realistic contingency planning, Kathleen Cavanaugh states:

As we know, simply put, the goal of all PMOs is to deliver projects on time and on budget. There is increased scrutiny placed on projects these days and many companies have enacted a protective governance gauntlet to help ensure the right projects come to fruition. With the time to market push/shove, it is easy for projects budgets and end dates to be estimated, even "promised" too soon much to the unease of the Project Manager. To help alleviate this potential heartbreak, the IT PMO at Zurich American Insurance Company has implemented a project contingency process for both duration and dollars.

The contingency process helps to mitigate the risk of known unknowns within the scope of a project. Since we are part of the global Zurich Financial Services Group we have a rather strict governance process that takes time and money to navigate. Once you make it through the gauntlet you don't really want to go back for re-approvals, date extension authorizations, etc., so, proper planning for risks and project changes is imperative.

The contingency process uses a determination matrix that considers particular risk factors such as resources and technology complexity just to name a couple. It is designed to help the PM assign the appropriate amount of contingency needed for both dollars and duration. The concept is nothing new but is still not widely accepted as a necessity.

The goal is not only to protect us from time consuming project governance submissions, but also to move away from padding individual estimates. Before this contingency approach was introduced, some Project Managers buried contingency within their estimates. The major disadvantage of this is that because the contingency is "hidden," there is no systematic process to release funds back into the project funding pool as risks lessen throughout the project. Now, contingency is kept separate in the project plan so that we can have better insight into where and why original estimates were off. Only then can we find the root cause and improve upon the estimating approach.

One important thing to note is that the customer takes part in determining the need for contingency so they have a good understanding of why it is so critical to the success of the project. The process makes contingency transparent and once the customer understands that contingency dollars cannot be used without their acknowledgement, they are more open to understanding the inherent changes that occur during the life of a project.

Contingency should be actively managed as the project progresses. Each month when project risks are re-assessed and risk probability decreases, the amount of contingency should be adjusted accordingly in both budget and schedule. It is expected that contingency would be released from the project if it is determined to no longer be needed based on the updated risk assessment. This allows money to be available for other efforts in the company.

To summarize, the bullet points below are an outline of the steps taken to effectively use the contingency process.

Plan: The IT Project Manager works with the Business Project Manager and project team using the determination matrix to calculate the appropriate contingency percent when the project is ready to go for full funding.

---

3. Material provided by Kathleen Cavanaugh, PMP, Zurich—IT PMO Consultant.

Document/Communicate: PM updates the project plan and Usage Log to document and communicate the contingency figures.

Approve: The Sponsor is responsible for providing sign-off before contingency is used.

Manage: The Project Manager maintains the Usage Log as contingency is consumed and other contingency data is updated each month along with the risk assessment.

Release: Contingency funds are released back into project funding pool as risks decrease.

Overall, this process addresses the age old problem of project work starting before everything that needs to be known is known and gives PMs a fighting chance to deliver on-time and on-budget in this ever changing environment. Because after all, change happens.

5.4 CONVERGENT COMPUTING

The importance of integrated processes, especially quality, has become part of all project management methodologies. According to Colin Spence, Project Manager/Partner at Convergent Computing (CCO):

There are steps in the life cycle that ensure these processes are addressed throughout the delivery process. Quality is addressed throughout the life cycle, starting with the attention from the team in crafting a strategy for services delivery in phase 2 and then preparing the statement of work (SOW) document (which is reviewed in draft form with the client in phase 2 and then finalized in phase 3). The SOW contains a section that focuses on any risks to the project and on any special change control processes to be put in place.

Whenever a deliverable is to be provided to the client (e.g., a configured server or deliverable document), the consultant must review it and the project manager must approve it before it can be considered complete. Internal checkpoints are also scheduled by the project manager to allow the team to get together and review progress with the client. This is a critical element in the change control process, and the project manager typically reports project status and any changes required to the client after these meetings.

Typically the success of these additional processes is reviewed in the final customer satisfaction meeting, and the client stakeholders are asked for specific feedback on the success or failure of these processes.

5.5 TOTAL QUALITY MANAGEMENT

During the past decade, the concept of TQM has revolutionized the operations and manufacturing functions of many companies. Companies have learned quickly that project management principles and systems can be used to support and administer TQM programs and vice versa. Ultimately excellent companies have completely integrated the two complementary systems.

The emphasis in TQM is on addressing quality issues in total systems. Quality, however, is never an end goal. Total quality management systems run continuously and concurrently in every area in which a company does business. Their goal is to bring to market products of better and better quality and not just of the same quality as last year or the year before.

Total quality management was founded on the principles advocated by W. Edwards Deming, Joseph M. Juran, and Phillip B. Crosby. Deming is famous for his role in turning postwar Japan into a dominant force in the world economy. Total quality management processes are based on Deming's simple plan–do–check–act cycle.

The cycle fits completely with project management principles. To fulfill the goals of any project, first you plan what you're going to do, then you do it. Next, you check on what you did. You fix what didn't work, and then you execute what you set out to do. But the cycle doesn't end with the output. Deming's cycle works as a continuous-improvement system, too. When the project is complete, you examine the lessons learned in its planning and execution. Then you incorporate those lessons into the process and begin the plan–do–check–act cycle all over again on a new project.

Total quality management also is based on three other important elements: customer focus, process thinking, and variation reduction. Does that remind you of project management principles? It should. The plan–do–check–act cycle can be used to identify, validate, and implement best practices in project management.

One of the characteristics of companies that have won the prestigious Malcolm Baldrige Award is that each has an excellent project management system. Companies such as Motorola, Armstrong World Industries, General Motors, Kodak, Xerox, and IBM use integrated TQM and project management systems.

In the mid 1990s, during a live videoconference on the subject, "How to Achieve Maturity in Project Management," Dave Kandt, Group Vice President for Quality and Program Management at Johnson Controls, commented on the reasons behind Johnson Controls' astounding success:

> We came into project management a little differently than some companies. We have combined project management and TQC (total quality control) or total quality management. Our first design and development projects in the mid-1980s led us to believe that our functional departments were working pretty well separately, but we needed to have some systems to bring them together. And, of course, a lot of what project management is about is getting work to flow horizontally through the company. What we did first was to contact Dr. Norman Feigenbaum, who is the granddaddy of TQC in North America, who helped us establish some systems that linked together the whole company. Dr. Feigenbaum looked at quality in the broadest sense: quality of products, quality of systems, quality of deliverables, and, of course, the quality of projects and new product launches. A key part of these systems included project management systems that addressed product introduction and the product introduction process. Integral to this was project management training, which was required to deliver these systems.
>
> We began with our executive office, and once we had explained the principles and philosophies of project management to these people, we moved to the management of plants, engineering managers, analysts, purchasing people, and of course project managers. Only once the foundation was laid did we proceed with actual project management and with defining the role and responsibility so that the entire company would understand their role in project management once these people began to work. Just the understanding allowed us to move to a matrix organization and eventually to a stand-alone project

management department. So how well did that work? Subsequently, since the mid-1980s, we have grown from 2 or 3 projects to roughly 50 in North America and Europe. We have grown from 2 or 3 project managers to 35. I don't believe it would have been possible to manage this growth or bring home this many projects without project management systems and procedures and people with understanding at the highest levels of the company.

In the early 1990s we found that we were having some success in Europe, and we won our first design and development project there. And with that project, we carried to Europe not only project managers and engineering managers who understood these principles but also the systems and training we incorporated in North America. So we had a company wide integrated approach to project management. What we've learned in these last 10 years that is the most important to us, I believe, is that you begin with the systems and the understanding of what you want the various people to do in the company across all functional barriers, then bring in project management training, and last implement project management.

Of course, the people we selected for project management were absolutely critical, and we selected the right people. You mentioned the importance of project managers understanding business, and the people that we put in these positions are very carefully chosen. Typically, they have a technical background, a marketing background, and a business and financial background. It is very hard to find these people, but we find that they have the necessary cross-functional understanding to be able to be successful in this business.

At Johnson Controls, project management and TQM were developed concurrently. Dave Kandt was asked during the same videoconference whether companies must have a solid TQM culture in place before they attempt the development of a project management program. He said:

I don't think that is necessary. The reason why I say that is that companies like Johnson Controls are more the exception than the rule of implementing TQM and project management together. I know companies that were reasonably mature in project management and then ISO 9000 came along, and because they had project management in place in a reasonably mature fashion, it was an easier process for them to implement ISO 9000 and TQM. There is no question that having TQM in place at the same time or even first would make it a little easier, but what we've learned during the recession is that if you want to compete in Europe and you want to follow ISO 9000 guidelines, TQM must be implemented. And using project management as the vehicle for that implementation quite often works quite well.

There is also the question of whether or not successful project management can exist within the ISO 9000 environment. According to Dave Kandt:

Not only is project management consistent with ISO 9000, a lot of the systems that ISO 9000 require are crucial to project management's success. If you don't have a good quality system, engineering change system, and other things that ISO requires, the project manager is going to struggle in trying to accomplish and execute that project. Further, I think it's interesting that companies that are working to install and deploy ISO 9000, if they are being successful, are probably utilizing project management techniques. Each of the different elements of ISO requires training, and sometimes the creation of systems inside the company that can all be scheduled, teams that can be assigned, deliverables that can be established, tracked, and monitored, and reports that go to senior management. That's exactly how we installed TQC at Johnson Controls, and I see ISO 9000 as having a very similar thrust and intent.

While the principles of TQM still exist, the importance of Six Sigma concepts has grown. According to Eric Alan Johnson and Jeffrey Alan Neal[4]:

Total Quality Management   In addition to the TQM PDCA cycle, the continuous improvement DMAIC (Define, Measure, Analyze, Improve, and Control) model can be used to improve the effectiveness of project management. This model has been successfully employed for Six Sigma and Lean Enterprise process improvement, but the basic tenets of its structured, data enabled problem solving methodology can also be employed to improve the success of project management.

By assessing data collected on both project successes and root cause of project failures, the DMAIC model can be used to improve and refine both the management of projects and the ultimate quality of products produced.

In the define phase, specific project definition and requirements are based on data gathered from the customer and on historical project performance. Gathering as much information as possible in these areas allows the project manager to concentrate on what is truly important to the customer while reviewing past performance in order to avoid the problems of and continue to propagate the successes of past projects. In the define stage, available data on the people, processes, and suppliers is reviewed to determine their ability to meet the cost, quality and schedule requirements of the project. The define phase, in short, should assess not only the requirements of the customer, but should also assess the capability of your system to meet those requirements. Both of these assessments must be based on data gathered by a dedicated measurement system. Additionally, the define stage should establish the metrics to be used during projects execution to monitor and control project progress. These metrics will be continually evaluated during the measure and analysis phase (these DMAIC phase are concurrent with the PMI phases of project management).

The next phase of the DMAIC model, measure, data (the metrics identified in the define stage) from the measurement system is continually reviewed during project execution to ensure that the project is being effectively managed. The same data metrics used in the define stage should be updated with specific project data to determine how well the project is progressing. The continual assessment of project performance, based on data gathered during the execution phase, is the key to data enabled project management.

During the continual measurement of the progress of the project, it is likely that some of these key metrics will indicate problems either occurring (present issues) or likely to occur (leading indicators). These issues must be addressed if the project is to execute on time, and on budget to meeting requirements. This is where the analysis aspect of the DMAIC model becomes a critical aspect of project management. The analysis of data is

---

4. Eric Alan Johnson, Satellite Control Network Contract Deputy Program Director, AFSCN, and the winner of the 2006 Kerzner Project Manager of the Year Award; and Jeffrey Alan Neal, Blackbelt/Lean Expert and Lecturer, Quantitative Methods, University of Colorado, Colorado Springs.

an entire field onto itself. Numerous books and articles have addressed the problem of how to assess data, but the main objective remains. The objective of data analysis is to turn data into usable information from which to base project decisions.

The methods of data analysis are specific to the data type and to the specific questions to be answered. The first step (after the data have been gathered) is to use descriptive techniques to get an overall picture of the data. This overall picture should include a measure of central tendency (i.e. mean), and a measure of variation such as standard deviation. Additionally, graphical tools such as histograms and Pareto charts are useful in summarizing and displaying information. Tests of significance and confidence interval development are useful in determining if the results of the analysis are statistically significant and for estimating the likelihood of obtaining a similar result.

In the continual monitoring of processes, control charts are commonly used tools to assess the state of stability of processes and to determine if the variation is significant enough to warrant additional investigation. In addition, control charts provide a basis for determining if the type of variation is special cause or common cause. This distinction is critical in the determination of the appropriate corrective actions that may need to be taken.

To provide a basis for the identification of potential root causes for project performance issues, tools such Failure Modes and Effects Analysis and the Fishbone (also known as the Ishkawa) diagram can be used to initiate and document the organized thought process needed to separate main causes of non-conformities from contributing causes.

If the data meets the statistical condition required, such tests as Analysis of Variance (ANOVA) and regression analysis can be extremely useful in quantifying and forecasting process and project performance. Because ANOVA (the General Linear Model) can be used to test for mean differences of two or more factors or levels, ANOVA can be used to identify important independent variables for various project dependent variables. Various regression models (simple linear, multiple linear and binary) can be used to quantify the different effects of independent variable on critical dependent variable that are key to project success.

In short, this phase uses the data to conduct an in-depth and exhaustive root cause investigation to find the critical issue that was responsible for the project execution problem and effects upon the project if left uncorrected.

The next phase involves the process correction and improvement that addressed the root cause identified in the previous phase. This is corrective action (fix the problem you are facing) and preventive action (make sure it or one like it doesn't come back). So, once the root cause has been identified, both corrective and preventive process improvement actions can be taken to address current project execution and to prevent the reoccurrence of that particular issue in future projects. To insure that current projects do not fall victim to that problem recently identified and that future projects avoid the mistakes of the past, a control plan is implemented to monitor and control projects. The cycle is repeated for all project management issues.

The continuing monitoring of project status and metrics along with their continual analysis and correction is an ongoing process and constitutes the control phase of the project. During this phase, the key measurements instituted during the initiation phase are used to track project performance against requirements. When the root cause of each project problem is analyzed, this root cause and the subsequent corrective and preventive action are entered into a "lessons learned" database. This allows for consistent problem resolution actions to be taken. The database is also then used to identify potential project risks and institute a priori mitigation actions.

Risk/Opportunity Management Using Six Sigma Tools and Probabilistic Models

Risk/Opportunity Management is one of the critical, if not the most, tools in a project or program managers tool box—regardless of contract type. Typically projects/programs focus on the potential impact and/or probability of a risk occurrence. While these are very critical factors to developing a good risk mitigation plan, the adroit ability of the project team to detect the risk will have the greatest impact on successful project execution. If you can't detect the risk, then your ability to manage it, will always be reactive. The undetectable risk is a greater threat to execution, than the high probability or high impact factors. This is where using one of the six sigma tools Failure Modes and Effects Analysis (FMEA) can be very effective. The FMEA tool can help a project team evaluate—risk detection. Focusing on risk detection will help the team think "out of the box" in proposing, planning or executing a successful project.

Example: If your project/program has a risk that has a significant probability of occurrence, then it is probably not really a ris —it is an issue/problem. If the impact is great and the probability is low, then you will keep an eye on this, but not usually spend Management Reserve (MR) to mitigate. However, if the risk has a high impact or probability, but has a low level of detectability, the results could be devastating.

The other side of managing a project/program is the lack of focus on opportunity identification and management. If a project team is only risk management focused, they may miss looking at the projects potential opportunities. Opportunities need to be evaluated with the same rigor as risks. The same level of focus in the areas of impact, probability AND the ability to recognize the opportunity must occur for a project team. The FMEA is also. . . . very useful for opportunity recognition and management. Sometimes undetectable risks will occur, but the ability to recognize and realize opportunities can counter this risk impacts. The use of opportunity recognition can have the greatest impacts on Fix-Price projects where saving costs can increase the projects profit margin.

If a project has risk schedule, how can we quantify that risk? One method is through the use probabilistic modeling. Probabilistic modeling of your schedule can help you forecast the likelihood of achieving all your milestones within your period of performance. If the risk of achieving your schedule is too high, you can use these models to perform "what if" analysis until the risk factors, can be brought to acceptable levels. This analysis should be done BEFORE the project is baselined or (ideally) during the proposal phase.

The key to successful implementation of this strategy is a relational database of information which will allow you to build the most realistic probabilistic model possible. This information must be gathered on a wide variety of projects so that information on projects of similar size and scope/complexity can be evaluated. It must be integrated with "lessons learned" from these other projects in order to build the best probabilistic model to mitigate your schedule risks. Always remember that a model is only as good as the information used to build it.

5.6 CONCURRENT ENGINEERING

The need to shorten product development time has always plagued U.S. companies. During favorable economic conditions, corporations have deployed massive amounts of resources to address the problem of long development times. During economic downturns, however, not only are resources scarce, but time becomes a critical constraint. Today, the principles of concurrent engineering have been almost universally adopted as the ideal solution to the problem.

Concurrent engineering requires performing the various steps and processes in managing a project in tandem rather than in sequence. This means that engineering, research and development, production, and marketing all are involved at the beginning of a project, before any work has been done. That is not always easy, and it can create risks as the project is carried through. Superior project planning is needed to avoid increasing the level of risk later in the project. The most serious risks are delays in bringing product to market and costs when rework is needed as a result of poor planning. Improved planning is essential to project management, so it is no surprise that excellent companies integrate concurrent engineering and project management systems.

Chrysler (now DaimlerChrysler) Motors used concurrent engineering with project management to go from concept to market with the Viper sports car in less than three years. Concurrent engineering may well be the strongest driving force behind the increased acceptance of modem project management.

5.7 RISK MANAGEMENT

Risk management is an organized means of identifying and measuring risk and developing, selecting, and managing options for handling those risks. Throughout this book, I have emphasized that tomorrow's project managers will need superior business skills in assessing and managing risk. This includes both project risks and business risks. Project managers in the past were not equipped to quantify risks, respond to risks, develop contingency plans, or keep lessons-learned records. They were forced to go to senior managers for advice on what to do when risky situations developed. Now senior managers are empowering project managers to make risk-related decisions, and that requires a project manager with solid business skills as well as technical knowledge.

Preparing a project plan is based on history. Simply stated: What have we learned from the past? Risk management encourages us to look at the future and anticipate what can go wrong and then to develop contingency strategies to mitigate these risks.

We have performed risk management in the past, but only financial and scheduling risk management. To mitigate a financial risk, we increased the project's budget. To mitigate a scheduling risk, we added more time to the schedule. But in the 1990s, technical risks became critical. Simply adding into the plan more time and money is not the solution to mitigate technical risks. Technical risk management addresses two primary questions:

- Can we develop the technology within the imposed constraints?
- If we do develop the technology, what is the risk of obsolescence, and when might we expect it to occur?

To address these technical risks, effective risk management strategies are needed based upon technical forecasting. On the surface, it might seem that making risk management an integral part of project planning should be relatively easy. Just identify and address risk factors before they get out of hand. Unfortunately, the reverse is likely to be the norm, at least for the foreseeable future.

For years, companies provided lip service to risk management and adopted the attitude that we should simply live with it. Very little was published on how to develop a structure risk management process. The disaster with the Space Shuttle Challenger in January 1986 created a great awakening on the importance of effective risk management.[5]

Risk management today has become so important that companies are establishing separate risk management organizations within the company. However, many companies have been using risk management functional units for years, and yet this concept has gone unnoticed. The following is an overview of the program management methodology of the risk management department of an international manufacturer headquartered in Ohio. This department has been in operation for approximately 25 years.

> The risk management department is part of the financial discipline of the company and ultimately reports to the treasurer, who reports to the chief financial officer. The overall objective of the department is to coordinate the protection of the company's assets. The primary means of meeting that objective is eliminating or reducing potential losses through loss prevention programs. The department works very closely with the internal environmental health and safety department. Additionally, it utilizes outside loss control experts to assist the company's divisions in loss prevention.
>
> One method employed by the company to insure the entire corporation's involvement in the risk management process is to hold its divisions responsible for any specific losses up to a designated self-insured retention level. If there is a significant loss, the division must absorb it and its impact on their bottom-line profit margin. This directly involves the divisions in both loss prevention and claims management. When a claim does occur, risk management maintains regular contact with division personnel to establish protocol on the claim and reserves and ultimate resolution.
>
> The company does purchase insurance above designated retention levels. As with the direct claims, the insurance premiums are allocated to its divisions. These premiums are calculated based upon sales volume and claim loss history, with the most significant percentage being allocated to claim loss history.
>
> Each of the company's locations must maintain a business continuity plan for its site. This plan is reviewed by risk management and is audited by the internal audit and environmental health and safety department.
>
> Risk management is an integral part of the corporation's operations as evidenced by its involvement in the due diligence process for acquisitions or divestitures. It is involved at the onset of the process, not at the end, and provides a detailed written report of findings as well as an oral presentation to group management.
>
> Customer service is part of the company's corporate charter. Customers served by risk management are the company's divisions. The department's management style with its customers is one of consensus building and not one of mandating. This is exemplified by the company's use of several worker's compensation third-party administrators (TPAs) in states where it is self-insured. Administratively, it would be much easier to utilize one nationwide TPA. However, using strong regional TPAs with offices in states where divisions

---

5. The case study "The Space Shuttle Challenger Disaster" appears in H. Kerzner, *Project Management Case Studies,*, 3rd ed. Wiley, New York, 2009, p. 425.

operate provides knowledgeable assistance with specific state laws to the divisions. This approach has worked very well for this company that recognizes the need for the individual state expertise.

The importance of risk management is now apparent worldwide. The principles of risk management can be applied to all aspects of a business, not just projects. Once a company begins using risk management practices, the company can always identify other applications for the risk management processes.

For multinational companies that are project-driven, risk management takes on paramount importance. Not all companies, especially in undeveloped countries, have an understanding of risk management or its importance. These countries sometimes view risk management as an overmanagement expense on a project.

Consider the following scenario. As your organization gets better and better at project management, your customers begin giving you more and more work. You're now getting contracts for turnkey projects, or complete-solution projects. Before, all you had to do was deliver the product on time and you were through. Now you are responsible for project installation and startup as well, sometimes even for ongoing customer service. Because the customers no longer use their own resources on the project, they worry less about how you're handling your project management system.

Alternatively, you could be working for third-world clients who haven't yet developed their own systems. One hundred percent of the risk for such projects is yours, especially as projects grow more complex (see Figure 5–6). Welcome to the twenty-first century!

One subcontractor received a contract to install components in a customer's new plant. The construction of the plant would be completed by a specific date. After construction was completed, the contractor would install the equipment, perform testing, and then start up. The subcontractor would not be allowed to bill for products or services until after a successful startup. There was also a penalty clause for late delivery.

The contractor delivered the components to the customer on time, but the components were placed in a warehouse because plant construction had been delayed. The contractor now had a cash flow problem and potential penalty payments because of external dependencies that sat on the critical path. In other words, the contractor's schedule was being controlled by actions of others. Had the project manager performed business risk management rather than just technical risk management, these risks could have been reduced.

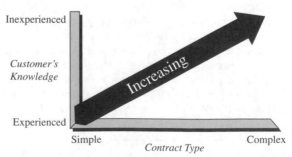

FIGURE 5–6.   Future risks.

For the global project manager, risk management takes on a new dimension. What happens if the culture in the country with which you are working neither understands risk management nor has any risk management process? What happens if employees are afraid to surface bad news or identify potential problems? What happens if the project's constraints of time, cost, and quality/performance are meaningless to the local workers?

5.8 EFFECTIVE RISK MANAGEMENT AT JEFFERSON COUNTY, COLORADO

Best practices in project risk management need not involve the expenditure of hundreds of thousands of dollars or involve a multitude of people. Applying the basic concepts of risk management can lead to best practices if done correctly. Such was the case at Jefferson County, Colorado. Joe Palmer, CIO/IT Services Director, Jefferson County, Colorado, describes the success that they had with risk management on a successful enterprise resource planning project. The remainder of the material in this section comes from Joe Palmer.

Background

Enterprise resource planning (ERP) systems standardize financial, human resource, procurement, and asset management processes and data models across the enterprise. As a result, they are high-risk projects. Risks are uncertain events or conditions that, if they occur, have an impact on the project's objectives. They could cause schedule or cost overruns or affect project quality or the ability to deliver on custom requirements.

The risk management plan utilized for the Jefferson County, Colorado ERP Project outlined risk categories, listed risks identified by the project team, including calculated risk factors, and outlined how the risks would be managed. The plan represented the cumulative input of project stakeholders, including the steering committee, business stakeholders, and the project team.

Risk Categories

Categories of risk include:

- *Technical Risk*—Areas of technical risk include hardware, software, middleware, third-party applications and interfaces, security administration, network, and database.
- *Leadership Risk*—Areas of leadership risk include politics, steering committee, project management, engagement management, and business stakeholders.
- *Team/Resource Risk*—Areas of team and resource risk include knowledge, time, focus, and commitment.
- *Business Alignment*—Areas of business alignment risk include training, requirements, end-user support/involvement, end-user acceptance, and business processes.
- *Project Management Risk*—Areas of triple constraint or project management risk include scope, time and schedule, and quality.

TABLE 5–1.   RISK FACTOR CALCULATIONS

Probability	Description
5%	Unlikely
25%	Possible
50%	Even chance
75%	Likely
95%	Very likely/certain

Risk probability was assessed using the guidelines in Table 5–1.
Risk impact was assessed using the guidelines in Table 5–2.

Risk Factor Calculations   Risk factors were calculated by multiplying the probability and impact assessments.

Risk Identification   The team, stakeholder, and governance committee collectively brainstormed project risks in a workshop.

Risk Management   Risk management encompasses the processes and techniques used to manage risk on the project. There are four possible risk management strategies:

- *Acceptance*—Choosing to suffer consequences of unmanaged risk
- *Avoidance*—Choosing to not proceed with a risk source or trigger
- *Transference*—Choosing to transfer the risk to a third party, usually at a cost
- *Mitigation*—Choosing to alter the approach to reduce or eliminate the risk source or trigger

A specific risk management technique was defined for every risk.

TABLE 5–2.   RISK FACTOR ASSESSMENTS

Impact	Description	Cost Impact	Schedule Impact
0.1	Minimal	≤$5000	≤1 week
0.3	Small	$5000–$9,999	1–2 weeks
0.5	Some	$10,000–$24,999	2–4 weeks
0.7	Significant	$25,000–$49,999	4–6 weeks
0.9	Large	≥$50,000	>6 weeks

Result   The result was that no previously unidentified risks arose during the course of the project. For every risk that arose, the selected risk management technique was applied. Risks were successfully managed such that the project was completed on time and under budget.

5.9 FAILURE OF RISK MANAGEMENT

There are numerous reasons why risk management can fail. Typical reasons might include:

- The inability to perform risk management effectively
- Not being able to identify the risks
- Not being able to measure the uncertainty of occurrence
- Not being able to predict the impact, whether favorable or unfavorable
- Having an insufficient budget for risk management work
- Having team members that do not understand the importance of risk management
- Fear that identification of the true risks could result in the cancellation of the project
- Fear that whoever identifies critical risks will get unfavorable recognition
- Peer pressure from colleagues and superiors that want to see this project completed regardless of the risks

All of these failures occur during the execution of the project. These failures we seem to understand and can correct with proper education and budget allocations for risk management activities. But perhaps the worst failures occur when people refuse to even consider risk management because of some preconceived notion about its usefulness of importance to the project. David Dunham discusses some of the reasons why people avoid risk management on new product development (NPD) projects[6]:

Discussing risk in new product development certainly seems to be a difficult thing to do. Despite the fact that the high-risk nature of new product development is built into the corporate psyche, many corporations still take a fatalistic approach toward managing the risk. Reasons for not being anxious to dwell on risk differ depending on the chair in which you are sitting.

Program Manager

- Spending time on risk assessment and management is counter to the action culture of many corporations. "Risk management does not create an asset," to quote one executive.
- Management feels that the learning can/should be done in the market.

---

6. D. J. Dunham, "Risk Management: The Program Manager's Perspective," in P. Belliveau, A. Griffin, and S. Somermeyer, *The PDMA Toolbook for New Product Development*, Wiley, Hoboken, NJ, 2002, p. 382.

Program Manager

- There is a natural aversion among developers to focus on the downside.
- Highlighting risk is counterintuitive for development teams who want to promote the opportunity when competing for NPD funding.

5.10 DEFINING MATURITY USING RISK MANAGEMENT

For years, project management maturity was measured by how frequently we were able to meet the project's triple constraints of time, cost, and performance or scope. Today, we are beginning to measure maturity in components, such as the areas of knowledge in the *PMBOK® Guide*. Maturity is now measured in stages and components, such as how well we perform scope management, time management, risk management, and other areas of knowledge. Gregory Githens believes that the way we handle risk management can be an indicator of organizational maturity[7]:

> Some firms have more capability to manage risk well, and these firms are the most consistent in their growth and profitability. Perhaps the simplest test for examining risk management maturity is to examine the level of authority given to the [New Product Development] NPD program [project] manager: If authority is high, then the organization is probably positioning itself well to manage risks, but if authority is low, then the blinders may be on. Another test is the use of checklists: if ticking off a checklist is the sole company response to risk, then organizational maturity is low. Risk management provides . . . [an] excellent lens by which to evaluate a firm's ability to integrate and balance strategic intent with operations.
>
> Many firms ignore risk management because they have not seen the need for it. They perceive their industry as stable and mostly focus on their competitive rivals and operational challenges. . . . By addressing risk at the project level, you encourage the organization to surface additional strategic concerns.
>
> Top NPD firms have a sophisticated capability for risk management, and they will "book" a project plan, pay attention to the details of product scope and project scope, use risk management tools such as computer simulations and principle-based negotiation, and document their plans and assumptions. These more mature firms are the ones that will consider risk in establishing project baselines and contracts. For example, Nortel uses a concept called "out of bounds" that provides the NPD program managers with the freedom to make trade-offs in time, performance, cost and other factors. Risk analysis and management is an important tool.
>
> Less mature firms typically establish a due date and pay attention to little else (and in my experience, this is the majority of firms). Firms that use the decision rule "Hit the launch date" default to passive acceptance—hiding the risk instead of managing it. Firefighting and crisis management characterize their organizational culture, and their

---

7. G. D. Githens, "How to Assess and Manage Risk in NPD: A Team-Based Approach," in P. Belliveau, A. Griffin, and S. Somermeyer, *The PDMA Toolbook for New Product Development*, Wiley, Hoboken, NJ, 2002, p. 208.

strategic performance is inconsistent. These firms are like the mythological character Icarus: They fly high but come crashing down because they ignored easily recognizable risk events.

5.11 BOEING AIRCRAFT COMPANY

As companies become successful in project management, risk management becomes a structured process that is performed continuously throughout the life cycle of the project. The two most common factors supporting the need for continuous risk management is how long the project lasts and how much money is at stake. For example, consider Boeing's aircraft projects. Designing and delivering a new plane might require 10 years and a financial investment of more than $5 billion.

Table 5–3 shows the characteristics of risks at Boeing. (The table does not mean to imply that risks are mutually exclusive of each other.) New technologies can appease customers, but production risks increase because the learning curve is lengthened with new technology compared to accepted technology. The learning curve can be lengthened further when features are custom designed for individual customers. In addition, the loss of suppliers over the life of a plane can affect the level of technical and production risk. The relationships among these risks require the use of a risk management matrix and continued risk assessment.

TABLE 5–3. RISK CATEGORIES AT BOEING

Type of Risk	Risk Description	Risk Mitigation Strategy
Financial	Up-front funding and payback period based upon number of planes sold	• Funding by life-cycle phases • Continuous financial risk management • Sharing risks with subcontractors • Risk reevaluation based upon sales commitments
Market	Forecasting customers' expectations on cost, configuration, and amenities based upon a 30- to 40-year life of a plane	• Close customer contact and input • Willingness to custom design per customer • Development of a baseline design that allows for customization
Technical	Because of the long lifetime for a plane, must forecast technology and its impact on cost, safety, reliability, and maintainability	• A structured change management process • Use of proven technology rather than high-risk technology • Parallel product improvement and new product development processes
Production	Coordination of manufacturing and assembly of a large number of subcontractors without impacting cost, schedule, quality, or safety	• Close working relationships with subcontractors • A structured change management process • Lessons learned from other new airplane programs • Use of learning curves

5.12 CHANGE MANAGEMENT

Companies use change management to control both internally generated changes and customer-driven changes in the scope of projects. Most companies establish a configuration control board or change control board to regulate changes. For customer-driven changes, the customer participates as a member of the configuration control board. The configuration control board addresses the following four questions at a minimum:

- What is the cost of the change?
- What is the impact of the change on project schedules?
- What added value does the change represent for the customer or end user?
- What are the risks?

The benefit of developing a change management process is that it allows you to manage your customer. When your customer initiates a change request, you must be able to predict immediately the impact of the change on schedule, safety, cost, and technical performance. This information must be transmitted to the customer immediately, especially if your methodology is such that no further changes are possible because of the life-cycle phase you have entered. Educating your customer as to how your methodology works is critical in getting customer buy-in for your recommendations during the scope change process.

Risk management and change management function together. Risks generate changes that, in turn, create new risks. For example, consider a company in which the project manager is given the responsibility for developing a new product. Management usually establishes a launch date even before the project is started. Management wants the income stream from the project to begin on a certain date to offset the development costs. Project managers view executives as their customers during new project development, but the executives view their customers as the stockholders who expect a revenue stream from the new product. When the launch date is not met, surprises result in heads rolling, usually executive heads first.

In the previous edition of the book, we stated that Asea, Brown and Boveri had developed excellent processes for risk management, so it is understandable that it also has structured change management processes. In companies excellent in project management, risk management and change management occur continuously throughout the life cycle of the project. The impact on product quality, cost, and timing is continuously updated and reported to management as quickly as possible. The goal is always to minimize the number and extent of surprises.

5.13 OTHER MANAGEMENT PROCESSES

Employee empowerment and self-directed work teams took the business world by storm during the early 1990s. With growing emphasis on customer satisfaction, it made sense to empower those closest to the customer—the order service people, nurses, clerks, and so on—to take action in solving customers' complaints. A logical extension of employee empowerment is the self-managed work team. A self-directed work team is a group of employees with given day-to-day responsibility for managing themselves and the work they perform. This includes the responsibility for handling resources and solving problems.

Some call empowerment a basis for the next industrial revolution, and it is true that many internationally known corporations have established self-directed work teams. Such corporations include Esso, Lockheed-Martin, Honeywell, and Weyerhauser. Time will tell whether these concepts turn out to be a trend or only a fad.

Reengineering a corporation is another term for downsizing the organization with the (often unfortunate) belief that the same amount of work can be performed with fewer people, at lower cost, and in a shorter period of time. Because project management proposes getting more done in less time with fewer people, it seems only practical to implement project management as part of reengineering. It still is not certain that downsizing executed at the same time as the implementation of project management works, but project-driven organizations seem to consider it successful.

Life-cycle costing was first used in military organizations. Simply stated, life-cycle costing requires that decisions made during the R&D process be evaluated against the total life-cycle cost of the system. Life-cycle costs are the total cost of the organization for the ownership and acquisition of the product over its full life.

5.14 EDS

Information technology enterprise management (ITEM) integrates all of the project management disciplines along with other IT disciplines such as engineering, testing, and model office. Many projects start with a charter and scope that make a project manager view the work with a definitive start and definitive end. (See Figure 5–7.) If this is applied to the relative stages of a release, they simply stretch the project management process groups across all of the release. That makes the suppliers of the project try to understand if designing and building are part of execution or if deployment of the project is the execution of the project.

As stated in the *PMBOK® Guide*, project management process groups should be repeated for each release stage. This promotes other project management strategies, such as roll wave planning and resource balancing. With this framework view, other capabilities can be applied to each stage, along with the project management capabilities (Figure 5–8).

5.15 EARNED-VALUE MEASUREMENT

An integral part of most project management methodologies is the ability to perform earned-value measurement. Earned-value measurement was created so that project managers would manage projects rather than merely monitor results. Even though some companies do not use earned-value measurement on a formal basis, core concepts such

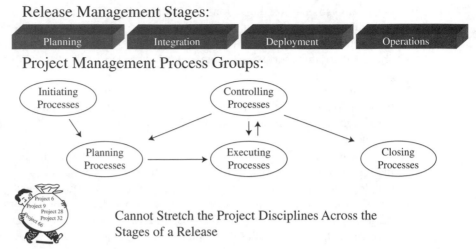

Release Management Stages:

| Planning | Integration | Deployment | Operations |

Project Management Process Groups:

FIGURE 5–7. Typical application of the *PMBOK® Guide*.

Cannot Stretch the Project Disciplines Across the
Stages of a Release

Release Management Stages:

| Planning | Integration | Deployment | Operations |

Project Management Process Groups:

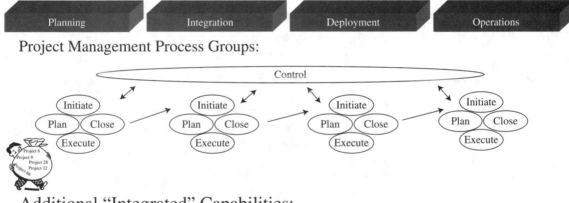

Additional "Integrated" Capabilities:

• Strategic Planning	• Hardware Design	• Site Assessments	• Service Desk
• Release Forecasting	• Application Design	• Procurement	• Incident Handling
• Release Impact Analysis	• Unit/System Tests	• Hardware Installation	• Change Management
• Client Feasibility Study	• Model Office	• User Acceptance Test	• Configuration Management
• Client Stage Gate	• Client Stage Gate	• Client Stage Gate	• Problem Management

FIGURE 5–8. Correct application of the *PMBOK® Guide*.

as variance analysis and reporting are being used. As an example, Keith Kingston, PMP®,
Manager of Program Management at Motorola, states:

> Variances to schedule performance are analyzed weekly or biweekly but not as part of a
> formal earned value approach. Greater than three days of schedule variation on any inter-
> nal deliverable requires analysis and a mitigation plan. Any variances that impact meeting
> a customer-required date requires analysis and a mitigation plan.

5.16 DTE ENERGY

One of the characteristics of integrated processes is that it must include an integration with a project management information system capable of reporting earned value measurement. Kizzmett Collins, PMP, Senior Project Manager, Software Engineering, Methods, and Staffing at DTE Energy, describes the integration with earned-value measurement:

DTE Energy's ITS Earned-Value Analysis Journey

Getting Started Beginning in 2001, the Information Technology Services' (ITS) project management office (PMO) sponsored several projects that helped to advance and develop earned-value analysis (EVA) understanding and knowledge among management and project managers.

The implementation of the Primavera Teamplay project management suite easily enabled the tracking of EVA metrics for individual projects and project portfolios. Among other sources of education, TeamPlay product training introduced ITS project managers and management to EVA metrics. The PMO developed processes and reports to aid project managers and management in the analysis and reporting of EVA metrics. ITS contracted Quentin W. Flemming to conduct EVA courses, which increased project managers' and ITS management's understanding of EVA.

Reporting EVA Metrics EVA metrics (such as SPI, CPI, CV, SV, and EAC) are reported weekly in project status reports and in the ITS project portfolio. The project manager provides additional status commentary for variances from target that exceed ± 10%.

During the ITS planning and management table (PMT) meetings, project mangers report project status, usually on a monthly basis. The PMT reviews EVA metrics and discusses variances and indicators as warranted. Other triggers that may necessitate a review of the projects' EVA metrics include:

- A project is at 20 percent of original estimated duration.
- A significant phase has ended.
- The project manager presents issues, risks, or changes.

Linking Rewards to EVA CPI Metric In 2003, the ITS Organization began linking rewards to the CPI metric. The CPI metric results are now included in both project manager performance reviews and the ITS Organizational Scorecard.

The ITS Organizational Scorecard is tied to the corporate Rewarding Employee Plan (REP). REP pays employees bonuses that are based on achieving the corporate and organizational goals. The ITS Organization's CPI metric is the aggregate of the CPI of each project in the project portfolio. As a result, all ITS employees have a monetary stake in the success of each project.

Project manager performance goals include the CPI metric for all projects within their area of responsibility. CPI results greater than 0.95 and less than 1.05 exceed performance expectations. Each project manager's performance review is linked to their merit increase.

Opportunities for Improvement   The ITS Organization introduced EVA to the corporation through the ITS scorecard. Both within ITS and across the corporation, further training is needed to expand our shared understanding of EVA. Internally, ITS can take the next step in further understanding what the EVA metrics indicate. For example, we have an opportunity to be more deliberate in allocating dollars elsewhere when a project's CPI indicates cost issues or problems.

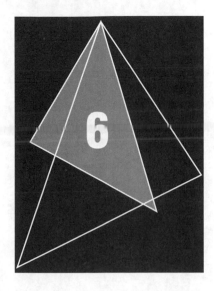

Culture

6.0 INTRODUCTION

Perhaps the most significant characteristic of companies that are excellent in project management is their culture. Successful implementation of project management creates an organization and culture that can change rapidly because of the demands of each project and yet adapt quickly to a constantly changing dynamic environment, perhaps at the same time. Successful companies have to cope with change in real time and live with the potential disorder that comes with it.

Change is inevitable in project-driven organizations. As such, excellent companies have come to the realization that competitive success can be achieved only if the organization has achieved a culture that promotes the necessary behavior. Corporate cultures cannot be changed overnight. Years are normally the time frame. Also, if as little as one executive refuses to support a potentially good project management culture, disaster can result.

In the early days of project management, a small aerospace company had to develop a project management culture in order to survive. The change was rapid. Unfortunately, the vice president for engineering refused to buy into the new culture. Prior to the acceptance of project management, the power base in the organization had been engineering. All decisions were either instigated or approved by engineering. How could the organization get the vice president to buy into the new culture?

The president realized the problem but was stymied for a practical solution. Getting rid of the vice president was one alternative, but not practical because of his previous successes and technical know-how. The corporation was awarded a two-year project that was strategically important to the company. The vice president was then temporarily assigned as the project manager and removed from his position as vice president for engineering. At the completion of the project, the vice president was assigned to fill the newly created position of vice president of project management.

6.1 CREATION OF A CORPORATE CULTURE

Corporate cultures may take a long time to create and put into place but can be torn down overnight. Corporate cultures for project management are based upon organizational behavior, not processes. Corporate cultures reflect the goals, beliefs, and aspirations of senior management. It may take years for the building blocks to be in place for a good culture to exist, but it can be torn down quickly through the personal whims of one executive who refuses to support project management.

Project management cultures can exist within any organizational structure. The speed at which the culture matures, however, may be based upon the size of the company, the size and nature of the projects, and the type of customer, whether it be internal or external. Project management is a culture, not policies and procedures. As a result, it may not be possible to benchmark a project management culture. What works well in one company may not work equally well in another.

Good corporate cultures can also foster better relations with the customer, especially external clients. As an example, one company developed a culture of always being honest in reporting the results of testing accomplished for external customers. The customers, in turn, began treating the contractor as a partner and routinely shared proprietary information so that the customers and the contractor could help each other.

Within the excellent companies, the process of project management evolves into a behavioral culture based upon multiple-boss reporting. The significance of multiple-boss reporting cannot be understated. There is a mistaken belief that project management can be benchmarked from one company to another. Benchmarking is the process of continuously comparing and measuring against an organization anywhere in the world in order to gain information that will help your organization improve its performance and competitive position. Competitive benchmarking is where one benchmarks organizational performance against the performance of competing organizations. Process benchmarking is the benchmarking of discrete processes against organizations with performance leadership in these processes.

Since a project management culture is a behavioral culture, benchmarking works best if we benchmark best practices, which are leadership, management, or operational methods that lead to superior performance. Because of the strong behavioral influence, it is almost impossible to transpose a project management culture from one company to another. What works well in one company may not be appropriate or cost-effective in another company.

Strong cultures can form when project management is viewed as a profession and supported by senior management. A strong culture can also be viewed as a primary business differentiator. Strong cultures can focus on either a formal or informal project management approach. However, with the formation of any culture, there are always some barriers that must be overcome.

According to a spokesperson from AT&T:

> Project Management is supported from the perspective that the PM is seen as a professional with specific job skills and responsibilities to perform as part of the project team. Does the PM get to pick and choose the team and have complete control over budget

allocation? No. This is not practical in a large company with many projects competing for funding and subject matter experts in various functional organizations.

A formal Project Charter naming an individual as a PM is not always done, however, being designated with the role of Project Manager confers the power that comes with that role. In our movement from informal to more formal, it usually started with Project Planning and Time Management, and Scope Management came in a little bit later.

In recent memory PM has been supported, but there were barriers. The biggest barrier has been in convincing management that they do not have to continue managing all the projects. They can manage the project managers and let the PMs manage the projects. One thing that helps this is to move the PMs so that they are in the same work group, rather than scattered throughout the teams across the company, and have them be supervised by a strong proponent of PM. Another thing that has helped has been the PMCOE's execution of their mission to improve PM capabilities throughout the company, including impacting the corporate culture supporting PM.

Our success is attributable to a leadership view that led to creating a dedicated project management organization and culture that acknowledges the value of Project Management to the business. Our vision: Establish a global best in class Project Management discipline designed to maximize the customer experience and increase profitability for AT&T.

In good cultures, the role and responsibilities of the project manager is clearly identified. It is also supported by executive management and understood by everyone in the company. According to Enrique Sevilla Molina, Corporate PMO Director at Indra:

Based on the historical background of our company and the practices we set in place to manage our projects, we found out that the project manager role constitutes a key factor for project success. Our project management theory and practice has been built to provide full support to the project manager when making decisions and, consequently, to give him (or her) full responsibility for project definition and execution.

We believe that he or she is not just the one that runs the project or the one that handles the budget or the schedule, but the one that "understand and look at their projects as if they were running their own business", as our CEO use to say, with an integrated approach to his/her job.

Our culture sets the priority on supporting the project managers in their job, helping them in the decision making processes, and providing them with the needed tools and training to do their job. This approach allow for a certain degree of a not so strict formal processes. This allows the Project manager's responsibility and initiative to be displayed, but always under compliance with the framework and set of rules that allows for a solid accounting and results reporting.

We can say that project management has always been supported throughout the different stages of evolution of the company, and throughout the different business units, although some areas have been more reluctant in implementing changes in their established way of performing the job. One of the main barriers or drawbacks is the ability to use the same project management concepts for the different types of projects and products. It is still a major concern in our training programs to try to explain how the framework and the methodology is applied to projects with a high degree of definition in scope and to projects with a lesser degree of definition (fuzzy projects).

6.2 CORPORATE VALUES

An important part of the culture in excellent companies is an established set of values that all employees abide by. The values go beyond the normal "standard practice" manuals and morality and ethics in dealing with customers. Ensuring that company values and project management are congruent is vital to the success of any project. In order to ensure this congruence of values, it is important that company goals, objectives, and values be well understood by all members of the project team.

Successful project management can flourish within any structure, no matter how terrible the structure looks on paper, but the culture within the organization must support the four basic values of project management:

- Cooperation
- Teamwork
- Trust
- Effective communication

6.3 TYPES OF CULTURES

There are different types of project management cultures based upon the nature of the business, the amount of trust and cooperation, and the competitive environment. Typical types of cultures include:

- *Cooperative Cultures:* These are based upon trust and effective communication, not only internally but externally as well.
- *Noncooperative Cultures:* In these cultures, mistrust prevails. Employees worry more about themselves and their personal interests than what is best for the team, company, or customer.
- *Competitive Cultures:* These cultures force project teams to compete with one another for valuable corporate resources. In these cultures, project managers often demand that the employees demonstrate more loyalty to the project than to their line manager. This can be disastrous when employees are working on multiple projects at the same time.
- *Isolated Cultures:* These occur when a large organization allows functional units to develop their own project management cultures. This could also result in a culture-within-a-culture environment. This occurs within strategic business units.
- *Fragmented Cultures:* Projects where part of the team is geographically separated from the rest of the team may result in a fragmented culture. Fragmented cultures also occur on multinational projects, where the home office or corporate team may have a strong culture for project management but the foreign team has no sustainable project management culture.

Cooperative cultures thrive on effective communications, trust, and cooperation. Decisions are made based upon the best interest of all of the stakeholders. Executive sponsorship is more passive than active, and very few problems ever go up to the executive levels for resolution. Projects are managed more informally than formally, with minimum documentation, and often with meetings held only as needed. This type of project management culture takes years to achieve and functions well during both favorable and unfavorable economic conditions.

Noncooperative cultures are reflections of senior management's inability to cooperate among themselves and possibly their inability to cooperate with the workforce. Respect is nonexistent. Noncooperative cultures can produce a good deliverable for the customer if one believes that the end justifies the means. However, this culture does not generate the number of project successes achievable with the cooperative culture.

Competitive cultures can be healthy in the short term, especially if there exists an abundance of work. Long-term effects are usually not favorable. An electronics firm would continuously bid on projects that required the cooperation of three departments. Management then implemented the unhealthy decision of allowing each of the three departments to bid on every job. Whichever department would be awarded the contract, the other two departments would be treated as subcontractors.

Management believed that this competitiveness was healthy. Unfortunately, the long-term results were disastrous. The three departments refused to talk to one another and the sharing of information stopped. In order to get the job done for the price quoted, the departments began outsourcing small amounts of work rather than using the other departments, which were more expensive. As more and more work was being outsourced, layoffs occurred. Management now realized the disadvantages of a competitive culture.

The type of culture can be impacted by the industry and the size and nature of the business. According to Eric Alan Johnson and Jeffrey Alan Neal[1]:

> *Data orientated culture*: The data orientated culture (also known as the data driven culture and knowledge based management) is characterized by leadership and project managers basing critical business actions on the results of quantitative methods. These methods include various tools and techniques such as descriptive and inferential statistics, hypothesis testing and modeling. This type of management culture is critically dependent on a consistent and accurate data collection system specifically designed to provide key performance measurements (metrics). A robust measurement system analysis program is needed to insure the accuracy and ultimate usability of the data.
>
> This type of culture also employs visual management techniques to display key business and program objects to the entire work population. The intent of a visual management program is not only to display the progress and performance of the project, but to instill a sense of pride and ownership in the results with those who are ultimately responsible for project and program success . . . the employees themselves.
>
> Also critical to the success of this type of management culture is the training required to implement the more technical aspects of such a system. In order to accurately collect,

---

1. Eric Alan Johnson, Satellite Control Network Contract Deputy Program Director, AFSCN, and the winner of the 2006 Kerzner Project Manager of the Year Award; and Jeffrey Alan Neal, Blackbelt/Lean Expert and Lecturer, Quantitative Methods. University of Colorado, Colorado Springs.

assess and enable accurate decision-making the diverse types of data (both nominal and interval data), the organization needs specialists skilled in various data analysis and interpretation techniques.

6.4 CORPORATE CULTURES AT WORK

Cooperative cultures are based upon trust, communication, cooperation, and teamwork. As a result, the structure of the organization becomes unimportant. Restructuring a company simply to bring in project management will lead to disaster. Companies should be restructured for other reasons, such as getting closer to the customer.

Successful project management can occur within any structure, no matter how bad the structure appears on paper, if the culture within the organization promotes teamwork, cooperation, trust, and effective communications.

Boeing

In the early years of project management, the aerospace and defense contractors set up customer-focused project offices for specific customers such as the Air Force, Army, and Navy. One of the benefits of these project offices was the ability to create a specific working relationship and culture for that customer.

Developing a specific relationship or culture was justified because the projects often lasted for decades. It was like having a culture within a culture. When the projects disappeared and the project office was no longer needed, the culture within that project office might very well disappear as well.

Sometimes, one large project can require a permanent cultural change within a company. Such was the case at Boeing with the decision to design and build the Boeing 777 airplane. The Boeing 777 project would require new technology and a radical change in the way that people would be required to work together. The cultural change would permeate all levels of management, from the highest levels down to the workers on the shop floor. Table 6–1 shows some of the changes that took place.[2]

As project management matures and the project manager is given more and more responsibility, project managers may be given the responsibility for wage and salary administration. However, even excellent companies are still struggling with this new approach. The first problem is that the project manager may not be on the management pay scale in the company but is being given the right to sign performance evaluations.

The second problem is determining what method of evaluation should be used for union employees. This is probably the most serious problem, and the jury hasn't come in yet on what will and will not work. One reason why executives are a little reluctant to implement wage and salary administration that affects project management is because of

---

2. The Boeing 777 case study appears in H. Kerzner, *Project Management Case Studies*, Wiley, Hoboken, NJ, 2009, p. 81.

TABLE 6–1.   CHANGES DUE TO BOEING 777 NEW AIRPLANE PROJECT

Situation	Previous New Airplane Projects	Boeing 777
• Executive communications	• Secretive	• Open
• Communication flow	• Vertical	• Horizontal
• Thinking process	• Two dimensional	• Three dimensional
• Decision-making	• Centralized	• Decentralized
• Empowerment	• Managers	• Down to factory workers
• Project managers	• Managers	• Down to nonmanagers
• Problem solving	• Individual	• Team
• Performance reviews (of managers)	• One way	• Three ways
• Human resources problem focus	• Weak	• Strong
• Meetings style	• Secretive	• Open
• Customer involvement	• Very low	• Very high
• Core values	• End result/quality	• Leadership/participation/customer satisfaction
• Speed of decisions	• Slow	• Fast
• Life-cycle costing	• Minimal	• Extensive
• Design flexibility	• Minimal	• Extensive

the union involvement. This dramatically changes the picture, especially if a person on a project team decides that a union worker is considered to be promotable when in fact his or her line manager says, "No, that has to be based upon a union criterion." There is no black-and-white answer for the issue, and most companies have not even addressed the problem yet.

Midwest Corporation (Disguised Company)

The larger the company, the more difficult it is to establish a uniform project management culture across the entire company. Large companies have "pockets" of project management, each of which can mature at a different rate. A large Midwest corporation had one division that was outstanding in project management. The culture was strong, and everyone supported project management. This division won awards and recognition on its ability to manage projects successfully. Yet at the same time, a sister division was approximately five years behind the excellent division in maturity. During an audit of the sister division, the following problem areas were identified:

- Continuous process changes due to new technology
- Not enough time allocated for effort
- Too much outside interference (meetings, delays, etc.)
- Schedules laid out based upon assumptions that eventually change during execution of the project
- Imbalance of workforce
- Differing objectives among groups
- Use of a process that allows for no flexibility to "freelance"
- Inability to openly discuss issues without some people taking technical criticism as personal criticism
- Lack of quality planning, scheduling, and progress tracking

- No resource tracking
- Inheriting someone else's project and finding little or no supporting documentation
- Dealing with contract or agency management
- Changing or expanding project expectations
- Constantly changing deadlines
- Last-minute requirements changes
- People on projects having hidden agendas
- Scope of the project is unclear right from the beginning
- Dependence on resources without having control over them
- Finger pointing: "It's not my problem"
- No formal cost-estimating process
- Lack of understanding of a work breakdown structure
- Little or no customer focus
- Duplication of efforts
- Poor or lack of "voice of the customer" input on needs/wants
- Limited abilities of support people
- Lack of management direction
- No product/project champion
- Poorly run meetings
- People do not cooperate easily
- People taking offense at being asked to do the job they are expected to do, while their managers seek only to develop a quality product
- Some tasks without a known duration
- People who want to be involved but do not have the skills needed to solve the problem
- Dependencies: making sure that when specs change, other things that depend on it also change
- Dealing with daily fires without jeopardizing the scheduled work
- Overlapping assignments (three releases at once)
- Not having the right personnel assigned to the teams
- Disappearance of management support
- Work being started in "days from due date" mode, rather than in "as soon as possible" mode
- Turf protection among nonmanagement employees
- Risk management nonexistent
- Project scope creep (incremental changes that are viewed as "small" at the time but that add up to large increments)
- Ineffective communications with overseas activities
- Vague/changing responsibilities (who is driving the bus?)

Large companies tend to favor pockets of project management rather than a companywide culture. However, there are situations where a company must develop a companywide culture to remain competitive. Sometimes it is simply to remain a major competitor; other times it is to become a global company.

6.5 SENTEL CORPORATION

Companies go to project management either because they have to or because they want to. Customers can often exert tremendous power over companies and mandate that project management become a way of life. When a company procrastinates in the implementation of project management, customers can become irate and remove contractors from approved bidding lists. The better choice is obviously to develop superior project management capability because you want to rather than waiting until the last minute and being forced into a situation that may be embarrassing. To do so requires strong executive leadership with an appreciation of the benefits of project management and a willingness to accept the lead role in its implementation. Executive-level leadership is critical. Such was the case with SENTEL Corporation.

Over the years, the U.S. government has become more actively involved in the way that contractors use project management. Some of the requirements imposed by certain government agencies include:

- Mandating that the contractors maintain a staff of PMP®s
- Mandating that the contractors have and use effectively a project management methodology that is compatible with government-required milestones
- Mandating that the contractors capture best practices both during the project and at completion and share the best practices and lessons learned with the government
- Convincing the government that the contractor has some degree of maturity in project management using assessment instruments of reasonably accepted project management maturity models

Companies that now wish to do business with the U.S. government must become project-centric companies, that is, a company with a culture that actively supports project management. When a company is a small- to medium-sized firm and reasonably transparent to its customers, the necessity for a superior project management culture exists. SENTEL Corporation recognized this early on in its growth and through strong executive leadership developed such a culture. According to Tresia Eaves, PMP® and Vice President, Technology Integration Group at SENTEL Corporation and winner of the 2007 Kerzner Project Manager of the Year Award:

> The culture within SENTEL Corporation, an approximately 350-person defense contracting company, is very supportive of project, program and portfolio management processes and techniques. Employees within the 3 diverse divisions of SENTEL share ideas with proposals, offer issue solutions and project referrals across "stove pipes" and discuss best practices through regularly scheduled meetings with Senior Management to include an annual PM Retreat sponsored by the CEO, Darrell Crapps. Mr. Crapps encourages the PMs to share ideas across projects and implements the sharing of best practices from the top down within the organization by having the Divisional VPs discuss their portfolios monthly in meetings with Corporate staff. Proposal teams are built across Division lines to get the most applicable technical and corporate knowledge included in responses, regardless of where the contractual work will be located. Many of SENTEL's clients demand

proven project, program and portfolio management processes be entrenched in the culture of the company and ask for referral clients who will attest to that fact. Mr. Crapps and other Executive Management, regularly travel to meet with customers and employees to provide assurance they have the commitment of everyone within the SENTEL in helping them meet their requirements and objectives. SENTEL is currently working on building a "best practices" intranet page which will include blogs, discussion threads, and an internal wiki page to share ideas across groups, projects, processes, and product lines anywhere in the world.

Developing a corporate culture cannot be developed overnight. However, the development time can be accelerated through strong executive leadership that is clearly visible to the work force and by functioning as executive-level project sponsors. Visibility and sponsorship are two critical supportive functions. Consider the following comments from two executives at SENTEL:

SENTEL Corporation is very supportive of project management going so far as to allow each PM to run their contracts as they would their own business. SENTEL's CEO, Darrell Crapps believes that ultimately the relationships these PMs have with their clients keeps them coming back to SENTEL and allows them to share the good news with others who eventually use SENTEL's services or purchase their products. (Tresia Eaves, Vice President)

Company executives definitely work as PM sponsors by encouraging communication forums and listening to ideas that PMs have about their work and the company and enabling the best ideas to become reality. An example of this is when Tresia Eaves, PMP® and VP of SENTEL's Technology Integration Group saw that her PMs had a kind skewed idea about how their project fit into the overall company budget so she suggested to her leadership that the company implement project-based accounting. With a few minor process bumps along the way, SENTEL rolled this process out in one planning cycle and provided PMs the visibility from their projects all the way up to the company-wide budget so they could easily appreciate how their successes or failures had effects through the whole company. The experience from this type of budgeting also trains PMs on critical accounting techniques they might not otherwise learn. (Darrell Crapps, CEO)

When a company develops a culture for project management, a decision must be made as to how much freedom each project manager will be given and whether the project managers will be allowed to manage their projects more informally rather than formally. Informality can allow project managers the luxury of not having to follow the enterprise project management methodology. Although this can work, it can also lead to costly mistakes. Not all customers are supportive of an informal project management process, especially when a customer like the U.S. government has thousands of contractors. Some sort of formality and standardization may be necessary. According to Darrell Crapps, CEO, SENTEL:

SENTEL defines project management processes in very formal ways due to the contractual obligations it has with the company's government clients. Often times, very detailed project, program and portfolio management and process maturity requirements are included in the Request for Proposals from savvy customers and the subsequent responses to these requests by SENTEL and competitors are critical to the decision of who retains or wins the work. The government can be a very small world and a company's culture and abilities with regard to project management is usually well known and has much to do with how successful they can be at growing their business. The formality of well defined processes which are then backed up with years of lessons learned and metrics gives companies like SENTEL an advantage over competitors who have not committed to an acceptable level of maturity in their organizations for these skills.

As mentioned previously, government agencies are now requiring contractors to demonstrate some level of maturity in project management. Using formal project management practices, the assessment of a level of maturity or reaching a certain maturity level is significantly easier than using informal project management practices. As stated by Tresia Eaves, PMP® and Vice President, Technology Integration Group at SENTEL Corporation:

> SENTEL has supported project management informally from its beginnings in 1987. The more formal support from project management came when the leadership of SENTEL foresaw that the SEI Capability Maturity Model would eventually become mandatory for government contractors and decided to embark on the more formal journey making the sometimes pricey investments in the company to build the infrastructure and support processes necessary to build a permanent and successful program; one that would truly be a part of the corporate culture—not just a box to check in a Statement of Work. Also, SENTEL is currently working toward becoming ISO 9001:2000 certified to further enhance its processes in support of the project management function and customer satisfaction.

Conversion to a project management culture is not easy. Barriers must be overcome even in those companies with strong leadership and visible project management support from the top down. Darrell Crapps, CEO, SENTEL, states:

> The barriers to implementing formal and informal project management techniques within SENTEL had much to do with our clients as they defined how much we would be able to do for them in this area. Often times, if a client has not worked within a supportive project management culture, they will have reservations but as soon as they discover the benefits and reap accolades from their managers, the support rolls in and trust is established. One of our customers has seen the power of agility that solid project, program and portfolio management processes gives them in an environment that is subject to legislative changes and cannot always plan for every contingency. Most of our customers believe in the value of the project management profession because they've seen the results by working with SENTEL. Tresia Eaves, PMP® and Vice President of the SENTEL Technology Integration Group, saw barriers with the transparency that good project managers promote. She said, "One of our customers was initially wary of published schedules, work breakdown structures, and estimates to actuals measurements but once we established trust and delivered on our promises, the customer relaxed and

the improvements started compounding to the point that they now push their government employees to get their PMP® and implement Earned Value Measurement techniques for all of their projects."

6.6 VITALIZE CONSULTING SOLUTIONS, INC.

When you work within your own company, time is often a luxury for changing the culture to be more supportive of project management. But when you work as a consultant and must perform miracles at a client that has a culture that refuses to support project management, life can be quite difficult. Marc Hirshfield, PMP, Director, Project Management Office at Vitalize Consulting Solutions, Inc., states:

> As a project management consultant working at a client site, the culture of the company many times does not typically support project management. Usually, the initial reaction to a PM is skepticism until they have a PM lead a project team and implement a successful project. Once they realize the importance of project management, including a well defined workplan, status reports, issues list, etc., their opinions change. In most cases, the culture allows project management to work on an informal basis. However, a PM can be successful by slowly introducing more formal methodologies to a project that create standards for future projects.
>
> One specific cultural barrier to an organization's acceptance of the project management function is the perception that project management is too time-consuming, requires too much detail, and that use of the project management tool methodologies overwhelm the project. With too much emphasis on the tool, this perception can become a reality. Projects work best, and are better accepted by management, if the PM focuses on communication as the goal, with the tool used as simply a way to keep large amounts of tasks organized and tracked into manageable pieces.
>
> Another large cultural challenge in our industry is scope creep. Key stakeholders, including the management staff, may request new scope to the project, which puts the project at risk. One way to overcome this challenge is to have a scope document prepared and approved before the project begins. If anyone attempts to add scope, the PM can remind the group of the original plan (scope statement) and how adding new work would impact the budget and the timeline.
>
> Lack of commitment by key stakeholders is also a cultural challenge in our industry. Sometimes people want to be included in the decision making process, but do not wish to attend the meetings. This causes unnecessary delays, which can impact the project's completion. One way to overcome this challenge is to meet with the key stakeholder(s) and personally ask them to attend the meeting, check their schedules in advance to ensure they are available or arrange to meet with them weekly to review all decisions and obtain their approval before proceeding with the project. Although, this may be additional work on the PM, it will ensure that unnecessary delays are avoided and support is provided.

6.7 DFCU FINANCIAL

There is no question concerning the importance of a good corporate culture in helping companies mature in project management. Too often, companies provide just lip service to the importance of culture and overemphasize the importance of tools and techniques. However, when companies focus on implementing the appropriate and needed cultural changes, good things can happen, and reasonably quickly. With a good culture, the organization recognizes quickly what works well and what needs improvement. Cultures can allow a needed change to occur many times with the idea that the change is in the best interest of the company. Such was the case with the culture at DFCU Financial[3]:

At $1.7 billion in assets, DFCU Financial Federal Credit Union is the largest credit union in Michigan and among the top 40 largest in the nation. With a 246% increase in net income since 2000, DFCU Financial has never done better, and effective project implementation has played a key role. At the root of this success story is a lesson in how to leverage what is best about your corporate culture.

Rolling back the clock to late 1997, I had just volunteered to be the Y2K project manager—the potential scope, scale and risk associated with this project scared most folks away. And with some justification—this was not a company known for its project successes. We made it through very well, however, and it taught me a lot about the DFCU Financial culture. We did not have a fancy methodology. We did not have business unit managers who were used to being formally and actively involved in projects. We did not even have many IT resources who were used to being personally responsible for specific deliverables. What we did have, however, was a shared core value to outstanding service—to doing whatever was necessary to get the job done well. It was amazing to me how effective that value was when combined with a well-chosen sampling of formal project management techniques.

Having tasted project management success, we attempted to establish a formal project management methodology—the theory being that if a little formal project management worked well, lots more would be better. In spite of its bureaucratic beauty, this methodology did not ensure a successful core system conversion in mid-2000. We went back to the drawing board to revisit project management since we were facing a daunting list of required projects.

With the appointment of a new president in late 2000, DFCU Financial's executive team began to change. It did not take long for the new team to assess the cultural balance sheet. On the debit side, we faced several cultural challenges directly affecting project success:

- Lack of accountability for project execution
- Poor strategic planning and tactical prioritization
- Projects controlled almost exclusively by IT
- Project management overly bureaucratic
- Limited empowerment

On the plus side, our greatest strength was still our strong service culture. Tasked with analyzing the company's value proposition in the market place, former senior vice president of marketing, Lee Ann Mares, made the following observations:

Through the stories that surfaced in focus groups with members and employees, it became very clear that this organization's legacy was extraordinary service. Confirming that the DFCU brand was all about service was the easy part. Making that generality accessible and actionable was tough. How do you break a high-minded concept like *outstanding*

---

3. The remainder of this section was provided by Elizabeth M. Hershey, PMP, Vice President, Delivery Channel Support at DFCU Financial.

service into things that people can relate to in their day-to-day jobs? We came up with three crisp, clear Guiding Principles: Make Their Day; Make It Easy; and Be An Expert. Interestingly enough, these simple rules have not only given us a common language, but have helped us to keep moving the bar higher in so many ways. We then worked with line employees from across the organization to elaborate further on the Principles. The result was a list of 13 Brand Actions—things each of us can do to provide outstanding service (Table 6–2).

While we were busy defining our brand, we were also, of course, executing projects. Since 2000, we have improved our operational efficiency through countless process improvement projects. We have replaced several key sub-systems. We have launched new products and services. We have opened new branches. We have also gotten better and better at project execution, due in large part to several specific changes we made in how we handle projects. When we look closer at what these changes were, it is striking how remarkably congruent they are with our Guiding Principles and Brand Actions. As simple as it may sound, we have gotten better at project management by truly living our brand.

Brand Action—Responsibility Project control was one of the first things changed. Historically, the IT department exclusively controlled most projects. The company's project managers even reported to the CIO. As chief financial officer, Eric Schornhorst comments, "Most projects had weak or missing sponsorship on the business

TABLE 6–2.   DFCU FINANCIAL BRAND ACTIONS

Make Their Day	Make It Easy	Be An Expert
Voice	We recognize team members as the key to the company's success, and each team member's role, contributions, and voice are valued.	
Promise	Our brand promise and its guiding principles are the foundation of DFCU Financial's uncompromising level of service. The promise and principles are the common goals we share and must be known and owned by all of us.	
Goals	We communicate company objectives and key initiatives to all team members, and it is everyone's responsibility to know them.	
Clarity	To create a participative working environment, we each have the right to clearly defined job expectations, training, and resources to support job function and a voice in the planning and implementation of our work.	
Teamwork	We have the responsibility to create a teamwork environment, supporting each other to meet the needs of our members.	
Protect	We have the responsibility to protect the assets and information of the company and our members.	
Respect	We are team members serving members, and as professionals, we treat our members and each other with respect.	
Responsibility	We take responsibility to own issues and complaints until they are resolved or we find an appropriate resource to own them.	
Empowerment	We are empowered with defined expectations for addressing and resolving member issues.	
Attitude	We will bring a positive, "can do" attitude to work each day—it is my job!	
Quality	We will use service quality standards in every interaction with our members or other departments to ensure satisfaction, loyalty and retention.	
Image	We take pride in and support our professional image by following dress code guidelines.	
Pride	We will be ambassadors for DFCU Financial by speaking positively about the company and communicate comments and concerns to the appropriate source.	

side. To better establish project responsibility, we moved the project managers out of IT, and we now assign them to work with a business unit manager for large-scale projects only. The project managers play more of an administrative and facilitating role, with the business unit manager actually providing project leadership." Our current leadership curriculum, which all managers must complete, includes a very basic project management course, laying the foundation for further professional development in this area.

Guiding Principle—Make It Easy

With project ownership more clearly established, we also simplified our project planning and tracking process. We now track all large corporate and divisional projects on a single spreadsheet that the executive team reviews monthly (see Table 6–3 for the report headers). Project priority is tied directly to our strategic initiatives. Our limited resources are applied to the most impactful and most critical projects. Eric Schornhorst comments, "Simplifying project management forms and processes has enabled us to focus more on identifying potential roadblocks and issues. We are much better at managing project risk."

Brand Action—Goals

Chief information officer Vince Pittiglio recalls the legacy issue of IT over-commitment. "Without effective strategic and tactical planning, we used to manage more of a project wish list than a true portfolio of key projects. We in IT would put our list of key infrastructure projects together each year. As the year progressed, individual managers would add new projects to our list. Often, many of these projects had little to do with what we were really trying to achieve strategically. We had more projects than we could do effectively, and to be honest, we often prioritized projects based on IT's convenience, rather than on what was best for the organization and our members." Focusing on key initiatives has made it possible to say "no" to low priority projects that are non-value addition or simply not in our members' best interest.

TABLE 6–3. DFCU FINANCIAL CORPORATE PROJECTS LIST REPORT HEADERS

Priority	1 = Board reported and/or top priority 2 = High priority 3 = Corporate priority, but can be delayed 4 = Business unit focused or completed as time permits
Project	Project name
Description	Brief entry, especially for new initiatives
Requirements document status	R = Required Y = Received N/A = Not needed
Status	Phase (discovery, development, implementation) and percentage completed for current phase
Business owner	Business unit manager who owns the project
Project manager	Person assigned to this role
Projected delivery time	The year/quarter targeted for delivery
Resources	Functional areas or specific staff involved
Project notes	Brief narrative on major upcoming milestones or issues

In addition, the current measuring stick for project success is not merely whether the IT portion of the project was completed, but rather that the project met its larger objectives and contributed to the company's success as a whole.

Brand Action—Teamwork   Historically, DFCU Financial was a strong functional organization. Cross-departmental collaboration was rare and occurred only under very specific conditions. This cultural dynamic did not provide an optimal environment for projects. The monthly project review meeting brings together the entire executive team to discuss all current and upcoming projects. The team decides which projects are in the best interest of the organization as a whole. This critical collaboration has contributed to building much more effective, cross-functional project teams. We are developing a good sense of when a specific team or department needs to get involved in a project. We also have a much better understanding of the concept that we will succeed or fail together. We are working together better than ever.

Brand Action—Empowerment   As chief operating officer Jerry Brandman points out, "our employees have always been positive and pleasant. However, our employees were never encouraged to speak their minds, especially to management. This often had a direct negative impact on projects—people foresaw issues, but felt it was not their place to sound the alarm. A lot of the fear related to not wanting to get others 'in trouble.' We have been trying to make it comfortable for people to raise issues. If the emperor is naked, we want to hear about it! To make people visualize the obligation they have to speak up, I ask them to imagine they are riding on a train and that they believe they know something that could put the trip in jeopardy. They have an obligation to pull the cord and stop the train. This has not been easy for people, but we are making headway every day."

Brand Action—Quality   At DFCU Financial project implementation in the past followed more of the big bang approach—implement everything all at once to everyone. When the planets aligned, success was possible. More often than not, however, things were not so smooth. Comments Jerry Brandman, "You have to have a process for rolling things out to your public. You also need to test the waters with a small-scale pilot whenever possible. This allows you to tweak and adjust your project in light of real feedback." Employees all have accounts at DFCU Financial, so we have a convenient pilot audience. Recent projects such as ATM to Debit Card conversions and the introduction of E-statements have all been piloted with employees prior to launching them to the entire membership.

Bottom line, the most significant best practice at DFCU Financial has been to be true to our core cultural value of providing extraordinary service. As we were working on defining this value and finding ways to make it actionable, we were also making changes to the way we approach project management that were very well aligned with our values. Our commitment to living our brand has helped us:

● Move project responsibility from IT to the business units
● Simplify project management forms and processes

- Use project review meetings to set priorities and allocate resources more effectively
- Break down organizational barriers and encourage input on projects from individuals across the organization
- Improve project success through pilots and feedback

As president and CEO, Mark Shobe summarizes, "good things happen when you have integrity, when you do what you say you are going to do. The improvements we have made in handling projects have rather naturally come out of our collective commitment to really live up to our brand promise. Have we made a lot of progress in how we manage projects? Yes. Is everything where we want it to be? Not yet. Are we moving in the right direction? You bet. And we have a real good road map to get there."

How Good Is the Road Map?

The preceding material was written in early 2005. By objective measures, the last four years have been good ones for DFCU Financial (see Table 6–4). With over $2 billion in assets, the credit union ranks in the top 10 among its peers in the most important key measures.

So, from a purely financial perspective, DFCU is doing very well, especially given the global economic climate as 2009 begins and the fact that DFCU largely serves members associated with the auto industry sector that is so much a part of the notoriously troubled Michigan economy.

These solid financials are the result of the current administration's efforts over the last eight years to streamline and improve operations, clarify DFCU's brand and value proposition and initiate effective project selection and execution processes.

While these efforts were underway, the executive team and board of directors were evaluating a troubling metric—one that could undermine the company's ability to sustain its recent successes: DFCU's membership has been nearly flat for many years—a trend affecting nearly all U.S. credit unions.

TABLE 6–4. DFCU FINANCIAL RESULTS

		For Quarter Ending September 30, 2008			
		Result		Ranking	
Metric	DFCU	National Peer[1] Average	Regional Peer[2] Average	National Peers	Regional Peers
Return on assets	1.94%	0.42%	0.77%	1	1
Return on equity	13.98%	4.23%	7.19%	2	2
Efficiency ratio	49.57%	65.41%	69.31%	5	3
Capital/assets	13.91%	9.82%	10.69%	2	9
Total assets	$2.0B	$3.4B	$1.0B	39	5

1 50 largest credit unions as measured by total assets.

2 Credit unions with at least $500 million in total assets in the states of Michigan, Pennsylvania, Ohio, Indiana, Illinois, Wisconsin, and Minnesota. The peer totals 56 credit unions.

The most recent years of the current administration have therefore been focused on the critical, strategic issue of growth. In addition, it is evident that DFCU's brand and guiding principles have helped shape the results.

Brand Actions—Voice and Quality; Guiding Principle—Make Their Day

While DFCU's executive team and board of directors explored several growth options, work on selection of a new core processing system began in mid-2005. The core system conversion project was viewed as a strategic imperative for growth, regardless of DFCU's operating structure. The prior system conversion in 2000 suffered from poor execution that left behind lingering data and process issues that needed to be addressed. At the start of the conversion project, which began formally in January 2006, the conversion was targeted for October of that same year. From the outset, however, we ran into difficulties with the system vendor. The vendor was going through one of its largest client expansions ever and was having a hard time satisfying all of the demands of the conversion projects in its pipeline. The impact to us was noticeable—high turnover in key vendor project team members; poor quality deliverables; lack of responsiveness to conversion issues. Due to the poor quality of the data cuts, the project conversion date was at serious risk by June.

Due to its scope, the system conversion project was the only corporate project commissioned in 2006, and all attention was on it. It was not an easy task, therefore, to deliver the message that the project was in trouble. "Mark was well aware that we were having difficulties when we sat down to discuss whether we would have a smooth October conversion," commented chief information officer, Vince Pittiglio. "I have had to deliver bad news before to other bosses, but the talk I had with Mark was a lot different than those I had before." Our stated objective of this conversion was basically to do no harm. We all agreed that we could not put our members or our employees through the same type of conversion we experienced in July 2000. It had to be as close to a non-event as possible. According to Pittiglio, "I laid out the key issues we were facing and the fact that none of us on the project believed they could be resolved by the October date. If we kept to the original date, we believed we would negatively impact member experience." But our CEO Mark Shobe was very clear—he insisted that this project be a quality experience for both members and employees, as we all had agreed at the outset, and he was willing to go to the board to bump the date and to put other key initiatives on hold to ensure the conversion's success. The project team agreed upon a revised conversion date in early June 2007. According to conversion project manager and senior vice president, Martha Peters, "Though the team continued to face difficulties with the vendor, we worked hard and got it done without any major issues. It made a real difference to know that our feedback was taken seriously by the chiefs and the board. In the end, it really was a non-event for most of our members and employees, exactly as we all wanted it to be. It was a tough decision to postpone the conversion, one that many companies are not willing to make. But it was the right decision to make—we really try to live our brand."

Brand Action—Clarity and Teamwork

By the time the system conversion was completed in June 2007, we had really only engaged one project in the previous two years, albeit a large-scale project of strategic importance—which was consciously delayed by eight months. This project drained our resources, so little else was accomplished in those two years. "Coming out of the system conversion was this huge, pent-up project demand. And everyone thought that the issues facing their division were, of course, the most pressing," commented chief financial officer, Eric Schornhorst. "We found out very quickly that our handy project tracking list and monthly corporate project meeting were insufficient tools for prioritizing how to deploy our scarce project resources." A small team was quickly assembled to put together a process for initiating and approving projects more effectively and consistently. A key objective for this team was to minimize bureaucracy, while trying to establish some useful structure, including a preliminary review of all new requests by the IT division. The output was a simple flow diagram that made the steps in the request and approval process clear to everyone (see Figure 6–1) and a form that integrated the instructions for each section. As Marsha Hubbard, senior vice president of human resources, related, "My group was one of the first to use the new process. It was surprisingly well put together and easy to use. We made the pitch to replace our learning management system with a more robust, outsourced solution. It was one of the projects that made it on the 2008 list. To be honest, we were really at the point in our company's history where we needed a bit more discipline in this process. In years past, we independently advocated for our projects at budget time with our division heads. If we received budgetary approval, we viewed our project as 'on the list.' When it came time to actually execute, however, we often had trouble lining up the resources from all the different areas that needed to be involved, especially those in IT."

The new process is contributing not only to clarity regarding corporate projects, but to teamwork as well. According to Schornhorst, "We recently reviewed the projects requested for 2009 using the new process. While we haven't completely addressed the backlog of projects created by the system conversion, we also have another project likely to take over the majority of resources in 2009. This is not good news for areas that have not yet seen their projects addressed. What's interesting though is how little contention there was as we reviewed the docket for 2009—and we had to put many important things on the back burner. I think that when you have everyone review the facts together in lock-step, it's easier to get to a set of priorities that makes sense for the organization and is mutually supported regardless of personal interest. It helps to bring the best out in all of us."

So Where Is the Road Headed?

At the time of this writing, in early 2009, DFCU Financial is poised for membership growth. After a review of charter options, the board and executive team decided to remain a credit union, but to pursue other avenues for growth. To that end, as 2009 began, the board put forth for vote a proposal to DFCU's membership for the merger with a credit union with branch locations in the lower central and western areas of Michigan. According to chief operating officer, Jerry Brandman, "We considered many different options to address our strategic vision for growth and expansion in membership—from mergers to various internal growth strategies. While we in the credit union industry currently face the same challenges as other financial institutions, we also are an industry known for service. And here at DFCU Financial we not only talk about service—we deliver. Our employees are happy and enthusiastic. They treat our members very well. We have been rated in the top 101 companies to work for in southeastern Michigan for 5 years in a row, based on feedback from our employees. And we have a member shops program that shows us how our service stacks up to industry benchmarks. We have consistently performed at the highest level of service relative to our peers. A pivotal event was a few years back when the employees of Honda Federal Credit Union sent us a huge card, congratulating us for surpassing them on member service scores! Merging

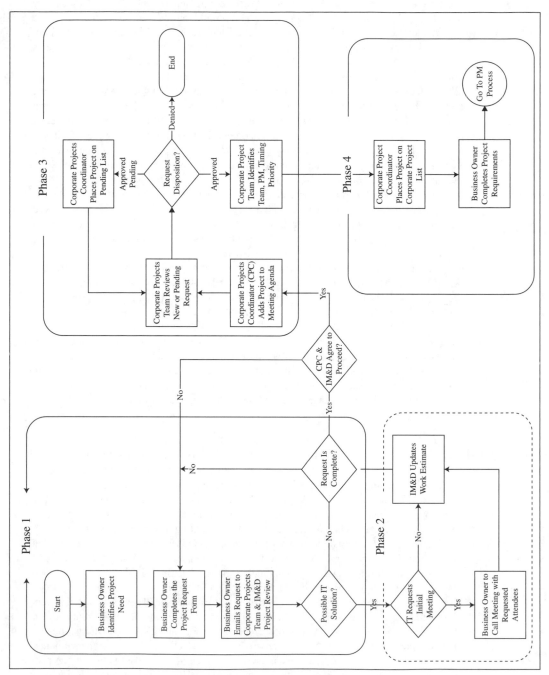

FIGURE 6–1. DFCU Financial's project initiation process.

with this other Michigan credit union and shifting to a community charter will provide us the growth we need to ensure a bright future for our members and our employees. It will allow us to spread the DFCU brand to other geographic areas. We believe this proposal will be appealing to our members and will be successful. We believe that people want to be a part of our organization."

Also in early 2009, DFCU Financial paid out a $17 million patronage dividend to its members for the third consecutive year, for a total of more than $50 million being paid out since inception, during one of the worst economic times to hit the Motor City in decades. According to Keri Boyd, senior vice president of marketing, "Since we are committed to remaining a credit union, we have looked at ways we could improve our value proposition to existing members and attract new members. The patronage dividend is the cornerstone of our approach to growing the business as a credit union." As Mark Shobe sums it up, "The board and I did not want to begin the payout until we were confident that we could sustain it over the years. It took hard work, some tough decisions, excellent project execution and diligence in our day-to-day operations to be in the position to share our success with our membership. The simple truth is that the driving force behind our success is our collective commitment to our brand." So, with a 2009 schedule of agreed-upon projects, a potential merger in the offing and some very, very satisfied members, how is the DFCU road map working? "Quite well, thank you!" replies Mark.

6.8 ILLUMINAT (TRINIDAD & TOBAGO) LIMITED

On paper, the merger of two or more companies may look easy. But if the companies have different cultures, and each culture has a different view of the benefits of project management, the actual merger may not produce the benefits expected. The remainder of this section been provided by Cynthia James-Cramer, EMBA, PMP, Manager, Project Management Services, ILLUMINAT (Trinidad & Tobago) Limited.

Managing the PM Culture in a Merger   A merger of companies always brings some inherent challenges to the newly formed organization. The methods employed in rising to these challenges would determine not only the success of the merger but also how soon the operations can reach a level of efficient operations.

When three entities of the ITC group (Information Technology & Communications) of the Neal & Massy Holdings company merged in the year 2001, September 11, the then CEO of ILLUMINAT, Mr. Keith Thomas, embarked on a thrust to introduce the organization to a Professional Services Methodology (Profserv) in which a project management methodology was embedded along with organizational learning and organizational adoption.

Of the three entities, only one company previously had a culture that supported project management because of its IT focus and had introduced the concept of a project office just prior to the merger. The tasks before the CEO were now to not only implement the Profserv methodology but also to institutionalize a project management culture. This was essential as the new company's mission was now focused on providing solutions which spanned the offerings of all three entities distinct from their previous undertakings of providing products and services.

The Mandate

The CEO issued a mandate for adoption of the Profserv methodology to all employees inclusive of senior management who was held responsible for the success of its execution. A similar mandate was extended to a companywide adoption of a standard approach to project management.

The Approach

First all managers, team leaders, and key department resources were sensitized to a project management culture and also shown what fallouts can emerge when such a culture does not exist. The evidence was apparent from a research done on a sample of key projects, which were executed without a project management methodology that clearly demonstrated where things went wrong and the impact on the success of these projects. The response and feedback from the employees were tremendous, which resulted in some employees' eagerness on the training up ahead. The organization itself was restructured to a team-based approach to services delivery and, where relevant, the supervisor and manager roles were renamed team leader.

The project office was now refocused to a project management office (PMO) with a mandate to establish and roll out a common project management methodology and to act as a repository for disseminating project management information as well as for maturing the culture in the short and long term. The organization's leadership showed its commitment by investing in staff training and complementing the existing staff with suitable, experienced project management recruitment.

The Success

The merger challenges of mixed cultures, merged departments, merged functions, and varying levels of project managment maturity or the absence of it were now bridged and tackled as a single test. The emerging success has been rewarding as the team approach has taken off. There is now a standardized approach to the whole services delivery cycle from winning the opportunity to implementing the solution. There has also been evidence of an increase in repeat business from existing customers and the opportunity for sole tendering.

Recently another research was done with a similar sample of projects to the first research, this time with projects executed using the project management methodology. The results spoke myriads to improved project success.

Sustaining the PM Culture

Top management is continuing its investment in the project management culture by supporting the ongoing development of the PMO staff and its endeavors. The PMO, in keeping its visibility in the organization, has launched a team-based annual recognition award program and has started a quarterly newsletter, which is circulated through the entire organization. The newsletter reports on project inventory, project successes, and project management–focused titbits.

Along with these efforts, staff with a specific interest in the discipline is being encouraged to pursue the basic project management certification, few have already done so, and others are taking up the challenge. In addition, new employees are introduced to the PMO through the new-employee orientation program. Yet another thrust is planned internal training for all key delivery personnel to reinforce the project management methodology.

Conclusion

Adopting a PM culture in a merger may look insurmountable at first, but by following best business practices of top management buy-in, involvement, and support, half of the job gets down. The rest lies with the PMO to stay visible by mentoring, teaching, and providing the necessary information and support to all arms of the organization until the transfer of the culture has been embedded.

In the words of the then CEO:

> Getting started! The approach to adoption is perhaps the most critical aspect of any change process. Arguably more critical than the change itself, it must be very carefully thought through and implemented. What follows is the result of due deliberation on the part of management and feedback from staff focus groups.
>
> Some believe that Adults accept new approaches best when they understand why. Therefore, we would begin with organization-wide sessions on the WHY.
>
> This would be followed by the HOW.

6.9 MCELROY TRANSLATION

Project management cultures are most often designed around the desires and aspirations of senior management. However, the information necessary to transform an existing culture into a project management culture, especially a culture that supports and fosters excellence in project management, also requires input and understanding from all levels of the organization. Tina (Wuelfing) Cargile, PMP®, Project Manager at McElroy Translation Company, describes her experience at McElroy Translation:

> McElroy Translation had been a successful company for decades, primarily working with established clients on fairly routine projects (patent and article translations from foreign languages into English). Procedures that had been developed over the years were performed as they always had been. Deadlines, for the most part, were determined arbitrarily and many projects lingered well beyond their due dates or were subject to sudden acceleration with little explanation. Individual departments operated independently, often unaware of the activities of neighboring departments, and were generally not informed about the procedures, activities, and goals of the sales and operations divisions of the company.
>
> In the 1980s, a rudimentary project management culture began to emerge, involving tracking and oversight of work flow by project coordinators and independent planning and tracking methodologies adopted by individual departments. While this improved

internal efficiency within each department and gave management an overview of project activities, there was still little communication between divisions, commitments were made regarding project scope and turnaround without the participation of staff, and line managers were obliged to cope with the specifications and deadlines assigned as best they could.

In the late 1990s, the nature of the translation business began to change rapidly with the advent of globalization as a corporate necessity and subsequent increased client demand for sophisticated, full-service solutions and aggressive turnaround times. As sales in traditional markets began to sag in the early 2000s, the company began to analyze both the increased activity in requests for translation from English into foreign languages and expedited delivery requests and further analyzed internal processes and how the existing business model could be modified to better serve new customer demands.

It was determined that the complexity and variety of the translation and localization needs of clients were not well served by the current operation. Management and sales personnel discussed pending projects with clients in terms of what the client was request-ing rather than in terms of what the client needed. This sometimes resulted in the client receiving unusable deliverables (e.g., a translation delivered with encoding incompatible with the client's software or a translation localized for use in Mainland China rather than Taiwan). And while historical best practices had created a series of processes that worked fairly well for standard projects, employees were unable to adapt effectively to unusual requests, often because they had no information regarding the clients' business needs or because the technical requirements involved in providing a requested deliverable had been overlooked. For example, the demand for a cursory edit on a rush patent translation was met with resentment in the editing department and a feeling that management did not care about the integrity of the quality process, since they were unaware that the client planned to use the rough deliverable to decide whether to proceed with further translations of asso-ciated patents. Production staff would often be forced to scramble to find a solution when the requested software application and language were incompatible.

After soliciting input from upper managers, project managers, line managers, and employees, it was determined that dissemination of information and improved commu-nication company-wide (including communication between company representatives and clients) had to be improved in order to position McElroy to develop a competitive edge by offering clients solutions rather than products.

A rapid and radical shift in culture and approach took place as a result of this evaluation.

- The most important change was management's decision to invest in the development of proprietary software to enable real-time project activity tracking by all employees, including access to project information, customized reporting, and planning tools, which allowed line managers to more effectively coordinate their staff activities with those of other departments and enabled project managers to efficiently monitor activi-ties and shift focus from information gathering and micromanagement to front-end considerations.
- Making all project materials—including client communications, planning assumptions, estimate information, source files, and so on—available to all employees in electronic form on the shared network made it possible for line staff to independently clarify their understanding of the client's intent and needs. This resulted in an additional value add when remote access to the database system and the network was provided to employees,

allowing flexible scheduling to accommodate communication with global vendors and clients beyond the normal workday.

- Independent department network filing and folder structures were merged into a single structure to eliminate redundancy and to make all project materials easily accessible.

- Project managers were involved in the estimate stage to directly engage in problem solving with clients, bringing in technical expertise from line staff when needed, and to ensure that sales personnel had a clear understanding of the internal processes and challenges that should be considered when discussing projects with clients. Sales personnel were also instructed to probe service requests with questions designed to determine the client's true business need:

1. How would you describe the business problem you are trying to solve?
2. What is the current system in place for addressing translation needs?
3. Please describe content authoring.
4. How will the translations be used?
5. What is your desired output?
6. What resources will be available to the translators? Do you own existing translation memories or terminology glossaries?
7. What are your highest volume and top priority?
8. Please quantify historical and projected translation activity.

(For less complex projects, the simple question "What are you planning to do with the deliverable?" often yields important information that can result in a more effective project plan.)

- Lengthy team meetings on selected projects were replaced with a daily meeting, generally lasting no longer than 15 minutes, reviewing daily and weekly project status, and giving representatives of all departments the opportunity to make suggestions, ask questions, or engage others in problem solving. This activity proved to be particularly valuable when dealing with high-risk projects but also consistently identified solvable problems in more routine projects that previously would not have been discussed in a team setting.

- Project managers established an open-door policy and were encouraged to communicate freely with all staff regarding decisions and assumptions and to incorporate staff feedback into current or future project plans. Management's visible respect for the technical expertise of the employees in the trenches improved morale and trust and often allowed project managers to proactively renegotiate deadlines or staged deliverables.

- Deadlines were assigned to best accommodate client needs and internal workload. Projects were prioritized using a simple coding system, designed to indicate the relative urgency of deadlines and also to indicate where deadline flexibility was possible.

- A dedicated customer service department was created to address client issues, to collect and disseminate information regarding postdelivery problems, and to conduct project postmortems to improve understanding of client needs company-wide.

- Analysis of how information was communicated—between upper management, sales, project management, operations, clients, and accounting—led to the development of electronic shipping procedures and customized electronic reporting that eliminated much of the faxing, copying, shipping, and filing tasks that had previously required the services of a full-time office clerk.

The availability and sharing of information, the emphasis on project management as a company culture, along with upper management's commitment to encouraging teamwork and cooperation have transformed the company, which now enjoys niche status as a vendor of choice for extremely high-risk projects that many agencies must turn away due to turnaround requests, complexity, or the need for customized solutions.

Shelly Orr Priebe, General Manager, McElroy Translation Company, describes the view of the McElroy culture as seen through the eyes of the general manager:

Leading a company with a rich tradition of stability and financial strength offers undeniable benefits. But what of the potential pitfalls of this enviable market position? The principles of physics too easily apply. How easy it is for a body in motion to remain in motion without changing direction or for a body at rest to remain at rest. But market needs evolve, competition stiffens, trends in offshore sourcing emerge, and technology developments change everything. The challenge at McElroy is to offer clients and employees the best combination of (1) valuing long-term relationships and constancy and (2) competing with the new breed of tech-savvy, "slick" competitors that proliferated on the localization industry landscape in the 1990s.

It would be convenient and "sexy" to laud the many ways that McElroy Translation embraced technology to leapfrog past competitors under a new general manager in 1999. I could write volumes about proprietary technology that was developed and implemented and is continually being improved by internal programmers and systems specialists to better serve the needs of our clients. Technology is fashionable and, admittedly, it has been critical to our ongoing success.

But effective technology use is the by-product of what has mattered most—the emphasis on project management as a company culture, along with upper management's commitment to encouraging teamwork and cooperation. Ideas for change and improvement flow up the organizational chart as often, or more often, than they flow down. Seeing their own good ideas implemented, employees are more vested in achieving overall company goals, instead of just personal or departmental ones. For example, the translation coordination department recently proposed a system for radically decreasing turnaround time for a certain type of project. They were concerned that the "standard procedure" was not realistically addressing the need for a certain market segment. It is exceptional that this initiative for change was generated by the supply side of the company, that is, operations instead of sales or management. These are employees who view the big picture.

Project management must be structured and organized but also be flexible and far reaching. For 15 minutes a day department managers pull together to look each other in the eye and quickly run through the day's priorities and near-term work flow projections. That meeting replaces countless emails and clarifies group priorities, and new ideas are generated there. At that meeting everyone becomes a project manager. Therein lies the key to success.

6.10 DTE ENERGY

Tim Menke, PMP®, Senior Continuous Improvement Expert, DO Performance Management at DTE Energy, describes the culture at DTE Energy:

> While the level of support for project management varies across departments, generally it is increasing overall. Several years ago it was common to find experienced practitioners managing large, high visibility projects using both internal and external methodologies. Today practitioners can be found across the company applying the tools and techniques to a variety of projects. In some cases this is a result of stakeholder expectations. In other cases, this is driven by the person managing the project. The Continuous Improvement Project Management Methodology is used by all Lean Six Sigma Black Belt Candidates at DTE Energy as it is. . . . required of all projects submitted in support of certification.
>
> Both formal and informal project management approaches are used. Projects reporting progress to governance boards or stakeholder committees tend to be managed more formally. Most departments have implemented progress reporting templates to drive consistency. Projects managed within a Director's organization or below tend to be managed less formally. Recognizing all projects do not require the same level of project management rigor, our Continuous Improvement Project Management Methodology has been revised to "scale" to the particular project and stakeholder needs.
>
> Challenges to implementing project and program management we have experienced include:
>
> - Stakeholders driving scope creep
> - Project Managers with varied experience levels
> - Team members struggling to balancing operational and project needs
>
> Ways we have overcome these barriers include:
>
> - Peer-to-peer mentoring by experienced practitioners
> - Employee development plan activities to improve proficiency
> - Increased focus on managing the matrix reporting relationship

6.11 EDS

Doug Bolzman, Consultant Architect, PMP®, ITIL Service Manager at EDS, believes that:

> In many cases, instituting project management is part of the cultural change required by an organization. When major improvements are needed, the culture is improved by instituting project management and project management cannot depend on the culture being there.
>
> Implementing enterprise wide releases/projects requires the culture to move the organization from a functional to a matrix management, move the delivery from project centric to component centric, and move the planning from tactical (emotional) to strategic (analytical) planning. This level of cultural change needs to be identified, designed, and implemented.

6.12 CONVERGENT COMPUTING

Rich Dorfman, VP Sales and Services, provides a management perspective of Convergent Computing (CCO):

Culturally, CCO takes a commonsense and value-based approach to business, viewing project management from a similar perspective. Since we're always looking to maximize the consulting value that we provide to clients, we try to avoid overengineering and overprocessing activities associated with project management. Practicality speaking, this translates into doing what makes the most sense given situational/scenario-based thinking, therefore balancing our execution around project management (and process) from one situation to the next.

Whereas larger consulting organizations such as PWC, IBM Global Services, and Microsoft Consulting Services tend to have very structured disciplines for project management (often treating their methodology more as a "religion" than a guide), we recognize it's important to manage generic elements associated with project management in order to drive results, elements such as identifying stakeholders, starting with objectives, uncovering constraints, identifying the activities that are time critical, having clearly understood roles and responsibilities, and establishing effective communications. So, culturally we support project management as it relates to these elements; but in most cases, we favor a more informal and practical implementation of them, rather than following the traditional steps and reporting associated with most formalized project management methodologies.

Historically, our focus has its roots in engineering, and our consulting services have been focused more on providing advanced technology expertise and experience rather than project oversight. Management's support for project management was less from the standpoint of project management discipline and more from driving results—the principles inherent in project management. Too often management confused project management with project administration, oversimplifying the role to one that focused almost exclusively to managing project budget and schedule commitments. More often than not, management communicated (in actions and in words) that customers would not pay CCO for project management. So, even where project management time was supposed to be included as a part of CCO's service proposal, it was typically negotiated out of the proposal or not even proposed, because it was assumed the customer would not pay for it. Contributing to the gap, our project managers had a tendency toward thinking about project management in a more formalized way than is culturally our norm while our engineers tended toward focusing in on the technology itself, rather than results. As a result, in most cases the success of the project—project management—rested squarely on the shoulders of the lead consultant, rather than on a dedicated project manager. This history, especially for employees who have worked at CCO for three plus years, left in question management's support for project management.

Over the past year, our business has been steadily shifting from providing technology expertise to helping clients drive measurable business results from the technology they invest in. As our business has evolved, management has recognized that project management takes on increased importance, especially where some of our larger and more profitable clients have attributed project management to be a critical success factor. Yet, because of the past, many employees remain skeptical, and management's more "enlightened" view around project management needs to be continually reinforced and managed, typically on a case-by-case basis, rather than as a corporate commitment. And, because CCO's philosophy remains significantly different than many other (larger) "high"-tech consulting firms, the enhanced importance management now places on project management is not widely understood.

Because management rarely functions in the role of a project manager but almost always acts as the customer/project sponsor, this divergence in roles creates a challenge for how CCO gets its managers to more actively support project management. Our strategies for working this issue revolve around (1) enhancing team communications so management gets to work more closely with project managers and (2) broadening the responsibility of project management to incorporate program management, relationship building, and business development, in hopes this will do a better job of aligning customer and profit goals with project goals. In the few instances we have tried this approach, we have had very good success combining the roles of project manager and account manager. (3) In addition, where possible, we are trying to replace job titles with job descriptions, so instead of project management, we use a more detailed description of project management that associates value-based activities the project manager will do.

We will continue to look for opportunities to make these kinds of organizational structure changes in the future and increase our commitment to project management. The more we establish a value around project management, the more we will embrace it in the future.

6.13 BARRIERS TO IMPLEMENTING PROJECT MANAGEMENT IN EMERGING MARKETS

Growth in computer technology and virtual teams has made the world smaller. First world nations are flocking to emerging market nations to get access to the abundance of highly qualified human capital that is relatively inexpensive and want to participate in virtual project management teams. There is no question that there exists an ample supply of talent in these emerging market nations. These talented folks have a reasonable understanding of project management and some consider it an honor to work on virtual project teams.

But working on virtual project management teams may come with headaches. While the relative acceptance of project management appears at the working levels where the team members operate, further up in the hierarchy there might be resistance to the implementation and acceptance of project management. Because of the growth of project management worldwide, many executives openly provide "lip service" to its acceptance yet, behind the scenes, create significant barriers to prevent it from working properly. This creates significant hardships for those portions of the virtual teams in first world nations that must rely upon their other team members for support. The ultimate result might be frustrations stemming from poor information flow, extremely long decision-making processes, poor cost control, and an abundance of external dependencies that elongate schedules beyond the buyer's contractual dates. Simply stated, there are strong cultural issues that need to be considered. In this section, we will typically use the United States as an example of the first world nations.

Barriers to effective project management implementation exist worldwide, not merely in emerging market nations. But in emerging market nations, the barriers are more apparent. For simplicity sake, the barriers can be classified into four categories:

- Cultural barriers
- Status and political barriers
- Project management barriers
- Other barriers

Culture

A culture is a set of beliefs that people follow. Every company could have its own culture. Some companies may even have multiple cultures. Some cultures are strong while others are weak. In some emerging market nations, there exist national cultures that can be so strong that they dictate the corporate cultures. There are numerous factors that can influence the culture of an organization. Only those factors that can have an impact on the implementation and acceptance of project management are discussed here and include:

- Bureaucratic centralization of authority in the hands of a few
- Lack of meaningful or real executive sponsorship
- Importance of the organizational hierarchy
- Improper legal laws
- The potential for corruption

Centralization of Authority: Many countries maintain a culture in which very few people have the authority to make decisions. Decision-making rests in the hands of a few and it serves as a source of vast power. This factor exists in both privately held companies and governmental organizations. Project management advocates decentralization of authority and decision-making. In many countries, the senior-most level of management will never surrender their authority, power, or right to make decisions to project managers. In these countries an appointment to the senior levels of management is not necessarily based upon performance. Instead, it is based upon age, belonging to the right political party, and personal contacts within the government. The result can be executives that possess little knowledge of their own business or who lack the leadership capacity.

Executive Sponsorship: Project sponsorship might exist somewhere in the company but most certainly not at the executive levels. There are two reasons for this. First, senior managers know their limitations and may have absolutely no knowledge about the project. Therefore, they could be prone to making serious blunders that could become visible to the people that put them into these power positions. Second, and possibly most important, acting as an executive sponsor on a project that could fail could signal the end of the executive's political career. Therefore, sponsorship, if it exists at all, may be at a low level in the organizational hierarchy, and at a level where people are expendable if the project fails. The result is that project managers end up with sponsors who either cannot or will not help them in time of trouble.

Organizational Hierarchy: In the United States, project managers generally have the right to talk to anyone in the company to get information relative to the project. The intent is to get work to flow horizontally as well as vertically. In some emerging market nations, the project manager must follow the chain of command. The organizational hierarchy is sacred. Following the chain of command certainly elongates the decision-making process to the point where the project manager has no idea how long it will take to get access to needed information or for a decision to be made even though a sponsor exists. There is no mature infrastructure in place to support project management. The infrastructure exists to filter bad news from the executive levels and to justify the existence of each functional manager.

In the United States, the "buck" stops at the sponsor. Sponsors have ultimate decision-making authority and are expected to assist the project managers during a crisis. The role of the sponsor is clearly defined and may be described in detail in the enterprise project management methodology. But in some emerging market countries, even the sponsor might not be authorized to make a decision. Some decisions may need to go as high as a government minister. Simply stated, one does not know where and when the decision needs to be made and where it will be made. Also, in the United States reporting bad news ends up in the hands of the project sponsor. In some nations, the news may go as high as government ministers. Simply stated, you cannot be sure where project information will end up.

Improper Legal Laws: Not all laws in emerging market nations are viewed by other nations as being legal laws. Yet American project managers, partnering with these nations, must abide by these laws. As an example, procurement contracts may be awarded not to the most qualified supplier or to the lowest bidder but to any bidder that resides in a city that has a high unemployment level. As another example, some nations have laws that imply that bribes are an acceptable practice when awarding contracts. Some contracts might also be awarded to relatives and friends rather than the best qualified supplier.

Potential for Corruption: Corruption can and does exist in some countries and plays havoc on project managers that focus on the triple constraint. Project managers traditionally lay out a plan to meet the objectives and the triple constraint. Project managers also assume that everything will be done systematically and in an orderly manner, which assumes no corruption. But in some nations, there are potentially corrupt individuals or organizations that will do everything possible to stop or slow down the project, until they could benefit personally.

Status and Politics   Status and politics are prevalent everywhere and can have a negative impact on project management. In some emerging market nations, status and politics actually sabotage project management and prevent it from working correctly. Factors that can affect project management include:

- Legal formalities and government constraints
- Insecurity at the executive levels
- Status consciousness
- Social obligations
- Internal politics
- Unemployment and poverty
- Attitude toward workers
- Inefficiencies

- Lack of dedication al all levels
- Lack of honesty

Legal Formalities and Government Constraints: Here in the United States, we believe that employees that perform poorly can be removed from the project or even fired. But in some emerging market nations, employees have the right to hold a job even if their performance is substandard. Having a job and a regular paycheck is a luxury. There are laws that clearly state under what conditions a worker can be fired, if at all.

There are also laws on the use of overtime. Overtime may not be allowed because paying someone to work overtime could eventually end up creating a new social class. Therefore, overtime may not be used as a means to maintain or accelerate a schedule that is in trouble.

Insecurity: Executives often feel insecurity more so than the managers beneath them because their positions may be the result of political appointments. As such project managers may be seen as the stars of the future and may be viewed as a threat to executives. Allowing project managers who are working on highly successful projects to make presentations to the senior-most levels of management in the government could be mired. If the project is in trouble, then the project manager may be forced to make the presentation. Executives are afraid of project managers.

Status Consciousness: Corporate officers in emerging market nations are highly status conscious. They have a very real fear that the implementation of project management may force them to lose their status, yet they refuse to function as active project sponsors. Status often is accompanied by fringe benefits such as a company car and other special privileges.

Social Obligations: In emerging market nations, social obligations due to religious beliefs (and possibly superstitious beliefs) and politics may be more important than in first world nations. Social obligations are ways of maintaining alliances with those people that have put an executive or a project manager in power. As such, project managers may not be allowed to interface socially with certain groups. This could also be viewed as a threat to project management implementation.

Internal Politics: Internal politics exist in every company in the world. Before executives consider throwing their support behind a new approach such as project management, they worry about whether they will become stronger or weaker, have more or less authority, and have a greater or lesser chance for advancement. This is one of the reasons why only a small percentage of emerging market companies have PMOs. Whichever executive gets control of the PMO could become more powerful than other executives. In the United States, we have solved this problem by allowing several executives to have their own PMO. But in the emerging markets, this is viewed as excessive headcount.

Unemployment and Government Constraints: Virtually all executives understand project management and the accompanying benefits, yet they remain silent rather than visibly showing their support. One of the benefits of project management implementation is that it can make organizations more efficient to the point where fewer resources are needed to perform the required work. This can be a threat to an executive because, unless additional business can be found, efficiency can result in downsizing the company, reducing the executive's power and authority, increasing the unemployment level, and possibly increasing poverty in the community. Therefore, the increased efficiencies of project management could be looked upon unfavorably.

Attitude toward Employees: In some nations, employees might be viewed as stepping-stones to building an empire. Hiring three below-average workers to do the same work as two average workers is better for empire building, yet possibly at the expense of the project's budget and schedule. It is true, however, that finding adequate human resources may be difficult, but sometimes companies simply do not put forth a good effort in their search. Friends and family members may be hired first regardless of their qualifications. The problem is further complicated when one must find people with project management expertise.

Inefficiencies: Previously, we stated that companies might find it difficult to hire highly efficient people in project management. Not all people are efficient. Some people simply are not committed to their work even though they understand project management. Other people may get frustrated when they realize that they do not have the power, authority, or responsibility of their colleagues in first world countries. Sometimes, new hirers that want to be efficient workers are pressured by the culture to remain inefficient or else the individual's colleagues will be identified as poor workers. Peer pressure exists and can prevent people from demonstrating their true potential.

Lack of Dedication: It is hard to get people motivated when they believe they cannot lose their job. People are simply not dedicated to the triple constraint. Some people prefer to see schedules slip because it provides some degree of security for a longer period of time. There is also a lack of dedication for project closure. As a project begins to wind down, employees will begin looking for a home on some other project. They might even leave their current project prematurely, before the work is finished, to guarantee employment elsewhere.

Honesty: People working in emerging market countries have a tendency to hide things from fellow workers and project managers, especially bad news, either to keep their prestige or to retain their power and authority. This creates a huge barrier for project managers that rely upon timely information, whether good or bad, in order to manage the project successfully. Delays in reporting could waste valuable time when corrective action could have been taken.

Implementation of Project Management

While culture, status, and politics can create barriers for any new management philosophy, there are other barriers that are directly related to project management, including:

- Cost of project management implementation
- Risks of implementation failure
- Cost of training and training limitations
- Need for sophistication
- Lack of closure on projects
- Work ethic
- Poor planning

Cost of Implementation: There is a cost associated with the implementation of project management. The company must purchase hardware and software, create a project management methodology, and develop project performance reporting techniques. This requires a significant financial expenditure, which the company might not be able to afford, and also requires significant resources to be tied up in implementation for an extended period of time. With limited resources, and the fact that the better resources would be required for implementation and removed from ongoing work, companies shy away from project management even though they know the benefits.

Risk of Failure: Even if a company is willing to invest the time and money for project management implementation, there is a significant risk that implementation will fail. And even if implementation is successful but projects begin to fail for any number of reasons, blame will be placed upon faulty implementation. Executives may find that their position in the hierarchy is now insecure once they have to explain the time and money expended for no real results. This is why some executives either refuse to accept or visibly support project management.

Training Limitations: Implementation of project management is difficult without training programs for the workers. This creates three additional problems. First, how much money must be allocated for training? Second, who will provide the training and what are the credentials of the trainers? Third, should people be released from project work to attend training classes? It is time consuming and expensive to train people in project management, whether it is project managers or team members that need to be trained. Adding together the cost of implementation and the cost of training might frighten executives from accepting project management.

Need for Sophistication: Project management requires sophistication, not only with the limited technology or tools that may be available but also in the ability of people to work together. This teamwork sophistication is generally lacking in emerging market countries. People may see no benefit in teamwork because others may be able to recognize their lack of competencies and mistakes. They have not been trained to work properly in teams and are not rewarded for their contribution to the team.

Lack of Closure on Projects: Employees are often afraid to be attached to the project at closure when lessons learned and best practices are captured. Lessons learned and best practices can be based upon what we did well and what we did poorly. Employees may not want to see anything in writing that indicates that best practices were discovered from their mistakes.

Work Ethic: In some nations, the inability to fire people creates a relatively poor work ethic which is contrary to effective project management practices. There is a lack of punctuality in coming to work and attending meetings. When people do show up at meetings, only good news is discussed in a group whereas bad news is discussed one-on-one. Communication skills are weak as is report writing. There is a lack of accountability because accountability means explaining your actions if things go bad.

Poor Planning: Poor planning is paramount in emerging market nations. There exists a lack of commitment to the planning process. Because of a lack of standards, perhaps attributed to the poor work ethic, estimating duration, effort, and cost is very difficult. The ultimate result of poor planning is an elongation of the schedule. For workers that are unsure about their next assignment, this can be viewed as job security at least for the short term.

Other Barriers

There are other barriers that are too numerous to mention. However, some of the more important ones are shown below. These barriers are not necessarily universal in emerging market nations, and many of these barriers can be overcome.

- Currency conversion inefficiencies
- Inability to receive timely payments
- Superstitious beliefs
- Laws against importing and exporting intellectual property
- Lack of tolerance for the religious beliefs of virtual team partners
- Risk of sanctions by partners' governments
- Use of poor or outdated technologies

Recommendations

Although we have painted a rather bleak picture, there are great future opportunities in these nations. Emerging market nations have an abundance of talent that is yet to be fully harvested. The true capabilities of these workers are still unknown. Virtual project management teams might be the starting point for the full implementation of project management.

As project management begins to grow, senior officers will recognize and accept the benefits of project management and see their business base increase. Partnerships and joint ventures using virtual teams will become more prevalent. The barriers that impede successful project management implementation will still exist, but we will begin to excel in how to live and work within the barriers and constraints imposed on the continually emerging virtual teams.

Greater opportunities are seen for the big emerging market economies. They are beginning to see more of the value of project management and have taken strides to expand its use. Some of the rapidly developing economies are even much more aggressive in providing the support needed for breaking many of the barriers addressed above. As more success stories emerge, the various economies will strengthen, become more connected, and start to fully utilize project management for what it really is.

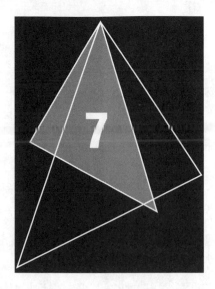

Management Support

7.0 INTRODUCTION

As we saw in Chapter 6, senior managers are the architects of corporate culture. They are charged with making sure that their companies' cultures, once accepted, do not come apart. Visible management support is essential to maintaining a project management culture. And above all, the support must be continuous rather than sporadic.

This chapter examines the importance of management support in the creation and maintenance of project management cultures. Case studies illustrate the vital importance of employee empowerment and the project sponsor's role in the project management system.

7.1 VISIBLE SUPPORT FROM SENIOR MANAGERS

As project sponsors, senior managers provide support and encouragement to the project managers and the rest of the project team. Companies excellent in project management have the following characteristics:

- Senior managers maintain a hands-off approach, but they are available when problems come up.
- Senior managers expect to be supplied with concise project status reports.
- Senior managers practice empowerment.
- Senior managers decentralize project authority and decision-making.
- Senior managers expect project managers and their teams to suggest both alternatives and recommendations for solving problems, not just to identify the problems.

However, there is a fine line between effective sponsorship and overbearing sponsorship. Robert Hershock, former vice president at 3M, said it best during a videoconference on excellence in project management:

> Probably the most important thing is that they have to buy in from the top. There has to be leadership from the top, and the top has to be 100 percent supportive of this whole process. If you're a control freak, if you're someone who has high organizational skills and likes to dot all the i's and cross all the t's, this is going to be an uncomfortable process, because basically it's a messy process; you have to have a lot of fault tolerance here. But what management has to do is project the confidence that it has in the teams. It has to set the strategy and the guidelines and then give the teams the empowerment that they need in order to finish their job. The best thing that management can do after training the team is get out of the way.

To ensure their visibility, senior managers need to believe in walk-the-halls management. In this way, every employee will come to recognize the sponsor and realize that it is appropriate to approach the sponsor with questions. Walk-the-halls management also means that executive sponsors keep their doors open. It is important that everyone, including line managers and their employees, feels supported by the sponsor. Keeping an open door can occasionally lead to problems if employees attempt to go around lower-level managers by seeking a higher level of authority. But such instances are infrequent, and the sponsor can easily deflect the problems back to the appropriate manager.

7.2 PROJECT SPONSORSHIP

Executive project sponsors provide guidance for project managers and project teams. They are also responsible for making sure that the line managers who lead functional departments fulfill their commitments of resources to the projects underway. In addition, executive project sponsors maintain communication with customers.

The project sponsor usually is an upper-level manager who, in addition to his or her regular responsibilities, provides ongoing guidance to assigned projects. An executive might take on sponsorship for several concurrent projects. Sometimes, on lower-priority or maintenance projects, a middle-level manager may take on the project sponsor role. One organization I know of even prefers to assign middle managers instead of executives. The company believes this avoids the common problem of lack of line manager buy-in to projects (see Figure 7–1).

In some large, diversified corporations, senior managers do not have adequate time to invest in project sponsorship. In such cases, project sponsorship falls to the level below corporate senior management or to a committee.

Some projects do not need project sponsors. Generally, sponsorship is required on large, complex projects involving a heavy commitment of resources. Large, complex projects also require a sponsor to integrate the activities of the functional lines, to dispel disruptive conflicts, and to maintain strong customer relations.

Consider one example of a project sponsor's support for a project. A project manager who was handling a project in an organization within the federal government decided that

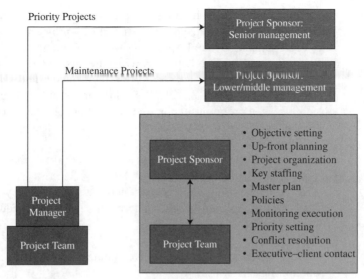

FIGURE 7–1. Roles of project sponsor. *Source:* Reprinted from H. Kerzner, *In Search of Excellence in Project Management,* Wiley, New York, 1998, p. 159.

another position would be needed on his team if the project were to meet its completion deadline. He had already identified a young woman in the company who fit the qualifications he had outlined. But adding another full-time-equivalent position seemed impossible. The size of the government project office was constrained by a unit-manning document that dictated the number of positions available.

The project manager went to the project's executive sponsor for help. The executive sponsor worked with the organization's human resources and personnel management department to add the position requested. Within 30 days, the addition of the new position was approved. Without the sponsor's intervention, it would have taken the organization's bureaucracy months to approve the position, too late to affect the deadline.

In another example, the president of a medium-size manufacturing company wanted to fill the role of sponsor on a special project. The project manager decided to use the president to the project's best advantage. He asked the president/sponsor to handle a critical situation. The president/sponsor flew to the company's headquarters and returned two days later with an authorization for a new tooling the project manager needed. The company ended up saving time on the project, and the project was completed four months earlier than originally scheduled.

Sponsorship by Committee As companies grow, it sometimes becomes impossible to assign a senior manager to every project, and so committees act in the place of individual project sponsors. In fact, the recent trend has been toward committee sponsorship in many kinds of organizations. A project sponsorship committee usually is made up of a representative from every function of the company: engineering, marketing, and production. Committees may be temporary, when a committee is brought together to sponsor one time-limited project, or permanent, when a standing committee takes on the ongoing project sponsorship of new projects.

For example, General Motors Powertrain had achieved excellence in using committee sponsorship. Two key executives, the vice president of engineering and the vice president of operations, led the Office of Products and Operations, a group formed to oversee the management of all product programs. This group demonstrated visible executive-level program support and commitment to the entire organization. Their roles and responsibilities were to:

- Appoint the project manager and team as part of the charter process
- Address strategic issues
- Approve the program contract and test for sufficiency
- Assure program execution through regularly scheduled reviews with program managers

EDS is also a frequent user of committee sponsorship. According to Gene Panter, Program Manager with EDS:

> EDS executives, in conjunction with key customers, tend to take the role of sponsor for large-scale projects or programs. Sponsors are asked to participate in various checkpoint reviews, assure the acquisition of necessary resources, and provide support in the elimination of barriers that may exist. Large programs tend to have guidance teams in place to provide necessary support and approvals for project teams.
>
> Line managers and functional managers tend to be responsible for working with the project or program manager to identify appropriate human resources for projects. Project or program managers provide the direct management function for the project. EDS is also widely utilizing the concept of a project or program office to provide standards for project tracking, oversight, management, and control support for projects.

Phases of Project Sponsorship

The role of the project sponsor changes over the life cycle of a project. During the planning and initiation phases, the sponsor plays an active role in the following activities:

- Helping the project manager establish the objectives of the project
- Providing guidance to the project manager during the organization and staffing phases
- Explaining to the project manager what environmental or political factors might influence the project's execution
- Establishing the project's priority (working alone or with other company executives) and then informing the project manager about the project's priority in the company and the reason that priority was assigned
- Providing guidance to the project manager in establishing the policies and procedures for the project
- Functioning as the contact point for customers and clients

During the execution phase of a project, the sponsor must be very careful in deciding which problems require his or her guidance. Trying to get involved with every problem that comes up on a project will result in micromanagement. It will also undermine the project manager's authority and make it difficult for the executive to perform his or her regular responsibilities.

For short-term projects of two years or less, it is usually best that the project sponsor assignment is not changed over the duration of the project. For long-term projects of five years, more or less, different sponsors could be assigned for every phase of the project, if necessary. Choosing sponsors from among executives at the same corporate level works best, since sponsorship at the same level creates a "level" playing field, whereas at different levels, favoritism can occur.

Project sponsors need not come from the functional area where the majority of the project work will be completed. Some companies even go so far as assigning sponsors from line functions that have no vested interest in the project. Theoretically, this system promotes impartial decision-making.

Customer Relations   The role of executive project sponsors in customer relations depends on the type of organization (entirely project-driven or partially project-driven) and the type of customer (external or internal). Contractors working on large projects for external customers usually depend on executive project sponsors to keep the clients fully informed of progress on their projects. Customers with multi-million-dollar projects often keep an active eye on how their money is being spent. They are relieved to have an executive sponsor they can turn to for answers.

It is common practice for contractors heavily involved in competitive bidding for contracts to include both the project manager's and the executive project sponsor's resumes in proposals. All things being equal, the resumes may give one contractor a competitive advantage over another.

Customers prefer to have a direct path of communication open to their contractors' executive managers. One contractor identified the functions of the executive project sponsor as:

- Actively participating in the preliminary sales effort and contract negotiations
- Establishing and maintaining high-level client relationships
- Assisting project managers in getting the project underway (planning, staffing, and so forth)
- Maintaining current knowledge of major project activities
- Handling major contractual matters
- Interpreting company policies for project managers
- Helping project managers identify and solve significant problems
- Keeping general managers and client managers advised of significant problems with projects

Decision-Making   Imagine that project management is like car racing. A yellow flag is a warning to watch out for a problem. Yellow flags require action by the project manager or the line manager. There is nothing wrong with informing an executive about a yellow-flag problem as long as the project manager is not looking for the sponsor to solve the problem. Red flags, however, usually do require the sponsor's direct involvement. Red flags indicate problems that may affect the time, cost, and performance parameters of the project. So red flags need to be taken seriously and decisions need to be made collaboratively by the project manager and the project sponsor.

Serious problems sometimes result in serious conflicts. Disagreements between project managers and line managers are not unusual, and they require the thoughtful intervention of the executive project sponsor. First, the sponsor should make sure that the disagreement could not be solved without his or her help. Second, the sponsor needs to gather information from all sides and consider the alternatives being considered. Then, the sponsor must decide whether he or she is qualified to settle the dispute. Often, disputes are of a technical nature and require someone with the appropriate knowledge base to solve them. If the sponsor is unable to solve the problem, he or she will need to identify another source of authority that has the needed technical knowledge. Ultimately, a fair and appropriate solution can be shared by everyone involved. If there were no executive sponsor on the project, the disputing parties would be forced to go up the line of authority until they found a common superior to help them. Having executive project sponsors minimizes the number of people and the amount of time required to settle work disputes.

Strategic Planning   Executives are responsible for performing the company's strategic planning, and project managers are responsible for the operational planning on their assigned projects. Although the thought processes and time frames are different for the two types of planning, the strategic planning skills of executive sponsors can be useful to project managers. For projects that involve process or product development, sponsors can offer a special kind of market surveillance to identify new opportunities that might influence the long-term profitability of the organization. Furthermore, sponsors can gain a lot of strategically important knowledge from lower-level managers and employees. Who else knows better when the organization lacks the skill and knowledge base it needs to take on a new type of product? When the company needs to hire more technically skilled labor? What technical changes are likely to affect their industry?

7.3 EXCELLENCE IN PROJECT SPONSORSHIP

Many companies have achieved excellence in their application of project sponsorship. Radian International depended on single-project sponsors to empower its project managers for decision-making. General Motors proved that sponsorship by committee works. Roadway demonstrated the vital importance of sponsorship training for both sponsorship by a single executive and sponsorship by a committee.

In excellent companies, the role of the sponsor is not to supervise the project manager but to make sure that the best interests of both the customer and the company are recognized. However, as the next two examples reveal, it is seldom possible to make executive decisions that appease everyone.

Franklin Engineering (a pseudonym) had a reputation for developing high-quality, innovative products. Unfortunately, the company paid a high price for its reputation:

a large R&D budget. Fewer than 15 percent of the projects initiated by R&D led to the full commercialization of a product and the recovery of the research costs.

The company's senior managers decided to implement a policy that mandated that all R&D project sponsors periodically perform cost–benefit analyses on their projects. When a project's cost–benefit ratio failed to reach the levels prescribed in the policy, the project was canceled for the benefit of the whole company.

Initially, R&D personnel were unhappy to see their projects canceled, but they soon realized that early cancellation was better than investing large amounts in projects that were likely to fail. Eventually, the project managers and team members came to agree that it made no sense to waste resources that could be better used on more successful projects. Within two years, the organization found itself working on more projects with a higher success rate but no addition to the R&D budget.

Another disguised case involves a California-based firm that designs and manufactures computer equipment. Let's call the company Design Solutions. The R&D group and the design group were loaded with talented individuals who believed that they could do the impossible and often did. These two powerful groups had little respect for the project managers and resented schedules because they thought schedules limited their creativity.

In June 1997, the company introduced two new products that made it onto the market barely ahead of the competition. The company had initially planned to introduce them by the end of 1996. The reason for the late releases: Projects had been delayed because of the project teams' desire to exceed the specifications required and not just meet them.

To help the company avoid similar delays in the future, the company decided to assign executive sponsors to every R&D project to make sure that the project teams adhered to standard management practices in the future. Some members of the teams tried to hide their successes with the rationale that they could do better. But the sponsor threatened to dismiss the employees, and they eventually relented.

The lessons in both cases are clear. Executive sponsorship actually can improve existing project management systems to better serve the interests of the company and its customers.

7.4 EMPOWERMENT OF PROJECT MANAGERS

One of the biggest problems with assigning executive sponsors to work beside line managers and project managers is the possibility that the lower-ranking managers will feel threatened with a loss of authority. This problem is real and must be dealt with at the executive level. Frank Jackson, formerly a senior manager at MCI, believes in the idea that information is power:

> We did an audit of the teams to see if we were really making the progress that we thought or were kidding ourselves, and we got a surprising result. When we looked at the audit, we found out that 50 percent of middle management's time was spent in filtering information up and down the organization. When we had a sponsor, the information went from the team to the sponsor to the operating committee, and this created a real crisis in our middle management area.

MCI has found its solution to this problem. If there is anyone who believes that just going and dropping into a team approach environment is an easy way to move, it's definitely not. Even within the companies that I'm involved with, it's very difficult for managers to give up the authoritative responsibilities that they have had. You just have to move into it, and we've got a system where we communicate within MCI, which is MCI mail. It's an electronic mail system. What it has enabled us to do as a company is bypass levels of management. Sometimes you get bogged down in communications, but it allows you to communicate throughout the ranks without anyone holding back information.

Not only do executives have the ability to drive project management to success, they also have the ability to create an environment that leads to project failure. According to Robert Hershock, former vice president at 3M:

Most of the experiences that I had where projects failed, they failed because of management meddling. Either management wasn't 100 percent committed to the process, or management just bogged the whole process down with reports and a lot of other innuendos. The biggest failures I've seen anytime have been really because of management. Basically, there are two experiences where projects have failed to be successful. One is the management meddling where management cannot give up its decision-making capabilities, constantly going back to the team and saying you're doing this wrong or you're doing that wrong. The other side of it is when the team can't communicate its own objective. When it can't be focused, the scope continuously expands, and you get into project creep. The team just falls apart because it has lost its focus.

Project failure can often be a matter of false perceptions. Most executives believe that they have risen to the top of their organizations as solo performers. It is very difficult for them to change without feeling that they are giving up a tremendous amount of power, which traditionally is vested in the highest level of the company. To change this situation, it may be best to start small. As one executive observed:

There are so many occasions where senior executives won't go to training and won't listen, but I think the proof is in the pudding. If you want to instill project management teams in your organizations, start small. If the company won't allow you to do it using the Nike theory of just jumping in and doing it, start small and prove to them one step at a time that they can gain success. Hold the team accountable for results—it proves itself.

It is also important for us to remember that executives can have valid reasons for micromanaging. One executive commented on why project management might not be working as planned in his company:

We, the executives, wanted to empower the project managers and they, in turn, would empower their team members to make decisions as they relate to their project or function. Unfortunately, I do not feel that we (the executives) totally support decentralization of decision-making due to political concerns that stem from the lack of confidence we have in our project managers, who are not proactive and who have not demonstrated leadership capabilities.

In most organizations, senior managers start at a point where they trust only their fellow managers. As the project management system improves and a project management culture develops, senior managers come to trust project managers, even though they do not occupy positions high on the organizational chart. Empowerment does not happen overnight. It takes time and, unfortunately, a lot of companies never make it to full project manager empowerment.

7.5 MANAGEMENT SUPPORT AT WORK

Visible executive support is necessary for successful project management and the stability of a project management culture. But there is such a thing as too much visibility for senior managers. Take the following case example, for instance.

Midline Bank

Midline Bank (a pseudonym) is a medium-size bank doing business in a large city in the Northwest. Executives at Midline realized that growth in the banking industry in the near future would be based on mergers and acquisitions and that Midline would need to take an aggressive stance to remain competitive. Financially, Midline was well prepared to acquire other small- and middle-size banks to grow its organization.

The bank's information technology group was given the responsibility of developing an extensive and sophisticated software package to be used in evaluating the financial health of the banks targeted for acquisition. The software package required input from virtually every functional division of Midline. Coordination of the project was expected to be difficult.

Midline's culture was dominated by large, functional empires surrounded by impenetrable walls. The software project was the first in the bank's history to require cooperation and integration among the functional groups. A full-time project manager was assigned to direct the project.

Unfortunately, Midline's executives, managers, and employees knew little about the principles of project management. The executives did, however, recognize the need for executive sponsorship. A steering committee of five executives was assigned to provide support and guidance for the project manager, but none of the five understood project management. As a result, the steering committee interpreted its role as one of continuous daily direction of the project.

Each of the five executive sponsors asked for weekly personal briefings from the project manager, and each sponsor gave conflicting directions. Each executive had his or her own agenda for the project.

By the end of the project's second month, chaos took over. The project manager spent most of his time preparing status reports instead of managing the project. The executives changed the project's requirements frequently, and the organization had no change control process other than the steering committee's approval.

At the end of the fourth month, the project manager resigned and sought employment outside the company. One of the executives from the steering committee then took over the project manager's role, but only on a part-time basis. Ultimately, the project was taken over by two more project managers before it was complete, one year later than planned. The company learned a vital lesson: More sponsorship is not necessarily better than less.

Contractco

Another disguised case involves a Kentucky-based company I'll call Contractco. Contractco is in the business of nuclear fusion testing. The company was in the process of bidding on a contract with the U.S. Department of Energy. The department required that the project manager be identified as part of the company's proposal and that a list of the project manager's duties and responsibilities be included. To impress the Department of Energy, the company assigned both the executive vice president and the vice president of engineering as cosponsors.

The DoE questioned the idea of dual sponsorship. It was apparent to the department that the company did not understand the concept of project sponsorship, because the roles and responsibilities of the two sponsors appeared to overlap. The department also questioned the necessity of having the executive vice president serve as a sponsor.

The contract was eventually awarded to another company. Contractco learned that a company should never underestimate the customer's knowledge of project management or project sponsorship.

Health Care Associates

Health Care Associates (another pseudonym) provides health care management services to both large and small companies in New England. The company partners with a chain of 23 hospitals in New England. More than 600 physicians are part of the professional team, and many of the physicians also serve as line managers at the company's branch offices. The physician-managers maintain their own private clinical practices as well.

It was the company's practice to use boilerplate proposals prepared by the marketing department to solicit new business. If a client were seriously interested in Health Care Associates' services, a customized proposal based on the client's needs would be prepared. Typically, the custom-designed process took as long as six months or even a full year.

Health Care Associates wanted to speed up the custom-designed proposal process and decided to adopt project management processes to accomplish that goal. In January 1994, the company decided that it could get a step ahead of its competition if it assigned a physician-manager as the project sponsor for every new proposal. The rationale was that the clients would be favorably impressed.

The pilot project for this approach was Sinco Energy (another pseudonym), a Boston-based company with 8600 employees working in 12 cities in New England. Health Care Associates promised Sinco that the health care package would be ready for implementation no later than June 1994.

The project was completed almost 60 days late and substantially over budget. Health Care Associates' senior managers privately interviewed each of the employees on the Sinco project to identify the cause of the project's failure. The employees had the following observations:

- Although the physicians had been given management training, they had a great deal of difficulty applying the principles of project management. As a result, the physicians ended up playing the role of invisible sponsor instead of actively participating in the project.
- Because they were practicing physicians, the physician sponsors were not fully committed to their role as project sponsors.
- Without strong sponsorship, there was no effective process in place to control scope creep.
- The physicians had not had authority over the line managers, who supplied the resources needed to complete a project successfully.

Health Care Associates' senior managers learned two lessons. First, not every manager is qualified to act as a project sponsor. Second, the project sponsors should be assigned on the basis of their ability to drive the project to success. Impressing the customer is not everything.

Sypris Electronics[1]

Management actively supports project management to the extent that Sypris Electronics has launched a transformation of the entire company.

The once functional-based silo'd organization is in the midst of an evolution to a value stream based organization based around project management. John Walsh, President of Sypris Electronics Group, sponsors this operational and cultural change from the top of the organization. Through a hoshin kanri policy deployment model (goal alignment philosophy) the entire company has embraced this change. All levels of management and the individual contributors, including direct labor employees, have targets to improve that support the top level results.

All levels of management function as project sponsors, from executive staff to manufacturing cell leaders. This encourages buy-in and fosters the desired cultural change. Line managers are primary change advocates at Sypris which support the project management based approach to running the company. This firm commitment stems from the leadership embracing and driving the change with a top down approach, as well as proper education. Sypris has brought in Dr. Harold Kerzner to speak to all levels in every discipline about what project management is and what successful project management can enable. This removes many misconceptions and underscores the importance of effective project management. Additionally, many of the salaried personnel are being trained to become certified PMP's (not just the project management office). Without everyone knowing their role, the right cultural change cannot occur. Finally, Sypris has selectively chosen other members for the executive staff based upon their bias for change that embrace this philosophy as well.

Vitalize Consulting Solutions, Inc.[2]

There are various levels of support from management regarding project management methodologies [at VCS]. Initially, management may not appreciate the value until they have a PM identify the deliverables, track the progress, provide real-time status reports, identify issues and risks and have someone actively oversee the project and maintain momentum.

---

1. Material on Sypris Electronics provided by Ryan Duran, Director, Program Management.
2. Material on Vitalize Consulting Solutions was provided by Marc Hirshfield, Director, Project Management Office.

Management support of project management in our industry is growing as the formal role of a project manager is becoming more important to meeting the needs of complex projects, but managers still need to embrace and empower the PM's. There are times the PM is not asked when changes occur (resource allocation as an example) or in some cases not supported as the person in charge of the initiative. Obviously culture and politics plays a factor in every industry, but in order for healthcare to meet the demands of the industry, management must continue to further invest and support the growth of their PMs.

At VCS, the entire leadership team (including our C-suite) is very supportive and actively involved in helping grow the PMO. The PMO resides as the central "hub" for providing our customer base a full service consulting service offering. This includes the ability to run a complex healthcare IT initiative with a project manager who has subject matter expertise on both the operations and technical side of the house along with a published methodology and process to follow. For VCS, this is a very big market differentiator and offers our customers great value in comparison to our competition.

Indra[3]

Executive management [at Indra] is highly motivated to support project management development within the company. They regularly insist upon improving our training programs for project managers as well as focusing on the need that the best project management methods are in place.

Sometimes the success of a project constitutes a significant step in the development of a new technology, in the launch in a new market, or for the establishment of a new partnership, and, in those cases, the managing directorate usually plays an especially active role as sponsors during the project or program execution. They participate with the customer in steering committees for the project or the program, and help in the decision making or risk management processes.

For a similar reason, due to the significance of a specific project but at a lower level, it is not uncommon to see middle level management carefully watching its execution and providing, for instance, additional support to negotiate with the customer the resolution of a particular issue.

Getting middle level management support has been accomplished using the same set of corporate tools for project management at all levels and for all kind of projects in the company. No project is recognized if it is not in the corporate system, and to do that, the line managers must follow the same basic rules and methods, no matter if it is a recurring, a non-recurring effort, or other type of project. A well developed WBS, a complete foreseen schedule, a risk management plan and a tailored set of earned value methods, may be applied to any kind of projects.

7.6 GETTING LINE MANAGEMENT SUPPORT

Management support is not restricted exclusively to senior management, as was shown in the previous section. Line management support is equally crucial for project management to work effectively. Line managers are usually more resistant to project management and often demand proof that project management provides value to the organization

3. Material on Indra provided by Enrique Sevilla Molina, PMP, Corporate PMO Director.

before they support the new processes. This problem was identified previously in Exel's journey to excellence in project management and also appeared at Motorola. According to a spokesperson at Motorola[4]:

> This (getting line management support) was tough at first. It took years of having PMs provide value to the organization.

When organizations become mature in project management, sponsorship at the executive levels and at middle management levels becomes minimal and integrated project teams are formed where the integrated or core team is empowered to manage the project with minimal sponsorship other than for critical decisions. These integrated or core teams may or may not include line management. The concept of core teams became a best practice at Motorola[5]:

> Most project decisions and authority resides in the project core team. The core team is made up of middle- to low-level managers for the different functional areas (marketing, software, electrical, mechanical, manufacturing, system test, program management, quality, etc.) and has the project ownership responsibility. This core team is responsible for reviewing and approving product requirements and committing resources and schedule dates. It also acts as the project change control board and can approve or reject project scope change requests. However, any ship acceptance date changes must be approved by senior management.

7.7 DTE ENERGY

Jason Schulist, Manager, Continuous Improvement, Operating Systems Strategy Group at DTE Energy, comments on project sponsorship:

> The champions of continuous improvement projects that use the 4-Gate/9-Step methodology [Figures 4-3 and 4-4] are primarily Vice Presidents and Directors of DTE Energy. These champions sign off at every gate to approve the results of the process. In some business units, a Master Black Belt also signs off at every gate to ensure that rigorous analysis and the DTE Energy Operating System methodology is followed.

The importance of this statement is that senior management should be the champions of the methodology in addition to becoming project champions. In one auto supplier, senior management undertook the role of champion for the development of a project management methodology. After the methodology was developed, the executive took on a different role in the company. Three years later, the company realized that there were no continuous improvement efforts on the methodology because there was no executive champion in place. The executive then returned to championing the methodology and continuous improvement efforts occurred.

---

4. H. Kerzner, *Project Management Best Practices: Achieving Global Excellence*, Wiley, 2006, p. 307.
5. Ibid., p. 308.

7.8 INITIATION CHAMPIONS AND EXIT CHAMPIONS

As project management evolved, so did the role of the executive in project management. Today, there are three roles that the executive plays:

- The project sponsor
- The project (initiation) champion
- The exit champion

The role of the executive in project management as a project sponsor has become reasonably mature. Most textbooks on project management have sections that are dedicated to the role of the project sponsor.[6] The role of the project champion however, is just coming of age. Stephen Markham defines the role of the champion:

> Champions are informal leaders who emerge in a somewhat erratic fashion. Championing is a voluntary act by an individual to promote a particular project. In the act of championing, individuals rarely refer to themselves as champions; rather, they describe themselves as trying to do the right thing for the right company. A champion rarely makes a single decision to champion a project. Instead, he or she begins in a simple fashion and develops increasing enthusiasm for the project. A champion becomes passionate about a project and ultimately engages others based upon personal conviction that the project is the right thing for the entire organization. The champion affects the way other people think of the project by spreading positive information across the organization. Without official power or responsibility, a champion contributes to new product development by moving projects forward. Thus, champions are informal leaders who (1) adopt projects as their own in a personal way, (2) take on risk by promoting the projects beyond what is expected of people in their position, and (3) promote the project by getting other individuals to support it.[7]

With regard to new product development (NPD) projects, champions are needed to overcome the obstacles in the "valley of death," as seen in Figure 7–2.[8] The valley of death is the area in NPD where recognition of the idea/invention and efforts to commercialize the product come together. In this area, good projects often fall by the wayside and projects with less value often get added into the portfolio of projects. According to Markham[9],

> Many reasons exist for the Valley of Death. Technical personnel (left side of Figure 7–2) often do not understand the concerns of commercialization personnel (right side) and vice versa. The cultural gap between these two types of personnel manifests itself in the results prized by one side and devalued by the other. Networking and contact management may be important to sales people but seen as shallow and self-aggrandizing by technical people. Technical people find value in discovery and pushing the frontiers of knowledge.

6. H. Kerzner, *Project Management: A Systems Approach to Planning, Scheduling and Controlling*, 10th ed., Wiley, Hoboken, NJ, 2009, Chapter 10. Also, H. Kerzner and F. Saladis, *What Executives Need to Know About Project Management*, Wiley, Hoboken, and IIL, New York, 2009.
7. S. K. Markham, "Product Champions: Crossing the Valley of Death," in P. Belliveau, A. Griffin, and S. Somermeyer (Eds.), *The PDMA Toolbook for New Product Development*, Wiley, Hoboken, NJ, 2002, p. 119.
8. Ibid., p. 120.
9. Ibid., p. 120.

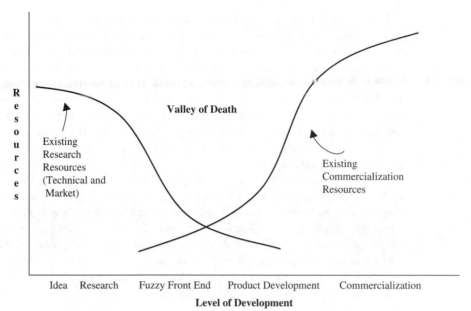

FIGURE 7–2. Valley of death.

Commercialization people need a product that will sell in the market and often consider the value of discovery as merely theoretical and therefore useless. Both technical and commercialization people need help translating research findings into superior product offerings.

As seen in Figure 7–2, the valley of death seems to originate somewhere near the fuzzy front end (FFE).

[FFE is] the messy "getting started" period of product development, which comes before the formal and well-structured product development process, when the product concept is still very fuzzy. It generally consists of the first three tasks (strategic planning, concept generation, and, especially, pre-technical evaluation) of the product development process. These activities are often chaotic, unpredictable, and unstructured. In comparison, the new product development process is typically structured and formal, with a prescribed set of activities, questions to be answered, and decisions to be made.[10]

Project champions are usually neither project managers nor project sponsors. The role of the champion is to sell the idea or concept until it finally becomes a project. The champion may not even understand project management and may not have the necessary skills to manage a project. Champions may reside much higher up in the organizational hierarchy than the project manager.

---

10. P. Belliveau, A. Griffin, and S. Somermeyer, *The PDMA Toolbook for New Product Development*, Wiley, Hoboken, NJ, 2002, p. 444.

Allowing the project champion to function as the project sponsor can be equally as bad as allowing them to function as the project manager. When the project champion and the project sponsor are the same person, projects never get canceled. There is a tendency to prolong the pain of continuing on with a project that should have been canceled.

Some projects, especially very long-term projects where the champion is actively involved, often mandate that a collective belief exist. The collective belief is a fervent, and perhaps blind, desire to achieve that can permeate the entire team, the project sponsor, and even the most senior levels of management. The collective belief can make a rational organization act in an irrational manner. This is particularly true if the project sponsor spearheads the collective belief.

When a collective belief exists, people are selected based upon their support for the collective belief. Champions may prevent talented employees from working on the project unless they possess the same fervent belief as the champion does. Nonbelievers are pressured into supporting the collective belief and team members are not allowed to challenge the results. As the collective belief grows, both advocates and nonbelievers are trampled. The pressure of the collective belief can outweigh the reality of the results.

There are several characteristics of the collective belief, which is why some large, high-technology projects are often difficult to kill:

- Inability or refusing to recognize failure
- Refusing to see the warning signs
- Seeing only what you want to see
- Fearful of exposing mistakes
- Viewing bad news as a personal failure
- Viewing failure as a sign of weakness
- Viewing failure as damage to one's career
- Viewing failure as damage to one's reputation

Project sponsors and project champions do everything possible to make their project successful. But what if the project champion, as well as the project team and sponsor, have blind faith in the success of the project? What happens if the strongly held convictions and the collective belief disregard the early warning signs of imminent danger? What happens if the collective belief drowns out dissent?

In such cases, an exit champion must be assigned. The exit champion sometimes needs to have some direct involvement in the project in order to have credibility, but direct involvement is not always a necessity. Exit champions must be willing to put their reputation on the line and possibly face the likelihood of being cast out from the project team. According to Isabelle Royer[11]:

> Sometimes it takes an individual, rather than growing evidence, to shake the collective belief of a project team. If the problem with unbridled enthusiasm starts as an unintended consequence of the legitimate work of a project champion, then what may be needed is a

---

11. I. Royer, "Why Bad Projects are So Hard to Kil," *Harvard Business Review*, February 2003, p. 11. Copyright © 2003 by the Harvard Business School Publishing Corporation. All rights reserved.

countervailing force—an exit champion. These people are more than devil's advocates. Instead of simply raising questions about a project, they seek objective evidence showing that problems in fact exist. This allows them to challenge—or, given the ambiguity of existing data, conceivably even to confirm—the viability of a project. They then take action based on the data.

The larger the project and the greater the financial risk to the firm, the higher up the exit champion should reside. If the project champion just happens to be the CEO, then someone on the board of directors or even the entire board of directors should assume the role of the exit champion. Unfortunately, there are situations where the collective belief permeates the entire board of directors. In this case, the collective belief can force the board of directors to shirk their responsibility for oversight.

Large projects incur large cost overruns and schedule slippages. Making the decision to cancel such a project, once it has started, is very difficult, according to David Davis[12]:

> The difficulty of abandoning a project after several million dollars have been committed to it tends to prevent objective review and recosting. For this reason, ideally an independent management team—one not involved in the projects development—should do the recosting and, if possible, the entire review. . . . If the numbers do not holdup in the review and recosting, the company should abandon the project. The number of bad projects that make it to the operational stage serves as proof that their supporters often balk at this decision.
> . . . Senior managers need to create an environment that rewards honesty and courage and provides for more decision making on the part of project managers. Companies must have an atmosphere that encourages projects to succeed, but executives must allow them to fail.

The longer the project, the greater the necessity for the exit champions and project sponsors to make sure that the business plan has "exit ramps" such that the project can be terminated before massive resources are committed and consumed. Unfortunately, when a collective belief exists, exit ramps are purposefully omitted from the project and business plans. Another reason for having exit champions is so that the project closure process can occur as quickly as possible. As projects approach their completion, team members often have apprehension about their next assignment and try to stretch out the existing project until they are ready to leave. In this case, the role of the exit champion is to accelerate the closure process without impacting the integrity of the project.

Some organizations use members of a portfolio review board to function as exit champions. Portfolio review boards have the final say in project selection. They also have the final say as to whether or not a project should be terminated. Usually, one member of the board functions as the exit champion and makes the final presentation to the remainder of the board.

---

12. D. Davis, "New Projects: Beware of False Economics," *Harvard Business Review*, March–April 1985, pp. 100–101. Copyright © 1985 by the President and Fellows of Harvard College. All rights reserved.

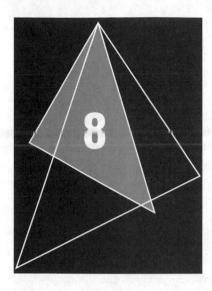

Training and Education

8.0 INTRODUCTION

Establishing project management training programs is one of the greatest challenges facing training directors because project management involves numerous complex and interrelated skills (qualitative/behavioral, organizational, and quantitative). In the early days of project management, project managers learned by their own mistakes rather than from the experience of others. Today, companies excellent in project management are offering a corporate curriculum in project management. Effective training supports project management as a profession.

Some large corporations offer more internal courses related to project management than most colleges and universities do. These companies include General Electric, General Motors, Kodak, the National Cryptological School, Ford Motor Company, and USAA. Such companies treat education almost as a religion. Smaller companies have more modest internal training programs and usually send their people to publicly offered training programs.

This chapter discusses processes for identifying the need for training, selecting the students who need training, designing and conducting the training, and measuring training's return on dollars invested.

8.1 TRAINING FOR MODERN PROJECT MANAGEMENT

During the early days of project management, in the late 1950s and throughout the 1960s, training courses concentrated on the advantages and disadvantages of various organizational forms (e.g., matrix, traditional, functional). Executives learned quickly, however, that any organizational structure could be made to work effectively and efficiently when basic project management is applied. Project management skills based in trust, teamwork, cooperation, and communication can solve the worst structural problems.

Starting with the 1970s, emphasis turned away from organizational structures for project management. The old training programs were replaced with two basic programs:

- Basic project management, which stresses behavioral topics such as multiple reporting relationships, time management, leadership, conflict resolution, negotiation, team building, motivation, and basic management areas such as planning and controlling.
- Advanced project management, which stresses scheduling techniques and software packages used for planning and controlling projects.

Today's project management training programs include courses on behavioral as well as quantitative subjects. The most important problem facing training managers is how to achieve a workable balance between the two parts of the coursework—behavioral and quantitative (see Figure 8–1). For publicly sponsored training programs, the seminar leaders determine their own comfort levels in the "discretionary zone" between technical and behavioral subject matter. For in-house trainers, however, the balance must be preestablished by the training director on the basis of factors such as which students will be assigned to manage projects, types of projects, and average lengths of projects (see Table 8–1).

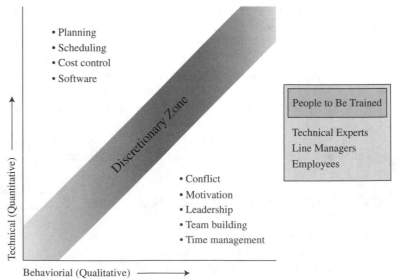

FIGURE 8–1. Types of project management training. *Source:* Reprinted from H. Kerzner, *In Search of Excellence in Project Management*, Wiley, New York, 1998, p. 174.

TABLE 8–1. EMPHASES IN VARIOUS TRAINING PROGRAMS

Type of Person Assigned for PM Training (PM Source)	Training Program Emphasis	
	Quantitative/ Technology Skills	Behavior Skills
Training Needed to Function as a Project Manager		
• Technical expert on short-term projects	High	Low
• Technical expert on long-term projects	High	High
• Line manager acting as a part-time project manager	High	Low
• Line manager acting as a full-time project manager	High	Average to high
• Employees experienced in cooperative operations	High	Average to high
• Employees inexperienced in cooperative operations	High	Average to high
Training Needed for General Knowledge		
• Any employees or managers	Average	Average

Source: Reprinted from H. Kerzner, *In Search of Excellence in Project Management,* Wiley, New York, 1998, p. 175.

8.2 NEED FOR BUSINESS EDUCATION

In the previous section, we discussed the importance of determining the right balance between quantitative skills and behavioral skills. That balance is now changing because of how we view the role of a project manager. Today, we have a new breed of project manager. Years ago, virtually all project managers were engineers with advanced degrees. These people had a command of technology rather than merely an understanding of technology. If the line manager believed that the project manager did in fact possess a command of technology, then the line manager would allow the assigned functional employees to take direction from the project manager. The result was that project managers were expected to manage people and provide technical direction.

Most project managers today have an understanding of technology rather than a command of technology. As a result, the accountability for the success of the project is now viewed as shared accountability between the project manager and all affected line managers. With shared accountability, the line managers must now have a good understanding of project management, which is why more line managers are now becoming PMP®s. Project managers are now expected to manage deliverables rather than people. Management of the assigned resources is more often than not a line function.

Another important fact is that project managers are treated as though they are managing part of a business rather than simply a project, and as such are expected to make sound business decisions as well as project decisions. Project managers must understand business principles. In the future, project managers may be expected to become externally certified by PMI® and internally certified by their company on the organization's business processes.

Now, when designing training courses, we determine the correct balance between quantitative skills, behavioral skills, and business skills. Soft skills and business acumen are crucial elements for a flawless project execution, says Benny Nyberg, formerly Group Assistant VP with responsibility for PM Methodologies and Talent Development at ABB:

After implementing the ABB PM Process as a common high level process throughout the company's project sales organizations as well as several product development organizations, one thing was very clear. In a technical company employing large numbers of highly skilled engineers, some of which are promoted to project management, the technical aspects of project management such as planning, scheduling and cost control are the least difficult to implement. Junior employees do need training in this area but the real challenge for reaching operational excellence in project management, a flawless project execution, desirable projects result and a high level of customer satisfaction lies in identifying and developing project managers with the right business acumen. Project management is a management position requiring excellent commercial, communications and leadership skills. A project manager must be very business minded, be able to communicate effectively with a variety of different stake holders and possess the ability to lead and motivate people. For delivery projects, a precise understanding of the contract, i.e. terms and conditions, scope and any promises made is crucial for being able to deliver just that, meet customer expectations and as such assure customer satisfaction and project success. Contract understanding is further a pre-requisite for maximizing financial outcome and recognizing up-selling opportunities as they occur.

The role of a project manager, especially for big contracts that take many months of several years to complete, is very close to the role of a key account manager. The following skills/abilities are among the most important for success; business mind, communication, negotiation, leadership, risk management, salesmanship.

In order to indentify and address training and other development activities required for the wide variety of competencies and skills, ABB have implemented a competency model. The model includes a definition of required competencies, questionnaires for self assessment, interview questionnaires plus development guides and last but not least a number of selectable training modules leading to appropriate level of certification.

8.3 INTERNATIONAL INSTITUTE FOR LEARNING

Given the importance of project management now and in the future, there will certainly be a continuing need for high-quality project management education. E. LaVerne Johnson, Founder, President and CEO of International Institute for Learning (IIL), comments on the growth of project management training. (For more information regarding IIL, you can visit the IIL website at www.iil.com.)

In IIL's nearly 20-year history, we've worked with thousands of companies and organizations around the world planning and delivering training to their project managers and helping them to create more effective project management environments. Our clients range across all industries and include large, world-class companies as well as smaller organizations trying to gain an edge through effective programs. With IIL serving our clients' day-to-day needs as well as those that are just emerging on the horizon, we've been in a unique position to participate as project management matured into a full-fledged profession. We have been "on the scene" as interest in project management developed into the need to take into account project, program, and portfolio management in order for companies to stay competitive.

From our perspective, courses that were sufficient just a few years ago now fall short—and that's a real sign of progress. The global market and expanding importance of project management has dictated whole new families of courses, much richer content, and a flexible range of delivery mechanisms allowing students to learn when and where they need to—in live and virtual classrooms, alone at their desks whether at home or in the office, self paced or instructor-led. It seems that right now learning and technology have come together to offer the project management profession extremely useful tools to keep on building the future. IIL takes pride in its Many Methods of Learning™ brand, which ensures that the education it provides serves a diversity of needs, styles, and interests.

Evolutionary Years: Learning Trends

Training courses during the 1980s were mostly geared to advancing the project manager's skills. The focus of training was on the basics: the fundamental methodology and the know-how required to pass PMI's Project Management Professional (PMP®) Certification Exam. In response, IIL launched training courses in project management fundamentals and established a comprehensive certification course that allowed individuals to prepare for and successfully pass PMI's PMP® exam. A small variety of books, traditional classroom courses, and software products were made available to those individuals responsible for managing projects in their companies.

Revolutionary Years: Marketplace Trends

In recent years, a far greater variety of companies and industries have recognized the business importance of managing projects more effectively and analyzing the ways in which projects meet overall corporate goals. Compared to previous years, a revolution is emerging. This has become evident in a number of trends:

- The volume of projects is burgeoning as more and more companies run their businesses via projects. Indeed some leading organizations undertake many hundreds of thousands of individual projects each year—some small and simple, others huge and complex.
- The ability to effectively manage projects has become critically important to business, and good project management skills have become a competitive advantage for leading companies.
- As a result of this revolutionary growth, the status and value of the project management professional has grown in importance—having this know-how allows a company to complete projects faster, at lower cost, with greater customer satisfaction, and with more desirable project outcomes.
- Knowledge that was once deemed "nice to have" is now considered "mandatory." A company's economic success and survival depend on its ability to determine which projects support its overall strategic objectives, and to enable it to sequence them in ways to achieve that success.
- Today, employees with project management skills expand beyond the PMP. Team members and middle and top management are developing expertise in the subject.
- The complexity and scope of project management methodologies grow to include new skills and new applications.
- Process development and improvement through direct, hands-on support or knowledge management solutions have become a requirement for economic survival in the most challenging of times. IIL's Unified Project Management® Methodology (UPMM™) Software Suite was developed to support consistency and quality in project, program, and portfolio management implementation.

- A large number of project management–related software programs have been developed to help manage projects (such as Microsoft Office® Project, Primavera, and dozens of others including a wide range of open source project management software).
- Project management certification has become an even more valuable asset to an individual's career path. As a result, there were more than 330,000 active PMPs and more than 300,000 PMI members as of early 2009—the latter a 14% increase over the previous year.
- Heretofore, the project manager's skill set has remained mostly technical. But today we are seeing project managers embrace additional skills: human resources, communications, information technology, managerial, and other advanced areas of knowledge.
- Strategic planning for project management has taken on importance. Organizations are now seeking systematic ways to better align project management with business objectives.
- More and more companies are establishing project and portfolio management offices.
- Approaches to project management within an organization remain relatively varied and nonstandardized. There is a need to work toward a more matured and common methodology in companies for more repeatable and predictable success.
- Companies and their project management offices are placing stronger emphasis on providing quality services, quality products, and improved processes using management tools such as Lean Six Sigma, Earned Valued, and Business Analysis.

Revolutionary Years: Learning Trends

In response to these trends, a far greater variety of courses are available to a broadening number of industries. New methods of learning have been introduced to meet the growing diversity of customer needs. Here are some examples of how IIL has responded to the burgeoning need and established best practices in training and education for project managers:

- IIL has dozens of different course titles in "advanced" areas of knowledge to increase the scope, application, and sophistication of the project manager. Such courses include advanced concepts in risk management, requirements management, the design and development of a project office, and how to manage multiple projects, just to name a few.
- Courses addressing the "softer" side of project management are now available to hone facilitation skills, interpersonal skills, leadership skills, and other nontechnical areas.
- As organizations are increasing in their levels of project management maturity, there is a need for training in the effective use of enterprise project management software. IIL worked with Microsoft to develop the Microsoft® Office Project 2007 Certification Curriculum, focused on the application of project and program management practices, using Microsoft® Office Project 2007 Enterprise Project Management Solution.
- More and more universities are recognizing project management as a part of their degree program. IIL has partnered with New York University (NYU) and the University of Southern California (USC) to offer project management certificate programs complete with university grade reports and certificates of completion where appropriate.
- The way we learn is changing. Employees have less time to devote to classroom study. As a result, in addition to traditional classroom training, IIL now offers innovative technology-based formats—Web-based, "self-paced" training; "virtual" instructor–led courses with synchronous communication; hands-on leadership simulation; and online mentoring. Our "virtual" classroom courses are available 24/7 to accommodate the busy professional's needs, budget, and schedule. Self-paced virtual courses using animation, streaming video, computer-aided simulations, and learning interactions engage and involve users in order to drive learning to maximum levels.

A Look into the Crystal Ball: Trends and Learning Responses

It's always a challenge to try and predict the future, but there are some emerging trends that allow us to take a reasonable stab at this. For each of these trends, there will be the need to develop the appropriate learning responses.

- A key competitive factor in companies will be their ability to undertake and effectively manage many, many projects (projects to develop new products and services, get to market faster, reduce costs, improve customer satisfaction, increase sales, and so on). The more well chosen the projects are, the more competitive a company will be in the marketplace (particularly regarding projects to improve products, processes, and customer satisfaction).
- We anticipate a blending of project management methodologies with other proven business strategies (such as Six Sigma, quality management, risk management, and business analysis). Training in these subjects will similarly become blended.
- Project, program, and portfolio management will continue to grow in business importance and ultimately become a strategic differentiating factor for remaining competitive.
- Senior management will become more knowledgeable and involved in project management efforts. This will require project management training that meets the unique needs of executives.
- Strategic planning for project management will become a way of life for leading organizations. The role of the "project office" will grow in importance and its existence will become commonplace and vital in companies. Membership will include the highest levels of executive management. Senior management will take leadership of the company's project management efforts.
- IIL will continue to partner with companies to offer dual certification for project managers. Project Managers will be required to be certified in project management and also certified in their internal business operations and processes.
- Today's executives will increasingly be involved in activities such as capacity planning, portfolio management, prioritization, business process improvement, supply chain management, and strategic planning specifically for project management. In fact, more and more executives are becoming certified Project and Program Management Professionals.
- An obstacle to participation by upper managers will be their limited experience and training in project, program, and portfolio management. An essential element will be to provide these managers with experience and training in how to manage project activities within their companies. This training must be tailored to be responsive to the unique needs and business responsibilities of upper management.
- The company's reward and recognition systems will change to stimulate and reinforce project management goals and objectives.
- Training in project management will be expanded to include all levels of the company hierarchy, including the non-PMP. Training will become responsive to the unique needs of this broad array of job functions, levels, and responsibilities.
- The status of the PMP and PgMP will grow significantly and the project manager's skills will be both technical and managerial.
- We will witness the establishment of a corporate-level project management executive (chief project management officer).
- Project benchmarking and continuous project improvement will become a way of life in leading organizations. The project management maturity models will play an important role in this regard, as they will help companies identify their strengths, weaknesses, and specific opportunities for improvement.

- The expanding importance of project, program, and portfolio management will require more individuals that are trained in project management. This in turn will necessitate the development of new and improved methods of delivery. Web-based training will play an increasingly important role.
- We will see an order-of-magnitude increase in the number of organizations reaching the higher levels of project management maturity.
- More and more colleges and universities will offer degree programs in project management and seek to align their courses with PMI standards and best practices.
- Project management will focus on providing the knowledge and best practices to support initiatives that offer sustainability. Project sustainability in the global economy through values, leadership, and professional responsibility will be the mandate of all project, program, and portfolio managers and sponsors.

8.4 IDENTIFYING THE NEED FOR TRAINING

Identifying the need for training requires that line managers and senior managers recognize two critical factors: first, that training is one of the fastest ways to build project management knowledge in a company and, second, that training should be conducted for the benefit of the corporate bottom line through enhanced efficiency and effectiveness.

Identifying the need for training has become somewhat easier in the past 10 years because of published case studies on the benefits of project management training. The benefits can be classified according to quantitative and qualitative benefits. The quantitative results include:

- Shorter product development time
- Faster, higher-quality decisions
- Lower costs
- Higher profit margins
- Fewer people needed
- Reduction in paperwork
- Improved quality and reliability
- Lower turnover of personnel
- Quicker "best practices" implementation

Qualitative results include:

- Better visibility and focus on results
- Better coordination
- Higher morale
- Accelerated development of managers
- Better control
- Better customer relations
- Better functional support
- Fewer conflicts requiring senior management involvement

Companies are finally realizing that the speed at which the benefits of project management can be achieved is accelerated through proper training.

8.5 SELECTING STUDENTS

Selecting the people to be trained is critical. As we have already seen in a number of case studies, it is usually a mistake to train only the project managers. A thorough understanding of project management and project management skills is needed throughout the organization if project management is to be successful. For example, one automobile subcontractor invested months in training its project managers. Six months later, projects were still coming in late and over budget. The executive vice president finally realized that project management was a team effort rather than an individual responsibility. After that revelation, training was provided for all of the employees who had anything to do with the projects. Virtually overnight, project results improved.

Dave Kandt, Group Vice President, Quality, Program Management and Continuous Improvement at Johnson Controls, explained how his company's training plan was laid out to achieve excellence in project management:

> We began with our executive office, and once we had explained the principles and philosophies of project management to these people, we moved to the managers of plants, engineering managers, cost analysts, purchasing people, and, of course, project managers. Only once the foundation was laid did we proceed with actual project management and with defining the roles and responsibilities so that the entire company would understand its role in project management once these people began to work. Just the understanding allowed us to move to a matrix organization and eventually to a stand-alone project management department.

8.6 FUNDAMENTALS OF PROJECT MANAGEMENT EDUCATION

Twenty years ago, we were somewhat limited as to availability of project management training and education. Emphasis surrounded on-the-job training in hopes that fewer mistakes would be made. Today, we have other types of programs, including:

- University courses
- University seminars
- In-house seminars
- In-house curriculums
- Distance teaming (e-learning)
- Computer-based training (CBT)

With the quantity of literature available today, we have numerous ways to deliver the knowledge. Typical delivery systems include:

- Lectures
- Lectures with discussion

- Exams
- Case studies on external companies
- Case studies on internal projects
- Simulation and role playing

Training managers are currently experimenting with "when to train." The most common choices include:

- *Just-in-Time Training:* This includes training employees immediately prior to assigning them to projects.
- *Exposure Training:* This includes training employees on the core principles just to give them enough knowledge so that they will understand what is happening in project management within the firm.
- *Continuous Learning:* This is training first on basic, then on advanced, topics so that people will continue to grow and mature in project management.
- *Self-Confidence Training:* This is similar to continuous learning but on current state-of-the-art knowledge. This is to reinforce employees' belief that their skills are comparable to those in companies with excellent reputations for project management.

8.7 DESIGNING COURSES AND CONDUCTING TRAINING

Many companies have come to realize that on-the-job training may be less effective than more formal training. On-the-job training virtually forces people to make mistakes as a learning experience, but what are they learning? How to make mistakes? It seems much more efficient to train people to do their jobs the right way from the start.

Project management has become a career path. More and more companies today allow or even require that their employees get project management certification. One company informed its employees that project management certification would be treated the same as a master's degree in the salary and career path structure. The cost of the training behind the certification process is only 5 or 10 percent of the cost of a typical master's degree in a business administration program. And certification promises a quicker return on investment (ROI) for the company. Project management certification can also be useful for employees without college degrees; it gives them the opportunity for a second career path within the company.

Linda Zaval, formerly a trainer with the International Institute for Learning, explained what type of project management training worked best in her experience during the early 1990s:

In our experience, we have found that training ahead of time is definitely the better route to go. We have done it the other way with people learning on the job, and that has been a rather terrifying situation at times. When we talk about training, we are not just talking about training. We want our project managers to be certified through the Project

Management Institute. We have given our people two years to certify. To that end there is quite a bit of personal study required. I do believe that training from the formal training end is great, and then you can modify that to whatever the need is in-house.

There is also the question of which are better: internally based or publicly held training programs. The answer depends on the nature of the individual company and how many employees need to be trained, how big the training budget is, and how deep the company's internal knowledge base is. If only a few employees at a time need training, it might be effective to send them to a publicly sponsored training course, but if large numbers of employees need training on an ongoing basis, designing and conducting a customized internal training program might be the way to go.

In general, custom-designed courses are the most effective. In excellent companies, course content surveys are conducted at all levels of management. For example, the R&D group of Babcock and Wilcox in Alliance, Ohio, needed a project management training program for 200 engineers. The head of the training department knew that she was not qualified to select core content, and so she sent questionnaires out to executive managers, line managers, and professionals in the organization. The information from the questionnaires was used to develop three separate courses for the audience. At Ford Motor Company, training was broken down into a 2-hour session for executives, a three-day program for project personnel, and a half-day session for overhead personnel.

For internal training courses, choosing the right trainers and speakers is crucial. A company can use trainers currently on staff if they have a solid knowledge of project management, or the trainers can be trained by outside consultants who offer train-the-trainer programs. Either way, trainers from within the company must have the expertise the company needs. Some problems with using internal trainers include the following:

- Internal trainers may not be experienced in all areas of project management.
- Internal trainers may not have up-to-date knowledge of the project management techniques practiced by other companies.
- Internal trainers may have other responsibilities in the company and so may not have adequate time for preparation.
- Internal trainers may not be as dedicated to project management or as skillful as external trainers.

But the knowledge base of internal trainers can be augmented by outside trainers as necessary. In fact, most companies use external speakers and trainers for their internal educational offerings. The best way to select speakers is to seek out recommendations from training directors in other companies and teachers of university-level courses in project management. Another method is contacting speakers' bureaus, but the quality of the speaker's program may not be as high as needed. The most common method for finding speakers is reviewing the brochures of publicly sponsored seminars. Of course, the brochures were created as sales materials, and so the best way to evaluate the seminars is to attend them.

TABLE 8–2.   COMMON PITFALLS IN HIRING EXTERNAL TRAINERS AND SPEAKERS

Warning Sign	Preventive Step
Speaker professes to be an expert in several different areas.	Verify speaker's credentials. Very few people are experts in several areas. Talk to other companies that have used the speaker.
Speaker's résumé identifies several well-known and highly regarded client organizations.	See whether the speaker has done consulting for any of these companies more than once. Sometimes a speaker does a good job selling himself or herself the first time, but the company refuses to rehire him or her after viewing the first presentation.
Speaker makes a very dramatic first impression and sells himself or herself well. Brief classroom observation confirms your impression.	Being a dynamic speaker does not guarantee that quality information will be presented. Some speakers are so dynamic that the trainees do not realize until too late that "The guy was nice but the information was marginal."
Speaker's résumé shows 10–20 years or more experience as a project manager.	Ten to 20 years of experience in a specific industry or company does not mean that the speaker's knowledge is transferable to your company's specific needs or industry. Ask the speaker what types of projects he or she has managed.
Marketing personnel from the speaker's company aggressively show the quality of their company, rather than the quality of the speaker. The client list presented is the company's client list.	You are hiring the speaker, not the marketing representative. Ask to speak or meet with the speaker personally and look at the speaker's client list rather than the parent company's client list.
Speaker promises to custom design his or her materials to your company's needs.	Demand to see the speaker's custom-designed material at least two weeks before the training program. Also verify the quality and professionalism of view graphs and other materials.

After a potential speaker has been selected, the next step is to check his or her rec-ommendations. Table 8–2 outlines many of the pitfalls involved in choosing speakers for internal training programs and how you can avoid them.

The final step is to evaluate the training materials and presentation the external trainer will use in the classes. The following questions can serve as a checklist:

- Does the speaker use a lot of slides in his or her presentation? Slides can be a problem when students do not have enough light to take notes.
- Does the instructor use transparencies? Have they been prepared professionally? Will the students be given copies of the transparencies?
- Does the speaker make heavy use of chalkboards? Too much chalkboard work usually means too much note taking for the trainees and not enough audiovisual preparation from the speaker.
- Does the speaker use case studies? If he or she does, are the case studies factual? It is best for the company to develop its own case studies and ask the speaker to use those so that the cases will have relevance to the company's business.
- Are role playing and laboratory experiences planned? They can be valuable aids to learning, but they can also limit class size.
- Are homework and required reading a part of the class? If so, can they be com-pleted before the seminar?

8.8 MEASURING RETURN ON INVESTMENT

The last area of project management training is the determination of the value earned on the dollars invested in training. Chapter 11 is devoted to this process of measuring ROI for training. It is crucial to remember that training should not be performed unless there is a continuous return on dollars for the company. Keep in mind, also, that the speaker's fee is only part of the cost of training. The cost to the company of having employees away from their work during training must be included in the calculation. Some excellent companies hire outside consultants to determine ROI. The consultants base their evaluations on personal interviews, on-the-job assessments, and written surveys.

One company tests trainees before and after training to learn how much knowledge the trainees really gained. Another company hires outside consultants to prepare and interpret posttraining surveys on the value of the specific training received.

The amount of training needed at any one company depends on two factors: whether the company is project-driven and whether it has practiced project management long enough to have developed a mature project management system. Figure 8–2 shows the amount of project management training offered (including refresher courses) against

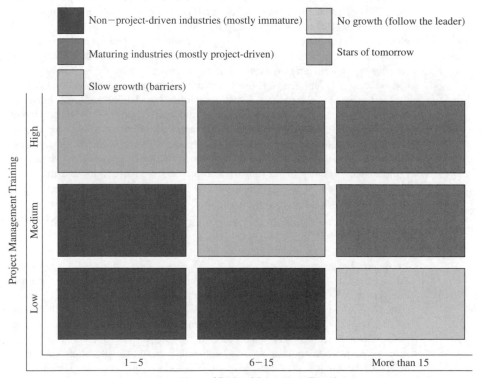

FIGURE 8–2. Amount of training by type of industry and year of project management experience. *Source:* Reprinted from H. Kerzner, *In Search of Excellence in Project Management,* Wiley, New York, 1998, p. 185.

the number of years in project management. Project-driven organizations offer the most project management training courses, and organizations that have just started implementing project management offer the fewest. That's no surprise. Companies with more than 15 years of experience in applying project management principles show the most variance.

8.9 PROJECT MANAGEMENT IS NOW A PROFESSION

For several years, project management was viewed as a part-time occupation and therefore all training was designed for one's primary job description, whatever that may be, rather than project management. As such, there was no need to develop job descriptions for project and program managers. Today, these job descriptions exist, project management is viewed as a profession, and training programs are provided based upon the job descriptions. When asked if AT&T had job descriptions, a spokesperson for AT&T responded "yes" to both project and program management:

Project Manager
Provides end-to-end project management throughout the lifecycle of a project by directing the efforts of project team(s) using dotted-line authority to deliver a completed product and/or service. Has full accountability for managing larger low complexity to high complexity projects, or projects within programs which may span multiple regions and/or multiple functions; multiple concurrent projects may be managed. Includes estimating, scheduling, coordinating, assigning resources, ensuring that project funding is secured, and assisting in recommending business solutions/alternatives for projects. Assesses, plans for, and manages project risks, issues, jeopardies, escalations and problem resolutions. Manages project scope, project budgeting and cost reporting, and ensures completion of projects while meeting quality, schedule and cost objectives using the organization's standard processes. Acts as project liaison between IT partners, client organizations and IT leadership. May assist in supplier management of existing vendors. May direct Associate Project Managers to provide support with project communications and tracking project progress. Does not include the management of extremely large and complex programs, with multiple sub-programs, requiring senior level oversight and extensive executive communications. Must spend 80% or more of time performing the project management duties described above.

Program Manager
Provides end-to-end project management and/or program management throughout the lifecycle of a project/program by directing the efforts of project/program team(s) using dotted-line authority to deliver a completed project and/or service. Has full accountability for managing concurrent high complexity projects and/or programs which may span multiple regions, functions and/or business units. Responsible for detailed planning including program/project structure & staffing, estimating, resource allocation and assignment, detailed scheduling, critical path analysis, consolidating project plans into an overall program plan and negotiating any sequencing conflicts. Directs project and/or program activities utilizing the organization's standard processes to ensure the timely delivery of stated business benefits, comparing actuals to plans and adjusting plans as necessary. Assesses,

plans for, and manages project/program risks including mitigation & contingency plans; manages issues, jeopardies, escalations and problem resolutions. Defines project/program scope and ensures changes to scope and deliverables are managed using the change control process. Manages large program or project budgets and cost reporting. Acts as liaison with client and IT leadership, providing communication and status regarding the progress of the project/program. May assist with RFP development, evaluation, and supplier selection, as well as ongoing relationships with suppliers or consultants. Utilizes knowledge of business, industry and technology to incorporate business process improvements into the organization and/or to develop business strategies and functional/business/technical architectures. May direct the efforts of project managers when they manage a project or sub-program over which the Senior Project/Program Manager has authority. May include the management of extremely large and complex programs, with multiple sub-programs, requiring senior level oversight and extensive executive communications. Must spend 80% or more of time performing the project management duties described above.

The recognition of project management as a profession has spread worldwide. According to Enrique Sevilla Molina, PMO Director, Indra:

Project management is considered the result of a specific blend of knowledge and experience, gained through dedication to achieve success in the projects under the project manager responsibility.

We have a set of management roles associated to the different levels of responsibilities and expertise to manage projects, programs and portfolios, and to develop business opportunities (i.e. Project Managers, Program Directors, etc). For each role, a specific set of skills in a certain degree is defined, so performance and achievement may be assessed. The yearly evaluation of personal performance is done based on the job descriptions, the role maturity achieved so far, the expected performance for the role and the actual performance, so the evolution in personal development may also be assessed.

8.10 COMPETENCY MODELS

Twenty years ago, companies prepared job descriptions for project managers to explain roles and responsibilities. Unfortunately, the job descriptions were usually abbreviated and provided very little guidance on what was required for promotion or salary increases. Ten years ago we still emphasized the job description, but it was now supported by coursework, which was often mandatory. By the late 1990s, companies began emphasizing core competency models, which clearly depicted the skill levels needed to be effective as a project manager. Training programs were instituted to support the core competency models. Unfortunately, establishing a core competency model and the accompanying training is no easy task.

Eli Lilly has perhaps one of the most comprehensive and effective competency models in industry today. Martin D. Hynes, III, Director, Pharmaceutical Projects Management (PPM), was the key sponsor of the initiative to develop the competency model. Thomas J. Konechnik, Operations Manager, Pharmaceutical Projects Management, was responsible for the implementation and integration of the competency model with other processes within the PPM group. The basis for the competency model is described below.

Lilly Research Laboratories project management competencies are classified under three major areas:

Scientific/Technical Expertise

- *Knows the Business:* Brings an understanding of the drug development process and organizational realities to bear on decisions.
- *Initiates Action:* Takes proactive steps to address needs or problems before the situation requires it.
- *Thinks Critically:* Seeks facts, data, or expert opinion to guide a decision or course of action.
- *Manages Risks:* Anticipates and allows for changes in priorities, schedules, and resources and changes due to scientific/technical issues.

Process Skills

- *Communicates Clearly:* Listens well and provides information that is easily understood and useful to others.
- *Attention to Details:* Maintains complete and detailed records of plans, meeting minutes, agreements.
- *Structures the Process:* Constructs, adapts, or follows a logical process to ensure achievement of objectives and goals.
- *Leadership*
- *Focuses on Results:* Continually focuses own and others' attention on realistic milestones and deliverables.
- *Builds a Team:* Creates an environment of cooperation and mutual accountability within and across functions to achieve common objectives.
- *Manages Complexity:* Organizes, plans, and monitors multiple activities, people, and resources.
- *Makes Tough Decisions:* Demonstrates assurance in own abilities, judgments, and capabilities; assumes accountability for actions.
- *Builds Strategic Support:* Gets the support and level of effort needed from senior management and others to keep project on track.

We examine each of these competencies in more detail below.

1. *Knows the Business:* Brings an understanding of the drug development process and organizational realities to bear on decisions.

   Project managers/associates who demonstrate this competency will:

   - Recognize how other functions in Eli Lilly impact the success of a development effort.
   - Use knowledge of what activities are taking place in the project as a whole to establish credibility.
   - Know when team members in own and other functions will need additional support to complete an assignment/activity.

- Generate questions based on understanding of nonobvious interactions of different parts of the project.
- Focus attention on the issues and assumptions that have the greatest impact on the success of a particular project activity or task.
- Understand/recognize political issues/structures of the organization.
- Use understanding of competing functional and business priorities to reality test project plans, assumptions, time estimates, and commitments from the functions.
- Pinpoint consequences to the project of decisions and events in other parts of the organization.
- Recognize and respond to the different perspectives and operating realities of different parts of the organization.
- Consider the long-term implications (pro and con) of decisions.
- Understand the financial implications of different choices.

Project managers/associates who do not demonstrate this competency will:

- Rely on resource and time estimates from those responsible for an activity or task.
- Make decisions based on what ideally should happen.
- Build plans and timelines by rolling up individual timelines and so on.
- Perceive delays as conscious acts on the part of other parts of the organization.
- Assume that team members understand how their activities impact other parts of the project.
- Focus attention on providing accurate accounts of what has happened.
- Avoid changing plans until forced to do so.
- Wait for team members to ask for assistance.

Selected consequences for projects/business of not demonstrating this competency are:

- Project manager or associate may rely on senior management to resolve issues and obtain resources.
- Proposed project timelines may be significantly reworked to meet current guidelines.
- Attention may be focused on secondary issues rather than central business or technical issues.
- Current commitments, suppliers, and so on, may be continued regardless of reliability and value.
- Project deliverables may be compromised by changes in other parts of Lilly.
- Project plans may have adverse impact on other parts of the organization.

2. *Initiates Action:* Takes proactive steps to address needs or problems before the situation requires it.

Project managers/associates who demonstrate this competency will:

- Follow up immediately when unanticipated events occur.
- Push for immediate action to resolve issues and make choices.

- Frame decisions and options for project team, not simply facilitate discussions.
- Take on responsibility for dealing with issues for which no one else is taking responsibility.
- Formulate proposals and action plans when a need or gap is identified.
- Quickly surface and raise issues with project team and others.
- Let others know early on when issues have major implications for project.
- Take action to ensure that relevant players are included by others in critical processes or discussions.

Project managers/associates who do not demonstrate this competency will:

- Focus efforts on ensuring that all sides of issues are explored.
- Ask others to formulate initial responses or plans to issues or emerging events.
- Let functional areas resolve resource issues on their own.
- Raise difficult issues or potential problems after their impact is fully understood.
- Avoid interfering or intervening in areas outside own area of expertise.
- Assume team members and others will respond as soon as they can.
- Defer to more experienced team members on how to handle an issue.

Selected consequences for projects/business of not demonstrating this competency are:

- Senior management may be surprised by project-related events.
- Project activities may be delayed due to "miscommunications" or to waiting for functions to respond.
- Effort and resources may be wasted or underutilized.
- Multiple approaches may be pursued in parallel.
- Difficult issues may be left unresolved.

3. *Thinks Critically:* Seeks facts, data, or expert opinion to guide a decision or course of action.

   Project managers/associates who demonstrate this competency will:

- Seek input from people with expertise or first-hand knowledge of issues and so on.
- Ask tough, incisive questions to clarify time estimates or to challenge assumptions and be able to understand the answers.
- Immerse self in project information to quickly gain a thorough understanding of a project's status and key issues.
- Focus attention on key assumptions and root causes when problems or issues arise.
- Quickly and succinctly summarize lengthy discussions.
- Gather data on past projects, and so on, to help determine best future options for a project.
- Push to get sufficient facts and data in order to make a sound judgment.
- Assimilate large volumes of information from many different sources.
- Use formal decision tools when appropriate to evaluate alternatives and identify risks and issues.

Project managers/associates who do not demonstrate this competency will:

- Accept traditional assumptions regarding resource requirements and time estimates.
- Rely on team members to provide information needed.
- Push for a new milestone without determining the reason previous milestone was missed.
- Summarize details of discussions and arguments without drawing conclusions.
- Limit inquiries to standard sources of information.
- Use procedures and tools that are readily available.
- Define role narrowly as facilitating and documenting team members' discussions.

Selected consequences for projects/business of not demonstrating this competency are:

- Commitments may be made to unrealistic or untested dates.
- High-risk approaches may be adopted without explicit acknowledgment.
- Projects may take longer to complete than necessary.
- New findings and results may be incorporated slowly only into current Lilly practices.
- Major problems may arise unexpectedly.
- Same issues may be revisited.
- Project plan may remain unchanged despite major shifts in resources, people, and priorities.

4. *Manages Risks:* Anticipates and allows for changes in priorities, schedules, resources, and changes due to scientific/technical issues.

   Project managers/associates who demonstrate this competency will:

   - Double-check validity of key data and assumptions before making controversial or potentially risky decisions.
   - Create a contingency plan when pursuing options that have clear risks associated with them.
   - Maintain ongoing direct contact with "risky" or critical path activities to understand progress.
   - Push team members to identify all the assumptions implicit in their estimates and commitments.
   - Stay in regular contact with those whose decisions impact the project.
   - Let management and others know early on the risks associated with a particular plan of action.
   - Argue for level of resources and time estimates that allow for predictable "unexpected" events.
   - Pinpoint major sources of scientific risks.

   Project managers/associates who do not demonstrate this competency will:

   - Remain optimistic regardless of progress.
   - Agree to project timelines despite serious reservations.

- Value innovation and new ideas despite attendant risks.
- Accept less experienced team members in key areas.
- Give individuals freedom to explore different options.
- Accept estimates and assessments with minimal discussion.

Selected consequences for projects/business of not demonstrating this competency are:

- Projects may take longer to complete than necessary.
- Project may have difficulty responding to shifts in organizational priorities.
- Major delays could occur if proposed innovative approach proves inappropriate.
- Known problem areas may remain sources of difficulties.
- Project plans may be subject to dramatic revisions.

5. *Communicates Clearly:* Listens well and provides information that is easily understood and useful to others.

Project managers/associates who demonstrate this competency will:

- Present technical and other complex issues in a concise, clear, and compelling manner.
- Target or position communication to address needs or level of understanding of recipient(s) (e.g., medical, senior management).
- Filter data to provide the most relevant information (e.g., does not go over all details but knows when and how to provide an overall view).
- Keep others informed in a timely manner about decision or issues that may impact them.
- Facilitate and encourage open communication among team members.
- Set up mechanisms for regular communications with team members in remote locations.
- Accurately capture key points of complex or extended discussions.
- Spend the time necessary to prepare presentations for management.
- Effectively communicate and represent technical arguments outside own area of expertise.

Project managers/associates who do not demonstrate this competency will:

- Provide all the available details.
- See multiple reminders or messages as inefficient.
- Expect team members to understand technical terms of each other's specialties.
- Reuse communication and briefing materials with different audiences.
- Limit communications to periodic updates.
- Invite to meetings only those who (are presumed to) need to be there or who have something to contribute.
- Rely on technical experts to provide briefings in specialized, technical areas.

Selected consequences for projects/business of not demonstrating this competency are:

- Individuals outside of the immediate team may have little understanding of the project.
- Other projects may be disrupted by "fire drills" or last-minute changes in plan.
- Key decisions and discussions may be inadequately documented.
- Management briefings may be experienced as ordeals by team and management.
- Resources/effort may be wasted or misapplied.

6. *Pays Attention to Details:* Systematically documents, tracks, and organizes project details.

   Project managers/associates who demonstrate this competency will:

   - Remind individuals of due dates and other requirements.
   - Ensure that all relevant parties are informed of meetings and decisions.
   - Prepare timely, accurate, and complete minutes of meetings.
   - Continually update or adjust project documents to reflect decisions and changes.
   - Check the validity of key assumptions in building the plan.
   - Follow up to ensure that commitments are understood.

   Project managers/associates who do not demonstrate this competency will:

   - Assume that others are tracking the details.
   - See formal reviews as intrusions and waste of time.
   - Choose procedures that are least demanding in terms of tracking details.
   - Only sporadically review and update or adjust project documents to reflect decisions and other changes.
   - Limit project documentation to those formally required.
   - Rely on meeting notes as adequate documentation of meetings.

   Selected consequences for projects/business of not demonstrating this competency are:

   - Coordination with other parts of the organization may be lacking.
   - Documentation may be incomplete or difficult to use to review project issues.
   - Disagreements may arise as to what was committed to.
   - Project may be excessively dependent on the physical presence of manager or associate.

7. *Structures the Process:* Constructs, adapts, or follows a logical process to ensure achievement of objectives and goals.

   Project managers/associates who demonstrate this competency will:

   - Choose milestones that the team can use for assessing progress.
   - Structure meetings to ensure agenda items are covered.
   - Identify sequence of steps needed to execute project management process.

- Maintain up-to-date documentation that maps expectations for individual team members.
- Use available planning tools to standardize procedures and structure activities.
- Create simple tools to help team members track, organize, and communicate information.
- Build a process that efficiently uses team members' time, while allowing them to participate in project decision; all team members should not attend all meetings.
- Review implications of discussion or decisions for the project plan as mechanism for summarizing and clarifying discussions.
- Keep discussions moving by noting disagreements rather than trying to resolve them there and then.
- Create and use a process to ensure priorities are established and project strategy is defined.

Project managers/associates who do not demonstrate this competency will:

- Trust that experienced team members know what they are doing.
- Treat complex sequences of activities as a whole.
- Share responsibility for running meetings, formulating agendas, and so on.
- Create plans and documents that are as complete and detailed as possible.
- Provide written documentation only when asked for.
- Allow team members to have their say.

Selected consequences for projects/business of not demonstrating this competency are:

- Projects may receive significantly different levels of attention.
- Project may lack a single direction or focus.
- Planning documents may be incomplete or out of date.
- Presentations and briefings may require large amounts of additional work.
- Meetings may be seen as unproductive.
- Key issues may be left unresolved.
- Other parts of the organization may be unclear about what is expected and when.

8. *Focuses on Results:* Continually focuses own and others' attention on realistic project milestones and deliverables.

   Project managers/associates who demonstrate this competency will:

   - Stress need to keep project-related activities moving forward.
   - Continually focus on ultimate deliverables (e.g., product to market, affirm/disconfirm merits of compound, value of product/program to Lilly) (manager).
   - Choose actions in terms of what needs to be accomplished rather than seeking optimal solutions or answers.
   - Remind project team members of key project milestones and schedules.
   - Keep key milestones visible to the team.
   - Use fundamental objective of project as means of evaluating option driving decisions in a timely fashion.

- Push team members to make explicit and public commitments to deliverables.
- Terminate projects or low-value activities in timely fashion.

Project managers/associates who do not demonstrate this competency will:

- Assume that team members have a clear understanding of project deliverables and milestones.
- Approach tasks and issues only when they become absolutely critical.
- Downplay or overlook negative results or outcomes.
- Keep pushing to meet original objectives in spite of new data/major changes.
- Pursue activities unrelated to original project requirements.
- Trust that definite plans will be agreed to once team members are involved in the project.
- Allow unqualified individuals to remain on tasks.
- Make attendance at project planning meetings discretionary.

Selected consequences for projects/business of not demonstrating this competency are:

- Milestones may be missed without adequate explanation.
- Functional areas may be surprised at demand for key resources.
- Commitments may be made to unreasonable or unrealistic goals or schedules.
- Projects may take longer to complete than necessary.
- Objectives and priorities may differ significantly from one team member to another.

9. *Builds a Team:* Creates an environment of cooperation and mutual accountability within and across functions to achieve common objectives.

Project managers/associates who demonstrate this competency will:

- Openly acknowledge different viewpoints and disagreements.
- Actively encourage all team members to participate regardless of their functional background or level in the organization.
- Devote time and resources explicitly to building a team identity and a set of shared objectives.
- Maintain objectivity; avoid personalizing issues and disagreements.
- Establish one-on-one relationship with team members.
- Encourage team members to contribute input in areas outside functional areas.
- Involve team members in the planning process from beginning to end.
- Recognize and tap into the experience and expertise that each team member possesses.
- Solicit input and involvement from different functions prior to their major involvement.
- Once a decision is made, insist that team accept it until additional data become available.
- Push for explicit commitment from team members when resolving controversial issues.

Project managers/associates who do not demonstrate this competency will:

- State what can and cannot be done.
- Assume that mature professionals need little support or team recognition.
- Limit contacts with team members to formal meetings and discussions.
- Treat issues that impact a team member's performance as the responsibility of functional line management.
- Help others only when explicitly asked to do so.
- Be openly critical about other team members' contributions or attitudes.
- Revisit decisions when team members resurface issues.

Selected consequences for projects/business of not demonstrating this competency are:

- Team members may be unclear as to their responsibilities.
- Key individuals may move onto other projects.
- Obstacles and setbacks may undermine overall effort.
- Conflicts over priorities within project team may get escalated to senior management.
- Responsibility for project may get diffused.
- Team members may be reluctant to provide each other with support or accommodate special requests.

10. *Manages Complexity:* Organizes, plans, and monitors multiple activities, people, and resources.

    Project managers/associates who demonstrate this competency will:

    - Remain calm when under personal attack or extreme pressure.
    - Monitor progress on frequent and consistent basis.
    - Focus personal efforts on most critical tasks: apply 80–20 rule.
    - Carefully document commitments and responsibilities.
    - Define tasks and activities to all for monitoring and a sense of progress.
    - Break activities and assignments into components that appear doable.
    - Balance and optimize workloads among different groups and individuals.
    - Quickly pull together special teams or use outside experts in order to address emergencies or unusual circumstances.
    - Debrief to capture "best practices" and "lessons learned."

    Project managers/associates who do not demonstrate this competency will:

    - Limit the number of reviews to maximize time available to team members.
    - Stay on top of all the details.
    - Depend on team members to keep track of their own progress.
    - Let others know how they feel about an issue or individual.
    - Rely on the team to address issues.
    - Assume individuals recognize and learn from their own mistakes.

Selected consequences for projects/business of not demonstrating this competency are:

- Projects may receive significantly different levels of attention.
- Projects may take on a life of their own with no clear direction or attainable outcome.
- Responsibility for decisions may be diffused among team members.
- Exact status of projects may be difficult to determine.
- Major issues can become unmanageable.
- Activities of different parts of the business may be uncoordinated.
- Conflicts may continually surface between project leadership and other parts of Lilly.

11. *Makes Tough Decisions:* Demonstrates assurance in own abilities, judgments, and capabilities; assumes accountability for actions.

Project managers/associates who demonstrate this competency will:

- Challenge the way things are done and make decisions about how things will get done.
- Force others to deal with the unpleasant realities of a situation.
- Push for reassessment of controversial decisions by management when new information/data become available.
- Bring issues with significant impact to the attention of others.
- Consciously use past experience and historical data to persuade others.
- Confront individuals who are not meeting their commitments.
- Push line management to replace individuals who fail to meet expectations.
- Challenge continued investment in a project if data suggest it will not succeed.
- Pursue or adopt innovative procedures that offer significant potential benefits even where limited prior experience is available.

Project managers/associates who do not demonstrate this competency will:

- Defer to the ideas of more experienced team members.
- Give others the benefit of the doubt around missed commitments.
- Hold off making decisions until the last possible moment.
- Pursue multiple options rather than halt work on alternative approaches.
- Wait for explicit support from others before raising difficult issues.
- Accept senior managers' decisions as "nonnegotiable."
- Rely on the team to make controversial decisions.
- Provide problematic performers with additional resources and time.

Selected consequences for projects/business of not demonstrating this competency are:

- Projects may take longer to complete than necessary.
- Failing projects may be allowed to linger.
- Decisions may be delegated upward.

- Morale of team may be undermined by nonperformance of certain team members.
- "Bad news" may not be communicated until the last minute.
- Key individuals may "bum out" in effort to play catch-up.

12. *Builds Strategic Support:* Gets the support and level of effort needed from senior management and others to keep projects on track.

   Project managers/associates who demonstrate this competency will:

   - Assume responsibility for championing the projects while demonstrating a balance between passion and objectivity.
   - Tailor arguments and presentations to address key concerns of influential decision-makers.
   - Familiarize self with operational and business concerns of major functions within Lilly.
   - Use network of contacts to determine best way to surface an issue or make a proposal.
   - Push for active involvement of individuals with the experience and influence needed to make things happen.
   - Pinpoint the distribution of influence in conflict situations.
   - Presell controversial ideas or information.
   - Select presenter to ensure appropriate message is sent.
   - Ask senior management to help position issues with other senior managers.

   Project managers/associates who do not demonstrate this competency will:

   - Meet senior management and project sponsors only in formal settings.
   - Propose major shifts in direction in group meetings.
   - Make contact with key decision-makers when faced with obstacles or problems.
   - Limit number of face-to-face contacts with "global" partners.
   - Treat individuals as equally important.
   - Avoid the appearance of "politicking."
   - Depend on other team members to communicate to senior managers in unfamiliar parts of Lilly.

   Selected consequences for projects/business of not demonstrating this competency are:

   - Viable projects may be killed without clear articulation of benefits.
   - "Cultural differences" may limit success of global projects.
   - Decisions may be made without the input of key individuals.
   - Resistance to changes in project scope or direction may become entrenched before merits of proposals are understood.
   - Key individuals/organizations may never buy in to a project's direction or scope.
   - Minor conflicts may escalate and drag on.

8.11 HARRIS CORPORATION

All too often, people attend seminars and courses leading to certification as a PMP and are overwhelmed with the knowledge presented in the course. They wonder how and why any corporation would perform all of the information presented in the course and why they must learn all of this information.

While it is true that many companies do not need or perform all of the activities that are covered, aerospace and defense contractors are required to perform all of these activities. Aerospace and defense contractors survive on how well they perform the project management processes. When companies like Harris Corporation become exceptionally good in project management and undergo continuous improvement in project management, they outperform their competitors. Harris Corporation has a history of success in project management. The remaining material in this section was provided by Alex Sadowski at Harris Corporation.[1] Perhaps after reading the remainder of this section, the reader will have a better understanding and appreciation of why this material is being taught and the complexities of performing in an industry whose life blood and survival rest upon maintaining superior project management capability.

1. The basic philosophy for Project Management applies to all projects no matter what size and what industry is involved. However, each industry has its own unique environment and culture and the application of Project Management philosophy can take different forms based upon this uniqueness. The Aerospace and Defense Contractor industry's unique environment can be characterized as follows:

- The environment is dynamic
- The schedule is most often aggressive
- Changing mission scenarios result in requirements changes
- Effective change management is absolutely necessary
- This is the first time a project of this type is being done
- In most cases the technology envelope is stretched
- To complete the project, the development of new technologies is often required.
- New methodologies have to be developed
- Risk management is of paramount importance
- Proprietary and/or classified information is involved
- This hampers open communication
- Security considerations affect all aspects of the project
- Being overly focused on new and challenging technologies can cause problems with managing the entire project

---

1. Alex Sadowski, PMP, is a program manager with the Government Communications Systems Division of Harris Corporation. He has over 30 years of experience in a large, diverse customer base including various civilian government and military organizations. His project management activities are concentrated in the aerospace and defense contractor business arena. He is also currently vice president of programs for the Space Coast Chapter of the Project Management Institute and is involved with initiatives for the Harris Program Management Executive Council and the Division Process Group.

- The demands of applying new and developing technologies can be very much overwhelming
- The trick is to meet the technological challenge without losing control of cost and schedule

2. As an Aerospace and Defense Contractor, Harris Corporation, has over the years developed detailed processes and procedures that cover all phases of project management from the identification of an opportunity all the way through to final completion, sell-off, and closure.
3. These processes and procedures are documented in great detail in the command media, and are regularly promulgated via seminars, courses, and general meetings as appropriate.
4. This process involves many gates (e.g., Pursue/No-Pursue Reviews, Bid/No-Bid Meetings, Proposal Red Teams, Pricing Reviews, Job Start-up Reviews, Regular Program Reviews, Systems Requirements Reviews, Preliminary Design Reviews, Critical Design Reviews, Peer Reviews, Production Readiness Reviews, Test Readiness Reviews, Final Customer Reviews, Contract Closure, etc.)
5. The processes and methodologies are quite extensive, but there are a few key aspects of this methodology that make everything fall into place and facilitate project success.
6. Through much experience, sometimes quite painful, we have found that detailed up-front planning is the best approach to achieving project success.
7. All too often there is the danger that the over-exuberance of the project team and the impatience of the customer results in the short circuiting of up-front planning. *This is a recipe for disaster!*
8. Proper planning has to be based upon understanding what the customer has contracted for, the price they have agreed to pay, and the date they expect to see completion.
9. A comprehensive Project Plan has to be developed before the full project team is engaged.
10. Effective planning depends on the following (see Figure 8–3):

- Work Breakdown Structure (WBS)
- Integrated Project Schedule (IPS)

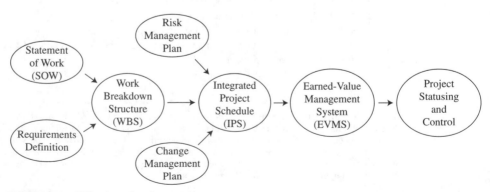

FIGURE 8–3. Effective planning steps.

- Earned Value Management System (EVMS)
- Change Management Planning
- Risk Management Planning

11. Both the Statement of Work (SOW) and the Systems Requirements Document (SRD) are used to develop the Work Breakdown Structure (WBS)

- The SOW identifies what is to be done, any constraints, any special customer gates, the deliverables and the timeframe for the project
- The SRD is the official requirements definition upon which the technical aspects of the project are based. This is used to make certain that the required functionality is achieved.

12. The WBS documents the details of what needs to be accomplished.

- It breaks down the project into its essential parts
- It provides the basis for assigning tasks and responsibilities
- It is a hierarchical structure that shows how each of the individual tasks contribute to the major task
- These individual tasks are essential in defining the schedule and in setting up the EVMS for the project.
- A WBS dictionary is developed that defines each task, and who is responsible for that task

13. Change Management
- Change is inevitable in any project
- Some changes have no overall impact on the project
- Some changes can have a dramatic effect on either the cost, schedule, or technical integrity of the effort. There are times when all three of these will be affected.
- A plan has to developed that allows
- for the identification of change,
- the impact of the change,
- how such changes should be addressed (accepted or rejected),
- how changes are negotiated with the customer
- how changes to the contract are made and by whom

14. Risk Management

- All projects have risks
- A risk is anything that can affect the successful outcome of a program
- A risk can impact
- Cost
- Schedule
- Technical Integrity (e.g., Functionality, Reliability, etc.)
- If a risk is not resolved it then becomes a problem
- At the very initial planning stages of a Project the risks have to be identified.
- A plan has to be developed that addresses
- Identification of risks
- Definition of their severity
- Definition of the probability of occurrence
- Definition of the impact to project success (i.e., cost, schedule, etc.)

- Risk Ranking
- Risk Mitigation
- Risk Retirement

15. The Integrated Project Schedule

- Details the schedule of activities of the project from beginning to end
- Shows the start and stop times of the discrete individual activities
- Shows the inter-relationships of all the individual activities
- Identifies the critical path
- Identifies all critical milestones

16. Earned Value Management System

- Facilitates the tracking of the progress of individual activities with respect to schedule and cost adherence
- Objectively shows which tasks are on schedule and whether they are within budget
- Provides a means of continually assessing whether or not a project is on schedule and within budget

17. Project Statusing & Control

- For any project to succeed, continual statusing and control is essential.
- Generally, reports from EVMS should be generated and reviewed every month. For problem projects or those of high risk, weekly statusing via EVMS would be most beneficial
- EVMS provides a convenient, accurate and ongoing methodology for determining the progress of a project

- The better the effort on the WBS and the IPS, the more accurate the information obtained from EVMS
- Realistic and meaningful milestones, sufficiently covering all tasks, are key to the successful use of EVMS
- Milestones must be tracked regularly (i.e., usually monthly, but as often as weekly if necessary)
- Once milestone accomplishments are accounted for then schedule and cost statusing can be accomplished.
- Generation of the Schedule Performance Index (SPI) provides an objective measure of whether or not the project is on schedule.
- Generation of the Cost Performance Index (CPI) provides an objective measure of whether or not the project is within budget
- A new Estimate to Complete (ETC) should be generated regularly (i.e., at monthly statusing) and compared with the project's Budgeted-cost at Completion (BAC)).
- Discrepancies between the ETC and BAC must be analyzed and reconciled. A result should be implementing a plan to get the program back on track

A well implemented EVMS is a most powerful management tool for the project manager.

- By considering the value of the SPI, the Project Manager knows if the project will meet schedule.

- If the value is less than 1, then the PM knows that the project is behind schedule, and must determine what tasks are falling behind and why, so that appropriate corrective action can be taken
- If the value is greater than 1, then it indicates that the project is ahead of schedule. Too much joy should not be taken at this time since this could be only a temporary situation. If the value is much greater than 1, then that could indicate a problem in the original estimation and planning. This requires detailed analysis.
- By considering the value of the CPI, the Project Manager knows if the project will stay within budget
- If the value is less than 1, then the PM knows that the project is over-running the budget, and must determine which tasks are exceeding cost and why, so that appropriate corrective action can be taken.
- If the value is greater than 1, then it means that the project is under budget. Caution must be taken here since this can be only a temporary situation. If the value is much greater than 1, then there is likely a problem with the original estimation and planning, and a detailed analysis must be done to determine why the discrepancy.
- When the project is planned, a budget is established. The budgeted cost of the project is what is expected to be expended at the conclusion of the project and is the BAC. As the project proceeds, actual costs are incurred. By reviewing the actual costs, and considering what still needs to be done, the PM can then arrive at an Estimate to Complete (ETC)
- By comparing the ETC and BAC, the Variance at Completion (VAC) can be calculated. The VAC indicates whether the project will over-run or under-run.
- If the VAC indicates an over-run then an investigation and analysis must be accomplished to identify the problem. Then a course of action can be taken to mitigate the problem

18. In the demanding and dynamic environment of Aerospace and Government projects, the use of a well defined and fully implemented Earned Value Management System can provide the necessary methodology to keep the project on schedule and within budget.

8.12 ALCATEL-LUCENT: RECOGNIZING THE VALUE OF A PMP

All too often, companies fail to capitalize on the intellectual property retained by PMPs and likewise cannot recognize the contribution they can make to the company. Some companies treat the PMP credentials as just something to put at the end of one's name on a business card or something to put into a proposal as part of competitive bidding efforts. Some companies even provide funding for training programs to obtain the credentials for fear that the workers will seek employment in other firms that provide employees with more opportunities for project management education.

But when companies like Alcatel-Lucent recognize the value that a PMP can make to the company, there can be a huge return-on-investment (ROI) on the PMPs. The ROI can be seen in the form of mentorship for project managers who aspire to become PMPs or wish to recertify their credentials, an advisory council or hotline for projects that are in trouble, or simply by establishing a global project management information network so that all employees worldwide understand current company developments and best practices with regard to project management.

Good things happen to companies like Alcatel-Lucent that recognizes the value of a PMP. According to Rich Maltzman, PMP, Senior Manager, Alcatel-Lucent Global PMO:

> Alcatel-Lucent has sought to increase PM maturity by establishing and deploying an integrated PM Development Framework including a dedicated career path where job descriptions are interlocked with a resource and skills management system connected directly to development options such as dedicated PM courseware. The company has also provided its PMs with a sophisticated Delivery Framework to help them implement and govern projects, measure and manage risk and to manage project financials. These frameworks provide the fundamental platform for achieving PM maturity. In particular, one focus group of PMPs has taken on a dynamic role to support soon-to-be PMPs and existing PMPs to facilitate the credentialing of PMs and to help existing PMPs not only to retain their PMP but to do so in a highly productive manner for their peers.
>
> The team, which gave itself the name PMP-CERT for PMP-Continuing Education and Recertification Team, meets weekly and focuses on several efforts to meet the goals spelled out in its title. Although an ad-hoc team, it works in concert with the Global PMO and Alcatel-Lucent University (a Registered Education Provider of the PMI), both of which are represented on the team. Here are some of PMP-CERT's main activities.

PMP Study Groups

Since the inception of this initiative, 24 PMP Study groups have been established. These are time-zone based groups, each with about 10 members, who study collaboratively for the PMP exam under the tutelage of an instructor. This instructor is typically a recently-credentialed PMP. Meeting at a frequency established by the team, and working in the language in which the majority of members expect to take the test, the Study Groups use the same materials and drill together on subject matter per a model syllabus established by the PMP-CERT team, but adapted for the needs of the particular group. A special PMP Study Hall website has been established specifically for these teams to ask questions, share findings and, of course, post the fact that they have passed the exam! Students receive 30 hours of contact time they need for the exam application, and the instructor is able to claim 10 valuable PDUs by virtue of their service as an instructor. Instructors and students have both reported that this has been an enriching experience, stating, for example, that the Study Groups, "allowed me to meet and learn experiences from other PMs with a background in different areas. This helped me become a stronger PM by using the knowledge gained from these various backgrounds."

Newsletter

For several years, the PMP-CERT team has published a newsletter periodically, but at least quarterly, with featured stories about project management, news of upcoming PDU opportunities, profiles of projects, and announcements of PM-related events around the world. It now has over 500 subscribers.

International PM Day Symposium

Alcatel-Lucent has decided to celebrate International PM Day by running an International PM Day Symposium. This is a web-based conference with connectivity by telephone and web-based video. It has been held since 2007, with attendance around 1,000 people at each. A theme is given to the Symposium and presenters submit proposals from which a PMP-CERT committee selects for use at the session. Past themes have included "Project Management that Matters", and "Project Management—a World of Difference" Presenters from about 6 0 different countries share lessons learned or tackle a project management knowledge area such as Risk Management. In addition, guest speakers, such as noted authors Dr. Harold Kerzner and Jean Binder (winner of PMI's David Cleland Award) or officials of PMI are featured. The Symposium has been registered with Alcatel-Lucent University, and students collect 6 PDUs for their verified attendance. These sessions are archived and are continuously available for replay through Alcatel-Lucent University.

Making PDUs Available—Learning Center

Again working with Alcatel-Lucent University, the PMP-CERT has helped established a PM Learning Center—a resource-rich website with tips on how to apply for the PMP exam, how to keep your credential once earned, suggested readings and courses, and links to the PMP Study Hall and to the established curricula of PM courseware.

Roundtables

PMP-CERT has sponsored five "Discussion Roundtables", 1-hour lunchtime telephone conversations focused on a particular topic and led by a subject-matter expert from inside or outside of Alcatel-Lucent. The sessions are casual and provide the student with 1 PDU.

Inspiration from PMP-CERT

The PMP-CERT team has also inspired other PM activities. Based partially on the success of PMP-CERT, Alcatel-Lucent has established a blog, "Scope Creep", covering general-interest PM topics, which has risen in the ranks of the over 100 corporate blogs to remain in the top 10 since its inception, receiving several thousand hits per year. Based on this success, the company is now starting to use blogs for "live" lessons learned. Projects can use the corporate blogging system with a special blog called SPEAK (Sharing Project Expertise and Knowledge). Key individuals from the project are given joint authoring rights to the SPEAK blog and post regularly with progress, findings, problems, and solutions to allow nearly real-time sharing of lessons-learned.

8.13 INTEGRATED PROJECT MANAGEMENT AT SATYAM

The term "integration" has several meanings whether it is used with regard to IT, engineering, sociology, or economics. Integration is a process of combining various elements and also includes management and control principles as well to assemble all of the components. Satyam has found an interesting use of integration for assembling all of the necessary elements to train and educate project managers. The remaining material in this section has been provided by Hirdesh Singhal, Satyam Learning World, Satyam Computer Services Limited.

Satyam manages IT service projects for customers around the globe. To manage the rapid growth and fill the demand–supply gap for skilled human resources in the organization, it was critical that Satyam rapidly grows Project Managers internally who could manage projects effectively and profitably meeting the expectations of various stakeholders.

A Project Management Competency Development Framework was available in Satyam that focused on making each of role holders in the delivery effective and preparing them to take up the next role. This was aligned to the Project Manager Competency Development Framework from PMI, USA.

With this background, the **Integrated Project Management (IPM)** practice was developed for senior Project Leads, aligned with Project Manager Competency Development Framework so that they are quickly prepared to take up the role of PM.

The practice was developed with following goals:

- Create Project Managers faster than competition
- Save cost by reducing external hiring of PMs
- Leverage the knowledge and skills of existing Project Leads to create PMs
- Reduce the cycle time to get a PM on board
- Make new PMs effective from day one in their job
- Prepare existing associates to take higher level responsibilities
- Motivate associates to stay with the company for longer duration

Some of the differentiators of Integrated Project Management practice include:

- **Holistic learning** to aspiring PMs which not only includes Project Management processes & tools but also, management & behavior, culture & languages and leadership also
- **Certification from the User Community**—Delivery Units leaders, Delivery Integrators/Program Managers are involved through the cycle and assess and certify participants on their readiness to take up PM role
- **Action Learning Projects** during the course provides deep understanding of delivery issues
- **Daily Learning Log Book leading** to Personal Development Plan after the completion of the program

The Design Process is shown in Figure 8–4.

Selection of Participants:

- Having more than 8 years of project experience
- Must have managed a team of at least 10 people
- Indication of abilities to take higher level responsibilities—a feedback from performance management system

We had ascertained that Project Management learning experience needs to be blended with behavioral, management and leadership learning experience to produce world class Project Managers. Participants could clearly understand the processes they need to follow, tools they need to use and various management techniques that can help them manage the project better. To improve the behavior, they did role plays and simulations to practice and demonstrate the change in behavior during the duration of the program. The practice leveraged various training programs and content that existed within the corporate university.

Satyam integrated the data in Tables 8–3 and 8–4.

One of the important facets of delivery excellence is to make the Project Manager productive from day one. They were not only learning during the day, at the end of the day they were also filling their Learning Log Books indicating what they could learn well and what they need to work on or practice more. This also helped them develop their long term development plan.

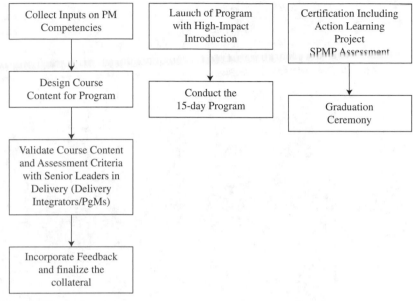

FIGURE 8–4. The design process.

TABLE 8–3. PROJECT MANAGEMENT TRAINING PROGRAMS

S No	Name of the Course	Duration	Delivery Mode
1	Satyam Project Management Practices	3 Days	ILT
2	PM Processes & Tools	2 Days	ILT
3	Software Estimations using Function Points	2 Days	ILT
4	Risk Management	5.5 hours	V-learning
5	Basic Concepts & Techniques of Estimation	6.5 hours	V-learning
6	Software Project Measurement & Metrics	4 hours	V-learning
7	Managing Project Full Lifecycle Businesses	1 Day	ILT

TABLE 8–4. MANGEMENT, BEHAVIORAL, AND LEADERSHIP PROGRAMS

S No	Name of the course	Duration	Delivery Mode
1	Business Challenge Simulation	2 days	ILT
2	Executive Presence	1 day	ILT
3	Conflict management Tools	1 day	ILT
4	Client Facing Skills	1 day	ILT

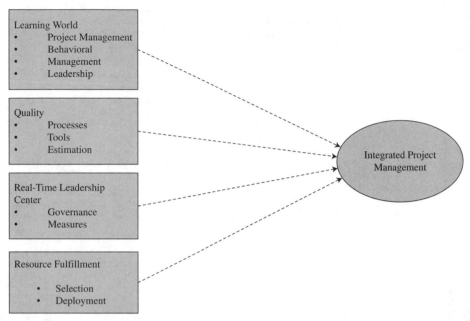

FIGURE 8–5. Integrated project management.

    The structure of the company is based on one key premise of collaboration. There was strong collaboration between the different functions of the company. See Figure 8–5.

- **Various schools of the Corporate university** came together to design and deliver the program covering Project Management, General Management, Behavioral and Leadership components.
- **Real Time Leadership Center** helped with insights on measures and attributes that a PM needs to focus on while managing projects.
- The **Quality function** provided insights on the PM processes and tools in the organization.
- The **Resource Fulfillment Team** deployed the participants by matching the demand and supply of PMs throughout the company.

Collaboration of Satyam Learning World with Other Functions of the Organization:

Key lessons:

1. Success of the practice is largely dependent on the selection of right participants for the program. Involvement of users and Business Units is crucial to right outcome.
2. This program helped participants learn all the aspects required to perform the role of PM.
3. Create a holistic program that prepares for all-around stakeholder management. Thus, include Management, Behavioral and leadership aspects.
4. Validation of the design and content of the program by the users helped us make the program effective from day one.
5. Action Learning Projects focused on ground level issues in the delivery of the projects. Issues that were not significant were removed. These topics were identified through a survey with the senior delivery leaders.

6. Strong support and sponsorship from senior executives for the practice is one of the key elements for the success.

7. Having pre-requisite programs before participants come for the program increase the effectiveness of the program and provides more time for the participants to practice together in the classroom and learn for each others.

8.14 HEWLETT-PACKARD

The quality of the project management training and education a company's employees receive is, along with executive buy-in, one of the most important factors in achieving success and ultimately excellence in project management. The training could be for both the employees of the company as well as for its suppliers who must interface with the customer's project management methodology. Let's look at some case examples of effective training programs.

Hewlett-Packard is clearly committed to program and project management development. According to Jim Crotty, Americas Program Management Profession Leader:

PM Development

HP Services has a comprehensive Program Management Development Program (PMDP) with courses that cover all aspects of program and project management training. A standard curriculum with over 100 courses is implemented throughout the world covering program/ project leadership, management, communication, risk management, contracting, managing business performance, scheduling and cost control, and quality. The courses are based on PMI's Project Management Body of Knowledge (*PMBOK® Guide*). The curriculum also encompasses specialized courses on key HP internal topics, such as the Program Methodology, as well as essential business and financial management aspects of projects.

All courses taught in HP's PMDP curriculum are registered in PMI's Registered Education Provider (REP) Program to ensure a consistent basis and oversight. PMDP won an Excellence in Practice award in career development and organizational learning from the American Society for Training & Development (ASTD), the world's leading resource on workplace learning and performance issues.

Even the most experienced HP PMs continue development activities to strengthen their knowledge and skills. HP Services sponsors the Project Management University (PMU) Program. PMU consists of five day symposiums in each major geography (Americas, Asia/Pacific and Europe) and one day events held in HP offices. These events provide project managers with an opportunity to devote concentrated time to study and to exchange knowledge and ideas with other HP project managers from around the world and in their local geography. PMU has been recognized for excellence by both ASTD and PMI.

PMP Certification

HP has a well-established program to encourage and support our project managers to achieve certification. HP Services has over 5000 individuals who have earned the PMP® (Project Management Professional) certification from PMI®.

PMI Support

HP actively supports the Project Management Institute (PMI®), a non-profit organization with more than 265,000 members. PMI® has set standards for Project Management excellence that are recognized by the industry and our customers worldwide.

HP employees participate on a number of PMI® boards and committees, including the Global Corporate Council. Global Accreditation Center, Research Program Membership Advisory Group, development of the Certified Associate in Project Management (CAPM), development of Certificates of Added Qualification (CAQ) in IT Systems, IT Networks, and Project Management Office, *PMBOK® Guide* review, and *PMBOK® Guide* Update Teams. Many HP employees hold leadership positions in PMI® Chapters and SIGs throughout the world.

8.15 EXEL

Training and education can accelerate the project management maturity process, especially if project management theoretical training is accompanied by training on the corporate project management methodology. This approach has been successfully implemented at Exel. According to Julia Caruso, PMP, Project Manager, Regional Project Management Group Americas:

> Project management (PM) training is currently not mandatory; however, all sectors are aware of the PM methodology and the training provided. The Regional Project Management Groups are responsible for scheduling and conducting the PM training courses in their respective regions. To date, over 7000 associates have been trained in Exel's PM methodology.
>
> Project Management 101 (PM101), Exel's introductory PM course, is a two-day, interactive course consisting of lecture, exercises, interactive team activities, and a comprehensive case study. Typically, the course is delivered by two facilitators in a tag-team fashion. An outside training consultant or certified PM101 trainer teaches general project management terms and theory & facilitates team exercises. A PM practitioner or member from the regional PM group supplements the training with instruction on Exel-specific PM tools and real-world examples.
>
> PM101 participants include members of all sectors, company-wide, who have involvement in projects as team members, project managers, sponsors, or business development.
>
> Presently, the PM training programs are trending toward a curriculum-based format. Originally, PM101 was the only, optional PM training offered in-house. Currently, we have expanded the program to include an on-line PM primer, PM201, DePICT® QuickStart, IT Supplemental and Project Management for Business Development and Sales Teams.
>
> The PM primer is a 30-minute, online tutorial on the basic concepts of PM and Exel's PM methodology. It is accessible on-demand, to all associates worldwide, via the internet.
>
> PM201—Techniques for Advanced Project Management is designed for experienced project managers who wish to enhance both 'hard' and 'soft' project management skills. Pre-work is assigned and the course includes topics such as managing critical path, risk management, conflict management, customer relationship management, and earned value.
>
> DePICT® QuickStart was developed after the acquisition by our parent company, DPWN, for those who were already trained in project management but just needed to

learn the terminology, processes, and tools of our proprietary methodology, DePICT®. This course is also offered by request to new project teams involving a customer, so that the customer stakeholders can learn our PM methodology that will be used to implement their business.

Lastly, IT Supplemental is for IT associates that covers specialized IT PM processes and tools, and PM for Business Development and Sales Team is a quick awareness program designed to provide our sales teams with the knowledge to promote the value of project management and Exel's PM capabilities.

Additionally, Exel is in the process of developing a case-study based PM training program—PM102. It is designed to apply the learnings from PM101 in a realistic project environment, and allow participants to experience typical issues and resolutions common across our projects.

Exel Course Listings

- Project Management On-line Primer
- Project Management 101
- Project Management 201
- DePICT® QuickStart
- IT Supplemental
- Project Management for Business Development and Sales Teams
- Project Management 102 (in development)

The PM training curriculum can also lead to certification training, as described by Todd Daily, PMP, Business Process Manager, International Supply Chain division:

The PM training curriculum does lead to certification training, in some cases. Not everyone who attends PM training or performs the role of project manager pursues formal certification. It is, however, becoming more prominent and recognized within the organization. Five years ago we had 16 certified Project Management Professionals (PMP®) in the U.S. Today we have almost 100 in the U.S., Mexico, and Brazil. We incorporate information about certification in the PM training courses and our course material follows the *PMBOK® Guide* approach.

The Regional Project Management Group Americas provides assistance to candidates within Exel who are interested in pursuing formal certification in project management. We have material and processes to guide potential candidates in their pursuit, and have sponsored in-house exam prep classes provided by an outside vendor.

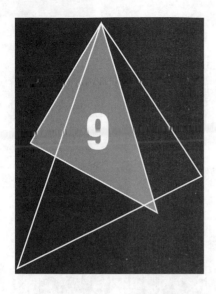

9 Informal Project Management

9.0 INTRODUCTION

Over the past 25 years, one of the most significant changes in project management has been the idea that informal project management does work. In the 1950s and 1960s, the aerospace, defense, and large construction industries were the primary users of project management tools and techniques. Because project management was a relatively new management process, customers of the contractors and subcontractors wanted evidence that the system worked. Documentation of the policies and procedures to be used became part of the written proposal. Formal project management, supported by hundreds of policies, procedures, and forms, became the norm. After all, why would a potential customer be willing to sign a $10 million contract for a project to be managed informally?

This chapter clarifies the difference between informal and formal project management, then discusses the four critical elements of informal project management.

9.1 INFORMAL VERSUS FORMAL PROJECT MANAGEMENT

Formal project management has always been expensive. In the early years, the time and resources spent on preparing written policies and procedures had a purpose: They placated the customer. As project management became established, formal documentation was created mostly for the customer. Contractors began managing more informally, while the customer was still paying for formal project management documentation. Table 9–1 shows the major differences between formal and informal project management. As you can see, the most relevant difference is the amount of paperwork.

TABLE 9–1.   FORMAL VERSUS INFORMAL PROJECT MANAGEMENT

Factor	Formal Project Management	Informal Project Management
Project manager's level	High	Low to middle
Project manager's authority	Documented	Implied
Paperwork	Exorbitant	Minimal

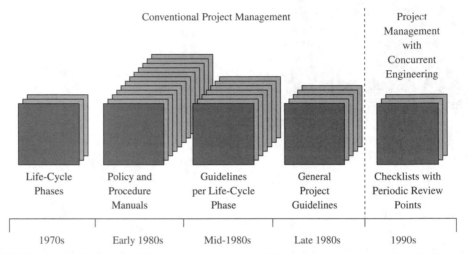

FIGURE 9–1. Evolution of policies, procedures, and guidelines. *Source:* Reprinted from H. Kerzner, *In Search of Excellence in Project Management,* Wiley, New York, 1998, p. 196.

Paperwork is expensive. Even a routine handout for a team meeting can cost $500–$2000 per page to prepare. Executives in excellent companies know that paperwork is expensive. They encourage project teams to communicate without excessive amounts of paper. However, some people are still operating under the mistaken belief that ISO 9000 certification requires massive paperwork.

Figure 9–1 shows the changes in paperwork requirements in project management. The early 1980s marked the heyday for lovers of paper documentation. At that time, the average policies and procedures manual probably cost between $3 million and $5 million to prepare initially and $1 million to $2 million to update yearly over the lifetime of the development project. Project managers were buried in forms to complete to the extent that they had very little time left for actually managing the projects. Customers began to complain about the high cost of subcontracting, and the paperwork boom started to fade.

Real cost savings did not materialize until the early 1990s with the growth of concurrent engineering. Concurrent engineering shortened product development times by taking activities that had been done in series and performing them in parallel instead. This change increased the level of risk in each project, which required that project management back away from some of its previous practices. Formal guidelines were replaced by less detailed and more generic checklists.

Policies and procedures represent formality. Checklists represent informality. But informality does not eliminate project paperwork altogether. It reduces paperwork requirements to minimally acceptable levels. To move from formality to informality

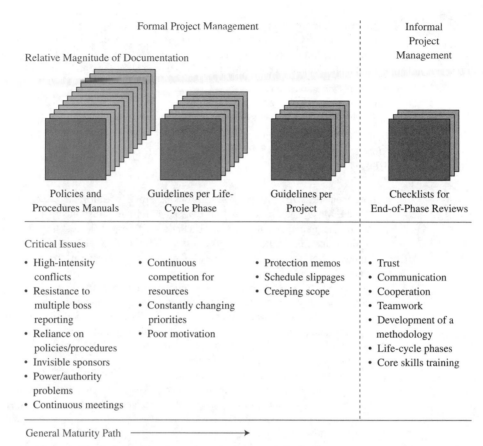

FIGURE 9–2. Evolution of paperwork and change of formality levels. *Source:* Reprinted from H. Kerzner, *In Search of Excellence in Project Management,* Wiley, New York, 1998, p. 198.

demands a change in organizational culture (see Figure 9–2). The four basic elements of an informal culture are these:

- Trust
- Communication
- Cooperation
- Teamwork

Large companies quite often cannot manage projects on an informal basis although they want to. The larger the company, the greater the tendency for formal project management to take hold. A former Vice President of IOC Sales Operations and Customer Service at Nortel Networks, believes that [1]:

> The introduction of enterprise-wide project process and tools standards in Nortel Networks and the use of pipeline metrics (customer-defined, industry standard measures) provides a

---

1. H. Kerzner, *Project Management Best Practices: Achieving Global Excellence*, Wiley, Hoboken, NJ, 2006, p. 329.

framework for formal project management. This is necessary given the complexity of tele-com projects we undertake and the need for an integrated solution in a short time frame. The Nortel Networks project manager crosses many organizational boundaries to achieve the results demanded by customers in a dynamic environment.

Most companies manage either formally or informally. However, if your company is project-driven and has a very strong culture for project management, you may have to manage either formally or informally based upon the needs of your customers. Carl Isenberg, Director of EDS Project Management Consulting, describes this dual approach at EDS [2]:

> Based on the size and scope of a project, formal or informal project management is utilized. Most organizations within EDS have defined thresholds that identify the necessary project management structure. The larger the project (in cost, schedule, and resource requirements), the more strictly the methodology is followed. Smaller projects (such as a modification to a report) require a more cursory approach to the management of that project.
>
> Larger projects require full development of a start-up plan (i.e., the plan for the plan), full-scale planning, execution of the plan and crucial closedown activities to ensure proper project completion, and documentation of lessons learned. These types of projects have a formal project workbook in place (typically in electronic format) to archive project documentation and deliverables. These archives are generally available to be leveraged on future projects that are similar in nature.
>
> The execution of EDS project management methodology is also sometimes dependent on the customer environment and relationship. The contractual nature and/or relationship EDS has with various customers sometimes affect whether project management is done formally or informally. The relationship also affects whose methodology will be used.

9.2 TRUST

Trusting everyone involved in executing a project is critical. You wake up in the morning, get dressed, and climb into your car to go to work. On a typical morning, you operate the foot pedal for your brakes maybe 50 times. You have never met the people who designed the brakes, manufactured the brakes, or installed the brakes. Yet you still give no thought to whether the brakes will work when you need them. No one broadsides you on the way to work. You do not run over anyone. Then you arrive at work and push the button for the elevator. You have never met the people who designed the elevator, manufactured it, installed it, or inspected it. But again you feel perfectly comfortable riding the elevator up to your floor. By the time you get to your office at 8 AM, you have trusted your life to uncounted numbers of people whom you have never even met. Still, you sit down in your office and refuse to trust the person in the next office to make a $50 decision.

---

2. Ibid., p. 330.

TABLE 9–2.   BENEFITS OF TRUST IN CUSTOMER–CONTRACTOR WORKING RELATIONSHIPS

Without Trust	With Trust
Continuous competitive bidding	Long-term contracts, repeat business, and sole source contracts
Massive documentation	Minimal documentation
Excessive customer–contractor team meetings	Minimal number of team meetings
Team meetings with documentation	Team meetings without documentation
Sponsorship at executive levels	Sponsorship at middle-management levels

Trust is the key to the successful implementation of informal project management. Without it, project managers and project sponsors would need all that paperwork just to make sure that everyone working on their projects was doing the work just as he or she had been instructed. And trust is also key in building a successful relationship between the contractor/subcontractor and the client. Let's look at an example.

Perhaps the best application of informal project management that I have seen occurred several years ago in the Heavy Vehicle Systems Group of Bendix Corporation. Bendix hired a consultant to conduct a three-day training program. The program was custom designed, and during the design phase the consultant asked the vice president and general manager of the division whether he wanted to be trained in formal or informal project management. The vice president opted for informal project management. What was the reason for his decision? The culture of the division was already based on trust. Line managers were not hired solely based on technical expertise. Hiring and promotions were based on how well the new manager would communicate and cooperate with the other line managers and project managers in making decisions in the best interests of both the company and the project.

When the relationship between a customer and a contractor is based on trust, numerous benefits accrue to both parties. The benefits are apparent in companies such as Hewlett-Packard, Computer Associates, and various automobile subcontractors. Table 9–2 shows the benefits.

9.3 COMMUNICATION

In traditional, formal organizations, employees usually claim that communication is poor. Senior managers, however, usually think that communication in their company is just fine. Why the disparity? In most companies, executives are inundated with information communicated to them through frequent meetings and dozens of weekly status reports coming from every functional area of the business. The quality and frequency of information moving down the organizational chart are less consistent, especially in more formal companies. But whether it is a problem with the information flowing up to the executive level or down to the staff, the problem usually originates somewhere upstairs. Senior managers are the usual suspects when it comes to requiring reports and meetings. And many of those reports and meetings are unnecessary and redundant.

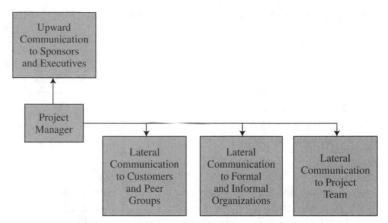

FIGURE 9–3. Internal and external communication channels for project management. *Source:* Reprinted from H. Kerzner, *In Search of Excellence in Project Management,* Wiley, 1998, New York, p. 200.

Most project managers prefer to communicate verbally and informally. The cost of formal communication can be high. Project communication includes dispensing information on decisions made, work authorizations, negotiations, and project reports. Project managers in excellent companies believe that they spend as much as 90 percent of their time on internal interpersonal communication with their teams. Figure 9–3 illustrates the communication channels used by a typical project manager. In project-driven organizations, project managers may spend most of their time communicating externally to customers and regulatory agencies.

Good project management methodologies promote not only informal project management but also effective communications laterally as well as vertically. The methodology itself functions as a channel of communication. A senior executive at a large financial institution commented on his organization's project management methodology, called Project Management Standards (PMS):

> The PMS guides the project manager through every step of the project. The PMS not only controls the reporting structure but also sets the guidelines for who should be involved in the project itself and the various levels of review. This creates an excellent communication flow between the right people. The communication of a project is one of the most important factors for success. A great plan can only go so far if it is not communicated well.

Most companies believe that a good project management methodology will lead to effective communications, which will allow the firm to manage more informally than formally. The question, of course, is how long it will take to achieve effective communications. With all employees housed under a single roof, the time frame can be short. For global projects, geographical dispersion and cultural differences may mandate decades before effective communication will occur. Even then, there is no guarantee that global projects will ever be managed informally.

Suzanne Zale, Global Program Manager of EDS, emphasized [3]:

> With any global project, communications becomes more complex. It will require much more planning up front. All constituents for buy-in need to be identified early on. In order to leverage existing subject matter, experts conversant with local culture, and suppliers, the need for virtual teams becomes more obvious. This increases the difficulty for effective communications.
>
> The mechanism for communication may also change drastically. Face-to-face conversations or meetings will become more difficult. We tend to rely heavily on electronic communications, such as video and telephone conferencing and electronic mail. The format for communications needs to be standardized and understood up front so that information can be sent out quickly. Communications will also take longer and require more effort because of cultural and time differences.

One of the implied assumptions for informal project management to exist is that the employees understand their organizational structure and their roles and responsibilities within both the organizational and project structure. Forms such as the linear responsibility chart and the responsibility assignment matrix are helpful. Communication tools are not used today with the same frequency as in the 1970s and 1980s.

For multinational projects, the organizational structure, roles, and responsibilities must be clearly delineated. Effective communications is of paramount importance and probably must be accomplished more formally than informally.

Suzanne Zale, Global Program Manager of EDS, stated:

> For any global project, the organizational structure must be clearly defined to minimize any potential misunderstandings. It is best to have a clearcut definition of the organizational chart and roles and responsibilities. Any motivation incentives must also contemplate cultural differences. The drivers and values for different cultures can vary substantially.

The two major communication obstacles that must be overcome when a company truly wants to cultivate an informal culture are what I like to call hernia reports and forensic meetings. Hernia reports result from senior management's belief that that which has not been written has not been said. Although there is some truth to such a belief, the written word comes with a high price tag. We need to consider more than just the time consumed in the preparation of reports and formal memos. There is all the time that recipients spend reading them as well as all the support time taken up in processing, copying, distributing, and filing them.

Status reports written for management are too long if they need a staple or a paper clip. Project reports greater than 5 or 10 pages often are not even read. In companies excellent in project management, internal project reports answer these three questions as simply as possible:

- Where are we today?
- Where will we end up?
- Are there any problems that require management's involvement?

---

3. Ibid., p. 332.

All of these questions can be answered on one sheet of paper.

The second obstacle is the forensic team meeting. A forensic team meeting is a meeting scheduled to last 30 minutes that actually lasts for more than 3 hours. Forensic meetings are created when senior managers meddle in routine work activities. Even project managers fall into this trap when they present information to management that management should not be dealing with. Such situations are an invitation to disaster.

9.4 COOPERATION

Cooperation is the willingness of individuals to work with others for the benefit of all. It includes the voluntary actions of a team working together toward a favorable result. In companies excellent in project management, cooperation is the norm and takes place without the formal intervention of authority. The team members know the right thing to do, and they do it.

In the average company (or the average group of any kind, for that matter), people learn to cooperate as they get to know each other. That takes time, something usually in short supply for project teams. But companies such as Ericsson Telecom AB, the General Motors Powertrain Group, and Hewlett-Packard create cultures that promote cooperation to the benefit of everyone.

9.5 TEAMWORK

Teamwork is the work performed by people acting together with a spirit of cooperation under the limits of coordination. Some people confuse teamwork with morale, but morale has more to do with attitudes toward work than it has to do with the work itself. Obviously, however, good morale is beneficial to teamwork.

In excellent companies, teamwork has these characteristics:

- Employees and managers share ideas with each other and establish high levels of innovation and creativity in work groups.
- Employees and managers trust each other and are loyal to each other and the company.
- Employees and managers are committed to the work they do and the promises they make.
- Employees and managers share information freely.
- Employees and managers are consistently open and honest with each other.

Making people feel that they are part of a team does not necessarily require a great deal of effort. Consider the situation at the Engineering and Construction Services Division of Dow Chemical Corporation several years ago. Dow Chemical had requested a trainer to develop a project management training course. The trainer interviewed several of the seminar participants before the training program to identify potential problem areas. The biggest problem appeared to be a lack of teamwork. This shortcoming was particularly evident in the drafting department. The drafting department personnel complained that too many changes were being made to the drawings. They simply could not understand the reasons behind all the changes.

The second problem identified, and perhaps the more critical one, was that project managers did not communicate with the drafting department once the drawings were complete. The drafting people had no idea of the status of the projects they were working on, and they did not feel as though they were part of the project team.

During the training program, one of the project managers, who was responsible for constructing a large chemical plant, was asked to explain why so many changes were being made to the drawings on his project. He said, "There are three reasons for the changes. First, the customers don't always know what they want up front. Second, once we have the preliminary drawings to work with, we build a plastic model of the plant. The model often shows us that equipment needs to be moved for maintenance or safety reasons. Third, sometimes we have to rush into construction well before we have final approval from the Environmental Protection Agency. When the agency finally gives its approval, that approval is often made contingent on making major structural changes to the work already complete." One veteran employee at Dow commented that in his 15 years with the company no one had ever before explained the reasons behind drafting changes.

The solution to the problem of insufficient communication was also easy to repair once it was out in the open. The project managers promised to take monthly snapshots of the progress on building projects and share them with the drafting department. The drafting personnel were delighted and felt more like a part of the project team.

9.6 COLOR-CODED STATUS REPORTING

The use of colors for status reporting, whether it be for printed reports or intranet-based visual presentations, has grown significantly. Color-coded reports encourage informal project management to take place. Colors can reduce risks by alerting management quickly that a potential problem exists. One company prepared complex status reports but color coded the right-hand margins of each page designed for specific audiences and levels of management. One executive commented that he now reads only those pages that are color coded for him specifically rather than having to search through the entire report. In another company, senior management discovered that color-coded intranet status reporting allowed senior management to review more information in a timely manner just by focusing on those colors that indicated potential problems. Colors can be used to indicate:

- Status has not been addressed.
- Status is addressed, but no problems exist.
- Project is on course.
- A potential problem might exist in the future.
- A problem definitely exists and is critical.
- No action is to be taken on this problem.

- Activity has been completed.
- Activity is still active and completion date has passed.

9.7 INFORMAL PROJECT MANAGEMENT AT WORK

Let's review two case studies that illustrate informal project management in action.

Polk Lighting

Polk Lighting (a pseudonym) is a $35 million company located in Jacksonville, Florida. The company manufactures lamps, flashlights, and a variety of other lighting instruments. Its business is entirely based in products and services, and the company does not take on contract projects from outside customers. The majority of the company's stock is publicly traded. The president of Polk Lighting has held his position since the company's start-up in 1985.

In 1994, activities at Polk centered on the R&D group, which the president oversaw personally, refusing to hire an R&D director. The president believed in informal management for all aspects of the business, but he had a hidden agenda for wanting to use informal project management. Most companies use informal project management to keep costs down as far as possible, but the president of Polk favored informal project management so that he could maintain control of the R&D group. However, if the company were to grow, the president would need to add more management structure, establish tight project budgets, and possibly make project management more formal than it had been. Also, the president would probably be forced to hire an R&D director.

Pressure from the company's stockholders eventually forced the president to allow the company to grow. When growth made it necessary for the president to take on heavier administrative duties, he finally hired a vice president of R&D.

Within a few years, the company's sales doubled, but informal project management was still in place. Although budgets and schedules were established as the company grew, the actual management of the projects and the way teams worked together remained informal.

Boeing Aerospace

Boeing was the prime contractor for the U.S. Air Force's new short-range attack missile (SRAM) and awarded the subcontract for developing the missile's propulsion system to the Thiokol Corporation.

It is generally assumed that communication between large customers and contractors must be formal because of the potential for distrust when contracts are complex and involve billions of dollars. The use of on-site representatives, however, can change a potentially contentious relationship into one of trust and cooperation when informality is introduced into the relationship.

Two employees from Boeing were carefully chosen to be on-site representatives at the Thiokol Corporation to supervise the development of the SRAM's propulsion system. The working relationship between Thiokol's project management office and Boeing's on-site representatives quickly developed into shared trust. Team meetings were held without the exchange of excessive documentation. And each party agreed to cooperate with the other. The Thiokol project manager trusted Boeing's representatives well enough to give them raw data from test results even before Thiokol's engineers could formulate their own opinions on the data. Boeing's representatives in turn promised that they would not relay the raw data to Boeing until Thiokol's engineers were ready to share their results with their own executive sponsors.

The Thiokol–Boeing relationship on this project clearly indicates that informal project management can work between customers and contractors. Large construction contractors have had the same positive results in using informal project management and on-site representatives to rebuild trust and cooperation.

Behavioral Excellence

10.0 INTRODUCTION

Previously, we saw that companies excellent in project management strongly emphasize training for behavioral skills. In the past it was thought that project failures were due primarily to poor planning, inaccurate estimating, inefficient scheduling, and lack of cost control. Today, excellent companies realize that project failures have more to do with behavioral shortcomings—poor employee morale, negative human relations, low productivity, and lack of commitment.

This chapter discusses these human factors in the context of situational leadership and conflict resolution. It also provides information on staffing issues in project management. Finally, the chapter offers advice on how to achieve behavioral excellence.

10.1 SITUATIONAL LEADERSHIP

As project management has begun to emphasize behavioral management over technical management, situational leadership has also received more attention. The average size of projects has grown, and so has the size of project teams. Process integration and effective interpersonal relations have also taken on more importance as project teams have gotten larger. Project managers now need to be able to talk with many different functions and departments. There is a contemporary project management proverb that goes something like this: "When researcher talks to researcher, there is 100 percent understanding. When researcher talks to manufacturing, there is 50 percent understanding. When researcher talks to sales, there is zero percent understanding. But the project manager talks to all of them."

Randy Coleman, former senior vice president of the Federal Reserve Bank of Cleveland, emphasizes the importance of tolerance:

> The single most important characteristic necessary in successful project management is tolerance: tolerance of external events and tolerance of people's personalities. Generally, there are two groups here at the Fed—lifers and drifters. You have to handle the two groups differently, but at the same time you have to treat them similarly. You have to bend somewhat for the independents (younger drifters) who have good creative ideas and whom you want to keep, particularly those who take risks. You have to acknowledge that you have some trade-offs to deal with.

A senior project manager in an international accounting firm states how his own leadership style has changed from a traditional to a situational leadership style since becoming a project manager:

> I used to think that there was a certain approach that was best for leadership, but experience has taught me that leadership and personality go together. What works for one person won't work for others. So you must understand enough about the structure of projects and people and then adopt a leadership style that suits your personality so that it comes across as being natural and genuine. It's a blending of a person's experience and personality with his or her style of leadership.

Many companies start applying project management without understanding the fundamental behavioral differences between project managers and line managers. If we assume that the line manager is not also functioning as the project manager, here are the behavioral differences:

- Project managers have to deal with multiple reporting relationships. Line managers report up a single chain of command.
- Project managers have very little real authority. Line managers hold a great deal of authority by virtue of their titles.
- Project managers often provide no input into employee performance reviews. Line managers provide formal input into the performance reviews of their direct reports.
- Project managers are not always on the management compensation ladder. Line managers always are.
- The project manager's position may be temporary. The line manager's position is permanent.
- Project managers sometimes are a lower grade level than the project team members. Line managers usually are paid at a higher grade level than their subordinates.

Several years ago, when Ohio Bell was still a subsidiary of American Telephone and Telegraph, a trainer was hired to conduct a three-day course on project management. During the customization process, the trainer was asked to emphasize planning, scheduling, and controlling and not to bother with the behavioral aspects of project management. At that time, AT&T offered a course on how to become a line supervisor that all of the seminar participants had already taken. In the discussion that followed between the trainer and the course content designers, it became apparent that leadership, motivation, and conflict reso lution were being taught from a superior-to-subordinate point of view in AT&T's course. When the course content designers realized from the discussion that project managers provide leadership, motivation, and conflict resolution to employees who do not report directly to them, the trainer was allowed to include project management–related behavioral topics in the seminar.

Organizations must recognize the importance of behavioral factors in working relationships. When they do, they come to understand that project managers should be hired for their overall project management competency, not for their technical knowledge alone. Brian Vannoni, formerly site training manager and principal process engineer at GE Plastics, described his organization's approach to selecting project managers:

> The selection process for getting people involved as project managers is based primarily on their behavioral skills and their skills and abilities as leaders with regard to the other aspects of project management. Some of the professional and full-time project managers have taken senior engineers under their wing, coached and mentored them, so that they learn and pick up the other aspects of project management. But the primary skills that we are looking for are, in fact, the leadership skills.

Project managers who have strong behavioral skills are more likely to involve their teams in decision-making, and shared decision-making is one of the hallmarks of successful project management. Today, project managers are more managers of people than they are managers of technology. According to Robert Hershock, former vice president at 3M:

> The trust, respect, and especially the communications are very, very important. But I think one thing that we have to keep in mind is that a team leader isn't managing technology; he or she is managing people. If you manage the people correctly, the people will manage the technology.

In addition, behaviorally oriented project managers are more likely to delegate responsibility to team members than technically strong project managers. In 1996, Frank Jackson, formerly a senior manager at MCI, said that:

> Team leaders need to have a focus and a commitment to an ultimate objective. You definitely have to have accountability for your team and the outcome of your team. You've got to be able to share the decision-making. You can't single out yourself as the exclusive holder of the right to make decisions. You have got to be able to share that. And lastly again, just to harp on it one more time, is communications. Clear and concise communication throughout the team and both up and down a chain of command is very, very important.

Some organizations prefer to have a project manager with strong behavioral skills acting as the project manager, with technical expertise residing with the project engineer.

Other organizations have found the reverse to be effective. Rose Russett, formerly the program management process manager for General Motors Powertrain, stated:

> We usually appoint an individual with a technical background as the program manager and an individual with a business and/or systems background as the program administrator. This combination of skills seems to complement one another. The various line managers are ultimately responsible for the technical portions of the program, while the key responsibility of the program manager is to provide the integration of all functional deliverables to achieve the objectives of the program. With that in mind, it helps for the program manager to understand the technical issues, but they add their value not by solving specific technical problems but by leading the team through a process that will result in the best solutions for the overall program, not just for the specific functional area. The program administrator, with input from all team members, develops the program plans, identifies the critical path, and regularly communicates this information to the team throughout the life of the program. This information is used to assist with problem solving, decision-making, and risk management.

10.2 CONFLICT RESOLUTION

Opponents of project management claim that the primary reason why some companies avoid changing over to a project management culture is that they fear the conflicts that inevitably accompany change. Conflicts are a way of life in companies with project management cultures. Conflict can occur on any level of the organization, and conflict is usually the result of conflicting objectives. The project manager is a conflict manager. In many organizations, the project managers continually fight fires and handle crises arising from interpersonal and interdepartmental conflicts. They are so busy handling conflicts that they delegate the day-to-day responsibility for running their projects to the project teams. Although this arrangement is not the most effective, it is sometimes necessary, especially after organizational restructuring or after a new project demanding new resources has been initiated.

The ability to handle conflicts requires an understanding of why conflicts occur. We can ask four questions, the answers to which are usually helpful in handling, and possibly preventing, conflicts in a project management environment:

- Do the project's objectives conflict with the objectives of other projects currently in development?
- Why do conflicts occur?
- How can we resolve conflicts?
- Is there anything we can do to anticipate and resolve conflicts before they become serious?

Although conflicts are inevitable, they can be planned for. For example, conflicts can easily develop in a team in which the members do not understand each other's roles and responsibilities. Responsibility charts can be drawn to map out graphically who is responsible for doing what on the project. With the ambiguity of roles and responsibilities gone, the conflict is resolved or future conflict averted.

Resolution means collaboration, and collaboration means that people are willing to rely on each other. Without collaboration, mistrust prevails and progress documentation increases.

The most common types of conflict involve the following:

- Manpower resources
- Equipment and facilities
- Capital expenditures
- Costs
- Technical opinions and trade-offs
- Priorities
- Administrative procedures
- Schedules
- Responsibilities
- Personality clashes

Each of these types of conflict can vary in intensity over the life of the project. The relative intensity can vary as a function of:

- Getting closer to project constraints
- Having met only two constraints instead of three (e.g., time and performance but not cost)
- The project life cycle itself
- The individuals who are in conflict

Conflict can be meaningful in that it results in beneficial outcomes. These meaningful conflicts should be allowed to continue as long as project constraints are not violated and beneficial results accrue. An example of a meaningful conflict might be two technical specialists arguing that each has a better way of solving a problem. The beneficial result would be that each tries to find additional information to support his or her hypothesis.

Some conflicts are inevitable and occur over and over again. For example, consider a raw material and finished goods inventory. Manufacturing wants the largest possible inventory of raw materials on hand to avoid possible production shutdowns. Sales and marketing wants the largest finished goods inventory so that the books look favorable and no cash flow problems are possible.

Consider five methods that project managers can use to resolve conflicts:

- Confrontation
- Compromise

- Facilitation (or smoothing)
- Force (or forcing)
- Withdrawal

Confrontation is probably the most common method used by project managers to resolve conflict. Using confrontation, the project manager faces the conflict directly. With the help of the project manager, the parties in disagreement attempt to persuade one another that their solution to the problem is the most appropriate.

When confrontation does not work, the next approach project managers usually try is compromise. In compromise, each of the parties in conflict agrees to trade-offs or makes concessions until a solution is arrived at that everyone involved can live with. This give-and-take-approach can easily lead to a win–win solution to the conflict.

The third approach to conflict resolution is facilitation. Using facilitation skills, the project manager emphasizes areas of agreement and deemphasizes areas of disagreement. For example, suppose that a project manager said, "We've been arguing about five points, and so far we've reached agreement on the first three. There's no reason why we can't agree on the last two points, is there?" Facilitation of a disagreement does not resolve the conflict. Facilitation downplays the emotional context in which conflicts occur.

Force is also a method of conflict resolution. A project manager uses force when he or she tries to resolve a disagreement by exerting his or her own opinion at the expense of the other people involved. Often, forcing a solution onto the parties in conflict results in a win–lose outcome. Calling in the project sponsor to resolve a conflict is another form of force project managers sometimes use.

The least used and least effective mode of conflict resolution is withdrawal. A project director can simply withdraw from the conflict and leave the situation unresolved. When this method is used, the conflict does not go away and is likely to recur later. Personality conflicts might well be the most difficult conflicts to resolve. Personality conflicts can occur at any time, with anyone, and over anything. Furthermore, they can seem almost impossible to anticipate and plan for.

Let's look at how one company found a way to anticipate and avoid personality conflicts on one of its projects. Foster Defense Group (a pseudonym) was the government contract branch of a Fortune 500 company. The company understood the potentially detrimental effects of personality clashes on its project teams, but it did not like the idea of getting the whole team together to air its dirty laundry. The company found a better solution. The project manager put the names of the project team members on a list. Then he interviewed each of the team members one on one and asked each to identify the names on the list that he or she had had a personality conflict with in the past. The information remained confidential, and the project manager was able to avoid potential conflicts by separating clashing personalities.

If at all possible, the project manager should handle conflict resolution. When the project manager is unable to defuse the conflict, then and only then should the project sponsor be brought in to help solve the problem. Even then, the sponsor should not come in and force a resolution to the conflict. Instead, the sponsor should facilitate further discussion between the project managers and the team members in conflict.

10.3 STAFFING FOR EXCELLENCE

Project manager selection is always an executive-level decision. In excellent companies, however, executives go beyond simply selecting the project manager. They use the selection process to accomplish the following:

- Project managers are brought on board early in the life of the project to assist in outlining the project, setting its objectives, and even planning for marketing and sales. The project manager's role in customer relations becomes increasingly important.
- Executives assign project managers for the life of the project and project termination. Sponsorship can change over the life cycle of the project, but not the project manager.
- Project management is given its own career ladder.
- Project managers given a role in customer relations are also expected to help sell future project management services long before the current project is complete.
- Executives realize that project scope changes are inevitable. The project manager is viewed as a manager of change.

Companies excellent in project management are prepared for crises. Both the project manager and the line managers are encouraged to bring problems to the surface as quickly as possible so that there is time for contingency planning and problem solving. Replacing the project manager is no longer the first solution for problems on a project. Project managers are replaced only when they try to bury problems.

A defense contractor was behind schedule on a project, and the manufacturing team was asked to work extensive overtime to catch up. Two of the manufacturing people, both union employees, used the wrong lot of raw materials to produce a $65,000 piece of equipment needed for the project. The customer was unhappy because of the missed schedules and cost overruns that resulted from having to replace the useless equipment. An inquisition-like meeting was convened and attended by senior executives from both the customer and the contractor, the project manager, and the two manufacturing employees. When the customer's representative asked for an explanation of what had happened, the project manager stood up and said, "I take full responsibility for what happened. Expecting people to work extensive overtime leads to mistakes. I should have been more careful." The meeting was adjourned with no one being blamed. When word spread through the company about what the project manager did to protect the two union employees, everyone pitched in to get the project back on schedule, even working uncompensated overtime.

Human behavior is also a consideration in assigning staff to project teams. Team members should not be assigned to a project solely on the basis of technical knowledge. It has to be recognized that some people simply cannot work effectively in a team environment. For example, the director of research and development at a New England company had an employee, a 50-year-old engineer, who held two master's degrees in engineering disciplines. He had worked for the previous 20 years on one-person projects. The director reluctantly assigned the engineer to a project team. After years of working alone, the engineer trusted no one's results but his own. He refused to work cooperatively with the other members of the team. He even went so far as redoing all the calculations passed on to him from other engineers on the team.

To solve the problem, the director assigned the engineer to another project on which he supervised two other engineers with less experience. Again, the older engineer tried to do all of the work by himself, even if it meant overtime for him and no work for the others.

Ultimately, the director had to admit that some people are not able to work cooperatively on team projects. The director went back to assigning the engineer to one-person projects on which the engineer's technical abilities would be useful.

Robert Hershock, former vice president at 3M, once observed:

> There are certain people whom you just don't want to put on teams. They are not team players, and they will be disruptive on teams. I think that we have to recognize that and make sure that those people are not part of a team or team members. If you need their expertise, you can bring them in as consultants to the team but you never, never put people like that on the team.
>
> I think the other thing is that I would never, ever eliminate the possibility of anybody being a team member no matter what the management level is. I think if they are properly trained, these people at any level can be participators in a team concept.

In 1996, Frank Jackson, formerly a senior manager at MCI, believed that it was possible to find a team where any individual can contribute:

> People should not be singled out as not being team players. Everyone has got the ability to be on a team and to contribute to a team based on the skills and the personal experiences that they have had. If you move into the team environment one other thing that is very important is that you not hinder communications. Communications is the key to the success of any team and any objective that a team tries to achieve.

One of the critical arguments still being waged in the project management community is whether an employee (even a project manager) should have the right to refuse an assignment. At Minnesota Power and Light, an open project manager position was posted, but nobody applied for the job. The company recognized that the employees probably did not understand what the position's responsibilities were. After more than 80 people were trained in the fundamentals of project management, there were numerous applications for the open position.

It's the kiss of death to assign someone to a project manager's job if that person is not dedicated to the project management process and the accountability it demands.

10.4 VIRTUAL PROJECT TEAMS

Historically, project management was a face-to-face environment where team meetings involved all players meeting together in one room. Today, because of the size and complexity of projects, it is impossible to find all team members located under one roof. Duarte and Snyder define seven types of virtual teams.[1] These are shown in Table 10–1.

---

1. D. L. Duarte and N. Tennant Snyder, *Mastering Virtual Teams.* Jossey-Bass, 2001, San Francisco, CA, p. 10. Reproduced by permission of John Wiley & Sons.

TABLE 10–1.   TYPES OF VIRTUAL TEAMS

Type of Team	Description
Network	Team membership is diffuse and fluid; members come and go as needed. Team lacks clear boundaries within the organization.
Parallel	Team has clear boundaries and distinct membership. Team works in the short term to develop recommendations for an improvement in a process or system.
Project or product development	Team has fluid membership, clear boundaries, and a defined customer base, technical requirement, and output. Longer-term team task is nonroutine, and the team has decision-making authority.
Work or production	Team has distinct membership and clear boundaries. Members perform regular and outgoing work, usually in one functional area.
Service	Team has distinct membership and supports ongoing customer network activity.
Management	Team has distinct membership and works on a regular basis to lead corporate activities.
Action	Team deals with immediate action, usually in an emergency situation. Membership may be fluid or distinct.

Source: D. L. Duarte and N. Tennant Snyder, *Mastering Virtual Teams*. Jossey-Bass, 2001, San Francisco, CA, p. 10. Reproduced by permission of John Wiley & Sons.

TABLE 10–2.   TECHNOLOGY AND CULTURE

Cultural Factor	Technological Considerations
Power distance	Members from high-power-distance cultures may participate more freely with technologies that are asynchronous and allow anonymous input. These cultures sometimes use technology to indicate status differences between team members.
Uncertainty avoidance	People from cultures with high uncertainty avoidance may be slower adopters of technology. They may also prefer technology that is able to produce more permanent records of discussions and decisions.
Individualism–collectivism	Members from highly collectivistic cultures may prefer face-to-face interactions.
Masculinity–femininity	People from cultures with more "feminine" orientations are more prone to use technology in a nurturing way, especially during team startups.
Context	People from high-context cultures may prefer more information-rich technologies, as well as those that offer opportunities for the feeling of social presence. They may resist using technologies with low social presence to communicate with people they have never met. People from low-context cultures may prefer more asynchronous communications.

Source: D. L Duarte and N. Tennant Snyder, *Mastering Virtual Teams,* Jossey-Bass, San Francisco, CA, 2001, p. 60.

Culture and technology can have a major impact on the performance of virtual teams. Duarte and Snyder have identified some of these relationships in Table 10–2. .

The importance of culture cannot be understated. Duarte and Snyder identify four important points to remember concerning the impact of culture on virtual teams. The four points are[2]:

1. There are national cultures, organizational cultures, functional cultures, and team cultures. They can be sources of competitive advantages for virtual teams that know

---

2. Ibid., p. 70.

how to use cultural differences to create synergy. Team leaders and members who understand and are sensitive to cultural differences can create more robust outcomes than can members of homogeneous teams with members who think and act alike. Cultural differences can create distinctive advantages for teams if they are understood and used in positive ways.

2. The most important aspect of understanding and working with cultural differences is to create a team culture in which problems can be surfaced and differences can be discussed in a productive, respectful manner.

3. It is essential to distinguish between problems that result from cultural differences and problems that are performance based.

4. Business practices and business ethics vary in different parts of the world. Virtual teams need to clearly articulate approaches to these that every member understands and abides by.

10.5 REWARDING PROJECT TEAMS

Today, most companies are using project teams. However, there still exist challenges in how to reward project teams for successful performance. The importance of how teams are rewarded is identified by Parker, McAdams, and Zielinski[3]:

> Some organizations are fond of saying, "We're all part of the team, but too often it is merely management-speak. This is especially common in conventional hierarchical organizations; they say the words but don't follow up with significant action. Their employees may read the articles and attend the conferences and come to believe that many companies have turned collaborative. Actually, though, few organizations today are genuinely team-based.
>
> Others who want to quibble point to how they reward or recognize teams with splashy bonuses or profit-sharing plans. But these do not by themselves represent a commitment to teams; they're more like a gift from a rich uncle. If top management believes that only money and a few recognition programs ("team of year" and that sort of thing) reinforce teamwork, they are wrong. These alone do not cause fundamental change in the way people and teams are managed.
>
> But in a few organizations, teaming is a key component of the corporate strategy, involvement with teams is second nature, and collaboration happens without great thought or fanfare. There are natural work groups (teams of people who do the same or similar work in the same location), permanent cross-functional teams, ad hoc project teams, process improvement teams, and real management teams. Involvement just happens.

Why is it so difficult to reward project teams? To answer this question, we must understand what a team is and is not[4]:

> Consider this statement: an organizational unit can act like a team, but a team is not necessarily an organizational unit, at least for describing reward plans. An organizational unit

---

3. G. Parker, J. McAdams, and D. Zielinski, *Rewarding Teams,* Jossey-Bass, San Francisco, CA, 2000, p. 17. Reproduced by permission of John Wiley & Sons.
4. Ibid., p. 17.

is just that, a group of employees organized into an identifiable business unit that appears on the organizational chart. They may behave in a spirit of teamwork, but for the purposes of developing reward plans they are not a "team." The organizational unit may be a whole company, a strategic business unit, a division, a department, or a work group.

A "team" is a small group of people allied by a common project and sharing performance objectives. They generally have complementary skills or knowledge and an interdependence that requires that they work together to accomplish their project's objective. Team members hold themselves mutually accountable for their results. These teams are not found on an organization chart.

Incentives are difficult to apply because project teams may not appear on an organizational chart. Figure 10–1 shows the reinforcement model for employees.[5] For project teams, the emphasis is the three arrows on the right-hand side of Figure 10–1.

Project team incentives are important because team members expect appropriate rewards and recognition[6]:

> Project teams are usually, but not always, formed by management to tackle specific projects or challenges with a defined time frame—reviewing processes for efficiency or cost-savings recommendations, launching a new software product, or implementing enterprise resource planning systems are just a few examples. In other cases, teams self-form around specific issues or as part of continuous improvement initiatives such as team-based suggestion systems.
>
> Project teams can have cross-functional membership or simply be a subset of an existing organizational unit. The person who sponsors the team—its "champion" typically

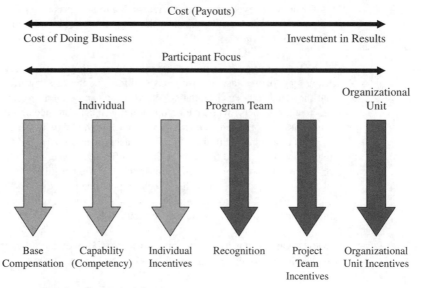

FIGURE 10–1. Reinforcement model.

5. Ibid., p. 29.
6. Ibid., pp. 38–39.

creates an incentive plan with specific objective measures and an award schedule tied to achieving those measures. To qualify as an incentive, the plan must include pre-announced goals, with a "do this, get that" guarantee for teams. The incentive usually varies with the value added by the project.

Project team incentive plans usually have some combination of these basic measures:

- Project Milestones: Hit a milestone, on budget and on time, and all team members earn a defined amount. Although sound in theory, there are inherent problems in tying financial incentives to hitting milestones. Milestones often change for good reason (technological advances, market shifts, other developments) and you don't want the team and management to get into a negotiation on slipping dates to trigger the incentive. Unless milestones are set in stone and reaching them is simply a function of the team doing its normal, everyday job, it's generally best to use recognition-after-the-fact celebration of reaching milestones—rather than tying financial incentives to it.

Rewards need not always be time-based, such that when the team hits a milestone by a certain date it earns a reward. If, for example, a product development team debugs a new piece of software on time, that's not necessarily a reason to reward it. But if it discovers and solves an unsuspected problem or writes better code before a delivery date, rewards are due.

- Project Completion: All team members earn a defined amount when they complete the project on budget and on time (or to the team champion's quality standards).
- Value Added: This award is a function of the value added by a project, and depends largely on the ability of the organization to create and track objective measures. Examples include reduced turnaround time on customer requests, improved cycle times for product development, cost savings due to new process efficiencies, or incremental profit or market share created by the product or service developed or implemented by the project team.

One warning about project incentive plans: they can be very effective in helping teams stay focused, accomplish goals, and feel like they are rewarded for their hard work, but they tend to be exclusionary. Not everyone can be on a project team. Some employees (team members) will have an opportunity to earn an incentive that others (non-team members) do not. There is a lack of internal equity. One way to address this is to reward core team members with incentives for reaching team goals, and to recognize peripheral players who supported the team, either by offering advice, resources, or a pair of hands, or by covering for project team members back at their regular job.

Some projects are of such strategic importance that you can live with these internal equity problems and non-team members' grousing about exclusionary incentives. Bottom line, though, is this tool should be used cautiously.

Some organizations focus only on cash awards. However, Parker et al. have concluded from their research that noncash awards can work equally well, if not better, than cash awards[7]:

Many of our case organizations use non-cash awards because of their staying power. Everyone loves money, but cash payments can lose their motivational impact over time.

---

7. Ibid., pp. 190–191.

However, non-cash awards carry trophy value that has great staying power because each time you look at that television set or plaque you are reminded of what you or your team did to earn it. Each of the plans encourages awards that are coveted by the recipients and, therefore, will be memorable.

If you ask employees what they want, they will invariably say cash. But providing it can be difficult if the budget is small or the targeted earnings in an incentive plan are modest. If you pay out more often than annually and take taxes out, the net amount may look pretty small, even cheap. Non-cash awards tend to be more dependent on their symbolic value than their financial value.

Non-cash awards come in all forms: a simple thank-you, a letter of congratulations, time off with pay, a trophy, company merchandise, a plaque, gift certificates, special services, a dinner for two, a free lunch, a credit to a card issued by the company for purchases at local stores, specific items or merchandise, merchandise from an extensive catalogue, travel for business or a vacation with the family, and stock options. Only the creativity and imagination of the plan creators limit the choices.

10.6 KEYS TO BEHAVIORAL EXCELLENCE

There are some distinguishing actions that project managers can take to ensure the successful completion of their projects. These include:

- Insisting on the right to select key project team
- Negotiating for key team members with proven track records in their fields
- Developing commitment and a sense of mission from the outset
- Seeking sufficient authority from the sponsor
- Coordinating and maintaining a good relationship with the client, parent, and team
- Seeking to enhance the public's image of the project
- Having key team members assist in decision-making and problem solving
- Developing realistic cost, schedule, and performance estimates and goals
- Maintaining backup strategies (contingency plans) in anticipation of potential problems
- Providing a team structure that is appropriate, yet flexible and flat
- Going beyond formal authority to maximize its influence over people and key decisions
- Employing a workable set of project planning and control tools
- Avoiding overreliance on one type of control tool
- Stressing the importance of meeting cost, schedule, and performance goals
- Giving priority to achieving the mission or function of the end item
- Keeping changes under control
- Seeking ways to assure job security for effective project team members

Earlier in this book, I claimed that a project cannot be successful unless it is recognized as a project and gains the support of top-level management. Top-level management must be willing to commit company resources and provide the necessary administrative support so that the project becomes part of the company's day-to-day routine of doing business. In addition, the parent organization must develop an atmosphere conducive to good working relationships among the project manager, parent organization, and client organization.

There are actions that top-level management should take to ensure that the organization as a whole supports individual projects and project teams as well as the overall project management system:

- Showing a willingness to coordinate efforts
- Demonstrating a willingness to maintain structural flexibility
- Showing a willingness to adapt to change
- Performing effective strategic planning
- Maintaining rapport
- Putting proper emphasis on past experience
- Providing external buffering
- Communicating promptly and accurately
- Exhibiting enthusiasm
- Recognizing that projects do, in fact, contribute to the capabilities of the whole company

Executive sponsors can take the following actions to make project success more likely:

- Selecting a project manager at an early point in the project who has a proven track record in behavioral skills and technical skills
- Developing clear and workable guidelines for the project manager
- Delegating sufficient authority to the project manager so that she or he can make decisions in conjunction with the project team members
- Demonstrating enthusiasm for and commitment to the project and the project team
- Developing and maintaining short and informal lines of communication
- Avoiding excessive pressure on the project manager to win contracts
- Avoiding arbitrarily slashing or ballooning the project team's cost estimate
- Avoiding "buy-ins"
- Developing close, not meddlesome, working relationships with the principal client contact and the project manager

The client organization can exert a great deal of influence on the behavioral aspects of a project by minimizing team meetings, rapidly responding to requests for information, and simply allowing the contractor to conduct business without interference. The positive actions of client organizations also include:

- Showing a willingness to coordinate efforts
- Maintaining rapport
- Establishing reasonable and specific goals and criteria for success
- Establishing procedures for making changes
- Communicating promptly and accurately

- Committing client resources as needed
- Minimizing red tape
- Providing sufficient authority to the client's representative, especially in decision-making

With these actions as the basic foundation, it should be possible to achieve behavioral success, which includes:

- Encouraging openness and honesty from the start from all participants
- Creating an atmosphere that encourages healthy competition but not cutthroat situations or liar's contests
- Planning for adequate funding to complete the entire project
- Developing a clear understanding of the relative importance of cost, schedule, and technical performance goals
- Developing short and informal lines of communication and a flat organizational structure
- Delegating sufficient authority to the principal client contact and allowing prompt approval or rejection of important project decisions
- Rejecting buy-ins
- Making prompt decisions regarding contract okays or go-aheads
- Developing close working relationships with project participants
- Avoiding arm's-length relationships
- Avoiding excessive reporting schemes
- Making prompt decisions on changes

Companies that are excellent in project management have gone beyond the standard actions as listed previously. These additional actions for excellence include the following:

- The outstanding project manager has these demonstrable qualities:
  - Understands and demonstrates competency as a project manager
  - Works creatively and innovatively in a nontraditional sense only when necessary; does not look for trouble
  - Demonstrates high levels of self-motivation from the start
  - Has a high level of integrity; goes above and beyond politics and gamesmanship
  - Is dedicated to the company and not just the project; is never self-serving
  - Demonstrates humility in leadership
  - Demonstrates strong behavioral integration skills both internally and externally
  - Thinks proactively rather than reactively
  - Is willing to assume a great deal of risk and will spend the appropriate time needed to prepare contingency plans
  - Knows when to handle complexity and when to cut through it; demonstrates tenaciousness and perseverance
  - Is willing to help people realize their full potential; tries to bring out the best in people
  - Communicates in a timely manner and with confidence rather than despair

- The project manager maintains high standards of performance for self and team, as shown by these approaches:
  - Stresses managerial, operational, and product integrity
  - Conforms to moral codes and acts ethically in dealing with people internally and externally
  - Never withholds information
  - Is quality conscious and cost conscious
  - Discourages politics and gamesmanship; stresses justice and equity
  - Strives for continuous improvement but in a cost-conscious manner
- The outstanding project manager organizes and executes the project in a sound and efficient manner:
  - Informs employees at the project kickoff meeting how they will be evaluated
  - Prefers a flat project organizational structure over a bureaucratic one
  - Develops a project process for handling crises and emergencies quickly and effectively
  - Keeps the project team informed in a timely manner
  - Does not require excessive reporting; creates an atmosphere of trust
  - Defines roles, responsibilities, and accountabilities up front
  - Establishes a change management process that involves the customer
- The outstanding project manager knows how to motivate:
  - Always uses two-way communication
  - Is empathetic with the team and a good listener
  - Involves team members in decision-making; always seeks ideas and solutions; never judges an employee's idea hastily
  - Never dictates
  - Gives credit where credit is due
  - Provides constructive criticism rather than making personal attacks
  - Publicly acknowledges credit when credit is due but delivers criticism privately
  - Makes sure that team members know that they will be held accountable and responsible for their assignments
  - Always maintains an open-door policy; is readily accessible, even for employees with personal problems
  - Takes action quickly on employee grievances; is sensitive to employees' feelings and opinions
  - Allows employees to meet the customers
  - Tries to determine each team member's capabilities and aspirations; always looks for a good match; is concerned about what happens to the employees when the project is over
  - Tries to act as a buffer between the team and administrative/operational problems
- The project manager is ultimately responsible for turning the team into a cohesive and productive group for an open and creative environment. If the project manager succeeds, the team will exhibit the following behaviors:
  - Demonstrates innovation
  - Exchanges information freely
  - Is willing to accept risk and invest in new ideas

- Is provided with the necessary tools and processes to execute the project
- Dares to be different; is not satisfied with simply meeting the competition
- Understands the business and the economics of the project
- Tries to make sound business decisions rather than just sound project decisions

10.7 CONVERGENT COMPUTING

One of the biggest challenges facing companies today is making project management a career path. There are two major issues that need to be considered. First, project management must be regarded as a profession. Second, job descriptions must be prepared that differentiate between pay grades of project managers, and this is the difficult part in deciding what should go into each job description. Anne Walker, Enterprise Project Manager at Convergent Computing (CCO), describes her experiences at CCO:

> Project management is regarded as a profession and particularly in our business; IT project management is a specialization. We need to have project managers who understand IT even if only at a high level to be able to manage the teams and projects we are engaged in. This initially has been a difficult combination skill-set to find; however, that is changing. We have dedicated project managers on staff and continue to see their efforts pay off.
>
> I have noticed a change in the past five years with many of our clients. Initially, project management was something they did not want to pay for and expected the engineer or consultant to "manage" the project. While many of our engineers and consultants successfully managed projects and delivered what the client was looking for, the focus was always on the technology or solution being utilized rather than successful project hand-offs, communication, or deliverables such as status reports and documentation. As CCO began implementing more and more project methodology into our projects, clients began to see the benefits of a more structured approach and their expectations about how projects are handled changed. Clients now want to ensure there is a communication plan for a project that is thought out and covers communication to all levels of the organization (if necessary). They want to see design sessions where features and functions are discussed and decisions jointly made about what will stay and what goes. Clients expect a structured testing plan that incorporates prototyping to ensure the technology fits with the business need. I know of a handful of our clients that are actively enrolling their employees in project management courses and looking to the employees to lead projects where in the past they participated but did not lead.
>
> Project managers help facilitate and coordinate the implementation of an IT project and are expected to obtain and maintain project manager certification. Project managers are involved in all phases of the project from obtaining requirements to roll-out and training of end users, all the while maintaining the budget integrity, time constraints, and objectives of the project. Major responsibilities include:
>
> - Validation of the allotted hours and task assignments with each project engineer for a project
> - Providing timelines to CCO's resource manager for engineer assignment(s)
> - Tracking project schedule

- Monitoring the project progress by comparing actual versus scheduled timeline
- Overseeing adherence to the role and responsibilities section within the scope
- Tracking budget progress
- Reviewing and monitoring project billable hours as recorded within Timesheet Professional
- Comparing the billed hours to the scoped hours contained in the project schedule
- Tracking scope
- Monitoring work progress with an emphasis on the following:
- Execution of tasks that are within the scope guidelines
- Managing "scope creep"
- Tracking customer satisfaction
- Insuring customer satisfaction utilizing major task sign-offs and customer surveys
- Monitoring ongoing customer satisfaction (throughout the project)
- Providing status updates and notification of major milestones and issues to account executives

10.8 EDS

Doug Bolzman, Consultant Architect, PMP®, ITIL Service Manager at EDS, discusses the approach at EDS for job descriptions:

> EDS project management roles are found in several job descriptions, along with the project management job family. To manage all of the required roles, each generic role is known as an agent. Table 10–3 provides a breakdown of how we define each agent (i.e., release manager, network engineer, financial manager). This is used by multiple organizations to manage the overall resource needs, resource balancing, and resource recruiting. As every organization uses the same listing, managing resources becomes more efficient and accurate.

10.9 PROACTIVE VERSUS REACTIVE MANAGEMENT

Perhaps one of the biggest behavioral challenges facing a project manager, especially a new project manager, is learning how to be proactive rather than reactive. Kerry Wills, discusses this problem.

Proactive Management Capacity Propensity[8]

In today's world Project Managers often get tapped to manage several engagements at once. This usually results in them having just enough time to react to the problems of the day that each project is facing. What they are not doing is spending the time to look ahead on each project to plan for upcoming work, thus resulting in more fires that need to be put out. There used to be an arcade game called "whack-a-mole" where the participant had a mallet and would hit each mole with it when one would pop up. Each time a mole was hit, a new mole would pop up. The cycle of

---

8. The remainder of this section was provided by Kerry R. Wills, PMP®. Reproduced by permission of Kerry R. Wills.

TABLE 10–3.   JOB DESCRIPTIONS

Section	Description
Agent name and version	The formal name of the role and the version number for that role. As frameworks improve and new technology is implemented, each role may need to be modified, which will be reflected in the version number. This allows for the changes to human resources to be reflected in the project plan, in terms of new training, new skills, new tools, etc.
Agent rational	The business justification for the role, this is where the role is integrated into the overall business environment.
Description	A paragraph that describes the role, authority, and accountability of the role.
Overall responsibilities	High-level responsibilities for the role. For project managers, this includes financial management, scope and schedule management, and resource management.
Component-specific responsibilities	Roles can be broken into specific components of the business environment. This section lists those responsibilities in detail. This allows for several processes to contribute to the development of a single role. These are measurable and the person fulfilling the role is measured against his or her conformance to these responsibilities
Work products	All the deliverables and products that are developed from the roles. These deliverables are integrated and assigned from the overall deliverable list, ensuring that each role is integrated by work products.
Aliases	Since many frameworks or work patterns may identify independent roles or agents, this allows to map this role to other best practices roles.
Comparable job codes	For human resources to function efficiently, this role is mapped to established job codes, so that if more people need to be hired into a role, those job codes can be used as qualified resources. This also helps to integrate the overall human resource family.
Qualifications	List of qualifications—years of experience, training, travel availability—that will contribute to a person successfully implementing the responsibilities of the role.
Skills	Specific skills that are required to fulfill the role, such as language skills, technology skills, business skills.

spending time putting out fires and ignoring problems that cause more fires can be thought of as "project whack-a-mole."

It is my experience that proactive management is one of the most effective tools that Project Managers can use to ensure the success of their projects. However, it is a difficult situation to manage several projects while still having enough time to look ahead. I call this ability to spend time looking ahead the "Proactive Management Capacity Propensity" (PMCP). This article will demonstrate the benefits of proactive management, define the PMCP, and propose ways of increasing the PMCP and thus the probability of success on the projects.

Proactive Management

Overview   Project Management involves a lot of planning up-front including work plans, budgets, resource allocations, and so on. The best statistics that I have seen on the accuracy of initial plans says there is a 30% positive or negative variance from the original plans at the end of a project. Therefore, once the plans have been made and the project has started, the Project Manager needs to constantly re-assess the project to understand the impact of the 60% unknowns that will occur.

The dictionary defines proactive as "acting in advance to deal with an expected difficulty." By "acting in advance" a Project Manager has some influence over the control of

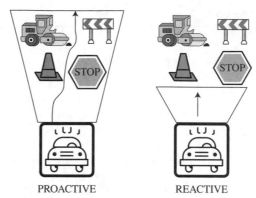

PROACTIVE   REACTIVE

FIGURE 10–2. Driving metaphor.

the unknowns. However, without acting in advance, the impacts of the unknowns will be greater as the Project Manager will be reacting to the problem once it has snowballed.

A Metaphor   When I drive into work in the morning, I have a plan and schedule. I leave my house, take certain roads and get to work in forty minutes. If I were to treat driving to work as a Project (having a specific goal with a finite beginning and end), then I have two options to manage my commute [Figure 10–2]:

By *proactively* managing my commute, I watch the news in the morning to see the weather and traffic. Although I had a plan, if there is construction on one of the roads that I normally take, then I can always change that plan and take a different route to ensure that my schedule gets met. If I know that there may be snow then I can leave earlier and give myself more time to get to work. As I am driving, I look ahead in the road to see what is coming up. There may be an accident or potholes that I will want to avoid and this gives me time to switch lanes.

A *reactive* approach to my commute could be assuming that my original plan will work fully. As I get on the highway, if there is construction then I have to sit in it because by the time I realize the impact, I have passed all of the exit ramps. This results in me missing my schedule goal. The same would happen if I walked outside and saw a foot of snow. I now have a chance to scope since I have the added activity of shoveling my driveway and car. Also, if I am a reactive driver, then I don't see the pothole until I have driven over it (which may lead to a budget variance since I now need new axles).

Benefits

The metaphor above demonstrates that reactive management is detrimental to projects because by the time that you realize that there is a problem it usually has a schedule, scope or cost impact. There are several other benefits to proactive management:

- Proactively managing a plan allows the Project Manager to see what activities are coming up and start preparing for them. This could be something as minor as setting up conference rooms for meetings. I have seen situations were tasks were not completed on time because of something as minor as logistics.
- Understanding upcoming activities also allows for the proper resources to be in place. Oftentimes, projects require people from outside of the project team and lining them

up are [*sic*] always a challenge. By preparing people in advance, there is a higher probability that they can be ready when needed.

The Relationship

- The Project Manager should constantly be replanning. By looking at all upcoming activities as well as the current ones, it can give a gage of the probability of success which can be managed rather than waiting until the day before something is due to realize that the schedule cannot be met.
- Proactive management also allows time to focus on quality. Reactive management usually is characterized by rushing to fix whatever "mole" has popped up as quickly as possible. This usually means a patch rather than the appropriate fix. By planning for the work appropriately, it can be addressed properly which reduces the probability of rework.
- As previously unidentified work arises, it can be planned for rather than assuming that "we can just take it on."

Proactive management is extremely influential over the probability of success of a project because it allows for replanning and the ability to address problems well before they have a significant impact.

Overview   I have observed a relationship between the amount of work that a project manager has and their ability to manage proactively. As Project Managers get more work and more concurrent projects, their ability to manage proactively goes down.

The relationship between Project Manager workload and the ability to manage proactively is shown in Figure 10–3 . . .. As Project Managers have increased work, they have less capacity to be proactive to and wind up becoming more reactive.

Not all Projects and Project Managers are equal. Some Project Managers can handle several projects well and some projects require more focus than others. I have therefore labeled this factor, the Project Management Capacity Propensity (PMCP). That is, the sum of those qualities that allow a Project Manager to proactively manage projects.

PMCP   There are several factors that make up the PMCP that I have outlined below.

Project Manager skill sets have an impact on the PMCP. Having good Time Management and organization techniques can influence how much a PM can focus on

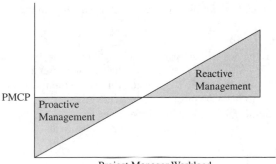

FIGURE 10–3. Proactivity graph.

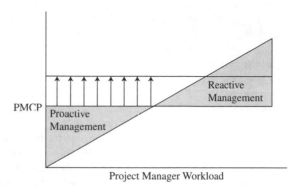

FIGURE 10–4. Increasing PMCP.

looking ahead. A Project Manager who is efficient with their time has the ability to review more upcoming activities and plan for them.

Project Manager expertise of the project is also influential to the PMCP. If the PM is an expert in the business or the project, this may allow for quicker decisions since they will not need to seek out information or clarification (all of which takes away time).

The PMCP is also impacted by team composition. If the Project Manager is on a large project and has several team leads who manage plans, then they have an increased ability to focus on replanning and upcoming work. Also, having team members who are experts in their field will require less focus from the Project Manager.

Increasing the PMCP   The good news about the PMCP is that it can be increased.

Project Managers can look for ways to increase their skillsets through training. There are several books and seminars on time management, prioritization, and organization. Attending these can build the effectiveness of the time spent by the PM on their activities.

The PM can also re-evaluate the team composition. By getting stronger team leads or different team members, the PM can offload some of their work and spend more time focusing on proactive management.

All of these items can increase the PMCP and result in an increased ability to manage proactively. The image . . . [in Figure 10–4] shows how a PMCP increase raises the bar and allows for more proactive management with the same workload.

Conclusion

To proactively manage a project is to increase your probability of being successful. There is a direct correlation between the workload that a PM has and their ability to look ahead. Project Managers do have control over certain aspects that can give them a greater ability to focus on proactive management. These items, the PMCP, can be increased through training and having the proper team.

Remember to keep your eyes on the road.

Measuring Return on Investment on Project Management Training Dollars

11.0 INTRODUCTION

For almost three decades, the 1960s through the 1980s, the growth and acceptance of project management were restricted to the aerospace, defense, and heavy construction industries. In virtually all other industries, project management was nice to have but not a necessity. There were very few project management training programs offered in the public marketplace, and those that were offered covered the basics of project management with weak attempts to customize the material to a specific company. The concept of measuring the return on investment (ROI) on training, at least in project management courses, was nonexistent. Within the past 10 years, there have been several studies on quantifying the benefits of project management with some pioneering work on the benefits of project management training.[1-4] There is still a great deal of effort needed, but at least we have recognized the need.

Today, our view of project management education has changed and so has our desire to evaluate ROI on project management training funds. There are several reasons for this:

- Executives realize that training is a basic necessity for companies to grow.
- Employees want training for professional growth and advancement opportunities.
- Project management is now viewed as a profession rather than a part-time occupation.

---

1. W. Ibbs and J. Reginato, *Quantifying the Value of Project Management*, Project Management Institute, Newton Square, PA, 2002.
2. W. Ibbs and Y-H. Kwak, *The Benefits of Project Management*, Project Management Institute, Newton Square, PA, 1997.
3. W. Ibbs, "Measuring Project Management's Value: New Directions for Quantifying PM/ROI®," in *Proceedings of the PMI Research Conference*, June 21–24, 2000, Paris, France.
4. J. Knutson, A three-part series in *PM Network*, January, February, and July 1999.

- The importance of becoming a PMP® has been increasing.
- There are numerous university programs available leading to MS, MBA, and PhD degrees in project management.
- There are certificate programs in advanced project management concepts where students are provided with electives rather than rigid requirements.
- The pressure to maintain corporate profitability has increased, resulting in less money available for training. Yet more and more training funds are being requested by the workers who desire to become PMP®s and then must accumulate 60 professional development units (PDUs) every three years to remain certified.
- Management realizes that a significant portion of training budgets must be allocated to project management education, but it should be allocated for those courses that provide the company with the greatest ROI. The concept of educational ROI is now upon us.

11.1 PROJECT MANAGEMENT BENEFITS

In the early years of project management, primarily in the aerospace and defense industries, studies were done to determine the benefits of project management. In a study by Middleton, the benefits discovered were[5]:

- Better control of projects
- Better customer relations
- Shorter product development time
- Lower program costs
- Improved quality and reliability
- Higher profit margins
- Better control over program security
- Improved coordination among company divisions doing work on the project
- Higher morale and better mission orientation for employees working on the project
- Accelerated development of managers due to breadth of project responsibilities

These benefits were identified by Middleton through surveys and were subjective in nature. No attempt was made to quantify the benefits. At that time, there existed virtually no project management training programs. On-the-job training was the preferred method of learning project management and most people learned from their own mistakes rather than from the mistakes of others.

Today the benefits identified by Middleton still apply and we have added other benefits to the list:

- Accomplishing more work in less time and with few resources
- More efficient and more effective performance

---

5. C. J. Middleton, "How to Set Up a Project Organization," *Harvard Business Review*, March–April 1967, pp. 73–82.

- Increase in business due to customer satisfaction
- A desire for a long-term partnership relationship with customers
- Better control of scope changes

Executives wanted all of the benefits described above and they wanted the benefits yesterday. It is true that these benefits could be obtained just by using on-the-job training efforts, but this assumed that time was a luxury rather than a constraint. Furthermore, executives wanted workers to learn from the mistakes of others rather than their own mistakes. Also executives wanted everyone in project management to look for continuous improvement efforts rather than just an occasional best practice.

Not every project management training program focuses on all of these benefits. Some courses focus on one particular benefit while others might focus on a group of benefits. Deciding which benefits you desire is essential in selecting a training course. And if the benefits can be quantified after training is completed, then executives can maximize their ROI on project management training dollars by selecting the proper training organizations.

11.2 GROWTH OF ROI MODELING

In the past several years, the global expansion of ROI modeling has taken hold. The American Society for Training and Development (ASTD) has performed studies on ROI modeling.[6] Throughout the world, professional associations are conducting seminars, workshops, and conferences dedicated to ROI on training. The Japan Management Association (JMA) has published case studies on the ROI process utilized at Texas Instruments, Verizon Communications, Apple Computers, Motorola, Arthur Anderson, Cisco Systems, AT&T, and the U.S. Office of Personnel Management.[7]

- A summary of the current status on ROI might be:
- The ROI methodology has been refined over a 25-year period.
- The ROI methodology has been adopted by hundreds of organizations in manufacturing, service, nonprofit, and government settings.
- Thousands of studies are developed each year using the ROI methodology.
- A hundred case studies are published on the ROI methodology.
- Two thousand individuals have been certified to implement ROI methodologies in their organizations.
- Fourteen books have been published to support the process.
- A 400-member professional network has been formed to share information.[8]

---

6. J. J. Phillips, *Return on Investment in Training and Performance Improvement Programs*, 2nd ed., Butterworth-Heinemann, Burlington, MA, 2003, Chapter 1. Reprinted with permission from Elsevier. In the opinion of this author, this is by far one of the best, if not the best, text on this subject. The bibliography is current and the examples are real world.
7. Ibid., p. 9.
8. Ibid., p. 11.

According to the 2001 *Training's* annual report, more than $66 billion was spent on training in 2001. It is therefore little wonder that management treats training with a business mindset, thus justifying the use of ROI measurement. But despite all of the worldwide commitment and documented successes, there is still a very real fear in many companies preventing the use of ROI modeling. Typical arguments are: "It doesn't apply to us"; "We cannot evaluate the benefits quantitatively"; "We don't need it"; "The results are meaningless"; and "It costs too much." These fears create barriers to the implementation of ROI techniques, but most barriers are myths that can be overcome.

In most companies, Human Resources Development (HRD) maintains the lead role in overcoming these fears and performing the ROI studies. The cost of performing these studies on a continuous basis could be as much as 4–5 percent of the HRD budget. Some HRD organizations have trouble justifying this expense. And to make matters worse, the HRD personnel may have a poor understanding of project management.

The salvation in overcoming these fears and designing proper project management training programs could very well be the project management office (PMO). Since the PMO has become the guardian of all project management intellectual property as well as designing project management training courses, the PMO will most likely take the lead role in calculating ROI on project management–related training courses. Members of the PMO might be required to become certified in educational ROI measurement the same way that they are certified as a PMP® or Six Sigma Black Belt.

Another reason for using the PMO is because of the enterprise project management (EPM) methodology. EPM is the integration of various processes such as total quality management, concurrent engineering, continuous improvement, risk management, and scope change control into one project management methodology that is utilized on a companywide basis. Each of these processes has measurable output that previously may not have been tracked or reported. This has placed additional pressure on the PMO and project management education to develop metrics and measurement for success.

11.3 THE ROI MODEL

Any model used must provide a systematic approach to calculating ROI. It should be prepared on a life-cycle basis or step-by-step approach similar to an EPM methodology. Just like with EPM, there is an essential criterion that must exist for any model to work effectively. A typical list of ROI criteria is[9]:

1. The ROI process must be *simple*, void of complex formulas, lengthy equations, and complicated methodologies. Most ROI attempts have failed with this requirement. In an attempt to obtain statistical perfection and use too many theories, some ROI models have become too complex to understand and use. Consequently, they have not been implemented.

---

9. Ibid., pp. 18–19.

2. The ROI process must be *economical* and must be implemented easily. The process should become a routine part of training and development without requiring significant additional resources. Sampling for ROI calculations and early planning for ROI are often necessary to make progress without adding new staff.

3. The assumptions, methodology, and techniques must be *credible*. Logical, methodical steps are needed to earn the respect of practitioners, senior managers, and researchers. This requires a very practical approach for the process.

4. From a research perspective, the ROI process must be *theoretically sound* and based on generally accepted practices. Unfortunately, this requirement can lead to an extensive, complicated process. Ideally, the process must strike a balance between maintaining a practical and sensible approach and a sound and theoretical basis for the process. This is perhaps one of the greatest challenges to those who have developed models for the ROI process.

5. The ROI process must *account for other factors* that have influenced output variables. One of the most often overlooked issues, isolating the influences of the HRD program, is necessary to build credibility and accuracy within the process. The ROI process should pinpoint the contribution of the training program when compared to the other influences.

6. The ROI process must be appropriate with a *variety of HRD* programs. Some models apply to only a small number of programs such as sales or productivity training. Ideally, the process must be applicable to all types of training and other HRD programs such as career development, organization development, and major change initiatives.

7. The ROI process must have the *flexibility* to be applied on a preprogram basis as well as a postprogram basis. In some situations, an estimate of the ROI is required before the actual program is developed. Ideally, the process should be able to adjust to a range of potential time frames.

8. The ROI process must be *applicable with all types of data*, including hard data, which are typically represented as output, quality, costs, and time; and soft data, which include job satisfaction, customer satisfaction, absenteeism, turnover, grievances, and complaints.

9. The ROI process must *include the costs of the program*. The ultimate level of evaluation is to compare the benefits with costs. Although the term ROI has been loosely used to express any benefit of training, an acceptable ROI formula must include costs. Omitting or underestimating costs will only destroy the credibility of the ROI values.

10. The actual calculation must use an *acceptable ROI formula*. This is often the benefits–cost ratio (BCR) or the ROI calculation, expressed as a percent. These formulas compare the actual expenditure for the program with the monetary benefits driven from the program. While other financial terms can be substituted, it is important to use a standard financial calculation in the ROI process.

11. Finally, the ROI process must have a successful *track record* in a variety of applications. In far too many situations, models are created but never successfully applied. An effective ROI process should withstand the wear and tear of implementation and should get the results expected.

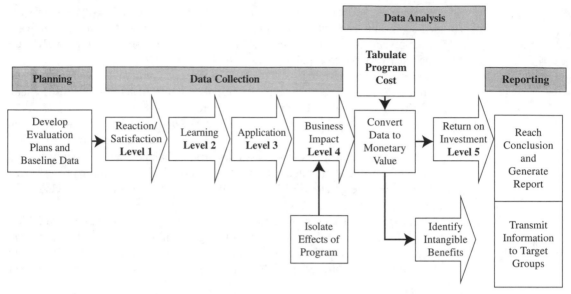

FIGURE 11–1. The ROI model. *Source:* Adapted from J. J. Phillips, *Return on Investment in Training and Performance Improvement Programs*, 2nd ed., Butterworth-Heinemann, Burlington, MA, 2003 p. 37.

TABLE 11–1. DEFINING LEVELS

Level	Description
1: Reaction/satisfaction	Measures the participants' reaction to the program and possibly creates an action plan for implementation of the ideas
2: Learning	Measures specific skills, knowledge, or attitude changes
3: Application	Measures changes in work habit or on-the-job performance as well as application and implementation of knowledge learned
4: Business impact	Measures the impact on the business as a result of implementation of changes
5: Return on investment	Compares monetary benefits with the cost of the training and expressed as a percentage

Because these criteria are considered essential, any ROI methodology should meet the vast majority of, if not all, criteria. The bad news is that most ROI processes do not meet these criteria. The good news is that the ROI process presented in Phillips' book meets all of these criteria. The model is shown in Figure 11–1. The definitions of the levels in Figure 11–1 are shown in Table 11–1.

11.4 PLANNING LIFE-CYCLE PHASE

The first life-cycle phase in the ROI model is the development of evaluation plans and baseline data. The evaluation plan is similar to some of the *PMBOK® Guide* knowledge

TABLE 11–2. TYPICAL PROGRAM OBJECTIVES

		Objectives	
Level	Description	Typical PMCP Training	Typical Best Practices Training Course
1	Reaction/satisfaction	Understand principles of *PMBOK® Guide*	Understand that companies are documenting their best practices
2	Learning	Demonstrate skills or knowledge in domain groups and knowledge areas	Demonstrate how best practices benefit an organization
3	Application	Development of EPM processes based upon *PMBOK® Guide*	Develop a best practices library or ways to capture best practices
4	Business impact	Measureement of customer and user satisfaction with EPM	Determine the time and/or cost savings from a best practice
5	Return on investment	Amout of business or customer satisfaction generated from EPM	Measure ROI for each best practice implemented

areas that require a plan as part of the first process step in each knowledge area. The evaluation plan should identify:

- The objective(s) of the program
- The way(s) the objective(s) will be validated
- The target audience
- Assumptions and constraints
- The timing of the program

Objectives for the training program must be clearly defined before ROI modeling can be completed. Table 11–2 identifies typical objectives. The objectives must be clearly defined for each of the five levels of the model. Column 3 in Table 11–2 would be representative of the objectives that a company might have when it registers a participant in a Project Management Certificate Program (PMCP) training course. In this example, the company funding the participant's training might expect the participant to become a PMP® and then assist the organization in developing an EPM methodology based upon the *PMBOK® Guide* with the expectation that this would lead to customer satisfaction and more business. Column 4 in Table 11–2 might be representative of a company that registers a participant in a course on best practices in project management. Some companies believe that if a seminar participant walks away from a training program with two good ideas for each day of the program and these ideas can be implemented reasonably fast, then the seminar would be considered a success. In this example, the objectives are to identify best practices in project management that other companies are doing and can be effectively implemented in the participant's company.

There can be differences in training objectives, as seen through the eyes of management. As an example, looking at columns 3 and 4 in Table 11–2, objectives might be:

- Learn skills that can be applied immediately to the job. In this case, ROI can be measured quickly. This might be representative of the PMCP course in column 3.

● Learn about techniques and advancements. In this case, additional money must be spent to achieve these benefits. ROI measurement may not be meaningful until after the techniques have been implemented. This might be representative of the best practices course in column 4.
● A combination of the above.

11.5 DATA COLLECTION LIFE-CYCLE PHASE

In order to validate that each level's objectives for the training course were achieved, data must be collected and processed. Levels 1–4 in Figure 11–1 make up the data collection life-cycle phase.

To understand the data collection methods, we revisit the course on best practices in project management, which covers best practices implemented by various companies worldwide. The following assumptions will be made:

● Participants are attending the course to bring back to their company at least two ideas that can be implemented in their company within six months.
● Collecting PDUs is a secondary benefit.
● The course length is two days.[10]

Typical data collection approaches are shown in Table 11–3 and explained below for each level.

TABLE 11–3. DATA COLLECTION

Level	Measures	Data Collection Methods and Instruments	Data Sources	Timing	Responsible Person
Reaction/ satisfaction	A 1–7 rating on end-of-course critique	Questionnaire	Participant (last day of program)	End of program	Instructor
Learning	Pretest, posttest, CD-ROM, and case studies	In-class tests and skill practice sets	Instructor	Each day of course	Instructor
Application	Classroom discussion	Follow-up session or questionnaire	Participant and/or PMO	Three months after program[a]	PMO
Business impact	Measurement of EPM continuous improvement efforts	Benefit–cost monitoring by the PMO	PMO records	Six months after program	PMO
Return on investment	Benefit–cost ratios	PMO studies	PMO records	Six months after program	PMO

[a] Usually for in-house program only. For public seminars, this may be done by the PMO within a week after completion of training.

---

10. Some companies have one-day, two-day, and even week-long courses on best practices in project management.

Level 1: Reaction and Satisfaction

Level 1 measures the participant's reaction to the program and possibly an action plan for implementation of the ideas. The measurement for level 1 is usually an end-of-course questionnaire where the participant rates the information presented, quality of instruction, instructional material, and other such topics on a scale of 1–7. All too often, the questionnaire is answered based upon the instructor's presentation skills rather than the quality of the information. While this method is most common and often serves as an indication of customer satisfaction hopefully leading to repeat business, it is not a guarantee that new skills or knowledge have been learned.

Level 2: Learning

This level measures specific skills, knowledge, or attitude changes learned during the course. Instructors use a variety of techniques for training, including:

- Lectures
- Lectures/discussions
- Exams
- Case studies (external firms)
- Case studies (internal projects)
- Simulation/role playing
- Combinations

For each training technique, a measurement method must be established. Some trainers provide a pretest at the beginning of the course and a posttest at the end. The difference in scores is usually representative of the amount of learning that has taken place. This is usually accomplished for in-house training programs rather than public seminars. Care must be taken in the use of pretests and posttests. Sometimes, a posttest is made relatively easy for the purpose of making it appear that learning has taken place. Out-of-class testing can also be accomplished using take-home case studies and CD-ROM multiple-choice questions.

Testing is necessary to validate that learning has taken place and knowledge has been absorbed. However, simply because learning has taken place is no guarantee that the information learned on best practices can or will be transferred to the company. The learning might simply confirm that the company is doing well and keeping up with the competitors.

Level 3: Application of Knowledge

This level measures changes in work habits or on-the-job performance as well as implementation of knowledge learned. Measurement at this level is normally done through follow-up sessions or follow-up questionnaires. However, for publicly offered courses with a large number of participants, it is impossible for the instructor to follow up with all participants. In such cases, the responsibility falls on the shoulders of the PMO. Participants may be required to prepare a short one- or two-page report on what they learned in the course and what best practices are

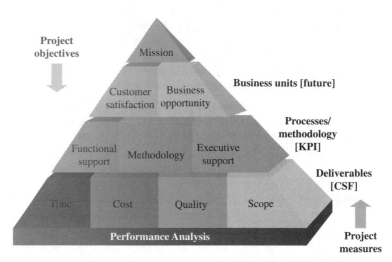

FIGURE 11–2. Postmortem pyramid. *Source*: From H. Kerzner, *Advanced Project Management: Best Practices in Implementation*, 2nd ed., Wiley, New York, p. 302.

applicable to the company. The report is submitted to the PMO that might have the final decision on the implementation of the ideas. Based on the magnitude of the best practices ideas, the portfolio management of projects may be impacted. However, there is no guarantee at this point that there will be a positive impact on the business.

Level 4: Business Impact   This level measures the impact on the business as a result of implementation of the changes. Typical measurement areas are shown in Figure 11–2.

The critical terms in Figure 11–2 are:

- *Critical Success Factor (CSF)*: This measures changes in the output of the project resulting from implementation of best practices. Hopefully, this will lead to improvements in time, cost, quality, and scope.
- *Key Performance Indicator (KPI)*: This measures changes in the use of the EPM system and support received from functional management and senior management.
- *Business Unit Impact:* This is measured by customer satisfaction as a result of the implementation of best practices and/or future business opportunities.

The measurement at level 4 is usually accomplished by the PMO. There are several reasons for this. First, the information may be company sensitive and not available to the instructor. Second, since there may be a long time span between training and the implementation of best practices, the instructor may not be available for support. And third, the company may not want anyone outside of the company talking to its customers about customer satisfaction. Although the implementation of best practices may have a favorable business impact, care must be taken that the implementation was cost effective.

As shown in Figure 11–1, an important input into level 4 is *isolate the effects of training*. It is often impossible to clearly identify the business impact that results directly from the training program. The problem is that people learn project management from multiple sources, including:

- Formal education
- Knowledge transfer from colleagues
- On-the-job experience
- Internal research on continuous improvements
- Benchmarking

Because of the difficulty in isolating the specific knowledge, this step is often overlooked.

11.6 DATA ANALYSIS LIFE-CYCLE PHASE

In order to calculate the ROI, the business impact data from level 4 must be converted to a monetary value. The information can come from interviews with employees and managers, databases, subject matter experts, and historical data. Very rarely will all of the information needed come from one source.

Another input required for data analysis is the cost of the training program. Typical costs that should be considered include:

- Cost of course design and development
- Cost of materials
- Cost of the facilitator(s)
- Cost of facilities and meals during training
- Costs of travel, meals, and lodgings for each participant
- Fully burdened salaries of participants
- Administrative or overhead cost related to the training course or approach of participants to attend training
- Possible cost (loss of income) of not having the participants available for other work during the time of training

Not all benefits can be converted to monetary values. This is the reason for the "identify intangible benefits" box in Figure 11–1. Some business impact benefits that are easily converted to monetary values include:

- Shorter product development time
- Faster, higher-quality decisions
- Lower costs
- Higher profit margins
- Fewer resources needed
- Reduction in paperwork

- Improved quality and reliability
- Lower turnover of personnel
- Quicker implementation of best practices

Typical benefits that are intangible and cannot readily be converted to monetary value include:

- Better visibility and focus on results
- Better coordination
- Higher morale
- Accelerated development of managers
- Better project control
- Better customer relations
- Better functional support
- Fewer conflicts requiring some management support

Despite the fact that these benefits may be intangible, every attempt should be made to assign monetary values of these benefits.

Level 5: Return on Investment

Two formulas are required for completion of level 5. The first formula is the BCR, which can be formulated as

$$BCR = \frac{program\ benefits}{program\ costs}$$

The second formula is the ROI expressed as a percent. The formula is based upon "net" program benefits, which are the benefits minus the cost. Mathematically, we can describe it as

$$ROI = \frac{net\ program\ benefits}{program\ costs} \times 100$$

To illustrate the usefulness of this level, we consider three examples all based upon the same training course. You attend a two-day seminar on best practices in project management. Your company's cost for attending the course is:

- Registration fee $ 475
- Release time (16 hr at $100/hr)   1600
- Travel expenses   800
                                                                   ―――
                                                                 $2875

When the seminar is completed, you come away with three best practices to recommend to your company. Your company likes all three ideas and assigns you as the project manager to implement all three best practices. Additional funds must be spent to achieve the benefits desired.

Example 1

During the seminar, you discover that many companies have adopted the concept of paperless project management by implementing a "traffic light" status-reporting system. Your company already has a Web-based EPM system but you have been preparing paper reports for status review meetings. Now, every status review meeting will be conducted without paper and with an LCD projector displaying the Web-based methodology with a traffic light display beside each work package in the work breakdown structure.

The cost of developing the traffic light system is:

Systems programming (240 hr at $100/hr)	$24,000
Project management (150 hr at $100/hr)	15,000
	$39,000

The benefits expressed by monetary terms are:

- Executive time in project review meeting (20 hr per project to 10 hr per project × 15 projects × 5 executives per meeting × $250/hr): $187,500
- Paperwork preparation time reduction (60 hr/project × 15 projects × $100/hr): $90,000
- Total additional benefit is therefore $275,500:

$$BCR = \frac{\$275,000 - \$39,000}{\$2875} = 82$$

$$ROI = \frac{\$275,000 - \$39,000 - \$2875}{\$2875} = 8109$$

This means that for every dollar invested in the training program, there was a return of $8109 in net benefits! In this example, it was assumed that workers were fully burdened at $100/hr and executives at $250/hr. The benefits were one-year measurements and the cost of developing the traffic light system was not amortized but expensed against the yearly benefits.

Not all training programs generate benefits of this magnitude. Lear in Dearborn, Michigan, has a project management traffic light reporting system as part of its Web-based EPM system. Lear has shown that in the same amount of time that it would review the status of one project using paper, it now reviewed the status of *all* projects using traffic light reporting.

Example 2

During the training program, you discover that other companies are using templates for project approval and initiation. The templates are provided to you during the training program and it takes a very quick effort to make the templates part of the EPM system and inform everyone about the update. The new templates will eliminate at least one meeting per week at a savings of $550:

$$Benefit = (\$500/meeting) \times (1 \ meeting/week) \times 50 \ weeks = \$27,500$$

$$BCR = \frac{\$27,500}{\$2875} = 9.56$$

$$\text{ROI} = \frac{\$27{,}500 - \$2875}{\$2875} = 8.56$$

In this example, for each $1 invested in the best practices program, a net benefit of $8.56 was recognized.

Example 3

During the training program, you learn that companies are extending their EPM systems to become more compatible with systems utilized by their customers. This should foster better customer satisfaction. The cost of updating your EPM system to account for diversified customer report generators will be about $100,000.

After the report generator is installed, one of your customers with whom you have four projects per year informs you they are so pleased with this change that they will now give you sole-source procurement contracts. This will result in a significant savings in procurement costs. Your company typically spends $30,000 preparing proposals:

$$\text{BCR} = \frac{(4 \text{ projects} \times \$30{,}000) - \$100{,}000}{\$2875} = 6.96$$

$$\text{ROI\%} = \frac{(4 \times \$30{,}000) - \$100{,}000 - \$2875}{\$2875} = 5.96$$

In this case, for every dollar invested in the best practices program there was a net benefit of $5.96 received.

Table 11–4 identifies typical ROI cases studies.[11] From Table 11–4, it should be obvious that the application of ROI on project management education could lead to fruitful results. To date, there have been very few attempts to measure ROI specifically on project management education. However, there have been some successes. In an insurance company, a $100 million project was undertaken. All employees were required to undergo project management training prior to working on the project. The project was completed 3 percent below budget.

Unsure of whether the $3 million savings was due to better project management education or poor initial estimating, the company performed a study on all projects where the employees were trained on project management prior to working on project teams. The result was an astounding 700 percent return on training dollars.

In another organization, the HRD people worked with project management to develop a computer-based project management training program. The initial results indicated a 900 percent ROI. The workers took the course on their own time rather than company time. Perhaps this is an indication of the benefits of e-learning programs. The e-learning programs may produce a much higher ROI than traditional courses because the cost of the course is significantly reduced with the elimination of the cost of release time.

11. P. P. Phillips, The Bottom Line on ROI, CEP Press, Atlanta, GA, 2002, p. 54,. The Center for Effective Performance, Inc., 1100 Johnson Ferry Road, Suite 150, Atlanta, GA 30342, 800-558-4237. Reprinted with permission.

TABLE 11–4.   ROI CASE STUDIES

Organization	Industry	Program	ROI (%)	Source
Office of Personnel Management	U.S. government	Supervisory training	150	1
Magnavox Electronic Systems Company	Electronics	Literacy training	741	1
Litton Guidance and Control Systems	Avionics	Self-directed work teams	650	1
Coca-Cola Bottling Company of San Antonio	Soft drinks	Supervisory training	1447	1
Commonwealth Edison	Electrical utility	Machine operator	57	2
Texas Instruments	Electronics	Sales training negotiation	2827	2
Apple Computer	Computer manufacturing	Process improvement	182	3
Hewlett-Packard Company	Computer support services	Sales training	195	3
First National Bank	Financial services	Sales training	555	3
Nassau County Police Department	Police department	Interpersonal skills training	144	3

Sources:
1. J. J. Phillips, Ed., *Measuring the Return on Investment,* Vol. 1, American Society for Training and Development, Alexandria, VA, 1994.
2. J. J. Phillips, Ed., *Measuring the Return on Investment,* Vol. 2, American Society for Training and Development, Alexandria, VA, 1998.
3. P. P. Phillips, Ed., *Measuring the Return on Investment,* Vol. 3, American Society for Training and Development, Alexandria, VA, 2001.

11.7 REPORTING LIFE-CYCLE PHASE

The final life-cycle phase in Figure 11–1 is reporting. The acceptance of the results could very well be based upon how the report is prepared. The report must be self-explanatory to all target groups. If assumptions are made concerning costs or benefits, then they must be justified. If the ROI numbers are inflated to make a training program look better than it was, then people may be skeptical and refuse to accept the results of future ROI studies. All results should be factual and supported by realistic data.

11.8 CONCLUSIONS

Because of the quantity and depth of available project management training programs, the concept of measuring ROI on training dollars can be expected to grow. Executives will recognize the benefits of this approach and its application to project management the same way it is applied to other training programs. Project management training organizations will be required to demonstrate expertise in ROI analysis. Eventually, PMI might even establish a special investigation group on ROI measurement.

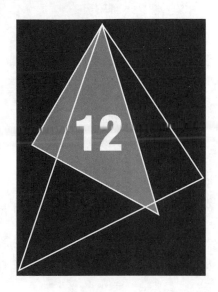

The Project Office

12.0 INTRODUCTION

As companies begin to recognize the favorable effect that project management has on profitability, emphasis is placed upon achieving professionalism in project management using the project office (PO) concept. The concept of a PO or project management office (PMO) could very well be the most important project management activity in this decade. With this recognition of importance comes strategic planning for both project management and the project office. Maturity and excellence in project management do *not* occur simply by using project management over a prolonged period of time. Rather, it comes through strategic planning for both project management and the PO.

General strategic planning involves the determination of where you wish to be in the future and then how you plan to get there. For PO strategic planning, it is often easier to decide which activities should be under the control of the PO than determining how or when to do it. For each activity placed under the auspices of the PO, there may appear pockets of resistance that initially view removing this activity from its functional area as a threat to its power and authority. Typical activities assigned to a PO include:

- Standardization in estimating
- Standardization in planning
- Standardization in scheduling
- Standardization in control
- Standardization in reporting
- Clarification of project management roles and responsibilities
- Preparation of job descriptions for project managers
- Preparation of archive data on lessons learned

- Benchmarking continuously
- Developing project management templates
- Developing a project management methodology
- Recommending and implementing changes and improvements to the existing methodology
- Identifying project standards
- Identifying best practices
- Performing strategic planning for project management
- Establishing a project management problem-solving hotline
- Coordinating and/or conducting project management training programs
- Transferring knowledge through coaching and mentorship
- Developing a corporate resource capacity/utilization plan
- Supporting portfolio management activities
- Assessing risks
- Planning for disaster recovery

In the first decade of the twenty-first century, the PO became commonplace in the corporate hierarchy. Although the majority of activities assigned to the PO had not changed, there was now a new mission for the PO:

- The PO now has the responsibility for maintaining all intellectual property related to project management and to actively support corporate strategic planning.

The PO was now servicing the corporation, especially the strategic planning activities for project management, rather than focusing on a specific customer. The PO was transformed into a corporate center for control of project management intellectual property. This was a necessity as the magnitude of project management information grew almost exponentially throughout the organization.

During the past 10 years, the benefits to executive levels of management of using a PO have become apparent. They include:

- Standardization of operations
- Company rather than silo decision-making
- Better capacity planning (i.e., resource allocations)
- Quicker access to higher-quality information
- Elimination or reduction of company silos
- More efficient and effective operations
- Less need for restructuring
- Fewer meetings which rob executives of valuable time
- More realistic prioritization of work
- Development of future general managers

All of the above benefits are either directly or indirectly related to project management intellectual property. To maintain the project management intellectual property, the PO must maintain the vehicles for capturing the data and then disseminating the data to the various stakeholders. These vehicles include the company project management intranet, project websites, project databases, and project management information systems. Since much of this information is necessary for both project management and corporate strategic planning then there must exist strategic planning for the PO.

The recognition of the importance of the PMO has now spread worldwide. Enrique Sevilla Molina, Corporate PMO Director for Indra, states:

> We have a PMO at corporate level and local PMOs at different levels throughout the company, performing a variety of functions. The PMO at corporate level provides directions on different project management issues, methodology clarifications, and tool use to local PMOs.
>
> Besides supporting the local PMOs and Project Managers as requested, the main functions of the corporate PMO include acting on the following areas:

- Maintenance and development of the overall project management methodology, including the extensions for Program and Portfolio levels
- Definition of the training material and processes for the PMs
- Management of the PMP certification process and candidates training and preparation
- Definition of the requirements for the corporate PM tools

> The Corporate PMO reports to the financial managing director.

A typical PMO does not have profit and loss responsibility on projects, nor does a typical PMO manage projects for external clients. According to Jim Triompo, Group Senior Vice President at ABB:

> The project office does not deliver projects. The projects managed by the project management office are limited to process/tools development, implementation, and training. The project management office is sometimes requested to perform reviews, participate in division-level risk reviews, and operational reviews in various countries.

Most PMOs are viewed as indirect labor and therefore subject to downsizing or elimination when a corporation is under financial stress. To minimize the risk, the PMO should set up metrics to show that the PMO is adding value to the company. Typical metrics include:

- Tangible measurements include:
  - CSF: Customer satisfaction
  - KPI: Projects at risk
  - KPI: Projects in trouble
  - KPI: The number of red lights that need recovery, and by how much added effort
- Intangible elements may also exist and these may not be able to be measured.
  - Early-on identification of problems
  - Quality and timing of information

12.1 SYPRIS ELECTRONICS

Every company has their own unique goals and objectives for a PMO. As such, the responsibilities of the PMO can vary from company to company. The following information was provided by Ryan Duran, Director, Program Management:

> Sypris Electronics has been in business for over 40 years in the defense and aerospace market. Until only recently has there been a project management office (PMO). Sypris is in the midst of a major transformation where both operational and cultural tides are shifting. Over 3.5M in investments has changed the mindset from a silo based functional organization to a value-stream based matrix organization centered on excellence in project management. With this change, a PMO has been created and it reports into the Director of continuous improvement who is part of the senior executive team.
>
> The PMO has a multitude of functions. Primarily, the PMO functions as a provider of services and organizational focus in project management. However, it also serves the other departmental areas as well. From a pure project focused area, it provides a means of consultation, mentorship, training, templates, forms, and checklists. It also serves as a repository for best practices. Below is a bulletized list of the main PMO functions:
>
> - Project management data archive
> - On-going training in project management
> - State-of-the-art procedures and guidelines
> - Promote consistency and uniformity in project management
> - Centralized point of reference for project management
> - Specialized skills and knowledge education
> - Best practices repository
> - Personnel bench strength
> - Mentor novice team members
> - Consultation and advisement

12.2 VITALIZE CONSULTING SOLUTIONS (VCS), INC.[1]

> VCS has a PMO which serves as the "home" for all of our project management consultants. The PMO reports directly to our Senior Vice President, who reports to our CEO. The group is matrixed into our functional or practice specific divisions. The PMO serves the company in two primary ways (internally and externally). Internally, we offer employees a career path into project management. This includes access to methodology, as well as standards and educational opportunities (i.e. PMI's PMP® credential) that are solely focused on project management. Externally, we offer our customers services with respect to healthcare information technology project management. This includes certified project managers with expertise in the industry (average at VCS is 14 years). We also offer our customers support with methodology and the tools to build the foundation of a PMO via

---

1. The material on VCS was provided by Marc Hirshfield, PMP, Director, Project Management Office at VCS.

our PMO JumpStart solution. VCS has provided project and program management support services to our customer base for approximately seven years.

VCS decided to invest and create the PMO to meet both internal and external market demands. Internally, the PMO allows VCS to offer a home and path for employees who want to focus solely on the career path of project management. It provided opportunities across different software vendor suites and offers a formal methodology and documented approach to running a healthcare IT initiative (project). Externally, it met the needs of our customer base who were looking for qualified project managers to help support their healthcare IT initiatives.

The biggest challenge with a PMO is following one methodology or standard across multiple initiatives which vary in size, scope and complexity. The key to success is making sure your PMO follows a consistent format, provides the tools needed to get the job done, but allows for scalability based on the individual PM's needs. A "one size fits all" mentality just does not work when running a PMO.

When we started the PMO at VCS, we had full executive (principal) support. Everyone was on-board with the value in the practice and this fact alone lead to our success. As we all know, having executive buy-in and support is not only critical to a successful project, but also integral in building a PMO.

Not only is the role of a PM at VCS considered a career path for our employees, but it is one of our solutions. Most healthcare organizations we provide services to are still learning the true value and ROI of a project manager. Some early adopter organizations that have invested and supported a PMO properly are seeing a large return on their investment. We have job descriptions for our project managers, which include levels one through three (three being our Senior PM).

Example of Forms in the PMO

The screen shots in Figures 12–1 and 12–2 shows our PMO Methodology and some examples of how our best practices are documented and outlined for our PM&'s to use in the field.

Shown . . . in Figure 12–3 is an example Charter Template from the VCS PMO Methodology.

12.3 CHURCHILL DOWNS INCORPORATED (CDI): ESTABLISHING A PMO _____

Deciding to implement a PMO is easy. Being able to do it requires that certain obstacles be overcome. Chuck Millhollan, Director of Program Management at CDI, discusses the chronology of events his organization went through and some of the obstacles that had to be overcome:

One of our primary barriers to implementing structured project, program and portfolio management processes was "familiarity." The Churchill Downs Incorporated PMO, chartered in April 2007, is the first in the thoroughbred horseracing industry. Our senior-most leadership understood the need for a structured, standardized approach to requesting, approving, and managing projects and maintaining the project portfolio; however, many of the organizational resources had never been exposed to formal project management concepts.

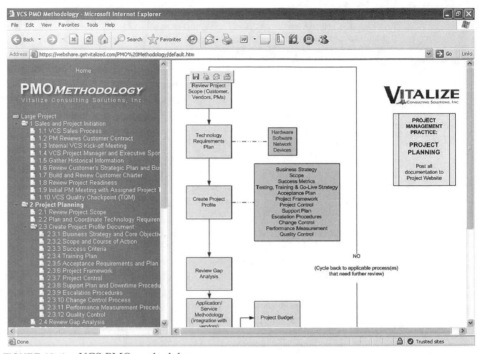

FIGURE 12–1. VCS PMO methodology.

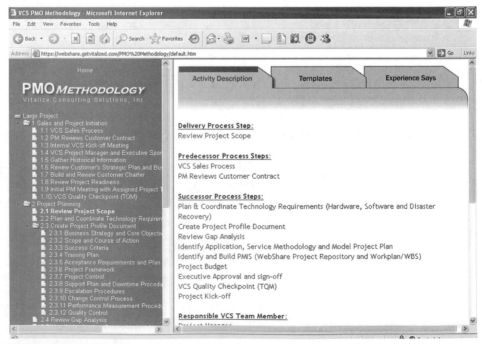

FIGURE 12–2. VCS PMO methodology.

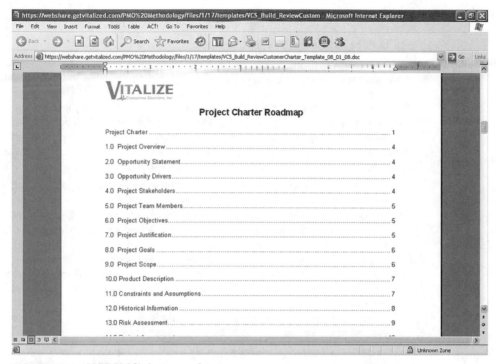

FIGURE 12–3. VCS PMO charter roadmap.

Our executives took an active role in the implementation process.

I would say this is one of the primary factors influencing the early success enjoyed by the Churchill Downs Incorporated PMO. We chartered our PMO with clearly defined vision and mission statements and business objectives. Our CEO signed the charter, granting authority to the PMO to expend organizational resources as related to managing capital projects.

i. Our PMO was chartered in April 2007.

ii. We developed a three-fold mission focused on the need identified by our senior leadership;

    1. Establish, facilitate, and manage the project portfolio selection and funding process.

    2. Create a foundation for consistent project success throughout the organization through development of a strong and pervasive Project Management discipline within CDI's project teams.

    3. Guide key projects to a successful conclusion by providing project management leadership, while improving the quality and repeatability of related processes.

iii. We defined the PMO's business objectives and linked progress to the PMO Director's compensation plan. Objectives included;

    1. Develop and Implement Standards for Project Selection.

    2. Develop and Implement a Standardized Project Management Methodology.

    3. Build Project Management Professionalism among CDI Staff.

    4. Manage the CDI Project Portfolio.

    5. Direct Project Management for Key Strategic Initiatives.

    6. Ensure Processes for Benefit Realization.

iv. We also conducted training classes on project management, team building, critical thinking, etc . . . to not only share our knowledge, but also to build relationships with project team members and other stakeholders.

v. The PMO has facilitated a book club (also chartered with clearly defined objectives) for the last year. This process received recognition throughout the organization and has directly contributed to developing relationships between different departments. Our Book Club membership includes representatives from 9 different departments, ranging from Vice President level members to individual contributors.

    1. Objective 1: Personal growth through completing chosen books and active involvement in discussions.

    2. Objective 2: Explore creative ideas and ways of addressing real-world business issues through practical application of concepts and shared learning as related to Churchill Downs and respective teams.

    3. Objective 3: Promote interaction among different functional areas within the Churchill team by active participation in book club discussions and sharing opportunities for addressing real-world work-related issues in a safe, confidential environment.

    4. Objective 4: Share learning within respective teams through intradepartmental discussion and implementation of learning related concepts.

vi. The primary driving factors behind Churchill Downs Incorporated's decision to staff a PMO were challenges with defining and managing the scope of projects, effectively allocating resources amongst multiple projects, and brining projects to a defined closure.

12.4 CHURCHILL DOWNS INCORPORATED (CDI): MANAGING SCOPE CHANGES

Mature PMOs either participate directly in the scope changes above a certain dollar level or set up processes for controlling scope changes. Chuck Millhollan, Director of Program Management at CDI, identifies six steps necessary for scope definition and change control[2]:

Step 1: Be Lean   Trying to introduce any type of structure or control in an organization or environment that has been absent of controls can present a significant challenge. Before a project management organization can address scope change control, they must implement a process to define scope. Getting organizational decision makers to accept the project management precepts is not overly difficult, but changing organizational behavior to leverage these principles is another matter altogether. The more change we attempt to introduce into an environment, the more difficulty that environment has in adapting to, accepting, and embracing that change. To avoid the natural resistance to

---

2. C. Millhollan, "Scope Change Control: Control Your Projects or Your Project Will Control You!." Copyright ©2008 by Chuck Millhollan. Reproduced by permission of Chuck Millhollan.

excessive change, a logical approach is to limit the scope of change and focus on immediate needs. Focus on the foundation and basics. Why have a complex, highly mature process if you are not consistently performing the basics well?

Step 2: Define Preliminary Scope

The immediate need for an organization without processes for capturing the business objectives associated with project requests is to define a structured approach for documenting, evaluating, and approving the preliminary scope of work. Note that *approving* the scope of work involves more than shaking heads, shaking hands, or a casual agreement on broad, subjective criteria. Approvals, in project management, imply documented endorsements More simply, signatures that provide evidence of agreement and a foundation to build upon. It is important to emphasize to stakeholders and sponsors unfamiliar with our profession's structured approach to managing projects that accepting a preliminary scope of work does not mean that you are locked in for the remainder of execution. Nothing could be farther from the truth. Instead, you are protecting their interests by beginning to set boundaries upon which effective planning can begin. In other words, you are increasing the probability (remember the research) that the project will be successful.

Step 3: Develop Understanding of What Final Acceptance Means to Project Sponsor or Sponsors

How do we know when we have arrived at a destination? When traveling, we know our trip is complete when we reach our intended destination. Likewise, we know that a project is complete when we have delivered on the business objectives identified in the project charter, right? Well, yes . . . and then some. The "and them some" is the focus of scope change control. How does your organization define final sponsor acceptance? The recommended approach is to define sponsor acceptance for stakeholders using plain language. Sponsor acceptance is the formal recognition that the objectives defined in the original agreed upon scope of work have been met, plus the objectives agreed upon in all of the formally approved change requests. This plain definition helps to avoid the differing perceptions around what was wanted versus what was documented.

Step 4: Define, Document, and Communicate a Structured Approach to Requesting, Evaluating, and Approving Change Requests

What is a change request? Some schools of thought suggest that changes are limited to requests for additional features, deliverables, or work. While this paper is focused on these types of change requests, or scope change requests, it is important to note that any change that has the potential to impact expectations should follow a formalized change request, approval, and communication process. Remember, aggressively managing expectations is our best opportunity to influence our stakeholder's perception of value. Scope, budget, schedules, and risks are typically interdependent and directly influence our stakeholder perceptions. Also, remember that the most effective change control processes include risk assessments that evaluate the potential risks of either approving or disapproving a change request.

Keep in mind that too much bureaucracy, too much analysis, or too much unnecessary paperwork will give stakeholders an incentive to circumvent your process. If you want your stakeholders to avoid, ignore or completely by-pass your process, include a great deal of administrivia. Administrivia is the new word for "trivial administrative process." (As the author, I reserve the right to add to the English language.) Remember, our profession's focus is on delivery and business results, not just adherence to a predefined process. Taking a lean approach to scope change request documentation can help influence acceptance of this sometimes painful, but vital, process for capturing change.

Process tip: Determine early (either as an enterprise standard or for your specific project) what the tripwires and associated levels of authority are for approving a requested change. What level of change can be approved internally? For example, a change with an impact of less than one week schedule delay or budget impact of under $10,000 may be approved by the project manager. What needs to be escalated to the project sponsor, what needs to be reviewed by a change control board or governance council? Determining these decision points in advance can remove a great deal of the mystique around how to manage change.

Ensure that everyone understands the difference between the natural decomposition process and identifying new work that must be accomplished to deliver on a previously agreed upon business objective and work associated with new or modified deliverables. Remember that omissions and errors in planning may lead to schedule and budget changes, but are usually not scope changes.

Step 5: Document and Validate Full Scope of Work (Create Work Breakdown Structure)

A great approach for defining all of the work required to complete a project is to start with the desired end state and associated expected benefits. What work is required to provide those benefits? What work is required to reach the approved end state goals (or business objectives)? Plan to the level of detail necessary to effectively manage the work. Decomposing work packages beyond the level required for effective management is considered Administrivia. Note that defining and communicating the processes for final sponsor acceptance and requesting changes both come before traditional decomposition. Why? Terrific question! The natural planning processes that we follow in breaking down business objectives into definable work packages can be a catalyst for change requests. We want to communicate up front that change is not free and that additional requests will need to be formally requested, documented, agreed upon, and approved before being included in the project scope of work.

Step 6: Manage Change

Your foundation is laid, you have documented the preliminary scope, you have defined processes for sponsor acceptance, you have defined and documented scope change request processes, and you have developed your WBS, now the only thing left is to manage according to your policies and plan. Almost forgot . . . you have to manage the change requests that are guaranteed to come too! Scope change control protects the project manager, and the performing organization, from scope creep and contributes to managing stakeholder expectations.

A question that frequently comes up among practitioners is "what do I do when my leadership does not allow me to define, document, and manage change?" This is a real, practical question that deserves a response. The instinctive approach is to communicate the necessity for a structured approach to documenting and managing scope. As our peers will confess, this is not always sufficient to get the support we need to set organizational policy. We can attempt to implement these processes without formalization, or just "do it anyway." This can be an effective approach for demonstrating the value, but can also be perceived as a self-protective measure instead of a process used to increase the likelihood of project success. People can be leery of someone else documenting requests, justifications, etc . . . for their needs. Ensure that you share the information and provide an explanation as to why this approach is designed to ensure you are managing to their expectations. In general, people have difficulty not accepting altruistic approaches to meeting their needs.

Learn from Other's Lessons:
A Real-World Application

Leveraging experience, best practices, and lessons learned, the Churchill Program Management Office began with the basics; they chartered their PMO. The three-fold mission of the newly founded PMO was to establish, facilitate, and manage the project portfolio selection and funding process; create a foundation for consistent project success throughout the organization through development of a strong and pervasive Project Management discipline; and to guide key projects to a successful conclusion by providing project management leadership, while improving the quality and repeatability of related processes. Sounds fairly standard, right? The mission was then broken down into specific objectives and successful completion of these objectives was tied to the PMO Director's compensation.

PMO Objectives Included

1. Develop and implement standard processes for project requests, evaluation, and funding to ensure that approved projects were aligned with Churchill Downs Incorporated's business goals and objectives.
2. Develop and implement a standardized project management methodology, to include policy, standards, guidelines, procedures, tools and templates.
3. Build project management professionalism by providing mentorship, training, and guidance to project teams as they learn and adopt project management processes and best practices.
4. Manage the Churchill Downs Incorporated project portfolio by ensuring required documentation is in place and that stakeholders are properly informed about the ongoing progress of the project portfolio through effective reporting of key performance indicators.
5. Direct project management for key strategic initiatives.
6. Ensure benefit realization by using processes for clearly defining business cases and the associated metrics for measuring project success. Facilitate post-implementation benefit measurement and reporting.

As related to change control, we wanted to ensure that the process was lean, that our stakeholders understood the importance of the process, and finally . . . arguably most important . . . communicated in a way that our stakeholders understood and could follow the change request processes. Here is a thought provoking question for our practitioners: "Why do we expect our stakeholders to learn and understand our vernacular?" To aid in understanding and training, we developed visual tools documenting our overall project management processes in a language that they understood. For example, the project "race track" (see Figure 4–10) demonstrated to our leadership and project team members what we, in our profession, take for granted as universally understood; that projects have a defined start, a defined finish, and require certain documentation throughout the planning and execution processes to ensure everyone understands expectations and that we will realize the intended benefit from the investment.

For Churchill Downs Incorporated, scope change control begins with the foundation of a completed Investment Request Worksheet (or business case) and an agreed to scope of work as outlined in a signed charter. The work is then decomposed to a level of detail required to control the effort and complete the work necessary to deliver on the requested and approved objectives as detailed in that charter and approved scope change requests. A scope change request consists of a simple to understand, fill in the blank template, and the process is facilitated by the project manager. More important, the scope change request form is used to document the business objectives for a change request, the metrics needed to ensure the change's benefits are realized, the impacts on schedule and costs, the funding source, and the necessary approvals required for including the request in the overall scope of work.

Some of the benefits that Churchill Downs Incorporated has realized to date from this structured approach to documenting and controlling scope include:

1. Retroactively documenting scope for legacy projects, which resulted in canceling projects that were plagued with uncontrolled change to the point that the final product would no longer deliver the benefits presented in the business case.
2. Denying scope change requests based on factual return on investment and impact analysis.
3. Ensuring that requested scope changes would contribute to the business objectives approved by the investment council.
4. Empowering project team members to say "no" to informal change requests that may or may not provide a quantifiable benefit.
5. Demonstrating that seemingly great ideas might not stand up to a structured impact analysis.

12.5 TYPES OF PROJECT OFFICES

There exist three types of POs commonly used in companies.

- *Functional PO:* This type of PO is utilized in one functional area or division of an organization such as information systems. The major responsibility of this type of PO is to manage a critical resource pool, that is, resource management. Many companies maintain an IT PMO, which may or may not have the responsibility for actually managing projects.
- *Customer Group PO:* This type of PO is for better customer management and customer communications. Common customers or projects are clustered together for better management and customer relations. Multiple customer group POs can exist at the same time and may end up functioning as a temporary organization. In effect, this acts like a company within a company and has the responsibility for managing projects.
- *Corporate (or Strategic) PO:* This type of PO services the entire company and focuses on corporate and strategic issues rather than functional issues. If this PMO does manage projects, it is usually projects involving cost reduction efforts.

As will be discussed later, it is not uncommon for more than one type of PMO to exist at the same time. For example, American Greetings maintained a functional PMO in IT and a corporate PMO at the same time. As another example, consider the following comments provided by a spokesperson for AT&T·

Client Program Management Office (CPMO) represents an organization (e.g., business unit, segment) managing an assigned set of Portfolio Projects and interfaces with:

- Client Sponsors and Client Project Managers for their assigned projects
- Their assigned DPMO (see DPMO Roles and Responsibilities for contacts)
- Their assigned Portfolio Administration Office (PAO) representative
- CPO-Resource Alignment (RA) organization Factories

The Department Portfolio Management Office (DPMO) supports their client organization's Executive Officer, representing their entire department Portfolio. It serves as the primary point of contact between the assigned CPMOs within their Client organization and the PAO for management of the overall departmental Portfolio in the following areas:

○ Annual Portfolio Planning
○ Capital and Expense Funding within portfolio capital and expense targets
○ In Plan List Change Management and Business Case Addendums
○ Departmental Portfolio Project Prioritization

The PMO is led by an Executive Director that is a peer to the line Project Management Executive Directors. All Executive Directors report to the Vice President—Project Management.

The functions of the PMO include: Define, document, implement, and continually improve project management processes, tools, management information, and training requirements to ensure excellence in the customer experience. The PMO establishes and maintains:

- Effective and efficient project management processes and procedures across the project portfolio.
- Systems and tools focused on improving efficiency of project manager's daily activities while meeting external and internal customer needs.
- Management of information that measures customer experience, project performance, and organizational performance.
- Training/certification curriculum supporting organizational goals.

12.6 STARTING UP A PMO AND CONSIDERATIONS[3]

Starting a PMO in any organization is not a task for the faint-hearted. While there are PMO-in-a-box type solutions on the market, there needs to be an understanding by those who are undertaking the task of implementing a PMO within the organization that moving

---

3. Material provided by Ben Stivers, PMP, Project Management Consultant.

toward a PMO for a virtual PMO is an *organizational change* program rather than a project that installs project management methodologies and processes.

Those involved in organizing the PMO should make a sniff-test about the future of the PMO by examining a single fact—is the PMO a part of the organization's strategic plan. If it is not, then it should be. The PMO, when properly placed, designed, and utilized in an organization, should be a *strategic asset*. That is, the PMO should provide a capability to the organization that gains it space in the marketplace that the organization would not ordinarily have if the PMO did not exist.

Then the PMO should define its *core purpose*—why does the PMO exist? This should be completely clear, and if you meet with project managers and then the supporting executives who are sponsoring the PMO, you might find that those two groups of participants do not agree. Project managers tend to believe that the PMO exists to produce processes for project managers, or provide metrics, or even deliver a project on time and on schedule. An executive, however, may be looking for a particular business outcome from the PMO (results), such as a 50 percent increase in ROI for projects or a 22 percent reduction in cost overruns. The outcomes are numerous, but it is imperative that those building the PMO and those who sponsor the PMO are on the same page about the core purpose of the PMO.

With the core purpose in place, the PMO may adopt the corporation's core values, or it may develop core values of its own that compliment the organization.

12.7 COMPUTER SCIENCES CORPORATION (CSC)

Consider for a moment the following situation. Your company wants to start up a PMO to function as the guardian of project management intellectual property for your company. You hire a person who has done this for a previous company with some success. What this person brings to your company are the successes and failures that he or she encountered at a previous company. This individual will make decisions based upon his or her experiences. This approach can work, and can work well. But there may be a better way to accomplish the goal of implementation of a PMO.

More companies today appear to be hiring consulting companies to help them implement a PMO. Consulting companies maintain files of successes and failures among all of their clients and the consultant(s) you hire can draw upon the experiences provided within the consulting company's knowledge repository. If the consultant has issues where help is needed, the consultant can call upon other company consultants for advice and guidance. Simply stated, the consultant brings to the table perhaps hundreds of man-years of documented experiences and best practices whereas the person you hired above may bring to the table only 5 or 10 years of personal experience.

Another problem occurs if the person you hire is unfamiliar with your industry. Consulting companies often maintain files by industry classification. This can significantly benefit the onsite consultant. With consulting companies, it is often easier to find industry-specific consultants. When starting up a PMO in a health care environment, this is essential.

To illustrate how hiring a consulting company can be beneficial, consider CSC[4]:

CSC's purpose is to deliver innovative business and technology solutions that help our commercial and government customers worldwide achieve what they want most—results. Focus is placed upon the uniqueness of each customer around the globe. CSC has over 91,000 employees in 80 countries worldwide and has been in business for over 49 years. CSC focuses upon a wide variety of industries and provides consulting, implementation, outsourcing and integration services at all levels.

Best Practices/Culture/ Governance Structure/ Methodology

In the above description of CSC, it is important to understand how critical the words "uniqueness of each customer" Are. When starting up a PMO in a health care environment, where improving and maintaining health are of prime concern, setting up a PMO based upon an engineering, construction, or manufacturing company may not be advisable. Nani Sadowski-Alvarez continues:

Best practices were utilized extensively during the incorporation of a PMO (Project Management Office) at a large 9,000 employee based Health Care System located in Baton Rouge, Louisiana entitled Franciscan Missionaries of our Lady. This implementation was conducted through a partnership with CSC over a duration of one year in order to formulate a streamlined, fully operational, optimized, and mature PMO team capable of facilitating and leading a large and balanced portfolio consisting of a blend of technical, clinical and business focused projects, programs and extended service requests. The Best Practices were established via a variety of experiences and lessons learned during prior PMO implementations at other client locations as well as through lessons learned shared via professional affiliations within the health care industry as well as a cross section of industries, project management based statistics, industry trends, and healthcare based trends and regulations. In addition, client surveys (both formal and informal) were conducted to determine and unveil:

- Clients' perception of areas within their organization that have proven to be operating in an efficient and effective manner
- Target areas that are recognized as being in need of improvement
- Areas and processes requiring focused attention that would enhance overall needs, mission/vision, desired organizational direction as well as provide desired emphasis upon the creation of positive and productive outcomes.
- What and who are the perceived drivers of this change.
- Overall organizational "pulse" and view surrounding the implementation of project methodology.
- Areas that require fine tuning and revisions.
- Internal employee views and feedback on the audits and assessments conducted by other consulting and audit agencies.
- The "understood" organizational strategic goals and initiatives.

---

4. The material on CSC in this section was provided by Nani Sadowski-Alvarez, PMP, CPHIMS Senior Management Consultant, Enterprise Program Management & Architecture, CSC.

- Overall perception and view of the change and the transformation underway.
- Environment and tone of interaction between organizational facilities and departments.

With the fact being that the implementation was driven/lead by consultants who closely partnered with the Health Care System, there are several additional factors to be taken into consideration with utilization of Best Practices as a guiding foundation. Following is a list of these *"Challenges and Opportunities as a Consultant Surrounding Incorporation/Utilization of Best Practices"*:

- Accurate interpretation of the organizational culture
- Listening effectively to all signs provided by the organization (verbal and nonverbal) encircling their needs
- Customization of Best Practices to a level that provides the most beneficial results.
- Obtaining and encouraging ample feedback and input from the leaders and end users to ensure that the first impression of the Best Practices is as positive as possible.
- Delivery of effective communications that target the impact that the initiative/project and the incorporated Best Practices will have.
- Providing adequate answers to the Who, What, When, Where, Why and How surrounding the need for the transformation/change underway.
- Instilling a balanced and flexible approach to use the proven Best Practices as a foundation with the addition of adjustments that will ensure they blend with the organization acquiring and implementing them.
- Creating the proper balance to ensure silos are removed allowing for more transparent and fluid communication and utilization of agreed upon Best Practices.

Best Practices Defined—Overall Best Practices are processes, procedures, standards, methodologies, techniques, activities, etc. that have proven to be more effective and successful in providing measurable efficiencies and results than other comparable processes, procedures, standards, activities, and techniques.

Factors and questions considered in the evaluation of best practices are:

- What is it? A process, procedure, template, methodology, action?
- Are they repeatable?
- Is it something that results in efficiencies?
- Is there impact to clinical care? How and what regulations (Patient Safety Goals, JCAHO—The Joint Commission on Accreditation of Healthcare Organizations, HIPAA Compliance Regulations) are met or impacted by this?
- Is there an impact to Business Initiatives and Technological Advancements?
- What "trial and error" processes were underway to test a sample set of options and determine what the "best route" would be to proceed with?
- What evaluation parameters were implemented in order to assess and determine if a process/procedure/template/methodology/action were in fact meeting the needs of the impacted areas and were there efficiencies related to it?
- Was there clear and definable collaboration involved by stakeholders to ensure that all factors were considered?

Samples of Best Practices
with Health Care PMOs

- Structuring all projects in a consistent manner with a core methodology in place (that is required to be adhered to)
- Undergoing periodic random project audits from an external department (either Legal Services or The Quality and Audit Department) that are made available to all internal requesting resources
- Conducting formalized project initiation and closing presentations with all stakeholders being invited to participate as well as Core Team member's attendance being required. These events/milestones are then followed-up and communicated via internal marketing communications accessible to all employees.
- Ensuring all projects (clinical, technical, and business) receive input and scoring/rating from an experienced clinical executive representative.
- Validating that project scoring/prioritizing contains a section directly related to the organizations strategic initiatives as well as process flows and quality of care.
- Creating detailed project metrics that are available on a weekly basis for all stakeholders and employees within the organization.
- Creating transparency with all project related documentation in order to ensure that any interested parties are able to assess the current action items, risks, issues, change requests, etc. with the ability to then provide suggestions/opportunities for improvement if they feel so inclined.
- Instilling that the PMs are parties to the contract review and negotiation process and that they facilitate the completion of a contract checklist.
- All Project/Program and Extended Service Request goals must be compiled in alignment with the SMART Goal philosophy in order to validate that the goals are measurable.
- All Projects/Programs and Extended Service Requests must undergo formal Operational Turnover prior to the Project being flagged as "closed." This turnover occurs with participating of Core Team Members, impacted areas, and key stakeholders and requires sign off from Management at all areas receiving support responsibilities.
- Projects must undergo the formal resource request process in order to validate that the Resource Managers will be able to support the dedication required to make a project successful.

Storage and Indication
of Best Practices

Potential Best Practices are assessed and evaluated throughout the duration of the project life cycle. These are then documented in the Lessons Learned template and made accessible to all organizational employees/Team Members. Project documentation is stored in SharePoint with view access provided to all Core Team Members, approving Committee Members, and Key Stakeholders as well as requesting parties.

Determined Best Practices are recorded as standards and often created into Policies/Procedures with the need for tiered sign off approval in order to ensure that supporting and enforcing levels of authority have the opportunity to provide feedback and input.

Methodology

The PMO was created at this Healthcare System (XYZ Health System) under the direction of the organization as well as the full-time support of the CIO/CMIO, Dr. Stephanie Mills and the Director of the PMO, Claudia Blackburn, PMP, CPHIMS. Nani Sadowski-Alvarez, PMP, CPHIMS was brought in from FCG/CSC to lead the organizational transformation Track focused upon the implementation. A blend of methodologies was utilized. The core foundation of methodology resided upon PMI based practices with a blend of ISO 9000 (for segments on the technology driven projects) lean project management principals from Six Sigma and best practices already established within the organization. With the foundation being PMI and *PMBOK® Guide* focused, the core phases of project management (Initiation, Planning, Executing/Controlling, and Closing) were infiltrated throughout all project implementations. Supporting policies and procedures were developed along with standards and detailed process flows that targeted all stakeholders and explained the methodology step-by-step. Each of these tools provided input surrounding templates that would be utilized within the various phases as well as a listing of what the outputs and results would be. The key with all of these, was to keep the concepts and explanations direct and simple and to provide details surrounding the resource responsible for executing the particular steps in the process, anticipated turnaround times, outcomes, and reference to related and interdependent processes. All templates surrounding the methodology, processes, procedures, standards, workflows were standardized and available to anyone within the organization as well as requesting partner vendors. This creation of a transparent department enabled all stakeholders to have access to pertinent project details and methodology at any point in time.

Governance

The transformation of the organization was divided into tracks with 1 Tracks main focus upon creation and Implementation of the PMO, another Tracks focus upon IT Governance, another tracks focus upon implementation of a PPM and internal IT Resource and Capacity Management tool, and the 4th track focused upon organizational governance. The key to success with this was that all tracks had to work collectively in order to ensure collaboration and cohesiveness of the undertakings of each track. The governance for the healthcare system was designed to ensure buy in from all key areas—Business, Clinical, and Technical. The PM (in partnership with the PMO and the Project Program Sponsor) was responsible for facilitation of the process and ensuring that the Business, Clinical, and Technical committees had the opportunity to review the Project/Program Charter (and provide feedback—suggestions) prior to delivery to the Steering Council. From there, the project may or may not go to the Executive Council (Pending project Cost, size, impact to resources as indicated in Governance Policies) as well as be presented to the board. The parameters were all clearly defined and created with participation from key resources throughout the organization. This ensured that all areas had equal input and that there was a greater likelihood of rapid buy in and support for the governance structure.

Project/Program Success

Project success is measured based upon the delivery of the projects on-time and within budget along with their ability to meet the project mission/vision as well as the SMART Goals. Now, with that being said, it is accurate to state that Change Management methodology is solidly in place and acceptable in re-aligning the project and its scope with adherence to the triple constraints and agreement that change requests must be signed by the Project/Program Sponsor. It is the PMs Responsibility to update project documentation accordingly and to adhere to the project change management Policy, procedure, standard, and process flow. Project/Program Status is available 24/7 to all requesting parties. It is encouraged that vendors participate in the status updates as well. Critical Success factors are clearly detailed within the project scope document (that is also signed by key stakeholders) and when changes are 25K or above they involve approval via the Governance Structure. PMs (Program and Project Managers) are required to conduct milestone updates and to report on them via the status report. PMs are also monoured and evaluated based upon their ability to achieve the role based performance standards/indicators. These indicators align with their daily duties, project based duties as well as methodology perspectives and organizational initiatives.

Overall, project success is defined by the organization at large. The project charters contain full detail surrounding project mission/vision and the project goals. A balanced representation is present at Approving Council meetings (from the Financial/Business side, clinical and technological side of the house to include lead Executives, CFO's, Physicians, Technology Directors, Departmental Leads, SMEs, etc.) to ensure that the checks and balances are in place prior to voting a project to the approved state. Once it is approved the detailed scope is reviewed (containing critical success factors, known issues/risks, detailed process flows, communication matrix, etc.) and open for suggestion/revision prior to sign off. Once signoff has occurred, the formal Change Management methodology is adhered to. This methodology does contain a 'break the glass' procedure in the situation that an item/opportunity for change arises that is of urgent and immediate attention status. The execution of this process does require executive level approval due to the magnitude of individuals whom will be impacted.

Excellence is defined clearly by the E^3 theory of Excellence Exceeding Expectations. With any effort to exceed the initially set and approved expectations, there is the ability to achieve excellence where streamlining, enhancing, cost savings, patient and employee satisfaction are present.

Questions to Ask Surrounding Program/Project Success

- Have milestones been met?
- Has effective change management taken place?
- Were Risks efficiently mitigated?
- Were status reports provided per the policy and procedure guidelines throughout the duration of the project?
- Were there regularly scheduled Core Team Meetings?
- Were the Signed Project/Program Goals and Objectives achieved?
- Was the project methodology followed throughout the project life cycle?
- What are the Sponsor(s) and Key Stakeholder(s) view of the end results of the Project/Program?
- Have all participants had the ability to provide their view of project successes and opportunities for improvement?
- Is the project within Budget?
- Was the project delivered on time?
- What are the results of the project audit(s)?

NOTE: We have discovered and confirmed that templates need to be fine-tuned for all clients in order to ensure that they meet the needs and initiatives of that organization. The length of the templates can at points hinder the success of the project. As a result, the recommendation is for the PMO to continually review and discuss the utilization of the templates and to adjust them accordingly.

Project Management Offices (PMO)

The project management office that was created for the Health Care Facility consists of a Director PMO, 10 PM's (2 of whom are detailed as Senior PMs), 2 Extended Service Request Coordinators, and 2 technology engineers (who focus upon guidance with project technology assessments and evaluations as well as audits). This PMO reports to the CIO/CMIO as well as the Steering Council and other governing Councils. Programs (such as the implementation of a Heart Hospital) are also facilitated through the PMO office.

The vision of the PMO is to "provide consistent and standardized project management with focus upon excellence in a collaborative setting. All approved projects will contain measurable goals and objectives directly related to the goals and standards of the Health System with convergence towards achieving a comprehensive continuum of quality care."

There is a solid governance process in place as referenced in the governance section.

Management Support

Support for the PMO trickles directly from the C-Level Executives on down at this Health Care organization as well as the consistent view point of the value that the PMO adds as a whole. Prior to the implementation of the PMO, an organizational audit occurred with the suggestion of more formalized methodology and standards as well as enhanced transparency surrounding project focused endeavors and initiatives.

Sponsors receive a one-on-one session with the Director of the PMO and the assigned PM in order to review a PowerPoint presentation and several hand outs that clearly define what the role of a project sponsor is. In this conversation it is also assessed if they feel prepared to take on the role of a Sponsor. If not, then this realization is brought to the Steering Council for discussion and determination as to who an appropriate Sponsor(s) might be. During the initial session with the Sponsor, items such as Project Communication, frequency of status reporting, regular standing meetings between the PM and Sponsor, etc. are all covered and details are aligned. Project Sponsors learn to become supportive of their projects and to value the partnership that the PM and PMO form with them in order to ensure that their project is a success. A PMO marketing tri-fold is also presented to them that contains information surrounding the following:

- Process flow for the Project Approval Process
- Definitions of key roles on the project team and what their responsibility is
- Critical Success Factors for all projects
- Definition of Project vs. Program vs. Extended Service Request
- High-level communication Expectations (to include their role in assisting with removal of barriers to the projects success)
- High-level review of project documentation that will consistently be provided
- Brief overview of project phases and methodology

Sponsors are held accountable for the success of their projects and are requested to be present in the initial presentation to the Steering Council requesting approval. They speak to the "need" of their project (for the organization), the ROI, the SMART Goals, overall added value, impact to resources, duration, scoring, etc. along with the PM and participate in the question/answer segment. As major changes (exceeding 25K in value or significant impact to Resources, or project schedule) arise, the Sponsor and PM are requested to attend the Steering Council meeting to provide the details as well as discuss the options for resolution with the Council and obtain a vote of approval for the newly targeted "direction" to follow.

Sponsors are required to be management level and above. For Extended Service Requests (initiatives that are 500 hours or less in duration and require formal structure in order to execute) a Sponsor may reside at the Managerial level. For Projects and Programs, however, a Sponsor must at least reside at the Director level in order to ensure that they have the leverage and authority to assist with the barrier removal.

12.8 UNDERSTANDING THE NATURE OF A PMO

In the previous examples in this chapter, we saw companies that implemented a PMO after a well-thought-out process. But, too often, a company will rush into the implementation of a PMO well before it decides on the intended use of the PMO, what functions it will perform, who will be assigned, and potential internal political issues. Not every PMO is the same, even in companies that have multiple PMOs. Carl Manello, PMP, Solutions Lead—Program and Project Management, Slalom Consulting, provides us with his insights on PMOs:

What's in a name? More than most of us would consider. Far too often the three letter acronym "PMO" is bandied about with far too little consideration for what it refers to in its implementation. I have seen this carelessness occur with my clients as well as within consulting organizations that offer solutions in this arena. The over-simplification of project management organizations into a "PMO" has led to some disastrous results. What do "PMO's" do after all? Are they:

- Project Management Centers of Excellence (e.g., process owners)
- Governance Offices (i.e., overseeing compliance)
- Project manager resource pools
- Resource management centers charged with balancing supply and demand

I have seen of each of these implemented with the same name: PMO. "So what?", you may ask. "Why is it so important what we call the function?" A quick anecdote may illustrate the point well.

A CFO reads in the latest trade magazine about the much celebrated success of a peer company in their implementation of a PMO. "Fantastic," she thinks and proceeds to call the CIO. "Bill, we need to get your team to build a PMO. I know that they can make a difference, have a financial impact and deliver results. Go make it happen!" After hanging up the phone, Bill is not sure where to turn. Does the CFO want a Project Management Office, Program Management Office, Program Office, Enterprise Program Office or Governance Office?

As professional project managers we must help our companies distinguish the difference in functional role of each of these different organizations. We must also help to clarify the span of control differences—say, between a Program Office and an Enterprise Program Office. I coach my colleagues and clients that we should stay away from a "one name fits all" reference of the mythical "PMO" and instead talk about operational functions.

For example, Figure 12–4 illustrates an IT Governance Office which has a variety of accountabilities. However, it is focused within IT and does not reach into the "business" side of the company. Focused solely within IT, Governance takes on many roles. This "PMO" focuses on providing assistance to project managers, aligning and monitoring the assignment of project managers, assisting the Office of the CIO with strategic planning, monitoring IT financial/project spend and overall IT reporting for leadership. By contrast, this "PMO" does not contain a pool of project managers to be assigned to projects throughout the information technology organization. Neither does this governance organization "PMO" have the same span of control as a "PMO" (Program Management Office) operating within the confines of a large business initiative.

To illustrate the difference in span of control, let's look at the following models. While not perfect solutions for every implementation, these figures show the range of possibilities and begin to break-down the view of a one-size-fits-all PMO. Figure 12–5 (I'm sure you'll align figure numbers later, just wanted to be sure you're pointing the right ones) provides one of the simplest models: The Project Office.[5] The Project Office provides oversight, monitoring and reporting and possibly even some governance functions over non-related projects. By contrast, the Program Management Office in Figure 12–6 is contained within a single large-scale business initiative. This PMO has accountability for driving a business program, keeping track of multiple associated projects and providing monitoring and reporting to the key business executives. In some implementations, the PMO is not just the project governing body of the program, but it is the home of the key executives running the initiative. At Sears, during the integration with K-Mart, and at Donnelley for the Year 2000 efforts, the PMO was the executive leadership team of the business focused effort. There were of course supporting teams of program and project managers on-board, but when the chairman referred to the "PMO" he was talking about leadership. Figure 12–7 illustrates the next wider governance organization: the Program Office. Much like the Project Office, the Program Office oversees non-related projects with the added responsibility for governance over programs. By nature of its oversight, this organization is usually set up with accountabilities across large functions (e.g., across business units). Program Office members work with their counter-parts in Program Management Offices to collect metrics, status and key updates, however, usually focused at a higher level. While the PMO works at a detailed level collecting, reporting and managing a single program, the Program Office summarizes cross-initiative information for consolidation at the executive level. The last variant is the Enterprise Program Office in Figure 12–8, which is typically implemented in very large organizations. Its role is to further roll-up, consolidate and coordinate initiatives across the enterprise. In addition to governance responsibilities, this group may be charged with portfolio management activities. With its view of the entire enterprise, the EPO is well suited to operate with the goals of the whole organization in mind, as opposed to siding with one business unit over another. As part of the first EPO at R. R. Donnelley & Sons in 1997, we worked on portfolio management for the whole enterprise, master schedule

---

5. The symbol in the upper right-hand corner of Figure 12–4 illustrates that each of the successive PMO structures in Figures 12–5 through 12–8 leverages the same project management capabilities in PM*f* that was discussed in Figure 4–7.

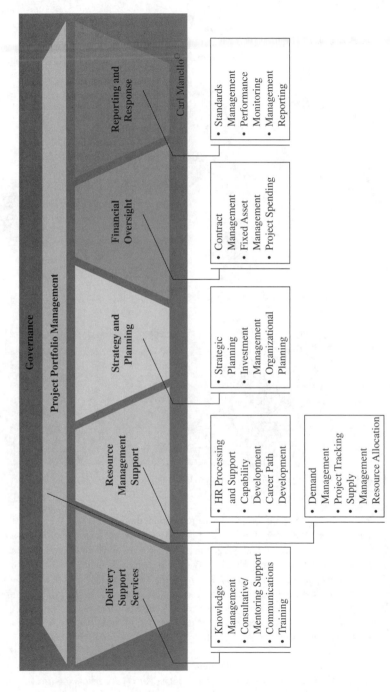

FIGURE 12-4. Example functions of a PMO.

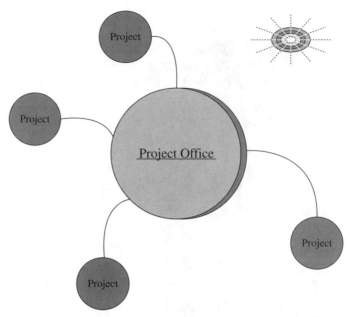

FIGURE 12–5. Project office that oversees independent projects.

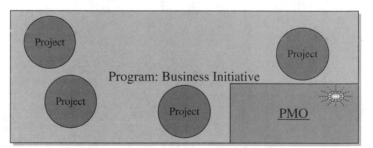

FIGURE 12–6. Program management office internal to a program.

development, creation of project management methods and processes and we provided project support. We worked across corporate functions, Research & Development, all the business units (e.g., Book and Magazine) and were involved in several key business programs. Given that there are such a variety of project management organization incarnations, with varying spans of control, it is no surprise that there are also a wide variety of reporting accountabilities; a project organization does not always report to the same function in every company. Typically, I have seen what I identify as *Project Offices* put in place within the Information Technology organization. Many if not all IT initiatives report into or have a link to the IT Project Office, however, initiatives that do not involve IT as a major component are typically not tied to this governing body. My R. R. Donnelley & Sons *EPO* reported into the head of the Research & Development organization (at that time, Donnelley did not have a full-time CIO controlling the IT organization). My *IT Governance Office* at Zurich North America reported into the Vice President of IT while the Sears implementation of a *Program*

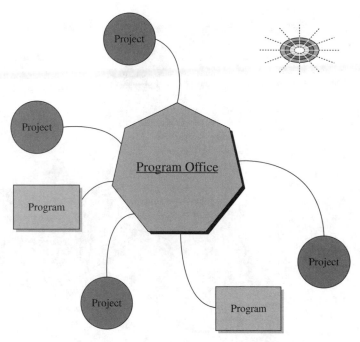

FIGURE 12–7. Program office that oversees programs and projects.

Management Office was accountable to the Office of the Chairman. The Sears' PMO was a special reporting relationship and was due to the importance and the visibility of the businesses program we were leading. However, most other Program Management Offices usually report directly into the executive champion in charge of the business initiative.

While the illustrations I have provided are certainly not exhaustive, I think it clear that there are a number of different types of organization. There are also a variety of span of control or reporting accountabilities. Given this variability of implementation, it should be clear to the reader that a generic label of "PMO," is wholly inadequate and may in fact be short-sighted.

As mentioned in an earlier chapter, customers are now *requiring* their contractors to assess their organization and determine their level of maturity in project management. This assessment process, whether customer-driven or internally driven, most often requires PMO involvement. Carl Manello continues:

With so many different types of "PMO's" (as discussed above), what is the right one to build? Put another way, how can an organization assess where it is at in the project management maturity continuum, in order to determine which type of "PMO" to implement? The answer is a bit more complex than one might first assume. One needs first to understand the environment into which the project management organization will be housed. For to plan a project management organization that is neither linked to the organization's current state (i.e., competencies and capabilities), nor tied to a vision for growth and development is a recipe for a sub-optimal solution.

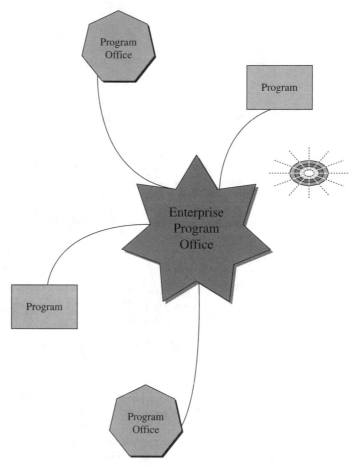

FIGURE 12–8. EPO that spans programs and POs.

There are scores of models for assessing project management maturity. Each organization must carefully consider the implications of the model they choose and understand where the implementation based on the model will lead them. One maturity model is illustrated in the diagrams below.

Much like the CMMI provides a stepped maturity model for software development, the model in Figure 12–9 . . . creates a maturity model for project management organizations. The various steps are loosely aligned as follows:

- Building Basics—establishing the Project Management Functions (PM*f*)
- Deliverable Focus—implementing a Project Office to coordinate/oversee discrete projects
- Business Focus—putting a Program Management Office in an accountable role within key business initiatives

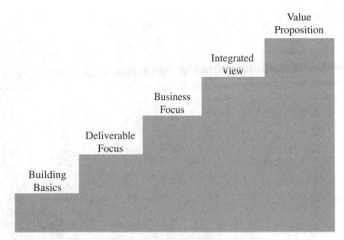

FIGURE 12–9. The project value maturity model.

- Integrated View—the Program Office, bringing together disparate projects and programs into a holistic view for the organization
- Value Proposition—realization of the Enterprise Program Office

While there is no absolute rule about progressing through maturity one step at a time, I think any organization that attempts to realize the value proposition of implementing an *EPO* before it has any experience with the less mature organizations, is looking for trouble. Just as one cannot jump maturity levels in CMMI, that same view should be considered when developing project management organizations. Another parallel to CMMI comes from the assessed performance level of an organization. To be "CMMI" in the truest sense, an organization must be assessed by a third-party against an international standard. While there is no such international standard to assess the maturity level of PM organizations, companies need to be brutally honest about their capabilities and competencies.

Too many of the organizations that I have worked for have had an over-optimistic view of where they lie on the CMMI scale (many companies thinks that they are at Level 3 or *higher*!). Similarly, companies that jump levels by over-optimistically assessing their PM maturity may actually diminish the value proposition of the "PMO" and in fact negatively impact the enterprise.

Figure 12–10 is a related view of the maturity progression, focusing on six project management organizational challenges, aligned with the six maturity levels. Note that this particular model was developed to illustrate the developmental challenges of an IT project management organization. Depending on the type of "PMO," the model will be different.

- Business Alignment—the movement of the IT organization from separate business unit to service provider to the rest of the organization
- Project Focus—the growth from discrete IT projects to aligned initiatives focused on the business
- Governance—moving from no oversight to a metrics and measures based management organization

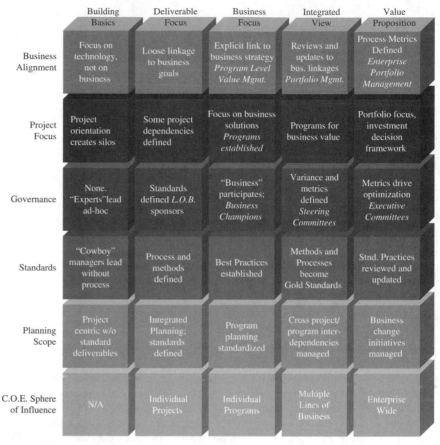

FIGURE 12–10. Maturity matrix.

- Standards—evolving from the "wild west" of project management to a organization run with standard practices
- Planning Scope—evolving from independent operating projects to ones planned in a standard repeatable way and integrated as change initiatives for the business
- Community of Expertise Sphere of Influence—parallels the growth of "PMO" organizations outlined above, from individual oversight to enterprise-wide influence

12.9 DTE ENERGY

Although functional POs can be developed anywhere in an organization, they are most common in an information systems environment. DTE Energy maintains an information systems PMO. According to Tim Menke, PMP®, Senior Continuous Improvement Expert, DO Performance Management:

> At DTE Energy the Customer Service Project Management Office (PMO) is a function with the Customer Service Program Management (CSPM) group reporting to the Vice President

of Customer Service. The Customer Service Program Management group consists of the PMO, the process management team, and the customer contact channel management team.

The PMO provides project management support for the continuous improvement initiatives within the department. These initiatives are designed to achieve the Customer Service Strategy.

Specific functions of the PMO include:

- Developing and maintaining the project management methodology
- Maintaining the portfolio of Customer Service continuous improvement projects
- Collecting and disseminating project data and metrics
- Providing project management tools and templates

The Customer Service portfolio contains projects from the five departments within Customer Service (Consumer Affairs, Customer Care, Billing, Data Acquisition, and Customer Service Program Management). The projects improve customer service, increase operational efficiency, and/or achieve savings from operations.

The PMO enables and facilitates the application of the project management process. The employees in the PMO have extensive backgrounds in project management and act as consultants, liaisons, and coaches for their respective projects.

As stated previously, the PMO can also participate in the portfolio management of projects. This is common in companies that wish to make maximum use of the talent in their PMO. Tim Menke explains:

> We select projects at the enterprise level based on various indicators including Return On Investment (ROI), Internal Rate of Return (IRR) and Net Present Value (NPV). This annual process involves the highest levels of organizational leadership and is integral to the prioritization and budgeting process.
>
> Our Project Management Office (PMO) engages with the project manager on "approved" projects. Our PMO aggregates projects into portfolios aligned by business unit. This approach allows us to analyze "trade offs" between projects within a business unit in an effort to elevate performance of the portfolio.
>
> As successes from this approach mount, our interest in performing portfolio management across business units increases. Our future focus includes a greater emphasis on resource allocation in accordance with enterprise strategies as opposed to business unit strategies.

12.10 EXEL[6]

For multinational companies, there can exist several POs that must function in a coordinated effort. According to Stan Krawczyk, PMP®, Director, Regional Project Management Group, Americas for Exel:

> Exel's Global PMO group serves the global organization as a project management center of excellence supporting project managers from all regions and sectors. The mission

---

6. Material on Exel Corporation was graciously provided by Julia Caruso, PMP®, Project Manager, Regional PM Group, Americas; Todd Daily, PMP®, Business Process Manager, DHL Global Forwarding, International Supply Chain; Stan Krawczyk, PMP®, Director, Regional PM Group, Americas; and David Guay, Global VP for Effective Operations.

of this group is to provide thought leadership and training of Exel's project management tools, techniques, and methodology. It is also responsible for the development of a strategic business plan that will leverage strengths of both project and resource management disciplines. The group provides a single-source solution for Exel's internal customers (sector PMOs and project managers). The Exel Global PMO is responsible for the following:

- Center of excellence, supporting project management tools and techniques
- Creation, deployment and maintenance of an enterprise-wide project management methodology
- Project management training for all sectors
- Consulting and mentoring of project managers
- Facilitation and support for the establishment of Regional PMOs in North America, Latin America, Asia, and Europe/U.K.
- Visibility to Exel's project portfolio and resource capacity across the organization
- Executive-level strategic reporting for PM initiatives
- Coordination between various functions utilizing PM methodology, including IT

The regional PMOs, established in the Americas, U.K./Middle East/Africa, Mainland Europe, and Asia/Pacific-Exel's primary theaters, provide dedicated project management in their respective regions. In addition, there are PMOs that focus on the IT segment as well as the DHL's Global Forwarding business (DGF). Regional PMOs (as shown in Figure 12–11) are groupings of project management associates (project managers, team members, etc.) who perform project management duties within specific regional or industry-specific areas. Primary PMO responsibilities are:

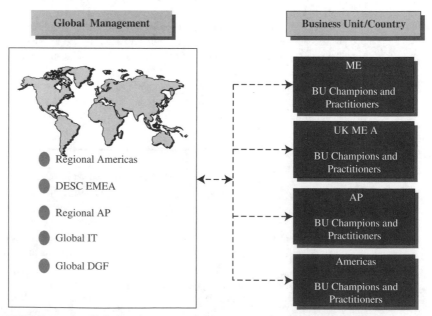

FIGURE 12–11. Global project management organization structure—matrix model.

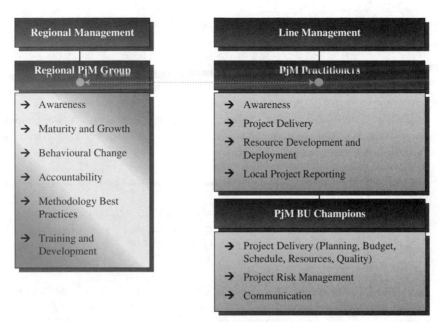

FIGURE 12–12. Global project management organization structure—responsibilities.

- Promotion of Exel's project management methodology
- Promote use of project management tools
- Project execution and delivery
- Subject matter expertise
- PM training delivery

Today, a globally consistent project management organization has been created to drive continuity, collaboration, and global visibility to all project management activity. (See Figure 12–12.) This organizational structure follows a matrix model and allows for the scalability of the methodology to accommodate cultural differences. Each project management group has specified roles and responsibilities that allow the project management community to mature while supporting the individualized needs of the internal business community and Exel's growing customer base. The global perspective of project management within Exel continues to evolve. According to Stan Krawczyk, PMP, Director, Regional PM Group, Americas,

> As Exel is now a truly global company in conjunction with the DHL brand, we now need to look even more at our project management community on a global basis across multiple business units. As project managers from different regions will be working together on large projects, we need to ensure that our standards for PM knowledge and expertise are consistent. Having DePICT® as our global methodology is a fine foundation, but we need to continue to drive commonality in our training curriculum and career paths for our community. This is one of our biggest challenges for the near future.

The following acronyms are used in the Exel figures:

- DePICT®—acronym for the five phases of Exel's project management methodology: define, plan, implement, control, transition.
- EMEA—acronym for Exel's Europe, Middle East, Africa region.
- APAC—acronym for Exel's Asia-Pacific region.
- PjM—acronym for "project management" used as an identifier within the Exel Way program, as PM could have been confused with performance management.

12.11 HEWLETT-PACKARD

Another company that has recognized the importance of global project management is Hewlett-Packard. According to Sameh Boutros, PMP, Director of Program and Project Management Practice at Hewlett-Packard:

> For large, global companies the need for project management standardization is essential in order to deliver higher value services at competitive costs. At Hewlett-Packard in the HP-EDS business group, there is a network of Program and Project Management (PPM) Practices in the Americas, Europe, and Asia Pacific regions. The mission of these practices are:
>
> To provide PM Services to HP Clients through Account PMO Leaders and PMs that lead IT services projects. The PPM Practice achieves its objectives when PMs consistently deliver projects on time, on budget, and to the client satisfaction, using disciplined and mature best practices. The PPM Practice supports the business objectives of efficient use of resources, profitability, growth, and customer satisfaction. It also provides profession leadership to ensure that the PMs are prepared to meet the needs of the business and have the opportunity to develop and grow their careers.

Project management development involves formal training and certification as well as informal development. Project management is a core skill and competency for HP Services. The award winning Project Management Development Program is organized by core project management courses, advanced project management topics, courses specific to HP Services practices, and professional skills training. Other activities that support project management development include:

- Driving project management certification programs
- Updating and managing the formal training curriculum in coordination with workforce development
- Driving and participating in major events like PMI congresses and regional project management training/networking events
- Encouraging informal communication and mentoring
- Providing mentorship to field project managers

The Global Method for Program Management provides project managers with methodologies and a standardized approach using industry best practices and incorporating the added value of HP's experience. This is shown in Figure 12–13.

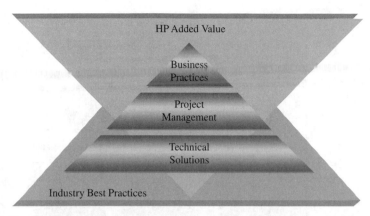

FIGURE 12–13. Global method, program methodology: standardized approach using industry best practices with company added value.

12.12 EDS

Doug Bolzman, Consultant Architect, PMP®, ITIL Service Manager at EDS, discusses the PMO approach at EDS:

> Most organizations have a PMO established and this was generated from the view that their individual projects required oversight. This is a significant jump for many organizations that 10 years ago did not see value in project managers and are now funding a PMO. But most of them are paying the price to staff the PMO but still do not see the value; they see it as a necessary evil. In other words, things would probably be worse if we did not staff the PMO.
>
> Major functions include project oversight, status reporting, and project conformance. Since release frameworks were not in place, companies had the situation where their main supplier organizations simply threw the solution over the fence to the next supplier. The PMO was created to facilitate these transactions. (See Figure 12–14.)
>
> The problem with the implementation of this approach is that there never was a single model developed for this type of framework and the PMO would add additional constraints, bureaucracy, or workloads. The PMO was looked at to plan the direction of the company though the implementation of individual projects. Instead, another model was

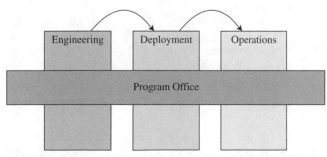

FIGURE 12–14. Using the PMO for facilitation.

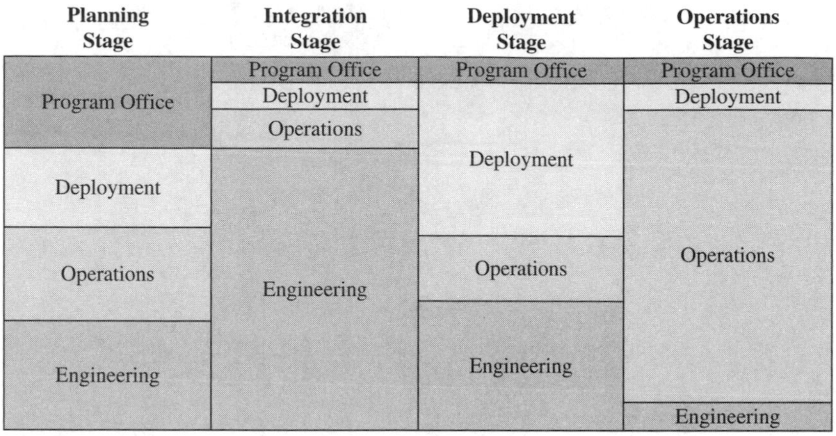

Planning Stage	Integration Stage	Deployment Stage	Operations Stage
Program Office	Program Office	Program Office	Program Office
	Deployment		Deployment
	Operations	Deployment	
Deployment	Engineering		Operations
Operations		Operations	
Engineering		Engineering	
			Engineering

FIGURE 12–15. Mapping the PMO to functionality.

developed to have all of the suppliers contribute to every stage of a release, which shares the accountability of planning and designing, while providing the PMO the proper level of functionality. (See Figure 12–15.)

12.13 STAR ALLIANCE[7]

Star Alliance is the world's first and largest airline alliance with 21 carriers and in the process of integrating 4 more (Air India, Continental, TAM and Brussels Airlines). Overall, the Star Alliance network offers more than 18,100 daily flights to 975 destinations in 162 countries. Its members carried a total of 405.7 million passengers with a turnover of US$95.3 billion in 2006. Each member of Star Alliance has a PMO.

- Air Canada
- Air China
- Air New Zealand
- ANA
- Asiana Airlines
- Austrian
- bmi
- EgyptAir
- LOT Polish Airlines
- Lufthansa
- Scandinavian Airlines
- Shanghai Airlines
- Singapore Airlines
- South African Airways
- Spanair
- SWISS

7. Information on Star Alliance has been provided by John Donohoe, PMP, Director, Project Management Office, Star Alliance Services GmbH.

- TAP Portugal
- THAI
- Turkish Airlines
- United
- US Airways
- Regional Members
  - Adria Airways
  - Blue1
  - Croatia Air

The Star PMO does not act as a "super-PMO" to member carriers PMO offices. The Star Alliance PMO provides project management services across the Star enterprise. The Star PMO carries out for the business units include such topics as information technology, marketing, sales, products, services, and frequent-flier programs, as well as common sourcing projects, which are projects that use the combined purchasing power of all carriers to jointly purchase common commodities (spare parts, in-flight services, etc).

Star Alliance projects are projects that are aimed at providing a common travel experience across all carriers or those that leverage our size to develop common IT apps, common networks, common lounges, check-in services or seamless Frequent Flyer upgrades across carriers. Project team members are normally business experts from member carriers distributed world-wide. We need to be very good in cultural awareness and consensus building.

Additionally, the Star Alliance PMO will assist and coordinate several member carriers to a Star Alliance Common IT Platform. The Star Alliance Common IT Platform is a strategic programme, focused on the effort to better serve the customer, markedly lower IT costs and significantly increase the speed of delivering new products to market. Once implemented, it will enable participating member airlines to improve customer services and enhance operational capabilities. It is based on Amadeus' pioneering new-generation Customer Management Solution portfolio which consists of Altéa Reservation, Altéa Inventory and Altéa Departure Control solutions.

Finally Star facilitates two separate groups for its global member carriers. The first is a Portfolio Management Advisory group with the objective . . . to identify, evaluate and prioritize future business initiatives that represent opportunities for <u>joint knowledge sharing, sourcing or development</u> with other Star carriers having similar interests. The second is a Project Management Workgroup with the objective to provide a common project management knowledge and collaboration repository to allow members to exchange project management best practices.

12.14 IMPORTANCE OF A PMO IN LOCAL GOVERNMENT

Most people seem to believe that the implementation of a PMO is restricted to large corporations only. Nothing could be further from the truth. The size of a company is irrelevant. The PMO concept in recent years has spread to all levels of government as well. According to Bill Dayton, IT Project Manager, IT Services, Jefferson County, Colorado:

Jefferson County's PMO has played a large role facilitating project success for Information Technology Services projects. The goal of the Project Management Office is to ensure that the standards and processes of Project Management are in place and are followed to ensure the success of projects that exist in the organization.

The PMO reports to the CIO/IT Services Director and consists of 3 Business Analysts and 6 Project Managers.

The PMO follows the model published by Project Management Institute's (PMI) Project Management Body of Knowledge (*PMBOK® Guide*). The PMO is responsible for creating and publishing procedures and processes as well as categorizing and prioritizing incoming projects within the organization.

Jefferson County's PMO currently has over 35 projects in progress with well over 100 proposed projects being considered. Projects types include software upgrades, design and development of software solutions, implementation of COTS solutions and analysis and improvement of business processes. Mobile applications, infrastructure upgrades, financial application upgrades, and document management projects are all on this list.

Project success is measured on a number of factors including adherence to Time and Budget constraints, calculation of earned value and customer feedback throughout the project.

12.15 PROJECT AUDITS AND THE PMO

In recent years, the necessity for a structured independent review of various parts of a business, including projects, has taken on a more important role. Part of this can be attributed to the Sarbanes-Oxley Law compliance requirements. These audits are now part of the responsibility of the PMO.

These independent reviews are audits that focus on either discovery or decision-making. They also can focus on determining the health of a project. The audits can be scheduled or random, and can be performed by in-house personnel or external examiners.

There are several types of audits. Some common types include:

- *Performance Audits:* These audits are used to appraise the progress and performance of a given project. The project manager, project sponsor, or an executive steering committee can conduct this audit.
- *Compliance Audits:* These audits are usually performed by the PMO to validate that the project is using the project management methodology properly. Usually the PMO has the authority to perform the audit but may not have the authority to enforce compliance.
- *Quality Audits:* These audits ensure that the planned project quality is being met and that all laws and regulations are being followed. The quality assurance group performs this audit.
- *Exit Audits:* These audits are usually for projects that are in trouble and may need to be terminated. Personnel external to the project, such as an exit champion or an executive steering committee, conduct the audits.
- *Best Practices Audits:* These audits can be conducted at the end of each life-cycle phase or at the end of the project. Some companies have found that project managers may not be the best individuals to perform the audit. In such situations, the company may have professional facilitators trained in conducting best practices reviews.

Checklists and templates often are the best means of performing audits and health checks. Nani Sadowski-Alvarez, PMP, CPHIMS Senior Management Consultant, Enterprise Program Management & Architecture at Computer Sciences Corporation (CSC), shares with us a template for auditing a project (see Table 12–1).

TABLE 12–1. TEMPLATE FOR AUDITING A PROJECTS

| Project Sponsor:
Project Go
Live Date: | | | Project Manager
Project Final
Audit Date: | |

Validation		Document/Item to be validated	Rating			Comments
Yes ☐	No ☐	**Council Approvals** (i.e., meeting minutes, sign off, etc. indicating that the project was approved and has been signed off on for execution and implementation).	☐	☐	☐	
Yes ☐	No ☐	**Signed Project Scope** (with original signatures and/or faxed/ electronic signatures attached).	☐	☐	☐	
Yes ☐	No ☐	**Project Kickoff Presentation and Kickoff Meeting Agenda** (to include date of kickoff).	☐	☐	☐	
Yes ☐	No ☐	Current/up-to-date **Project Capital and Operating Expense Cost Sheet.**	☐	☐	☐	
Yes ☐	No ☐	All project-specific **Change Requests** (with detail in relation to triple constraints— schedule/budget/resource), with all corresponding change request sign-off approvals attached.	☐	☐	☐	
Yes ☐	No ☐	Signed (by project sponsor) **Project Acceptance—Closure Letter and Operational Turnover.**	☐	☐	☐	
Yes ☐	No ☐	**Project Closure Presentation** along with agenda and date of closure meeting.	☐	☐	☐	
Yes ☐	No ☐	**Vendor Contracts and Statements of Work (SOW)** signed [by all impacted parties—i.e., vendor, sponsor(s), legal, etc.]	☐	☐	☐	
Yes ☐	No ☐	Completed **Project Closure Checklist** upon closure of the project.	☐	☐	☐	

Validation		Document/Item to be validated	Rating			Comments
		Initiation				
Yes ☐	No ☐	**Project Charter and Project Score**	☐	☐	☐	
Yes ☐	No ☐	**Finalized Approved Project Budget**	☐	☐	☐	
Yes ☐	No ☐	Early phase **Workflows** (where applicable) created to demonstrate project need, streamlining of work effort, etc. NOTE: If project is strictly surrounding hardware (HW)/ equipment—diagram provided can be an equipment or netware (NW) diagram.	☐	☐	☐	

Validation		Document/Item to be validated	Rating			Comments
		Planning				
Yes ☐	No ☐	**Signed Contracts** from Vendor along with SOW.	☐	☐	☐	
Yes ☐	No ☐	**Project Schedule** that has been created and maintained via Clarity and Work Bench. (NOTE: Initial signed copy should be included as an addendum to the project scope. PM will show current schedule to auditor in Clarity/WorkBench upon request).	☐	☐	☐	

(Continued)

TABLE 12–1.   *(Continued)*

Validation		Document/Item to be validated	Rating			Comments
		Planning				
Yes ☐	No ☐	**Signed Project Scope**—full document to include project organization chart, communication plan, high-level milestones, technical/business/clinical assessments, as well as project financial materials (when detailed financials are incorporated into the roll-out of projects) etc.	☐	☐	☐	
Yes ☐	No ☐	**Technical Assessment and Relating Diagrams**	☐	☐	☐	
Yes ☐	No ☐	**Provisioning Assessment** (where applicable) to include cost estimates, forecasted equipment, etc.	☐	☐	☐	
Yes ☐	No ☐	**Security Assessment** as established by financial information security officer (FISO).	☐	☐	☐	
Yes ☐	No ☐	**Finalized Approved Project Budget**	☐	☐	☐	
Validation		Document/Item to be validated	Rating			Comments
		Executing/Controlling				
Yes ☐	No ☐	**Design/Build**—Any applicable design/build documentation.	☐	☐	☐	
Yes ☐	No ☐	**Risks and Issues**—PM should document issues and risks via the issues/risk/change request tab in Clarity and upload the affiliated documents to the specific risk or issue.	☐	☐	☐	
Yes ☐	No ☐	**Change Requests and Affiliated Sign-Off's**—The PM should document change requests for their projects, via the issues/risk/change request tab. Change requests will also be uploaded into SharePoint under the applicable project folder.	☐	☐	☐	
Yes ☐	No ☐	**Revised and Enhanced Process Flows and Procedures**	☐	☐	☐	
Yes ☐	No ☐	**Testing**—All project testing related documentation (testing scripts, testing plan, etc.).	☐	☐	☐	
Yes ☐	No ☐	**Training**—All project training related documentation (training plan/schedule, course information, etc.).	☐	☐	☐	
Yes ☐	No ☐	Appropriate measures and forms in place surrounding **Provisioning** (to include any necessary processes for obtaining sign-off for security access).	☐	☐	☐	
Yes ☐	No ☐	**Purchase Order Details** (purchase requests, cost estimates, etc.).	☐	☐	☐	
Yes ☐	No ☐	**All Invoices** (these will also be tracked via Clarity with the review and guidance of the IS accountant).	☐	☐	☐	
Yes ☐	No ☐	**Pre-Go Live Security Assessment** as preformed by FISO.	☐	☐	☐	
Yes ☐	No ☐	**Pre-Go Live Process Flows** (where applicable) detailing current state and what the forecasted state is following the implementation of the project; if the project involves equipment.	☐	☐	☐	

| Yes No | Overview Assessment with IS Accountant | □ □ □ |
| □ □ | | |

| Yes No | **Activation/Go Live Plan** to include all necessary details for the core team and other impacted parties to achieve a successful and efficient project go live. | □ □ □ |
| □ □ | | |

Validation	Document/Item to be validated	Rating	Comments
	Closing		

Yes No	**Project Budget**—Approved budget details vs. final actual closing budget.	□ □ □	
□ □			
Yes No	**Lessons Learned**—As compiled by the PM and the entire project core team.	□ □ □	
□ □			
Yes No	**Operational Turnover** documentation.	□ □ □	
□ □			
Yes No	**Scanned Project Acceptance (Closure) Letter with Sponsor(s) Signature**	□ □ □	
□ □			

Validation	Document/Item to be validated	Rating	Comments
	Agendas/Minutes		

Yes No	All project-related **Agendas** (saved with date format in the title to ensure ease of use for reference).	□ □ □	
□ □			
Yes No	All project-related **Meeting Notes** (saved with date format in title to ensure ease of use for reference).	□ □ □	
□ □			

Validation	Document/Item to be validated	Rating	Comments
	Presentations		

Yes No	**Kickoff PowerPoint Presentation**	□ □ □	
□ □			
Yes No	**Vendor-Related Demos/Presentations**	□ □ □	
□ □			
Yes No	**Internal Presentations Utilized for Project Approval** (if applicable).	□ □ □	
□ □			
Yes No	**Project Closure PowerPoint Presentation**	□ □ □	
□ □			

Summary Scores			
Hard Copies of Documentation			
Electronic Copies			

12.16 PROJECT HEALTH CHECKS

Quite often, projects undergo health checks, but by the wrong people. The concept of performing a health check is sound practice provided the right people are performing the health check and the right information is being discussed. The purpose of the health check should be to provide constructive criticism and evaluate alternative approaches as necessary. Too often, the meetings end up being a personal attack on the project team. Executive-level reviews and likewise reviews by the PMO may not provide the project manager with the constructive information that the project manager desires. Mark Gray and Eric Maurice of NXP have identified an innovative way of doing this. They title this approach: *If two heads are better than one, then why not use three or four?*[8]

In our drive to increase the probability of success for projects, one mechanism that is often used is peer-reviews whereby we get other experts to critically analyse our project and give their judgement and advice. The main problem with this approach is the fact that it is often a one-shot, very brief review of the project management documentation with little or no insight to the actual mechanics of the project.

Eric Maurice, a project manager in the R&D area of NXP Semiconductors has come up with a novel way of getting real value out of the approach—the *multi-brained project manager!* (This is sometimes referred to as a Hydra project manager.)

Triggered by the findings of an MBTI (Myers Briggs Type Indicator) exercise done at the team level, Eric realised that there is a significant danger of becoming too biased towards one perspective of the project with the possibility of overlooking what could be obvious issues. This is further exacerbated by the level of complexity of the projects and the number of (sometimes conflicting) data that the project manager has to consider at the start-up.

Eric then approached several project management peers (from across the organisation) via the local network and asked them to become a part of a neural network—sharing ideas, concepts and viewpoints in the context of this particular project. The reason for using this approach rather than a simple peer review was to overcome the constraint of just having a one-shot input while also showing the added value opportunity.

Of course, to make this work required some preparation and starting conditions:

- Given that the peers were from vastly different project backgrounds, quite a lot of preparation had to go into explaining the project context to the group.
- Ground rules for mutual trust, openness, honesty and constructive criticism were needed (although not formally stated). This was specially helpful in identifying potential weaknesses in the risk plan, and helping to face the (sometimes brutal) truth.
- Overcoming one's own barriers to showing weak points is never easy—this again relies on a good level of trust and cooperation in the group.
- The frequency needs to be reasonably regular—in this case it was once per month (on a 20 month project). This is in order to secure the shared view of the project is kept current.

---

8. The material in this section was provided by Mark Gray, Senior Project Manager at NXP Semiconductors, and Eric Maurice, PMP, Project Manager at NXP Semiconductors.

From the experience we have the following observations and outcomes:

- A tangible outcome was a reduction in the risk level for the project—a review of the risk plan helped to secure the content and response planning.
- Strong ties were built between the peer project managers that in fact remained in operation outside of this project context. This also helped to reinforce the value of networking in the organisation.
- The feedback from the participants was also positive, with appreciation shown for the opportunity to share and to learn from each other.
- An element that was seen as adding to the success was the decision to focus on just one specific topic area for each session (planning, risk . . .). This setting of a "theme" allowed the peers to apply their knowledge (or learn from the others) on a specific focus area in the context of a real and understood project.
- The small but dynamic group (between 3 and 6 people was seen as the ideal size) also served as a real incubator for new ideas, as well as an excellent conduit for lessons learned to be transmitted between projects and across the organisation.

In conclusion we can safely say that the usage of the multi-brained project manager approach has a clear value-added in getting to project execution excellence, much more so than either formal project reviews or the normal peer-review "snapshots". This gain is not only for the project, but also for the participants and the organisation as a whole!

Some Recommendations— And a Health Warning!

Since the process of setting up and running Hydra sessions takes a non-trivial investment in time, some consideration should be given to when this would be appropriate. We have some suggestions for when (and when not) this may be an appropriate approach:

- This process would have a good return on investment where either the project has a very high level of perceived risk, or where the desire is to use it as an opportunity for mentoring (either the project manager is new on the job or the peers have an opportunity to learn from an "adept").
- We would not recommend using this on very short-term projects (a few months duration) as this reduces the possibility of traction, or on projects with a low-level of strategic importance as this will reduce the level of interest from the peers.
- It's not a good idea to have project managers involved in multiple Hydra sessions—not just from point of view of the time required but also this would dilute the focus too much.
- Putting a Hydra approach in place on a project should come from the project manager themselves; forcing it turns it into a chore, or worse still indicates a lack of trust in the project manager.

Some would say that this should be the domain of the PMO (where one exists) but here we would like to give a health warning: The PMO should of course be the person(s) that help put it in place, set it up, and support the capturing of results—but the true value comes from having the peers really involved in the project under scrutiny. In the authors opinion if the Hydra becomes the domain of the PMO it has the risk of becoming the monster of Greek legend . . .

This approach is not intended to become just another "monitoring and controlling" tool—the real benefit is the shared learning and the multiple perspectives on the day-to-day functioning of the project.

12.17 CRITICAL RATIO AS AN INDICATOR OF THE HEALTH OF A PROJECT _____

People that attend training programs for preparation for the PMP Certification Exam are often overwhelmed with the amount of equations that they must learn as part of the earned-value module under the cost control area of knowledge. While some companies use all of the equations, many companies use only those equations that are directly pertinent to their projects. When the formulas are used correctly, they can form the basis for a health check on a project. According to Donald L. Redmond, PMP, Senior Project Manager, IT Services, Jefferson County, Colorado:

> Earned value is a much overlooked method for assessing the health of a project, especially in small organizations such as local governments. But, when you look at the whole suite of factors used in earned value calculations, it can be overwhelming—intimidating, even. As demonstrated by these two equations:

$$EV = \sum_{Start}^{Current} PV \ (Completed)$$

$$IEAV = \sum AC + \frac{BAC - \sum EV}{CPI} = \frac{BAC}{CPI}$$

> It appears to be very complicated and, in reality, it can be. Such complexity is often justified in very large efforts (multimillion dollar construction projects for example), but what about the smaller projects? If only there was a way to apply earned value using a simple method. Well, there is.
>
> The method involves the use of Critical Ratio (critical ratio being the product of Schedule Performance Index and Cost Performance Index) and an enterprise project management tool such as Microsoft Office Project Server©. By setting the thresholds of this value for different levels of project health (e.g., Green, Yellow, and Red), MS Project© will calculate the health for you.
>
> Critical Ratio is easy to determine and, with some simple formulas, rules, and thresholds that you set, can provide a valuable measure of the overall health of a particular project and, indeed, for an entire portfolio of projects when applied in a server environment.

12.18 PMO OF THE YEAR AWARD _____

Some people contend that the most significant change in project management in the first decade of the twenty-first century has been the implementation of the PMO concept. As such, it is no big surprise that The Center for Business Practices initiated the "PMO of the Year" award.[9]

---

9. The material in this section has been provided by The Center for Business Practices, Rockwell Automation, and Alcatel-Lucent. For additional information on the Center for Business Practices and The PMO of The Year Award, visit their website: www.cbponline.com.

Award Criteria   The PMO of the Year Award is presented to the PMO that best
                                      illustrates—through an essay and other documentation—their
project management improvement strategies, best practices, and lessons learned. Additional
support documentation—such as charts, graphs, spreadsheets, brochures, etc.—could not
exceed five documents. While providing additional documentation was encouraged, each
eligible PMO clearly demonstrated its best practices and lessons learned in the awards essay.
Judges reviewed the essays to consider how the applicant's PMO linked project management
to their organization's business strategies and played a role in developing an organizational
project management culture. The essays were judged on validity, merit, accuracy, and consistency in addition to the applicant PMO's contribution to project and organizational success.
     Types of best practices judges looked for include:

- Practices for integrating PMO strategies to manage projects successfully
- Improvements in project management processes, methodologies, or practices lead-
  ing to more efficient and/or effective delivery of the organization's projects
- Innovative approaches to improving the organization's project management capability
- Practices that are distinctive, innovative, or original in the application of project
  management
- Practices that promote an enterprisewide use of project management standards
- Practices that encourage the use of performance measurement results to aid
  decision-making
- Practices that enhance the capability of project managers

Best practice outcomes included:

- Evidence of realized business benefits—customer satisfaction, productivity,
  budget performance, schedule performance, quality, ROI, employee satisfaction,
  portfolio performance, strategic alignment
- Effective use of resources
- Improved organizational project management maturity
- Executive commitment to a project management culture expressed in policies and
  other documentation
- A PMO that exhibits an organizational business results focus
- Effective use of project management knowledge and lessons learned
- Individual performance objectives and potential rewards linked to measurement of
  project success
- Project management functions applied consistently across the organization

Completing the Essay   The essay comprised three sections. Incomplete submissions were
                                      disqualified.

Section 1: Background of the PMO In no more than 1000 words, the applicants
described their PMO, including background information on its scope, vision and mission,
and organizational structure. In addition, they described:

- How long the PMO has been in place
- Their role within the PMO

- How the PMO's operation is funded
- How the PMO is structured (staff, roles and responsibilities, enterprisewide, departmental, etc.)
- How the PMO uses project management standards to optimize its practices

Section 2: PMO Innovations and Best Practices   In no more than 1500 words, the applicants addressed the challenges their organization encountered prior to implementing the new PMO practices and how they overcame those challenges. They described clearly and concisely the practices implemented and their effect on project and organizational success.

Section 3: Impact of the PMO and Future Plans   In no more than 500 words, the applicants described the overall impact of the PMO over a sustained period (e.g., customer satisfaction, productivity, reduced cycle time, growth, building or changing organizational culture, etc.). If available, the applicants provided quantitative data to illustrate the areas in which the PMO had the greatest business impact. Finally, they briefly described their PMO's plans for 2009 and how those plans will potentially impact their organization.

Two of the companies discussed in this book competed for the award: Rockwell Automation, which won the award of the 2009 PMO of the Year, and Alcatel-Lucent, which was recognized as one of the finalists for the award. Both of their profiles are discussed below.

Rockwell Automation: 2009 PMO of the Year Winner

> Software Program Management Office
> Type of organization: manufacturing
> Headquarters: Milwaukee, Wisconsin, USA
> Number of full-time employees (FTEs): 21,000 +
> PMO FTEs: 30
> PMO annual operating budget: $3.2 million
> PMO Director: James C. Brown, Director, A&S Program Management Office
> Presenting challenge: Rolling out a consistent product and project management practice across 16 businesses
> Business benefits: Increased predictability and productivity; faster pace of innovation; delivery of a major release comprising 20 + projects on time and under cost for the first time in company history.
> Website: http://www.rockwellautomation.com/

Rockwell Automation: From "Clean Slate" to Global Innovator in Under Five Years

Rockwell Automation was formed by bringing two major automation companies, Allen-Bradley and Reliance Electric, together in the late 1980s. Over the years, Rockwell Automation has continued to acquire other leading automation suppliers as a growth strategy and also as a way to bring new advanced automation technologies into the company. In 2005, as Rockwell Automation was planning the rollout of a new SAP business system, the company recognized the need for a new, common product development (CPD) process that would be based on company best practices combined with industry best practices for product development. This effort resulted in a CPD process that allowed for enterprisewide adoption. All 16 different product businesses ranging from high-volume component suppliers to complex continuous process control systems solution suppliers now use the same high-level process framework for their new product developments.

Since project managers are instrumental in the execution of a product development process, it was quickly realized that introducing an end-to-end process to a company built from many related but very different product businesses would require consistent application of project management across all the product lines. To complicate matters, each business segment was at a different maturity level relative to all aspects of product development. A formal project management organization, established in 2004, already existed and was capitalized upon to support this effort. Says PMO Director James C. Brown, "If consistency, transparency and risk mitigation are important to your business, and they are for us, then we believe that a formal well recognized and managed project management entity is paramount."

Brown, hired in 2004 to help implement the PMO, called the project management environment "a clean slate," other than those people who were already identified as project managers. The PMO is structured by function with program managers overseeing programs and project managers overseeing projects, assisted by an additional two resources supporting tools such as MSProject Server and Sharepoint.

Rockwell's new PMO ramped up quickly, getting everyone PMP certified within two months, which caught the attention of the senior VP for the division. From there they began establishing processes and methodologies, establishing scorecards and metrics, and deploying tools in support of new product development and services. They moved from a waterfall approach in driving projects to an agile approach, driving processes from 20+ pages to fewer than 5 pages and moving them from notebook binders to electronic media. As Brown says, "We moved from reporting on everything to reporting on exceptions only."

The PMO grew from 10 people to 30 over just four years, and its reach went from North American to global. The number of projects under its direction grew from 12–15 to over 50 concurrently. On the way, Rockwell Automation deployed a portfolio management process in their Architecture and Software Group. The goals and purpose of the process are to link investments to business strategy, maximize the value of the portfolio, achieve a desired balance (mix) of projects, and focus the organization's efforts. The portfolio management process links to related processes, such as idea management, strategy development, program and project management, and the recently deployed CPD process, and has become an integral part of the planning process. Brown focuses on the human side of project portfolio management (PPM)'s benefits: "It's about people reaching consensus using trusted data, and a common decision making framework."

Of course, company culture is hard to change, so governance is critical; and that requires management commitment. The driving force behind management's commitment to implement this new process was the vision of a common consistent methodology for new product development across the enterprise. This consistency was prioritized from the top (direct management involvement in stage gate reviews) down, in order to realize benefit as soon as possible.

All too often, businesses were forced to deny funding for strategic projects due to the never-ending incremental product improvements that just kept coming. By forcing business management to approve each project's passage from one phase to the next, the new processes pushed the visibility of every project, every resource, and every dollar up to the decision-makers who wrestled with trying to find dollars to fund the real game-changers.

This visibility also made it easier for the business owners to kill projects with questionable returns or to delay a project in order to free up critical resources. This contrasted with the old way of executing a project, where reviews were informal and haphazard. Teams were able to continue spending and even overspending without any real fear of cancellation. Under the new process, every dependent organization is represented at the appropriate review and given the chance to agree or disagree with the project manager that all deliverables are available. The intent is to have the go/no-go decision made by both the primary organization responsible for the deliverables during the previous phase and the primary organization responsible for the deliverables in the subsequent phase. Both these organizations are required at each stage gate review. In this way, the process helps Rockwell Automation avoid surprises during the later phases by ensuring transparency during the earlier phases.

All of this has been implemented with a light touch that has eased acceptance of the new processes. As Brown notes, "There is a fine line between rigor and burden, the trick is to push this line hard to insure rigorous implementation without slowing the progress of the project team down."

The PMO has been instrumental in Rockwell Automation's quest for increased predictability, productivity, and visibility. By delivering—for the first time in the company's history—a major release that contained in excess of four programs and 20+ projects, on time and under budget, the organization has proven its business value.

Alcatel-Lucent: 2009 PMO of the Year Finalist

Global Program Management Office
Type of Organization: telecom
Headquarters: Paris, France
Number of full-time employees (FTEs): 70,000
PMO FTEs: 10
PMO annual operating budget: $4.5 million
PMO senior manager: Rich Maltzman, PMP, Senior Manager, Learning and Professional
    Advancement
Presenting challenge: Combining the project management improvement efforts of two
    companies into one supercharged initiative
Business benefits: Improvements in a wide array of project metrics on projects that impact
    customer satisfaction
Website: http:www.alcatel-lucent.com

Alcatel-Lucent: Two Best-Practice Telecoms Unite Their PM Strengths

The Alcatel-Lucent Global Program Management Office (GPMO) combines the best project management practices of Alcatel and Lucent, both of which were already in the midst of major efforts to revitalize project management as a discipline at the time that Lucent merged with Alcatel in November 2006. Both organizations had already researched best practices in project management, and the discipline was given priority by the new company leadership. A core team was assigned to combining the project management efforts into one new initiative in late 2006. The initial focus of the enterprisewide GPMO was on the 2000+ customer-facing project managers who oversee the turnover of new solutions to customers. The GPMO focused on two major "frameworks"—a project delivery framework and a project management development framework.

The project delivery framework, dedicated to the methodologies and tools that project managers use across the company, brings a new level of project management maturity by offering practice consistency across business units and geographical regions. At its core is a gate-based methodology called the contract implementation process (CIP). The CIP points to a collection of tools that can be used as appropriate by the customer-facing project managers on a region-by-region and unit-by-unit basis. Each CIP methodology is traceable to a *PMBOK® Guide* process. The company has adopted Computer Associates' Clarity system as a standard for managing its telecom projects as a portfolio.

The project manager development framework is a nine-piece integrated model that recognizes the interconnectedness between such key project manager development elements such as a competency model, a career path, project manager training, industry certification, internal accreditation and recognition, and project manager skills management. The GPMO has set stringent targets for PMP® certification for its project managers over the next two years. Alcatel-Lucent was featured in PMI's *Leadership in Project Management* annual for its work in this area. The company's depth of commitment to providing a supportive environment for project managers is illustrated by a number of programs, including:

- *Project Management Professional Accreditation.* Alcatel-Lucent has its own program of accreditation, above and beyond the PMP® certification, the General Project Manager Accreditation, an accreditation above and beyond the PMP® that honors excellence in real-word deployment of external customer projects. It requires the completion of an extensive case-based set of advanced project management courseware and is subject to extremely strict criteria, including a formal jury board. This certification helps guarantee that project managers have not only the general project management wisdom needed for their work but also the particular experience and background in actual projects in the telecom field to support customers. Started with the project manager population (2000 people), professional accreditation was so successful that it is now being rolled out to all of the contributors in the services organization—approximately 18,000 people.
- *Competency Model.* The heart of the project management development framework is a competency model that takes the best from the heritage of Alcatel and Lucent. This is a living model, updated every year to keep up with changes in the project management discipline as well as the fast-paced telecom business.
- *RSMS.* The resource and skills management system facilitates project managers' ability to monitor their own progress in development, using their job profile as a basis in identifying skills gaps and suggesting development options to fill those gaps.

- *Alcatel-Lucent University.* As a PMI Registered Education Provider, Alcatel-Lucent University provides access to a wide array of Web-based and instructor-led training, some of which is highly customized and case based to allow project managers to learn from real successful projects.
- *PMP Study Groups.* The GPMO now has 27 PMP study groups, 8–12 individuals who pool their efforts in studying for the PMP exam. A PMP instructor guides the group, which meets at a frequency of their own choosing via teleconference, and completes the study using a recommended book and set of practice questions. The program also benefits the instructor, yielding 10 PDUs, as we have registered this program with PMI as part of Alcatel-Lucent University.
- *International Project Management Day Symposium.* The second annual International Project Management Day Symposium featured presentations from seven countries on project management topics, with speakers such as Dr. Harold Kerzner. About 1000 project managers participated in the symposium, which received excellent feedback scores.

Integration: A Best Practice in Itself

According to Rich Maltzman, PMP, Senior Manager, Learning and Professional Advancement—a role that focuses on the human side of project management, including the career path, training, internal recognition programs, and skills management, "We feel that the integrated nature of the PM Development Framework is a best practice. It forces the interaction between supporting elements of a successful Project Management career, and in turn a PM discipline that can best support customer projects and thus increase the company's financial position."

In addition, some of the primary best practice tools include:

- *Project Delivery Framework.* The CIP, the heart of the delivery framework, is the standardized process for managing the project life cycle in the company and is focused on the handoffs at key gates in that project life cycle. Recognizing that people make projects work, it provides responsibility matrices to show which role is responsible for which activities at each handoff point. The extensive number of tools and processes that it defines for project managers provides a uniform and effective means to manage projects.
- *Project Management Community Building.* The GPMO Web page provides news, executive messages, links to PMP exam preparation resources, CIP, and a frequently upated general project management blog. That blog, Scope Creep, now has almost 100 postings, receiving as many as 1500 unique views, putting and keeping it in the top 10 of the over 100 Alcatel-Lucent Center for Business Practices corporate blogs since its inception about a year ago. The GPMO is experimenting with blogging as a mechanism for project managers to provide "live" lessons learned rather than to rely on a static repository.
- *Project Team of the Year Award.* A dedicated program was established in January 2008 to recognize the most outstanding project teams in each region and in the entire company. The program was designed with peer project managers. Nominations for 32 teams were received and judged by a panel of project managers and executives. During the year, feature stories based on the nominations were placed on the

GPMO and other key corporate websites. Nine finalists were chosen, and from that, a single winner was selected. All of the finalists will receive a team dinner, and the winning team sends a representative to Paris to receive a special award from the Alcatel-Lucent CEO.

- *Maturity Assessment and Improvement.* In 2008, the GPMO began a deliberate measurement of project management maturity using a custom survey tool built from recommendations from the Software Engineering Institute (Carnegie Mellon Univeristy, Capability Maturity Meaurement Integrated, and the Project Management Office Executive Council (part of the Corporate Executive Board). The number of questions was limited to 25, divided among areas such as governance, project performance, resource management, and financials. The response rate was very high (700 respondents) and the answers have yielded mathematical data as well as verbatim feedback that provide a way to map out improvements. The GPMO plans to run the maturity assessment at least once per year, providing a way to track the improvements by aspect area as well as region.

Tracking the Benefits

Enrollment in Resource and Skills Management System (RSMS) has increased in the months since it was introduced to the point where—despite significant turnover in the workforce—well over 90% have actively started managing their skills using the dedicated skills programs customized for the four project manager job profiles. Over 100 new PMPs have been certified thanks to the establishments of targets and the use of the PMP study groups. In addition, 36 new general project manager accreditations were awarded in 2008, an increase of almost 30%.

The GPMO spearheads the dissemination of project management thought leadership by Alcatel-Lucent via:

- Presentations at PMI world congresses
- Article in PMI's *Leadership in Project Management* magazine feature
- Alcatel-Lucent presentations at the PMO Summit (Florida) and the PMO Symposium (Texas)
- Operations committee membership for the fourth edition of *PMBOK® Guide*

Finally, in terms of statistical results, all of the following measures demonstrated improvements over end-of-year 2007 or even within 2008:

- Baseline of projects under financial margin control is up 160 percent.
- Percent of projects with (approved) upscope has nearly doubled.
- Percent of projects with underruns has more than doubled.
- Percent of projects covered by the enterprise project management system went up from 52 to 87 percent within the year 2008.
- Percent of CFPMs with formal project management development plans has gone from 52 to over 80 percent.

Clearly, a dual focus on people and processes has served the company well and assisted the projects under the umbrella of the GPMO to sail through the merger period—a significant achievement in itself.

Six Sigma and the Project Management Office

13.0 INTRODUCTION

In the previous chapter, we discussed the importance of the PMO for strategic planning and continuous improvements. In some companies, the PMO was established specifically for the supervision and management of Six Sigma projects. Six Sigma teams throughout the organization would gather data and make recommendations to the PMO for Six Sigma projects. The Six Sigma project manager, and possibly the team, would be permanently assigned to the PMO.

Unfortunately, not all companies have the luxury of maintaining a large PMO where the Six Sigma teams and other supporting personnel are permanently assigned to the PMO. It is the author's belief that the majority of the PMOs have no more than four or five people permanently assigned. Six Sigma teams, including the project manager, may end up reporting "dotted" to the PMO and administratively "solid" elsewhere in the organization. The PMO's responsibility within these organizations is primarily for the evaluation, acceptance, and prioritization of projects. The PMO may also be empowered to reject recommended solutions to Six Sigma projects.

For the remainder of this chapter we will focus on organizations that maintain small PMO staffs. The people assigned to the PMO may possess a reasonable knowledge concerning Six Sigma but may be neither Green nor Black Belts in Six Sigma. These PMOs can and do still manage selected Six Sigma projects but perhaps not the traditional type of Six Sigma projects taught in the classroom.

13.1 PROJECT MANAGEMENT—SIX SIGMA RELATIONSHIP

Is there a relationship between project management and Six Sigma? The answer is definitely "yes." The problem is how to exchange the benefits such that the benefits of Six Sigma can be integrated into project management and, likewise, the benefits of project management can be integrated into Six Sigma. Some companies, such as EDS, have already recognized this important relationship, especially the input of Six Sigma principles to project management. Doug Bolzman, Consultant Architect, PMP®, ITIL Service Manager at EDS, discusses this relationship:

> We have incorporated the Information Technology Information Library (ITIL) into the information technology enterprise management (ITEM) framework to design the operational model required to maintain and support the release. ITIL operations components are evaluated and included within each release that requires an operational focus. Six Sigma models have been generated to assist the organization in understanding the capabilities of each release and how to manage the requirements, standards, and data for each of the established capabilities.

Today, there is a common belief that the majority of traditional, manufacturing-oriented Six Sigma failures are because of the lack of project management; nobody is managing the Six Sigma projects as projects. Project management provides Six Sigma with structured processes as well as faster and better execution of improvements.

From a project management perspective, problems with Six Sigma Black Belts include:

- Inability to apply project management principles to planning Six Sigma projects
- Inability to apply project management principles to the execution of Six Sigma projects
- Heavy reliance on statistics and minimum reliance on business processes
- Inability to recognize that project management is value added

If these problem areas are not resolved, then Six Sigma failures can be expected as a result of:

- Everyone plans but very few execute improvements effectively.
- There are too many projects in the queue and poor prioritization efforts.
- Six Sigma stays in manufacturing and is not aligned with overall business goals.
- Black Belts do not realize that executing improvements are projects within a project.

Six Sigma people are project managers and, as such, must understand the principles of project management, including statements of work, scheduling techniques, and so on. The best Six Sigma people know project management and are good project managers; Black Belts are project managers.

A possible solution to some of the Six Sigma failures is to require Six Sigma personnel to use the enterprise project management methodology. Jason Schulist, Manager—Continuous Improvement, Operating Strategy Group at DTE Energy, discusses this:

In 2002, DTE Energy developed an Operating System Framework to enable systemic thinking around continuous improvement. We blended a lean tool and Six Sigma implementation strategy to develop our current "Lean Sigma" systemic approach. This approach utilizes a 4-Gate/9-Step project management model.[1]

Members of the various business units submit ideas for projects. A review committee prioritizes the projects within each business unit using a project selection document. Once prioritized, each business unit allocates 1–2 percent of its organizational staff to full-time continuous improvement initiatives. Most of these resources are either Lean Sigma Black Belt certified or in training. These resources use the four-gate/nine-step project management model for all projects.

The 4-Gate/9-Step Management Model

In Steps 1–3, the project lead scopes the project opportunity, forms a team, and analyzes the current reality using rigorous data analysis techniques. By using the $y = f(x)$ tool, teams are able to quantify their metrics and scope their projects to the appropriate level. All Black Belt (BB) certified projects must achieve at least $250,000 in savings or have a significant impact in safety or customer service. The team develops a project charter, which the Champion signs along with the Gate 1 review form.

In Steps 4–6, the team defines the ideal state design that most effectively improves the metrics agreed to in Gate 1. The team identifies gaps and develops countermeasures [using the failure mode and effect analysis (FMEA)] to migrate the initiative from the current state to the ideal state. The team develops a master plan for implementing the changes and commits to targets for each of the metrics. The team measures both input and output metrics and measures the success of the project with respect to improving these metrics. The Champion signs off Gate 2.

In Step 7, the team implements its plan and course corrects as necessary in order to achieve the metrics. At Gate 3 the team reviews its performance to plan, addresses gaps and plan countermeasures, discusses progress to project targets, and ensures that outcomes will be achieved.

In Steps 8 and 9, the team measures project progress, sustains the goals, acknowledges the team, reflects on the project, and communicates the results. The team performs an after-action review (AAR) to inculcate the learning from the project and reduce mistakes in future implementations. The team must achieve the project metrics from Gate 2 before the sponsor signs off at Gate 4. If the team does not reach its targets, the team most likely returns to Gate 2 to redefine the ideal state and confirm that the original targets are still feasible.

The DTE Energy Sarah Sheridan award recognizes many successful completed projects using the 4-Gate/9-Step project management model. Operating systems improvements using the project methodology have saved DTE Energy over $40 million in 2003 and over $100 million in 2004.

---

1. The model is shown in Figures 4–3, 4–4, and 4–40.

13.2 INVOLVING THE PMO

The traditional PMO exists for business process improvements and supports the entire organization, including Six Sigma Black Belts, through the use of the enterprise project management methodology. Project managers, including Black Belts, focus heavily upon customer value-added activities, whether it be an internal or external customer. The PMO focuses on corporate value-added activities.

The PMO can also assist with the alignment of Six Sigma projects with strategy. This includes the following:

- Continuous reprioritization may be detrimental. Important tasks may be sacrificed and motivation may suffer.
- Hedging priorities to appease everyone may result in significant work being prolonged or disbanded.
- A cultural change may be required during alignment.
- Projects and strategy may be working toward cross-purposes.
- Strategy starts at the top whereas projects originate at the middle of the organization.
- Employees can recognize projects but may not be able to articulate strategy. Selecting the proper mix of projects during portfolio management of projects cannot be accomplished effectively without knowing the strategy. This may result in misinterpretation.
- "Chunking" breaks a large project into smaller ones to better support strategy. This makes it easier for revitalization or rejection.

The PMO can also assist in solving some of the problems associated with capturing Six Sigma best practices, such as:

- Introducing a best practice can "raise the bar" too soon and pressure existing projects to possibly implement a best practice that may not be appropriate at that time.
- Employees and managers are unaware of the existence of the best practices and do not participate in their identification.
- Knowledge transfer across the organization is nonexistent and weak at best.
- Falling prey to the superstitious belief that most best practices come from failures rather than from successes.

Simply stated, the marriage of project management with Six Sigma allows us to manage better from a higher level.

13.3 TRADITIONAL VERSUS NONTRADITIONAL SIX SIGMA

In the traditional view of Six Sigma, projects fall into two categories: manufacturing and transactional. Each category of Six Sigma is multifaceted and includes a management strategy, metric, and process improvement methodology. This is shown in Figure 13–1. Manufacturing Six Sigma processes utilize machines to produce products whereas transactional Six Sigma processes utilize people and/or computers to produce services. The process

FIGURE 13–1. Six Sigma categories (traditional view).

improvement methodology facet of Six Sigma addresses both categories. The only difference is what tools you will use. In manufacturing, where we utilize repetitive processes that make products, we are more likely to use advanced statistical tools. In transactional Six Sigma, we might focus more on graphical analysis and creative tools/techniques.

The traditional view of a Six Sigma project has a heavy focus on continuous improvement to a repetitive process or activity associated with manufacturing. This traditional view includes metrics, possibly advanced statistics, rigor, and a strong desire to reduce variability. Most of these Six Sigma projects fit better for implementation in manufacturing than in the PMO. Six Sigma teams manage these manufacturing-related projects.

Not all companies perform manufacturing and not all companies support the PMO concept. Companies without manufacturing needs might focus more on the transactional Six Sigma category. Companies without a PMO rely heavily upon the Six Sigma teams for the management of both categories of projects.

Those companies that do support a PMO must ask themselves the following three questions:

- Should the PMO be involved in Six Sigma projects?
- If so, what type of project is appropriate for the PMO to manage even if the organization has manufacturing capability?
- Do we have sufficient resources assigned to the PMO to become actively involved in Six Sigma project management?

PMOs that are actively involved in most of the activities described in Chapter 12 do not have the time or resources required to support all Six Sigma projects. In such a case, the PMO must be selective as to which projects to support. The projects selected are

FIGURE 13–2.   Six Sigma categories (nontraditional view).

commonly referred to as nontraditional projects that focus more on project management–related activities than manufacturing.

Figure 13–2 shows the nontraditional view of Six Sigma. In this view, operational Six Sigma includes manufacturing activities and all other activities from Figure 13–1, and transactional Six Sigma now contains primarily those activities to support project management.

In the nontraditional view, the PMO can still manage both traditional and nontraditional Six Sigma projects. However, there are some nontraditional Six Sigma projects that are more appropriate for management by the PMO. Some of the projects currently assigned to the PMOs include enhancements to the enterprise project management methodology, enhancements to the PMO tool set, efficiency improvements, and cost avoidance/reduction efforts. Another project assigned to the PMO involves process improvements to reduce the launch of a new product and improving customer management. Experts in Six Sigma might view these as nontraditional types of projects. There is also some concern as to whether these are really Six Sigma projects or just a renaming of a continuous improvement project to be managed by a PMO. Since several companies now refer to these as Six Sigma projects, the author will continue this usage.

Strategic planning for Six Sigma project management is not accomplished merely once. Instead, like any other strategic planning function, it is a cycle of continuous improvements. The improvements can be small or large, measured quantitatively or qualitatively, and designed for either internal or external customers.

There almost always exists a multitude of ideas for continuous improvements. The biggest challenge lies in effective project selection and then assigning the right players. Both of these challenges can be overcome by assigning Six Sigma project management best practices to the project management office. It may even be beneficial having Six Sigma specialists with Green Belts or Black Belts assigned to the PMO.

13.4 UNDERSTANDING SIX SIGMA

Six Sigma is not about manufacturing widgets. It is about a focus on processes. And since the PMO is the guardian of the project management processes, it is only fitting that the PMO have some involvement in Six Sigma. The PMO may be more actively involved in identifying the "root cause" of a problem than in managing the Six Sigma solution to the problem.

Some people contend that Six Sigma has fallen short of expectations and certainly does not apply to activities assigned to a PMO. These people argue that Six Sigma is simply a mystique that some believe can solve any problem. In truth, Six Sigma can succeed or fail but the intent and understanding must be there. Six Sigma gets you closer to the customer, improves productivity, and determines where you can get the biggest returns. Six Sigma is about process improvement, usually repetitive processes, and reducing the margin for human and/or machine error. Error can only be determined if you understand the critical requirements of either the internal or external customer.

There are a multitude of views and definitions of Six Sigma. Some people view Six Sigma as merely the renaming of total quality management (TQM) programs as Six Sigma. Others view Six Sigma as the implementation of rigorous application of advanced statistical tools throughout the organization. A third view combines the first two views by defining Six Sigma as the application of advanced statistical tools to TQM efforts.

These views are not necessarily incorrect but are incomplete. From a project management perspective, Six Sigma can be viewed as simply obtaining better customer satisfaction through continuous process improvement efforts. The customer could be external to the organization or internal. The word "satisfaction" can have a different meaning whether we are discussing external or internal customers. External customers expect products and services that are a high quality and reasonably priced. Internal customers may define satisfaction in financial terms, such as profit margins. Internal customers may also focus on such items as cycle time reduction, safety requirements, and environmental requirements. If these requirements are met in the most efficient way without any non-value-added costs (e.g., fines, rework, overtime), then profit margins will increase.

Disconnects can occur between the two definitions of satisfaction. Profits can always be increased by lowering quality. This could jeopardize future business with the client. Making improvements to the methodology to satisfy a particular customer may seem feasible but may have a detrimental effect on other customers.

The traditional view of Six Sigma focused heavily on manufacturing operations using quantitative measurements and metrics. Six Sigma tool sets were created specifically for this purpose. Six Sigma activities can be defined as operational Six Sigma and transactional Six Sigma. Operational Six Sigma would encompass the traditional view and focus on manufacturing and measurement. Operational Six Sigma focuses more on processes, such as the enterprise project management methodology, with emphasis on continuous improvements in the use of the accompanying forms, guidelines, checklists, and templates. Some people argue that transactional Six Sigma is merely a subset of operational Six Sigma. While this argument has merit, project management and specifically the PMO spend the majority of their time involved in transactional rather than operational Six Sigma.

The ultimate goal of Six Sigma is customer satisfaction, but the process by which the goal is achieved can differ whether we are discussing operational or transactional Six

TABLE 13–1.   GOALS OF SIX SIGMA

Goal[a]	Method of Achievement
Understand and meet customer requirements (do so through defect prevention and reduction instead of inspection)	Improvements to forms, guidelines, checklists, and templates for understanding customer requirements
Improve productivity	Improve efficiency in execution of the project management methodology
Generate higher net income by lowering operating costs	Generate higher net income by streamlining the project management methodology without sacrificing quality or performance
Reduce rework	Develop guidelines to better understand requirements and minimize scope changes
Create a predictable, consistent process	Continuous improvement on the processes

[a] From *The Fundamentals of Six Sigma*, International Institute for Learning, New York, 2008, pp. 1–24.

TABLE 13–2.   GOALS VERSUS FOCUS AREAS

Executive Goals	PMO Focus Areas
Provide effective status reporting	• Identification of executive needs • Effective utilization of information • "Traffic light" status reporting
Reduce the time for planning projects	• Sharing information between planning documents • Effective use of software • Use of templates, checklists, and forms • Templates for customer status reporting • Customer satisfaction surveys • Extensions of the enterprise project management methodology into the customer's organization

Sigma. Table 13–1 identifies some common goals of Six Sigma. The left-hand column lists the traditional goals that fall more under operational Six Sigma, whereas the right-hand column indicates how the PMO plans on achieving the goals.

The goals for Six Sigma can be established at either the executive levels or the working levels. The goals may or may not be able to be completed with the execution of just one project. This is indicated in Table 13–2.

Six Sigma initiatives for project management are designed not to replace ongoing initiatives but to focus on those activities that may have a critical-to-quality and critical-to-customer-satisfaction impact in both the long and short terms.

Operational Six Sigma goals emphasize reducing the margin for human error. But transactional Six Sigma activities managed by the PMO may involve human issues such as aligning personal goals to project goals, developing an equitable reward system for project teams, and project career path opportunities. Fixing people problems is part of transactional Six Sigma but not necessarily of operational Six Sigma.

13.5 SIX SIGMA MYTHS[2]

Ten myths of Six Sigma are given in Table 13–3. These myths have been known for some time but have become quite evident when the PMO takes responsibility for project management transactional Six Sigma initiatives.

Works Only in Manufacturing   Much of the initial success in applying Six Sigma was based on manufacturing applications; however, recent publications have addressed other applications of Six Sigma. Breyfogle[3] includes many transactional/service applications. In GE's 1997 annual report, CEO Jack Welch proudly states that Six Sigma "focuses on moving every process that touches our customers—every product and *service* (emphasis added)—toward near-perfect quality."

Ignores Customer in Search of Profits   This statement is not myth but rather a misinterpretation. Projects worthy of Six Sigma investments should (1) be of primary concern to the customer and (2) have the potential for significantly improving the bottom line. Both criteria must be met. The customer is driving this boat. In today's competitive environment, there is no surer way of going out of business than to ignore the customer in a blind search for profits.

Creates Parallel Organization   An objective of Six Sigma is to eliminate every ounce of organizational waste that can be found and then reinvest a small percentage of those savings to continue priming the pump for improvements. With the large amount of downsizing that has taken place throughout the world during the past decade, there is no room or inclination to waste money through the duplication of functions. Many functions are understaffed as it is. Six Sigma is about nurturing any function that adds significant value to the customer while adding significant revenue to the bottom line.

TABLE 13–3.   TEN MYTHS OF SIX SIGMA

1. Works only in manufacturing
2. Ignores the customer in search of bottom-line benefits
3. Creates a parallel organization
4. Requires massive training
5. Is an add-on effort
6. Requires large teams
7. Creates bureaucracy
8. Is just another quality program
9. Requires complicated, difficult statistics
10. Is not cost effective

2. Adapted from F. W. Breyfogle III, J. M. Cupello, and B. Meadows, *Managing Six Sigma,* Wiley, New York, 2001, pp. 6–8.
3. F. W. Breyfogle, III, *Implementing Six Sigma; Smarter Solutions Using Statistical Methods,* Wiley, New York, 1999.

Requires Massive Training   Peter B. Vaill states:

> Valuable innovations are the positive result of this age (we live in), but the cost is likely to be continuing system disturbances owing to members' nonstop tinkering. Permanent white water conditions are regularly taking us all out of our comfort zones and asking things of us that we never imagined would be required. It is well for us to pause and think carefully about the idea of being continually catapulted back into the beginner mode, for that is the real meaning of being a continual learner. We do not need competency skills for this life. We need incompetency skills, the skills of being effective beginners.

Is an Add-On Effort   This is simply the myth "creates a parallel organization" in disguise. Same question, same response.

Requires Large Teams   There are many books and articles in the business literature declaring that teams have to be small if they are to be effective. If teams are too large, the thinking goes, a combinational explosion occurs in the number of possible communication channels between team members, and hence no one knows what the other person is doing.

Creates Bureaucracy   A dictionary definition of bureaucracy is "rigid adherence to administrative routine." The only thing rigid about wisely applied Six Sigma methodology is its relentless insistence that the customer needs to be addressed.

**Is Just Another Quality   Based upon the poor performance of untold quality programs during the
Program**   past three to five decades,[4] an effective quality program would be welcome. More to the point,[5] Six Sigma is "an entirely new way to manage an organization."

**Requires Complicated,   There is no question that a number of advanced statistical tools are
Difficult Statistics**   extremely valuable in identifying and solving process problems. We believe that practitioners need to possess an analytical background and understand the wise use of these tools but do not need to understand all the mathematics behind the statistical techniques. The wise application of statistical techniques can be accomplished through the use of statistical analysis software.

Is Not Cost Effective   If Six Sigma is implemented wisely, organizations can obtain a very high rate of return on their investment within the first year.

---

4. J. Micklethwait and A. Wooldridge, *The Witch Doctors of the Management Gurus,* Random House, New York, 1997.
5. T. Pyzdek, "Six Sigma Is Primarily a Management Program, *Quality Digest,* 1999, p. 26.

13.6 USE OF ASSESSMENTS

One of the responsibilities that can be assigned to a PMO is the portfolio management of projects. Ideas for potential projects can originate anywhere in the organization. However, ideas specifically designated as transactional Six Sigma projects may need to be searched out by the PMO.

One way to determine potential projects is through an assessment. An assessment is a set of guidelines or procedures that allows an organization to make decisions about improvements, resource allocations, and even priorities. Assessments are ways to:

- Examine, define, and possibly measure performance opportunities
- Identify knowledge and skills necessary for achieving organizational goals and objectives
- Examine and solve performance gap issues
- Track improvements for validation purposes

A gap is the difference between what currently exists and what it should be. The gaps can be in cost, time, quality, and performance or efficiency. Assessments allow us to pinpoint the gap and determine the knowledge, skills, and abilities necessary to compress the gap. For project management gaps, the assessments can be heavily biased toward transactional rather than operational issues, and this could easily result in behavior modification projects.

There are several factors that must be considered prior to performing an assessment. These factors might include:

- Amount of executive-level support and sponsorship
- Amount of line management support
- Focus on broad-based applications
- Determining who to assess
- Bias of the participants
- Reality of the answers
- Willingness to accept the results
- Impact on internal politics

The purpose of the assessment is to identify ways to improve global business practices first and functional business practices second. Because the target audience is usually global, there must exist unified support and understanding for the assessment process and that it is for the best interest of the entire organization. Politics, power, and authority issues must be put aside for the betterment of the organization.

Assessments can take place at any level of the organization. These can be:

- Global organizational assessments
- Business unit organizational assessments
- Process assessments
- Individual or job assessments

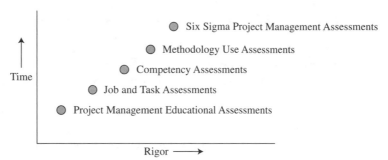

FIGURE 13–3. Time and effort expended.

● Customer feedback assessments (satisfaction and improvements)

There are several tools available for assessments. A typical list might include:

● Interviews
● Focus groups
● Observations
● Process maps

Assessments for Six Sigma project management should not be performed unless the organization believes that opportunities exist. The amount of time and effort expended can be significant, as shown in Figure 13–3.

The advantages of assessment can lead to significant improvements in customer satisfaction and profitability. However, there are disadvantages, such as:

● Costly process
● Labor intensive
● Difficulty in measuring which project management activities can benefit from assessments
● May not provide any meaningful benefits
● Cannot measure a return on investment from assessments

Assessments can have a life of their own. There are typical life-cycle phases for assessments. These life-cycle phases may not be aligned with the life-cycle phases of the enterprise project management methodology and may be accomplished more informally than formally. Typical assessment life-cycle phases include:

● Gap or problem recognition
● Development of the appropriate assessment tool set
● Conducting the assessment/investigation
● Data analyses
● Implementation of the changes necessary
● Review for possible inclusion in the best practices library

TABLE 13–4. SCALES

Strongly agree	Under 20%
Agree	Between 20 and 40%
Undecided	Between 40 and 60%
Disagree	Between 60 and 80%
Strongly disagree	Over 80%

Determining the tool set can be difficult. The most common element of a tool set is a focus on questions. Types of questions include:

- Open ended
  - Sequential segments
  - Length
  - Complexity
  - Time needed to respond
- Closed ended
  - Multiple choice
  - Forced choices (yes–no, true–false)
  - Scales

Table 13–4 illustrates how scales can be set up. The left-hand column solicits a qualitative response and may be subjective whereas the right-hand column would be a quantitative response and more subjective.

It is vitally important that the assessment instrument undergo pilot testing. The importance of pilot testing would be:

- Validate understanding of the instructions
- Ease of response
- Time to respond
- Space to respond
- Analysis of bad questions

13.7 PROJECT SELECTION

Six Sigma project management focuses on continuous improvements to the enterprise project management methodology. Identifying potential projects for the portfolio is significantly easier than getting them accomplished. There are two primary reasons for this:

- Typical PMOs may have no more than three or four employees. Based upon the activities assigned to the PMO, the employees may be limited as to how much time they can allocate to Six Sigma project management activities.
- If functional resources are required, then the resources may be assigned first to those activities that are mandatory for the ongoing business of the firm.

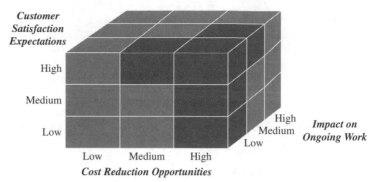

FIGURE 13–4. Project selection cube.

The conflict between ongoing business and continuous improvements occurs frequently. Figure 13–4 illustrates this point. The ideal Six Sigma project management activity would yield high customer satisfaction, high cost reduction opportunities, and significant support for the ongoing business. Unfortunately, what is in the best interest of the PMO may not be in the best, near-term interest of the ongoing business.

All ideas, no matter how good or how bad, are stored in the "idea bank." The ideas can originate from anywhere in the organization, namely:

- Executives
- Corporate Six Sigma champions
- Project Six Sigma champions
- Master Black Belts
- Black Belts
- Green Belts
- Team members

If the PMO is actively involved in the portfolio management of projects, then the PMO must perform feasibility studies and cost–benefit analyses on projects together with prioritization recommendations. Typical opportunities can be determined using Figure 13–5. In this figure, ΔX represents the amount of money (or additional money) being spent. This is the input to the evaluation process. The output is the improvement, ΔY, which is the benefits received or cost savings. Consider the following example.

Convex Corporation   Convex Corporation identified a possible Six Sigma project involving the streamlining of internal status reporting. The intent was to eliminate as much paper as possible from the bulky status reports and replace it with color-coded "traffic light" reporting using the company intranet. The PMO used the following data:

- Burdened hour at the executive level $240
- Typical number of project status review meetings per project 8
- Duration per meeting 2 hours

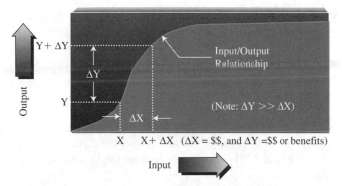

FIGURE 13–5. Six Sigma quantitative evaluation.

- Number of executives per meeting 5
- Number of projects requiring executive review 20

Using the above information, the PMO calculated the total cost of executives as:

(8 meetings) × (5 executives) × (2 hr/meeting) × ($240/hr) × (20 projects) = $384,000

Convex assigned one systems programmer (burdened at $100/hr) for four weeks. The cost for adding traffic light reporting to the intranet methodology was $16,000.

Six months after implementation, the number of meetings had been reduced to five per project for an average of 30 minutes in duration. The executives were now focusing on only those elements of the project that were color coded as a potential problem. On a yearly basis, the cost for the meetings on the 20 projects was now about $60,000. In the first year alone, the company identified a savings of $324,000 for one investment of $16,000.

13.8 TYPICAL PMO SIX SIGMA PROJECTS

Projects assigned to the PMO can be operational or transactional but mainly the latter. Typical projects might include:

- *Enhanced Status Reporting:* This project could utilize traffic light reporting designed to make it easier for customers to analyze performance. This could be intranet based. The intent is to achieve paperless project management. The colors could be assigned based upon problems, present or future risks, or title, level, and rank of the audience.
- *Use of Forms:* The forms should be user friendly and easy to complete. Minimal input by the user should be required and the data inputted into one form should service multiple forms if necessary. Nonessential data should be eliminated. The forms should be cross-listed to the best practices library.

- *Use of Checklists/Templates:* These documents should be comprehensive yet easy to understand. They should be user friendly and easy to update. The forms should be flexible such that they can be adapted to all situations.
- *Criteria of Success/Failure:* There must exist established criteria for what constitutes success or failure on a project. There must also exist a process that allows for continuous measurement against these criteria as well as a means by which success (or failure) can be redefined.
- *Team Empowerment:* This project looks at the use of integrated project teams, the selection of team members, and the criteria to be used for evaluating team performance. This project is designed to make it easier for senior management to empower teams.
- *Alignment of Goals:* Most people have personal goals that may not be aligned with goals of the business. This includes project versus company goals, project versus functional goals, project versus individual goals, project versus professional goals, and other such alignments. The greater the alignment between goals, the greater the opportunity for increased efficiency and effectiveness.
- *Measuring Team Performance:* This project focuses on ways to uniformly apply critical success factors and key performance indicators to team performance metrics. This also includes the alignment of performance with goals and rewards with goals. This project may interface with the wage and salary administration program by requiring two-way and three-way performance reviews.
- *Competency Models:* Project management job descriptions are being replaced with competency models. A competency criterion must be established, including goal alignment and measurement.
- *Financial Review Accuracy:* This type of project looks for ways of including the most accurate data into project financial reviews. This could include transferring data from various information systems such as earned-value measurement and cost accounting.
- *Test Failure Resolution:* Some PMOs maintain a failure-reporting information system that interfaces with FMEA. Unfortunately, failures are identified but there may be no resolution on the failure. This project attempts to alleviate this problem.
- *Preparing Transitional Checklists:* This type of project is designed to focus on transition or readiness of one functional area to accept responsibility. As an example, it may be possible to develop a checklist on evaluating the risks or readiness of transitioning the project from engineering to manufacturing. The ideal situation would be to develop one checklist for all projects.

This list is by no means comprehensive. However, the list does identify typical projects managed by the PMO. Some conclusions can be reached by analyzing this list. First, the projects can be both transactional and operational. Second, the majority of the projects focus on improvements to the methodology. Third, having people with Six Sigma experience (i.e., Green, Brown, or Black Belts) would be helpful.

When a PMO takes the initiative in Six Sigma project management, the PMO may develop a Six Sigma toolbox exclusively for the PMO. These tools most likely will not include the advanced statistics tools that are used by Black Belts in manufacturing but may be more process-oriented tools or assessment tools.

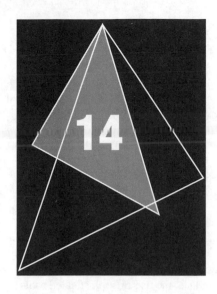

Project Portfolio Management

14.0 INTRODUCTION

Your company is currently working on several projects and has a waiting list of an additional 20 projects that they would like to complete. If available funding will support only a few more projects, how does a company decide which of the 20 projects to work on next? This is the project portfolio management process.It is important to understand the difference between project management and project portfolio management. Debra Stouffer and Sue Rachlin have made this distinction for IT projects[1]:

> An IT portfolio is comprised of a set or collection of initiatives or projects. Project management is an ongoing process that focuses on the extent to which a specific initiative establishes, maintains, and achieves its intended objectives within cost, schedule, technical and performance baselines.
>
> Portfolio management focuses attention at a more aggregate level. Its primary objective is to identify, select, finance, monitor, and maintain the appropriate mix of projects and initiatives necessary to achieve organizational goals and objectives.
>
> Portfolio management involves the consideration of the aggregate costs, risks, and returns of all projects within the portfolio, as well as the various tradeoffs among them. Of course, the portfolio manager is also concerned about the "health" and well being of each project that is included within the IT portfolio. After all, portfolio decisions, such as whether to fund a new project or continue to finance an ongoing one, are based on information provided at the project level.
>
> Portfolio management of projects helps determine the right mix of projects and the right investment level to make in each of them. The outcome is a better balance between ongoing and new strategic initiatives. Portfolio management is not a series of project-specific calculations such as ROI, NPV, IRR, payback period,

---

1. D. Stouffer and S. Rachlin, "A Summary of First Practices and Lessons Learned in Information Technology Portfolio Management," prepared by the Chief Information Officer (CIO) Council, Washington, DC, March 2002, p. 7.

and cash flow and then making the appropriate adjustment to account for risk. Instead, it is a decision-making process for what is in the best interest of the entire organization.

Portfolio management decisions are not made in a vacuum. The decision is usually related to other projects and several factors, such as available funding and resource allocations. In addition, the project must be a good fit with other projects within the portfolio and with the strategic plan.

The selection of projects could be based upon the completion of other projects that would release resources needed for the new projects. Also, the projects selected may be constrained by the completion date of other projects that require deliverables necessary to initiate new projects. In any event, some form of a project portfolio management process is needed.

14.1 WHY USE PORTFOLIO MANAGEMENT?

Not all companies assign the same degree of importance to portfolio management. In some companies, it is a manual process while in others it mandates the use of sophisticated tools. Some companies believe that portfolio management is part of the strategic planning efforts while others see it as a support function for capacity planning. Carl Manello, PMP, Solutions Lead—Program and Project Management, Slalom Consulting, believes that:

> Portfolio management is critical to ensure that companies are spending their limited resources in the best possible way. I have worked with companies at various levels to implement some type of portfolio management. In some cases, portfolio management is simply an extension of the annual budgetary process. In the best cases, it develops into a fully vetted assessment and on-going prioritization process.
>
> But there is no one best approach to portfolio management. The process of portfolio management has varying degrees to which it can be implemented. For example, one can choose to put [a] process in place with or without a sophisticated project portfolio management tool. Even the process by which prioritization is conducted may vary. Depending on the capability of the organization, the approach may be an inventory of current and proposed projects with a forced ranking created by the PMO. For the more sophisticated enterprise, a full blown process may engage senior and executive management into a negotiation to decide where to invest scarce corporate resources based on business unit, strategic objective and projected returns. Figure 14–1 illustrates a typical portfolio matrix. At one company, I worked to develop a relatively simplistic numerical algorithm for assessing projects. After criteria were established for *Business Importance* and *Complexity* (these were the two measures defined by the Enterprise Program Office and the Chief Financial Officer), each project was plotted on a simple two by two matrix. This visualization offered us the ability to compare and contrast dissimilar projects with very little effort.

14.2 INVOLVEMENT OF SENIOR MANAGEMENT, STAKEHOLDERS, AND THE PMO

The successful management of a project portfolio requires strong leadership by individuals who recognize the benefits that can be accrued from portfolio management. The commitment by senior management is critical. Stouffer and Rachlin comment on the role of senior management in an IT environment in government agencies[2]:

---

2. Ibid., p. 8.

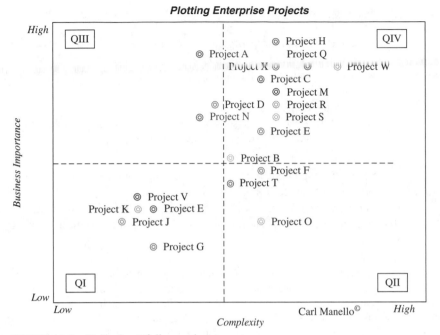

FIGURE 14–1. Typical portfolio matrix.

Portfolio management requires a business and an enterprise-wide perspective. However, IT investment decisions must be made both at the project level and the portfolio level. Senior government officials, portfolio and project managers, and other decision makers must routinely ask two sets of questions.

First, at the project level, is there sufficient confidence that new or ongoing activities that seek funding will achieve their intended objectives within reasonable and acceptable cost, schedule, technical, and performance parameters?

Second, at the portfolio level, given an acceptable response to the first question, is the investment in one project or a mix of projects desirable relative to another project or a mix of projects?

Having received answers to these questions, the organization's senior officials, portfolio managers, and other decision makers then must use the information to determine the size, scope, and composition of the IT investment portfolio. The conditions under which the portfolio can be changed must be clearly defined and communicated. Proposed changes to the portfolio should be reviewed and approved by an appropriate decision making authority, such as an investment review board, and considered from an organization-wide perspective.

Senior management is ultimately responsible for clearly defining and communicating the goals and objectives of the project portfolio as well as the criteria and conditions considered for the portfolio selection of projects. According to Stouffer and Rachlin, this includes[3]:

---

3. Ibid., p. 13.

- Adequately define and broadly communicate the goals and objectives of the IT portfolio.
- Clearly articulate the organization's and management's expectations about the type of benefits being sought and the rates of returns to be achieved.
- Identify and define the type of risks that can affect the performance of the IT portfolio, what the organization is doing to avoid and address risk, and its tolerance for ongoing exposure.
- Establish, achieve consensus, and consistently apply a set of criteria that will be used among competing IT projects and initiatives.

Senior management must also collect and analyze data in order to assess the performance of the portfolio and determine whether or not adjustments are necessary. This must be done periodically such that critical resources are not being wasted on projects that should be canceled. Stouffer and Rachlin provide insight on this through their interviews[4]:

> According to Gopal Kapur, President of the Center for Project Management, organizations should focus on their IT portfolio assessments and control meetings on critical project vital signs. Examples of these vital signs include the sponsor's commitment and time, status of the critical path, milestone hit rate, deliverables hit rate, actual cost versus estimated cost, actual resources versus planned resources, and high probability, high impact events. Using a red, yellow, or green report card approach, as well as defined metrics, an organization can establish a consistent method for determining if projects are having an adverse impact on the IT portfolio, are failing and need to be shut down.
>
> Specific criteria and data to be collected and analyzed may include the following:
>
> - Standard financial measures, such as return on investment, cost benefit analysis, earned value (focusing on actuals versus plan, where available), increased profitability, cost avoidance, or payback. Every organization participating in the interviews included one or more of these financial measures.
> - Strategic alignment (def.ined as mission support), also included by almost all organizations.
> - Client (customer) impact, as defined in performance measures.
> - Technology impact (as measured by contribution to, or impact on, some form of defined architecture).
> - Initial project and (in some cases) operations and schedules, as noted by almost all organizations.
> - Risks, risk avoidance (and sometimes risk mitigation specifics), as noted by almost all participants.
> - Basic project management techniques and measures.
> - And finally, data sources and data collection mechanisms also are important. Many organizations interviewed prefer to extract information from existing systems; sources include accounting, financial, and project management systems.

---

4. Ibid., p. 18.

One of the best practices identified by Stouffer and Rachlin for IT projects was careful consideration of both internal and external stakeholders[5]:

Expanding business involvement in portfolio management often includes the following:

- Recognizing that the business programs are critical stakeholders, and improving that relationship throughout the life cycle
- Establishing service level agreements that are tied to accountability (rewards and punishment)
- Shifting the responsibilities to the business programs and involving them on key decision making groups

In many organizations, mechanisms are in place to enable the creation, participation and "buy-in" of stakeholder coalitions. These mechanisms are essential to ensure the decision making process is more inclusive and representative. By getting stakeholder buy-in early in the portfolio management process, it is easier to ensure consistent practices and acceptance of decisions across an organization. Stakeholder participation and buy-in can also provide sustainability to portfolio management processes when there are changes in leadership.

Stakeholder coalitions have been built in many different ways depending on the organization, the process and the issue at hand. By including representatives from each major organizational component who are responsible for prioritizing the many competing initiatives being proposed across the organization, all perspectives are included. The approach, combined with the objectivity brought to the process by using pre-defined criteria and a decision support system, ensures that everyone has a stake in the process and the process is fair.

Similarly, the membership of the top decision making body is comprised of senior executives from across the enterprise. All major projects, or those requiring a funding source, must be voted upon and approved by this decision making body. The value of getting stakeholder participation at this senior level is that this body works toward supporting the organization's overall mission and priorities rather than parochial interests.

More and more companies today are relying heavily upon the PMO for support with portfolio management. Typical support activities include capacity planning, resource utilization, business case analysis, and project prioritization. The role of the PMO in this regard is to support senior management, not to replace them. Portfolio management will almost always remain as a prime responsibility for senior management, but recommendations and support by the PMO can make the job of the executive a little easier. In this role, the PMO may function as more of a facilitator. Chuck Millhollan, Director of Program Management at Churchill Downs Inc. (CDI), describes portfolio management in his organization:

Our PMO is responsible for the portfolio management process and facilitates portfolio reviews by our "Investment Council." We have purposefully separated the processes for

---

5. Ibid., pp. 22–23.

requesting and evaluating projects (having projects approved in principle) and authorizing work (entry into the active portfolio).

When asked to describe the PMO's relationship to portfolio management, Chuck Millhollan commented:

Investment Council: The Investment Council is comprised of senior (voting) members (CEO, COO, CFO, EVPs) and representatives from each business unit. There are regularly scheduled monthly meetings, facilitated by the PMO, to review and approve new requests and review the active portfolio. The Investment Council's goals and objectives include:

1. Prioritize & allocate capital to projects.
2. Approve/disapprove requested projects based on the merit of the associated Business Case.
3. Act individually and collectively as vocal and visible project champions throughout their representative organizations.
4. As necessary, take an active role in approving project deliverables, helping resolve issues and policy decisions, and providing project related direction and guidance.

Request, Evaluation & Approval: We use an "Investment Request Worksheet" to standardize the format in which projects (called investment requests) are presented to the Investment Council. Elements include request description, success criteria and associated metrics, a description of the current and future state, alignment to strategic goals, preliminary risk assessment, identification of dependent projects, preliminary resource availability and constraint assessment and a payback analysis for ROI and Cost-Out initiatives.

Work Authorization: If projects were approved during the annual operational planning processes and are capital investments that generate ROI or result in a Cost-Out, come back to the Investment Council for work authorization and addition to the portfolio of active projects. This can be done concurrently with request, evaluation and approval for projects that are initiated mid-planning cycle.

Portfolio Maintenance: We use a bi-weekly project status reporting process and only include projects that the Investment Council has identified as requiring portfolio review and/or oversight. The portfolio reports are provided bi-weekly and presented monthly during the Investment Council meetings.

When the PMO supports or facilitates the portfolio management process, the PMO becomes an active player in the strategic planning process and supports senior management by making sure that the projects in the queue are aligned with strategic objectives. The role might be support or monitoring and control. Enrique Sevilla Molina, PMP, Corporate PMO Director at Indra, discusses portfolio management in his organization:

Portfolio management is strongly oriented to monitor and control the portfolio performance, and to review its alignment with the strategic planning. A careful analysis of trends and forecasts is also periodically performed, so the portfolio composition may be assessed and reoriented if required.

Once the strategic targets for the portfolio have been defined and allocated through the different levels in the organization, the main loop of the process includes reporting, reviewing and taking actions on portfolio performance, problems, risks, forecasts and new contracts planning. A set of alerts, semaphores and indicators have been defined and automated in order to focus the attention on the main issues related with the portfolio management. Those projects or proposals marked as requiring specific attention are carefully followed by the management team, and a specific status reporting is provided for those.

One of the key tools used for [the] portfolio management process is our Projects Monitor. It is a web based tool that provides a full view of the status of any predefined set of projects (or portfolio), including general data, performance data, indicators and semaphores. It has also the capability to produce different kinds of reports, at single project level, at portfolio level, or a specialized risk report for the selected portfolio.

Besides the corporate PMO, major Business Units throughout the company use local PMOs in their portfolio management process. Some of them are in charge of risk status reporting for the major projects or programs in the portfolio. Others are in charge of an initial definition of the risk level for the projects and operations in order to provide an early detection of potential risk areas. And others play a significant role in providing the specific support to the portfolio managers when reporting the status to upper level management.

Our corporate level PMO defines the portfolio management processes in order to be consistent with the project management level and, in consequence, the requirements for the implementation of those processes in the company tools and information systems.

Some companies perform portfolio management without involvement by the PMO. This is quite common when portfolio management might include a large amount of capital spending projects. According to a spokesperson at AT&T:

> Our PMO is not part of portfolio management. We maintain a Portfolio Administration Office (PAO) which approves major capital spending projects and programs through an annual planning process. The PAO utilizes change control for any modifications to the list of approved projects. Each Project Manager must track the details of their project and update information in the Portfolio Administration Tool (PAT). The Corporate Program Office uses data in PAT to monitor the health and well being of the projects. Individual projects are audited to ensure adherence to processes and reports are prepared to track progress and status.

14.3 PROJECT SELECTION OBSTACLES[6]

Portfolio management decision-makers frequently have much less information to evaluate candidate projects than they would wish. Uncertainties often surround the likelihood of success for a project, the ultimate market value of the project, and its total cost to completion. This lack of an adequate information base often leads to another difficulty: the lack of a systematic approach to project selection and evaluation. Consensus criteria and

---

6. W. Souder, *Project Selection and Economic Appraisal*, Van Nostrand Reinhold, New York, 1984, pp. 2–3.

methods for assessing each candidate project against these criteria are essential for rational decision-making. Although most companies have established organizational goals and objectives, they are usually not detailed enough to be used as criteria for project portfolio management decision-making. However, they are an essential starting point.

Portfolio management decisions are often confounded by several behavioral and organizational factors. Departmental loyalties, conflicts in desires, differences in perspectives, and an unwillingness to openly share information can stymie the project selection, approval, and evaluation processes. Much project evaluation data and information is necessarily subjective in nature. Thus, the willingness of the parties to openly share and put trust in each other's opinions becomes an important factor.

The risk-taking climate or culture of an organization can also have a decisive bearing on the project selection process. If the climate is risk adverse, high-risk projects may never surface. Attitudes within the organization toward ideas and the volume of ideas being generated will influence the quality of the projects selected. In general, the greater the number of creative ideas generated, the greater the chances of selecting high-quality projects.

14.4 IDENTIFICATION OF PROJECTS

The overall project portfolio management process is a four-step approach, as shown in Figure 14–2. The first step is the identification of the ideas for projects and needs to help support the business. The identification can be done through brainstorming sessions, market research, customer research, supplier research, and literature searches. All ideas, regardless of merit, should be listed.

Because the number of potential ideas can be large, some sort of classification system is needed. There are three common methods of classification. The first method is to place the projects into two major categories, such as survival and growth. The sources and types of funds for these two categories can and will be different. The second method comes from typical R&D strategic planning models, as shown in Figure 14–3. Using this

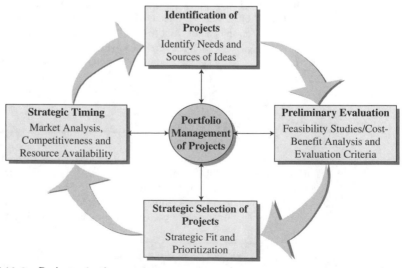

FIGURE 14–2. Project selection process.

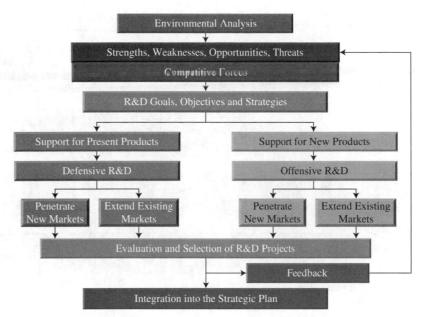

FIGURE 14–3. R&D strategic planning process.

approach, projects to develop new products or services are classified as either offensive or defensive projects. Offensive projects are designed to capture new markets or expand market share within existing markets. Offensive projects mandate the continuous development of new products and services.

Defensive projects are designed to extend the life of existing products or services. This could include add-ons or enhancements geared toward keeping present customers or finding new customers for existing products or services. Defensive projects are usually easier to manage than offensive projects and have a higher probability of success.

Another method for classifying projects would be:

● Radical technical breakthrough projects
● Next-generation projects
● New family members
● Add-ons and enhancement projects

Radical technological breakthrough projects are the most difficult to manage because of the need for innovation. Figure 14–4 shows a typical model for innovation. Innovation projects, if successful, can lead to profits that are many times larger than the original development costs. Unsuccessful innovation projects can lead to equally dramatic losses, which is one of the reasons why senior management must exercise due caution in approving innovation projects. Care must be taken to identify and screen out inferior candidate projects before committing significant resources to them.

There is no question that innovation projects are the most costly and difficult to manage. Some companies mistakenly believe that the solution is to minimize or limit the total number of ideas for new projects or to limit the number of ideas in each category. This could be a costly mistake.

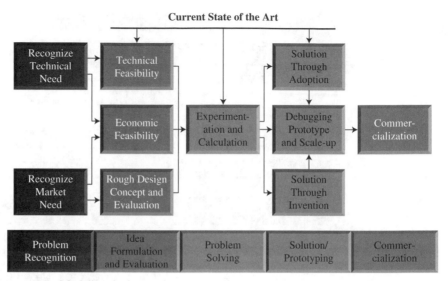

FIGURE 14–4. Modeling the innovation process.

In a study of the new-product activities of several hundred companies in all industries, Booz, Allen, and Hamilton[7] defined the new-product evolution process as the time it takes to bring a product to commercial existence. This process began with company objectives, which included fields of product interest, goals, and growth plans, and ended with, hopefully, a successful product. The more specifically these objectives were defined, the greater guidance would be given to the new-product program. This process was broken down into six manageable, fairly clear sequential stages:

- *Exploration:* The search for product ideas to meet company objectives.
- *Screening:* A quick analysis to determine which ideas were pertinent and merit more detailed study.
- *Business Analysis:* The expansion of the idea, through creative analysis, into a concrete business recommendation, including product features, financial analysis, risk analysis, market assessment, and a program for the product.
- *Development:* Turning the idea-on-paper into a product-in-hand, demonstrable and producible. This stage focuses on R&D and the inventive capability of the firm. When unanticipated problems arise, new solutions and trade-offs are sought. In many instances, the obstacles are so great that a solution cannot be found, and work is terminated or deferred.
- *Testing:* The technical and commercial experiments necessary to verify earlier technical and business judgments.
- *Commercialization:* Launching the product in full-scale production and sale; committing the company's reputation and resources.

7. *Management of New Products*, Booz, Allen & Hamilton, 1984, pp. 180–181.

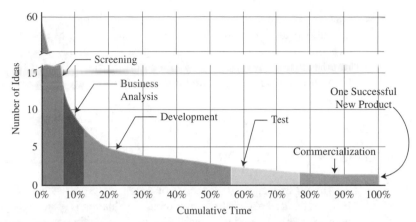

FIGURE 14-5. Mortality of new product ideas.

In the Booz, Allen & Hamilton study, the new-product process was characterized by a decay curve for ideas, as shown in Figure 14–5. This showed a progressive rejection of ideas or projects by stages in the product evolution process. Although the rate of rejection varied between industries and companies, the general shape of the decay curve is typical. It generally takes close to 60 ideas to yield just one successful new product.

The process of new-product evolution involves a series of management decisions. Each stage is progressively more expensive, as measured in expenditures of both time and money. Figure 14–6 shows the rate at which expense dollars are spent as time accumulates for the average project within a sample of leading companies. This information

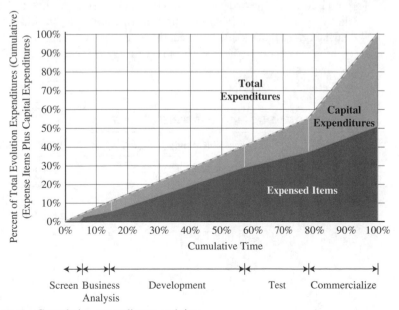

FIGURE 14–6. Cumulative expenditures and time.

was based on an all-industry average and is therefore useful in understanding the typical industrial new-product process. It is significant to note that the majority of capital
expenditures are concentrated in the last three stages of evolution. It is therefore very
important to do a better job of screening for business and financial analysis. This will
help eliminate ideas of limited potential before they reach the more expensive stages of
evolution.

14.5 PRELIMINARY EVALUATION

As shown in Figure 14–2, the second step in project selection is preliminary evaluation.
From a financial perspective, preliminary evaluation is basically a two-part process. First, the
organization will conduct a feasibility study to determine whether the project can be done.
The second part is to perform a benefit-to-cost analysis to see whether the company should
do it (see Table 14–1).

The purpose of the feasibility study is to validate that the idea or project meets feasibility of cost, technological, safety, marketability, and ease of execution requirements. It is
possible for the company to use outside consultants or subject matter experts (Smells) to
assist in both feasibility studies and benefit-to-cost analyses. A project manager may not be
assigned until after the feasibility study is completed because the project manager may not
have sufficient business or technical knowledge to contribute prior to this point in time.

If the project is deemed feasible and a good fit with the strategic plan, the project
is prioritized for development along with other approved projects. Once feasibility is
determined, a benefit-to-cost analysis is performed to validate that the project will, if executed correctly, provide the required financial and nonfinancial benefits. Benefit-to-cost
analyses require significantly more information to be scrutinized than is usually available
during a feasibility study. This can be an expensive proposition.

Estimating benefits and costs in a timely manner is very difficult. Benefits are often
defined as:

TABLE 14–1.   FEASIBILITY STUDENTS AND COST–BENEFIT ANALYSES

	Feasibility Study	Cost–Benefit Analysis
Basic question	Can we do it?	Should we do it?
Life-cycle phase	Preconceptual	Conceptual
Project manager selected	Not yet	Perhaps
Analysis	Qualitative	Quantitative
	• Technical	• NPV
	• Cost	• Discounted cash flow (DCF)
	• Quality	• IRR
	• Safety	• ROI
	• Legal	• Assumptions
	• Economical	• Reality
Decision criteria	Strategic fit	Benefits > cost

- Tangible benefits, for which dollars may be reasonably quantified and measured
- Intangible benefits, which may be quantified in units other than dollars or may be identified and described subjectively

Costs are significantly more difficult to quantify, at least in a timely and inexpensive manner. The minimum costs that must be determined are those that are used specifically for comparison to the benefits. These include:

- The current operating costs or the cost of operating in today's circumstances.
- Future period costs that are expected and can be planned for.
- Intangible costs that may be difficult to quantify. These costs are often omitted if quantification would contribute little to the decision-making process.

There must be careful documentation of all known constraints and assumptions that were made in developing the costs and the benefits. Unrealistic or unrecognized assumptions are often the cause of unrealistic benefits. The go or no-go decision to continue with a project could very well rest upon the validity of the assumptions.

14.6 STRATEGIC SELECTION OF PROJECTS

From Figure 14–2, the third step in the project selection process is the strategic selection of projects, which includes the determination of a strategic fit and prioritization. It is at this point where senior management's involvement is critical because of the impact that the projects can have on the strategic plan.

Strategic planning and the strategic selection of projects are similar in that both deal with the future profits and growth of the organization. Without a continuous stream of new products or services, the company's strategic planning options may be limited. Today, advances in technology and growing competitive pressure are forcing companies to develop new and innovative products while the life cycle of existing products appears to be decreasing at an alarming rate. Yet, at the same time, executives may keep research groups in a vacuum and fail to take advantage of the potential profit contribution of R&D strategic planning and project selection.

There are three primary reasons that corporations work on internal projects:

- To produce new products or services for profitable growth
- To produce profitable improvements to existing products and services (i.e., cost reduction efforts)
- To produce scientific knowledge that assists in identifying new opportunities or in "fighting fires"

Successful project selection is targeted, but targeting requires a good information system, and this, unfortunately, is the weakest link in most companies. Information systems are needed for optimum targeting efforts, and this includes assessing customer and market needs, economic evaluation, and project selection.

Assessing customer and market needs involves opportunity-seeking and commercial intelligence functions. Most companies delegate these responsibilities to the marketing group, and this may result in a detrimental effort because marketing groups appear to be overwhelmed with today's products and near-term profitability. They simply do not have the time or resources to adequately analyze other activities that have long-term implications. Also, marketing groups may not have technically trained personnel who can communicate effectively with the R&D groups of the customers and suppliers.

Most organizations have established project selection criteria, which may be subjective, objective, quantitative, qualitative, or simply a seat-of-the-pants guess. The selection criteria are most often based upon suitability criteria, such as:

- Similar in technology
- Similar marketing methods used
- Similar distribution channels used
- Can be sold by current sales force
- Will be purchased by the same customers as current products
- Fits the company philosophy or image
- Uses existing know-how or expertise
- Fits current production facilities
- Both research and marketing personnel enthusiastic
- Fits the company long-range plan
- Fits current profit goals

In any event, there should be a valid reason for selecting the project. Executives responsible for selection and prioritization often seek input from other executives and managers before moving forward. One way to seek input in a quick and reasonable manner is to transform the suitability criteria shown above into rating models. Typical rating models are shown in Figures 14–7, 14–8, and 14–9.[8] These models can be used for both strategic selection and prioritization.

Prioritization is a difficult process. Factors such as cash flow, near-term profitability, and stakeholder expectations must be considered. Also considered are a host of environmental forces, such as consumer needs, competitive behavior, existing or forecasted technology, and government policy.

Being highly conservative during project selection and prioritization could be a road map to disaster. Companies with highly sophisticated industrial products must pursue an aggressive approach to project selection or risk obsolescence. This also mandates the support of a strong technical base.

14.7 STRATEGIC TIMING

Many organizations make the fatal mistake of taking on too many projects without regard for the limited availability of resources. As a result, the highly skilled labor is assigned

---

8. W. Souder, *Project Selection and Economic Appraisal,* Van Nostrand Reinhold, New York, 1984, pp. 66–69.

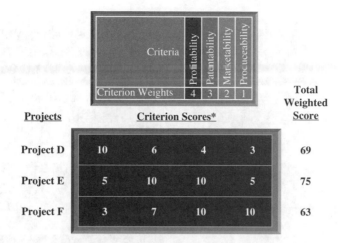

Projects	Criterion Scores*				Total Weighted Score
	Profitability	Patentability	Marketability	Procuceability	
Criterion Weights	4	3	2	1	
Project D	10	6	4	3	69
Project E	5	10	10	5	75
Project F	3	7	10	10	63

Total Weighted Score = Σ (Criterion Score × Criterion Weight)
* Scale: 10=Excellent;   1=Unacceptable

FIGURE 14–7. Scoring model.

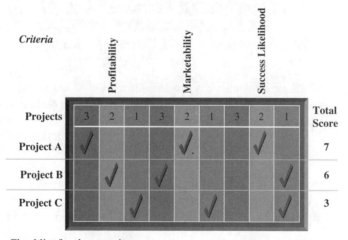

FIGURE 14–8. Checklist for three projects.

to more than one project, creating schedule slippages, lower productivity, less than antici-pated profits, and never-ending project conflicts.

The selection and prioritization of projects must be made based upon the availability of qualified resources. Planning models are available to help with the strategic timing of resources. These models are often referred to as *aggregate planning models*.

Another issue with strategic timing is the determination of which projects require the best resources. Some companies use a risk–reward cube, where the resources are assigned based upon the relationship between risk and reward. The problem with this approach is that the time required to achieve the benefits (i.e., payback period) is not considered.

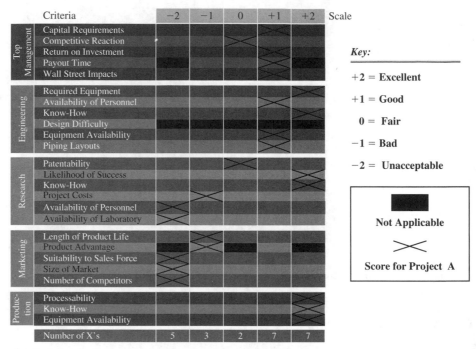

FIGURE 14–9. Scaling model for one project, project A.

Aggregate planning models allow an organization to identify the overcommitment of resources. This could mean that high-priority projects may need to be shifted in time or possibly be eliminated from the queue because of the unavailability of qualified resources. It is a pity that companies also waste time considering projects for which they know that the organization lacks the appropriate talent.

Another key component of timing is the organization's tolerance level for risk. Here, the focus is on the risk level of the portfolio rather than the risk level of an individual project. Decision makers who understand risk management can then assign resources effectively such that the portfolio risk is mitigated or avoided.

14.8 ANALYZING THE PORTFOLIO

Companies that are project-driven organizations must be careful about the type and quantity of projects they work on because of the resources available. Because of critical timing, it is not always possible to hire new employees and have them trained in time or to hire subcontractors that may end up possessing questionable skills.

---

9. This type of portfolio was adapted from the life-cycle portfolio model commonly used for strategic planning activities.

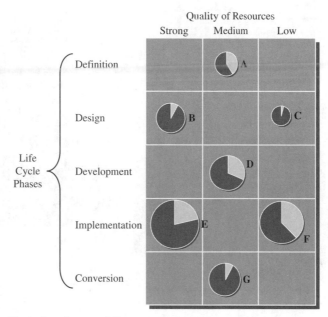

FIGURE 14–10. Typical project portfolio.

Figure 14–10 shows a typical project portfolio.[9] Each circle represents a project. The location of each circle represents the quality of resources and the life-cycle phase that the project is in. The size of the circle represents the magnitude of the benefits relative to other projects, and the pie wedge represents the percentage of the project completed thus far.

In Figure 14–10, project A has relatively low benefits and uses medium-quality resources. Project A is in the definition phase. However, when project A moves into the design phase, the quality of resources may change to low-quality or strong quality. Therefore, this type of chart has to be updated frequently.

Figures 14–11, 14–12, and 14–13 show three types of portfolios. Figure 14–11 represents a high-risk project portfolio where strong resources are required on each project. This may be representative of project-driven organizations that have been awarded large, highly profitable projects. This could also be a company in the computer field that competes in an industry that has short product life cycles and where product obsolescence occurs six months downstream.

Figure 14–12 represents a conservative, profit portfolio where an organization works on low-risk projects that require low-quality resources. This could be representative of a project portfolio selection process in a service organization or even a manufacturing firm that has projects designed mostly for product enhancement.

Figure 14–13 shows a balanced portfolio with projects in each life-cycle phase and where all levels of resources are being utilized, usually quite effectively. A very delicate juggling act is required to maintain this balance.

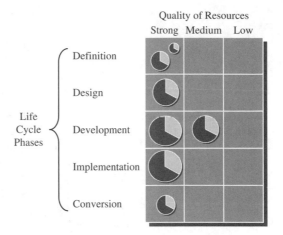

FIGURE 14–11. High-risk portfolio.

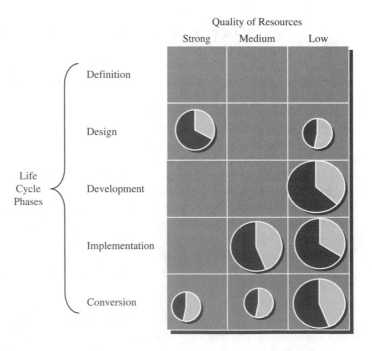

FIGURE 14–12. Profit portfolio.

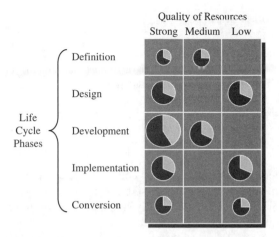

FIGURE 14–13. Balanced portfolio.

14.9 PROBLEMS WITH MEETING EXPECTATIONS

Why is it that, more often than not, the final results of either a project or an entire portfolio do not meet senior management's expectations? This problem plagues many corporations and the blame is ultimately (and often erroneously) rationalized as poor project management practices. As an example, a company approved a portfolio of 20 R&D projects for 2001. Each project was selected on its ability to be launched as a successful new product. The approvals were made following the completion of the feasibility studies. Budgets and timetables were then established such that the cash flows from the launch of the new products would support the dividends and the cash needed for ongoing operations.

Full-time project managers were assigned to each of the 20 projects and began with the development of detailed schedules and project plans. For eight of the projects, it quickly became apparent that the financial and scheduling constraints imposed by senior management were unrealistic. The project managers on these eight projects decided not to inform senior management of the potential problems but to wait awhile to see if contingency plans could be established. Hearing no bad news, senior management was left with the impression that all launch dates were realistic and would go as planned.

The eight trouble-plagued projects were having a difficult time. After exhausting all options and failing to see a miracle occur, the project managers then reluctantly informed senior management that their expectations would not be met. This occurred so late in the project life cycle that senior management became quite irate and several employees had their employment terminated, including some of the project sponsors.

Several lessons can be learned from this situation. First, unrealistic expectations occur when financial analysis is performed from "soft" data rather than "hard" data. In Table 14–1 we showed the differences between a feasibility study and a benefit-to-cost analysis. Generally speaking, feasibility studies are based upon soft data.

TABLE 14–2.   COST/HOUR ESTIMATES

Estimating Method	Generic Type	WBS Relationship	Accuracy	Time to Prepare
Parametric	ROM[a]	Top down	25% to + 75%	Days
Analogy	Budget	Top down	−10% to + 25%	Weeks
Engineering (grass roots)	Definitive	Bottom up	−5% to + 10%	Months

[a] Rough order of magnitude.

Therefore, critical financial decisions based upon feasibility study results may have significant errors. This can also be seen from Table 14–2, which illustrates the accuracy of typical estimates. Feasibility studies use top-down estimates that can contain significant error margins.

Benefit-to-cost analyses should be conducted from detailed project plans using more definitive estimates. Benefit-to-cost analysis results should be used to validate that the financial targets established by senior management are realistic.

Even with the best project plans and comprehensive benefit-to-cost analyses, scope changes will occur. There must be a periodic reestimation of expectations performed on a timely basis. One way of doing this is by using the rolling wave concept shown in Figure 14–14. The rolling wave concept implies that as you get further along in the project, more knowledge is gained, which allows us to perform more detailed planning and estimating. This then provides additional information from which we can validate the original expectations.

Continuous reevaluation of expectations is critical. At the beginning of a project, it is impossible to ensure that the benefits expected by senior management will be realized at project completion. The length of the project is a critical factor. Based upon project length, scope changes may result in project redirection. The culprit is most often changing economic conditions, resulting in invalid original assumptions. Also, senior management must be made aware of events that can alter expectations. This information must be made known quickly. Senior management must be willing to hear bad news and have the courage to possibly cancel a project.

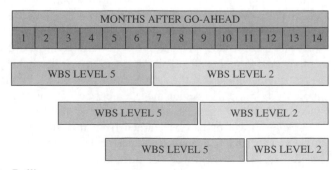

FIGURE 14–14.   Rolling wave concept.

Since changes can alter expectations, project portfolio management must be integrated with the project's change management process. According to Mark Forman, the Associate Director for IT and e-Government in the Office of Management and Budget[10]:

Many agencies fail to transform their process for IT management using the portfolio management process because they don't have change management in place before starting. IT will not solve management problems—re-engineering processes will. Agencies have to train their people to address the cultural issues. They need to ask if their process is a simple process. A change management plan is needed. This is where senior management vision and direction is sorely needed in agencies.

Although the comments here are from government IT agencies, the problem is still of paramount importance in nongovernment organizations and across all industries.

14.10 PORTFOLIO MANAGEMENT AT ROCKWELL AUTOMATION[11]

Rockwell Automation has deployed a portfolio management process in their Architecture and Software Group. The goals and purpose of the process are to link investments to our business strategy, maximize the value of the portfolio, achieve a desired balance (mix) of projects, and focus the organization's efforts. The Portfolio Management process puts strategic focus on how we manage our investment dollars, by becoming an integral part of our planning process. It is about people reaching consensus using trusted data, and a common decision making framework. The Portfolio Management Process links to related processes, such as Idea Management, Strategy Development, Program and Project Management and our recently deployed Common Product Development Process. (See Figure 14–15.)

Investment Proposals are qualified through a common Concept Scorecard, which is a dynamic spreadsheet that quantifies and scores the attractiveness of a concept through an Investment Proposal, and if approved is utilized to quantify a project in the stage gate process, which includes funding events. The Data build for decision-making starts with less early and builds to more later as the accuracy and certainty of estimates improve. The sum of all non-funded (proposals) and funded (projects) are managed through the Ranked Ordered Concept List, which is fed from the Concept Scorecard. (See Figure 14–16.)

14.11 SYNOVUS FINANCIAL PROJECT PORTFOLIO IMPACT ASSESSMENT

Most people seem to equate a best practice as the way we use a special form, guideline, template, or checklist. But best practices can also be the way the company uses a complete set of processes such as with an enterprise project management methodology. At Synovus

---

10. See note 1, p. 1.
11. This section on Rockwell Automation was provided by James C. Brown, PgMP, PMP, OPM3 AC, MPM, CIPM, CSP, CSSMBB, Director, A&S Enterprise Program Management Office; Karen Wojala, Manager, Business Planning; and Matt Stibora, Lean Enterprise Manager.

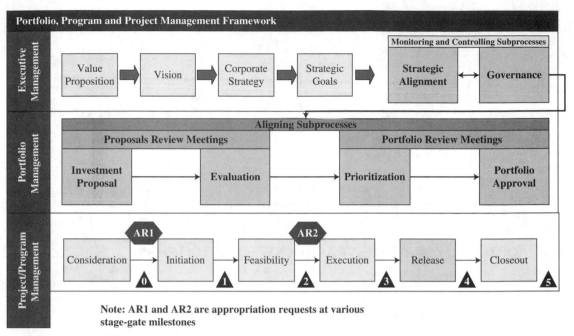

FIGURE 14–15. Portfolio management and a common product development stage gate process.

Financial Corporation, the entire Project Portfolio Impact Assessment Process is viewed as a best practice. Scot E. Hanley, recipient of the 2008 Kerzner Project Management of the Year Award, provides us with an insight of how this impact assessment process works at Synovus.[12]

> The Synovus Project Management Office was established in the company's Technology Division as a strategic initiative to improve project delivery to the business, manage risk and control project related costs. The initial implementation of the Project Management Office in the Synovus Technology Division involved many challenges, foremost of which was an understanding of the impact from new projects to the current projects underway. Lacking a formal estimation process for forecasting project schedules, and absent documented statements of work (SOWs), the Technology Division was left to make educated guesses on [a] timeline and updated Division forecasts, leading to missed customer expectations, cost overruns and other issues.

The PMO's solution was the creation of the "Project Portfolio Impact Assessment" with a key dependency of the SOW process with documented SOWs. A summary of the impact assessment process is presented below.

---

12. Material provided by Scot Hanley, formerly Director, Professional Services Organization, Synovus Financial Corporation. Reproduced by permission of Synovus Financial Corporation.

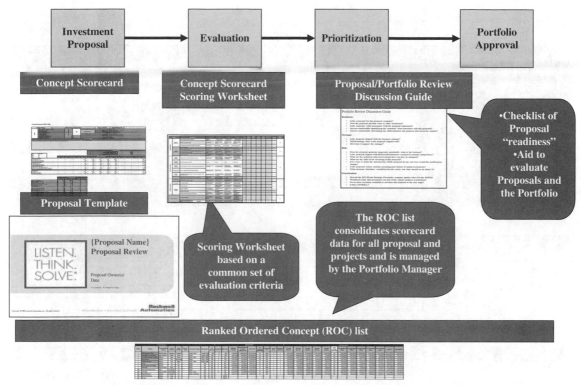

FIGURE 14–16. Project management process and common template overview.

Best Practice Summary

- The Synovus *Busines–IT Partner (BITP)* (a PMO department dedicated to working closely with the business to define new projects) receives notification from the business that there is a new project and consults with the business to document the project scope for the appropriate phase (concept, design, implementation, etc.)

- The BITP works with all pertinent stakeholders (including sponsors, resource managers, analysts, etc.) to flesh out high-level scope, assumptions, dependencies, resource estimates, timeline, and other factors

- The BITP submits the documented results in the form of a statement of work (SOW) for submission to the project management office (PMO). See Figure 14–17.

- The PMO validates the SOW and loads relevant data from the SOW into a portfolio management tool, Planview Enterprise, for review by the PMO. See Figure 14–18.

- Before and after analysis of all project schedules, resource allocation is conducted across hundreds of projects to determine any impacts to the current project portfolio forecast (see Figure 14–19). Impacted projects, staffing bottlenecks, revised costs, and forecasts are documented for review by the implementation steering commit-

STATEMENT OF WORK

Project Number:

General Information			
Project Name:		Date Submitted:	
Business Unit Name and CC#:		Project Manager:	
Business Unit Sponsor:		IT Contact:	
Accountability Manager:		Date Last Updated:	
Requested Finish Date:		Version:	

Statement of Work Phases:				
#	Name	Est.Start	Est.Finish	Notes/Comments
1	Concept/Overview			
2	Collaboration			
3	Design			
4	Implementation			
5	Close-out			

Project Attributes:			
Is Project TCO > $250,000 or requires more than 1000 Hours?	☒ YES ☐ NO	Are there Regulatory/Compliance Drivers?	☒ YES ☐ NO
Project Level Rating	☐ Level 1 ☐ Level 2 ☐ Level 3 ☐ Level 4		

NOTE: Total Cost of Ownership (TCO) estimates should take into consideration all direct and indirect expenses over the life (or amortization schedule) of the asset. You are not expected to include TCO figures for the Concept paper; however you should be able to provide an Order of Magnitude to Budget Estimate at the completion of the Collaboration Phase.

Divisions Involved: (Indicate which Divisions are involved in completing the project)			
☐ Accounting/Finance ☐ Banking Affiliates ☐ Card Services ☐ Compliance ☐ Corporate Services/CPD	☐ FMS Companies ☐ Innovative Solutions ☐ Insurance ☐ Leasing	☐ Legal ☐ Loan Administration ☐ Mortgage (SMC) ☐ Operations ☐ Sales, Mktg, Product Dev	☐ S-Link ☐ Team Services (HR) ☐ Technology Division ☒ All
Field Impact: (Indicate which areas will be impacted by the final implementation of the project)			
Branch Personnel ☐ Tellers ☐ Personal Bankers ☐ Branch Managers ☐ Lenders/Lending Staff	Bank Admin/Support ☐ CFO/Staff ☐ Compliance Officer/Staff ☐ Marketing ☐ Sales Managers ☐ Executive Group	FMS/Mortgage/Other ☐ Brokers ☐ Trust Officers ☐ Financial Planners ☐ Mortgage Lenders/Staff ☐ Leasing	☒ All Team Members

FIGURE 14–17. Two pages from the Synovus SOW illustrating (page 1) basic information regarding the project and (page 2) a table of deliverables including the resource name, estimated hours, duration, due date, and other pertinent information.

IT Departments Involved: (For ITSC use)			
☒ Business Analyst ☐ Business Intelligence ☐ Business Systems	☐ Channel Tech Solution ☐ Customer Support ☐ DBA ☐ Information Security	☐ IT Architecture ☐ Network Engineering ☐ Network Operations ☐ Network Services ☐ Portal Team ☒ PMO	☐ Procurement ☐ Production Control ☐ Quality Assurance ☐ Systems Engineering ☒ Web Development

In Scope Project Deliverables:

*NOTE: Information loaded into the below table should specifically identify the milestones and/or deliverables for the project and phase, the role/resource responsible assigned to the task, and estimated date the deliverable will be complete. In addition to other project, compliance, and/or policy related deliverables, **the below data should compliment the standard PHASE deliverables listed in the PSO Solutions Framework**. For projects with an existing project plan and/or a WBS, the data in this table should map to the milestones/deliverables in the plan.*

Deliverable	Resource	Billable	Estimated Hours	Estimated Due Date	Calendar Duration	Status	Remarks/Notes
1.		☐					
2.		☐					
3.		☐					
4.		☐					
5.		☐					
6.		☐					
7.		☐					
8.		☐					
9.		☐					
10.		☐					
11.		☐					
12.		☐					
13.		☐					
14.		☐					
15.		☐					

FIGURE 14–18. Planview Enterprise™ is used for loading and managing all project schedule information from the SOW.

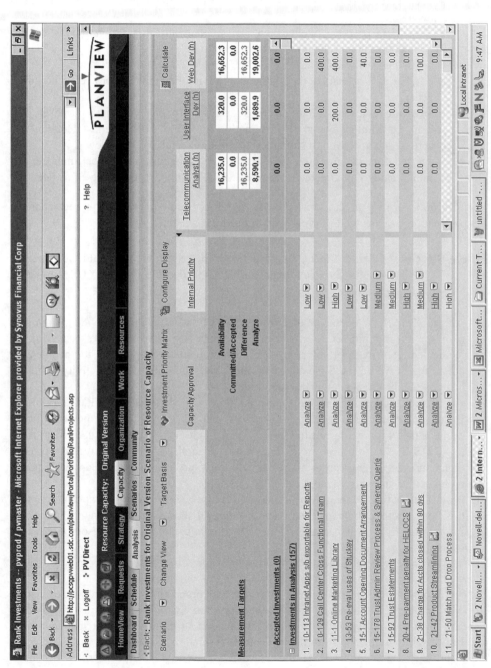

FIGURE 14–19. Planview Enterprise Portfolio Management (EPM) module used for analyzing impact across projects and conducting "what if" analysis.

tee (composed of Synovus business directors). Impacts that are identified include: Team members required to work on more than one project at the same time (a staffing bottleneck)

- Team members whose forecasted work exceeds the time they have available to work
- Changes to vendor timelines that are outside contractual obligations or require renegotiation on scheduling vendor resources
- Missing skill sets that are not available in the current resource pool
- "What If" scenarios that illustrate the implications of project scheduling changes
- Impacts are reviewed with the steering committee for change control purposes. The steering committee reviews the information listed above and business decisions are proposed and approved or submitted to senior executives for review.
- A revised forecast and project schedules are completed for project managers to communicate to all affected team members. The process repeats itself as needed.

It is noteworthy that the same process above is used for material project changes (scope, vendors etc.) as well as project termination.

Best Practice Critical Success Factors (CSFs)   In order for this best practice to be repeatable and yield the expected results, Synovus Financial has identified the following CSFs must be in place:

1. A formalized SOW document and process must be workable and provide meaningful planning information before an impact assessment best practice can be successful.
2. Dedicated team members must be organized to focus on supporting the business to create the SOW. In the Synovus experience, the separation of this activity from the business analyst responsibility is required.
3. An analytical tool, such as the Enterprise Portfolio Management (EPM) module of Planview Enterprise is required to enable "what if" analysis across the portfolio or projects (see Figure 14–19). In our experience, attempts to use nonspecialized tools (e.g., MS Excel) result in errors, delays, and generally an inability to analyze more than a half-dozen large projects in an acceptable time. Use of an EPM analyst ensures repeatability and uniform application of technique.
4. Project estimation techniques must become a competency of the organization attempting this best practice. Synovus's experience has been successful in leveraging historical information from project actuals combined with project manager and resource manager estimates.
5. The business must be engaged and active in the decision-making process.

Additionally, it is recommended that this best practice is integrated with existing corporate project governance bodies. For example, at Synovus, the results of this process are used by the Synovus IT steering committee, a group of senior managers that regularly examines the IT portfolio of projects, including capacity, forecast, projected ROI, and cost. Based upon the results of each impact assessment, informed decisions and recommendations are made by the committee for communication to the business and other governance bodies (such as the implementation leads mentioned above). These can include (1) changes to existing project schedules, (2) projects to be delayed or closed, (3) suggested scope changes, (4) staffing recommendations, and other items.

Tight integration and communication across all project governance teams ensure complete transparency of the information, thus driving accountability to the organization in timely decision-making aimed at creating the best results possible for the company.

. . .

Global Project Management Excellence

15.0 INTRODUCTION

In the previous chapters, we discussed excellence in project management, the use of project management methodologies, and the hexagon of excellence. Many companies previously described in the book have excelled in all of these areas. In this chapter we will focus on six companies, namely IBM, Computer Associates, Microsoft, Deloitte, Johnson Controls, and Siemens, all of which have achieved specialized practices and characteristics related to in-depth globalized project management:

- They are multinational.
- They sell business solutions to their customers rather than just products or services.
- They recognize that, in order to be a successful solution provider, they must excel in project management.
- They recognize that that they must excel in all areas of project management rather than just one area.
- They recognize that a global project management approach must focus more on a framework, templates, checklists, forms, and guidelines, rather than rigid policies and procedures, and that the approach can be used equally well in all countries and for all clients.
- They understand the necessity of having project management tools to support their project management approach.
- They understand that, without continuous improvement in project management, they could lose clients and market share.
- They regard project management as a strategic competency.

These characteristics can and do apply to all of the companies discussed previously but they are of the highest importance to multinational companies.

15.1 IBM[1]

Culture

In his book, *Who Says Elephants Can't Dance*, IBM's Chairman, Lou Gerstner wrote:

- The issue, in essence, is: How do we deal with the IBM matrix? How can we execute our strategies in such a complex company?
- . . . . IBM's unique value to customers is our ability to build integrated solutions. . . .
- . . . our ability to integrate products, skills and insight is our *only* sustainable competitive advantage.
- . . . there is no long-term, sustainable competitive advantage in technology.
- The question then is: what kind of management system supports our fundamental strategy of integration? . . . The classic chain-of-command system is not only too slow for the pace of this industry, but it opposes our ability to work across organizations. Hierarchy, which seems to support integration, actually fights it. Hierarchy erects vertical lines and boundaries and fosters the infamous "silo" mentality.
- . . . a well-managed matrix is highly fluid and adaptable. Roles change often. Teams form and disband. Decisions about which business unit will lead in any particular situation are not codified. This puts a premium on the judgment of leaders at every level.[2]

Industry dynamics such as worldwide competition, resource pressures, rapid change in customer segments, and technology drove IBM to a different organizational structure than its traditional hierarchical management approach. Additionally, IBM identified the lack of good project management as a major contributor to project failure and customer satisfaction issues across the corporation. These factors led IBM to decide to become a project-based enterprise, applying and integrating project management across its core businesses, processes and systems.

In November 1996, IBM consolidated its efforts to become a project-based enterprise and established the Project Management Center of Excellence (PM/COE). The Project Management Center of Excellence charter is to drive IBM's transition to and support of professional project management worldwide, a competency deemed necessary to effectiveness and success within a matrix enterprise.

The Charter: Raising Project Management to a Core Competence in IBM

IBM identified project management as key to its ability to reliably deliver its commitments. As in many businesses in the 1990s, the lack of good project management contributed to failure to deliver commitments on time, within budget, and with appropriate profit. Analysis of project management capabilities against industry best practices showed that IBM:

2. L. V. Gerstner, "Who Says Elephants Can't Dance?" Copyright ©2002, Harper Business Publishers, an Imprint of Harper Collins Publishers, Appendix A, pp. 313–314.

- Often did not recognize a work effort as a project and, therefore, did not apply the disciplines needed to deliver the work successfully;
- Did not have skilled project managers or a pipeline to cover project opportunities; and
- Had an organizational culture and systems (management and business) that were not project-based and, therefore, did not support, and sometimes actually impeded, project work.

IBM took definitive steps to change its culture and support systems to improve its business posture. The company took bold steps in how to organize, execute, and track work and began to organize work into projects that produced services, products, and solutions for our customers. This approach was applied to both external and internal development projects. Since its inception, the IBM Project Management Center of Excellence, working hand in hand with all business units worldwide, worked to ensure that IBM became a project-based enterprise that applies and integrates project management disciplines into all core business processes and systems. (See Figure 15–1.) The initial charter drove more consistent and broader use of project management disciplines, important to IBM's ability to meet customer needs. However, project management cannot be leveraged without pervasive organizational competence in project

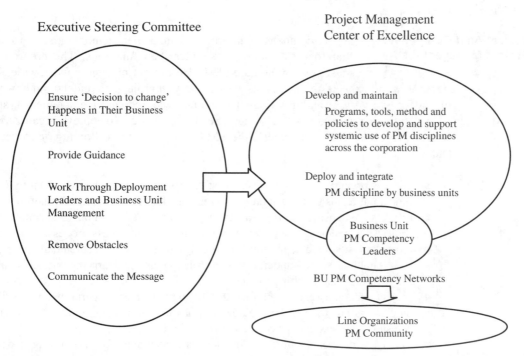

FIGURE 15–1. Involvement at all levels of the enterprise is critical to the systemic acceptance and use of project management disciplines.

Build Skills	Deliver Key Enablers	Make PM Systemic

Skills
Education
Qualification
Knowledge
    Sharing
	Methods
	Tools
	Assessments
	Financial Systems
		Executive Involvement
		Measurements
		Communications
		Integration:
		    Business, Cultural,
		    Organization

Individual ⟹ Enterprise

FIGURE 15–2. The roadmap to achieve project management competency reaches from the individual project management professional across the enterprise, ensuring organizational effectiveness and acceptance.

management skills. Actions drove the formalization of the position of project manager throughout IBM as well as focus on improving the project management skills of all IBM employees.

Organizational Competence in Project Management

A core competence in project management requires an environment that focuses on projects rather than functions. The environment must develop a cadre of skilled project managers worldwide. IBM encourages and expects all employees associated with projects to understand how their roles affect delivery. It requires financial, human resource, and management systems to support project management activities. (See Figure 15–2.) To achieve organizational competence in project management, the PM/COE drives the following action across its culture:

- Work is organized along project lines; efforts and initiatives are driven by cross-functional teams that focus on achieving specific goals in defined timeframes.
- Projects are consistently funded, staffed, and managed in all business units.
- Project management has a consistent role in overall business processes.
- Individuals involved in projects—including executives, project team members, and functional managers—understand their role in planning, carrying out, and evaluating individual and multiple projects.
- Project outcomes can be predicted with high degrees of certainty. When those predictions are wrong, the organization has the knowledge and ability to adjust decisions and take proper corrective action early.
- Timely, accurate, quantitative methods are available for monitoring progress, predicting results, and evaluating risk; systems are in place to support those techniques; and results are continually improved.

- Project managers are qualified and assigned according to stringent professional criteria.
- Project management is designated as a critical professional discipline with programs for continuous skill growth and professional development.

The Project Management Center of Excellence established the following actions to achieve such organizational competence and a project-based culture across IBM.

Action 1: IBM adopted one consistent approach for project management. This approach embodies core methodologies, practices, tools and techniques common to all IBM organizations, but flexible enough to accommodate individual business considerations.

Initially, there were multiple approaches to project management. Inclusive of different vocabularies, processes, and tools. Project managers were trained differently in different business units. Even the role of the project manager differed from one organization to the next.

An organizational competence requires a consistent approach. Consistency improves the timeliness and quality, and it reduces project costs. Communication is simplified and timelier with standardized tools and formats. Project managers across IBM benefit from the experience of others as they no longer waste time reinventing techniques already learned once through experience. Clearly, an identical detailed project management process will not be applicable to all business units and could even interfere with individual business processes and needs. However, there are key characteristics common to all successful projects regardless of their type or magnitude. We have developed a core process based on best practices within IBM and from our peers and competitors. Institutionalizing this core process ensures that the project management disciplines critical to success are now applied throughout the corporation.

Action 2: Selected, certified project managers manage significant projects while the management systems enable this process. IBM project managers meet appropriate requirements for professional education and experience, commensurate with position requirements.

IBM project managers must meet educational standards. A progressive qualification structure, including the requirement to pass independent industry examinations, ensures that IBM project managers obtain the requisite knowledge and experience as they move through their careers. A detailed competency study was used to map the experience and education critical to developing the best project managers and ensuring their professional vitality.

The career path and qualification approach is supported by an integrated human resource framework and professional development program. These approaches are aligned and have converged into a single IBM-wide approach to project management education and professional vitality. The approach is flexible so that it can also embrace new Human Resource programs, which ensure skill levels attained are consistent not only across project management roles but against other IBM job roles.

Equally important to project manager development and qualification is a refinement of the process by which they are assigned. Projects are assessed based on size, revenue implications, risk, customer significance, time urgency, market necessity, and other characteristics; qualified project managers are assigned to them based on required education and experience factors.

Action 3: IBM monitors project performance and measures progress in terms of technical, financial, and schedule performance. Project managers and their sponsoring executives are accountable for specific project results. To aid in these efforts, methods and tools necessary to analyze performance are also available.

Project management is too often thought of as simply managing tasks in a schedule. However, project success is also determined by the degree to which requirements are met and by the financial performance of the project. An effort that is completed on schedule but doesn't meet requirements or significantly overruns its budget projections may well end up costing the company clients, market share, or marketplace leadership as well as immediate profit and employee morale.

In evaluating project performance, all aspects of project management—technical, schedule, and financial performance—are consistently assessed. IBM processes and systems support this approach and provide timely and comprehensive project and program data. There are a number of systems available to support task and schedule management as well as to evaluate the financial performance of individual projects. The PM/COE has, since its inception, garnered historical databases that allow for the identification and evaluation of trends and historical comparisons. It continues to ensure that efforts underway to improve systems are supported and expanded to provide a project orientation for all work across the corporation.

Action 4: IBM established a project management Center of Competence, the mission of which is to support the practice of professional project management throughout the IBM Corporation.

IBM designated an executive sponsor for project management and then established and staffed a PM Center of Excellence. The PM/COE is the virtual source and clearinghouse for project management expertise across the corporation. The PM/COE is staffed with certified project managers on rotational assignments from all business units. It is modeled after the highly successful approach of the IBM Academy and utilizes the governance model developed by Global Services for their IBM Solutions Institute.

Specific PM/COE responsibilities encompass:

- Developing an IBM-wide strategy and plans for the development and support of project management as an organizational competency as well as an individual profession
- Driving the development of processes, practices, tools, and curricula for the achievement of that strategy and coordinating related efforts in all business units
- Providing subject matter expertise and assistance on project management across the corporation
- Maintaining a professional project management community within IBM and coordinating that community with other related internal and external communities

Action 5: IBM project managers further the advancement of project management as a professional discipline through "giveback" activities during or in between project assignments.

A professional community thrives and grows only through the contributions of its members as they help each other become more proficient in the practice of their discipline.

These efforts include mentoring, teaching, project assessment and assurance activities as well as the sharing of experience through lessons learned exercises and published papers. Such giveback serves to refresh and renew project managers, and it is considered an honor to contribute to one's profession in this way. However, such professional growth and enhancement can occur only with conscious organizational support for these activities.

According to Debi Dell, PMP, Manager of Communications and Operations for IBM's Project Management Center of Excellence,

> "It is only through giveback and similar activities that IBM benefits from its learning, growing project management profession. We have set an expectation that our best and brightest project managers will contribute to the PM community—for they truly have the most to give. Participation in giveback activities is built into the qualification requirements for project managers across the corporation. Leadership must continue to endorse the profession and these activities as valuable and directly contributing to IBM's effectiveness."

In conclusion, the PM/COE develops and implements corporate wide strategy and plans for achieving organizational competence in project management. It establishes and drives a consistency of approach, a network of knowledgeable practitioners, and supportive business processes and systems. (See Figure 15–3.) Finally, it establishes and maintains a professional project management community within IBM and acts as the interface between IBM's community and other internal and external professional communities. Over time, the PM/COE has driven the transformation and integration of project management into the fabric of IBM. IBM professionals are more experienced and capable in their abilities to get their work done using project management disciplines. In summary, project management disciplines have become systemic across IBM, resulting in successful projects and higher client value.

Management Support

Strong executive backing began at the top of IBM when the Corporate Executive Committee, led by Lou Gerstner (CEO), declared in November 1996 that IBM

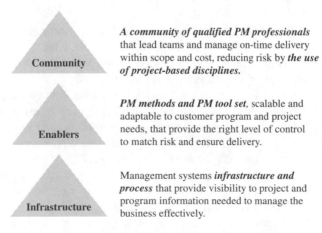

A community of qualified PM professionals that lead teams and manage on-time delivery within scope and cost, reducing risk by *the use of project-based disciplines.*

PM methods and PM tool set, scalable and adaptable to customer program and project needs, that provide the right level of control to match risk and ensure delivery.

Management systems *infrastructure and process* that provide visibility to project and program information needed to manage the business effectively.

FIGURE 15–3. A corporatewide set of objectives to achieve project management competency.

would become a project based enterprise with common methods, tools, and a recognized project management profession. This transformation continues to receive strong executive support across the corporation.

> "The shaping and deployment of IBM's Enterprise Project Management strategy is a glorious example of successful organizational transformation. Initiated by a CEO charter from Lou Gerstner in 1996 to make IBM into a project based business, IBM's Project Management Center of Excellence has directed the progressive phases of the program across these past thirteen years. Key to the PM/COE's success has been sponsorship by a senior executive committee, who acted as program champions for project management across a diverse set of global business units. The organizational benefits of this transformation that were realized over that time are readily evident in IBM's current business success."
> —**Steve DelGrosso, Director, IBM Project Management Center of Excellence**

Many executives actively support project management within their organizations and send strong messages throughout their business unit in this regard. This is critical for getting others within the organization to support project management. One such executive is Rodney Adkins, the Senior Vice President of Development & Manufacturing in IBM's System and Technology Group (STG). He published the following statement of support of project management within STG:

> "Project managers are a critical element of our end-to-end development and business execution model. Our goal is to have sound project management practices in place to provide better predictability in support of our products and offerings. As a team, you help us see challenges before they become gating issues and ensure we meet our commitments to STG and clients. . . .
>
> We continue to focus on project management as a career path for high-potential employees and we strongly encourage our project managers to become certified, not only PMI, but ultimately IBM certified. . . . End-to-end project management must become ingrained in the fabric of our business."

The STG example is just one of many. Management support must start at the top with a statement of strong executive support, which is communicated across and within the organization. This, in turn, is instrumental in getting middle management and line management to support the PM initiatives. The PM/COE works with the executives within business units without this top down support to increase their understanding of the importance of project management to achieving their business objectives and to subsequently articulate their support and messages to their organization.

Following the attainment of general executive support in an organization, education is provided to middle managers, line managers, and PM practitioners about IBM's PM initiatives and how they can fully embrace project management. This education takes many different forms, such as courses, monthly articles, periodic meetings, conferences, etc.

Finally, a critical component to the integration of PM disciplines within all business units is the PM Competency Network. A core team of Competency Network Leaders (CNLs) exist in most of IBM's major geographies, lines of business, and business units (BU) who are responsible for their project management community. These leaders and teams see that PM activities, tools, methods and education are implemented within their geography or business unit. Additionally, a Global Competency Council, a partnership between the PM/COE and CNLs provide direct feedback on new and changing programs and initiatives.

Project, Program, and Portfolio Management

IBM is structured around hardware, software, and services organizations globally. Implementing common processes, procedures, methods, tools, and skills that can cross over business units and geographies and be effective was initially difficult. However, by allowing organizations to tailor the approach to fit each organization's and geography's business model, there remains a consistency, commonality, and effectiveness across PM practices worldwide.

Within IBM's project-based enterprise, it is important for all levels of management to understand how projects, the basic construct for accomplishing work, related to programs and portfolios. This relationship is critical in order to apply the appropriate management approach, management authority, and management controls to ensure success. Definitions consistent across the business for these constructs include:

- Projects are temporary endeavors undertaken to produce a unique product or service. Projects are tactical endeavors as they are short term and have a defined scope or objectives.
- Programs are long term endeavors undertaken to implement a strategy or mission to meet business or organizational goals. A program is implemented through multiple interrelated projects and ongoing activities. Many times Complex projects should more correctly be managed as Programs. In IBM typical programs include contracts, product development.
- Project/Program Portfolios is a business view of projects and/or programs that share specific common characteristics and are viewed as a group for management purposes. In IBM typical portfolios are: IBM Global Services service offerings, projects under a principal, and programs under a sector.
- IBM developed corporate practices aimed at supporting the management of each of these areas. The practices are short, concise documents that are a good reference to use in verifying that a team is addressing all common areas that have been proven as success factors in a project based construct. The practices specifically address the following areas of PM discipline and form a common base for management in all business units (See Figure 15–4): The relationship of projects, programs and portfolios;
- The responsibilities of the management and stakeholders of projects, programs and portfolios; and
- Project and Program Management Practices.

Compliance to a practice is recommended, except in cases where a business unit policy or process makes compliance mandatory. By using an IBM Corporate Practice, individual business units can make the appropriate decision as to compliance to the practice relevant to their business objectives.

Strategic Endeavors

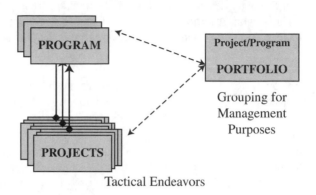

Tactical Endeavors

IBM Corporate Practices:

Projects:
              C-P 0-0145-030
Programs:
              C-P 0-0145-040
Portfolios:
              C-P 0-0145-050

FIGURE 15–4. The relationship between projects, programs and portfolios and supporting materials.

Enterprise Project Management (Enterprise PM) is the capstone in IBM's evolution to a project based enterprise and encapsulates the project, program and portfolio constructs. See Figure 15–5.) It is the next logical step to making project management systemic to all levels of the management structure. Enterprise PM is a management approach that applies project management disciplines to the management of all functions of an organization. An Enterprise PM solution includes a project based infrastructure (methods, tools, processes), skilled and experienced PM professionals, and the use of project management disciplines in the day-to-day management of the business. It makes PM systemic to how IBM does business, it is woven into the fabric of its culture, it is global, and it results in business value.

Project Management Offices (PMO)

As part of its project-based mission, the IBM PM/COE focuses on the understanding and implementation of Project Management Offices (PMOs) to assist in the PM transformation process.

PMOs, defined as a coordinated group consisting of one or more individuals within an organization, are established within IBM to perform project management functions for a single project or for a portfolio of projects to make the organization more effective. In fact, **the PM/COE is structured as a PMO.**

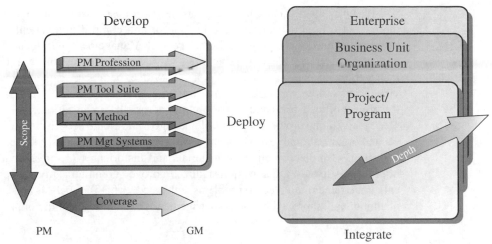

FIGURE 15–5. Enterprise Project Management.

Within IBM, two types of PMOs primarily exist:

● An internal project office that provides project management support to a single project or program; and
● A centralized project office that provides project management expertise to multiple projects, programs and organizations or that has a charter beyond executing PM tasks.

The PM/COE developed and defined the Project Office program by providing a body of knowledge on the effective design and implementation of project office capabilities, both within IBM and on behalf of clients. The Project Office program provides assistance to project offices through a number of tools including the Project Office Guide, Setup Guide, and Staffing Roles Guidance.

Additionally, members of the PM/COE present the value of PMOs across the business, encouraging their use when appropriate. In today's world of highly complex solution delivery, a Project Management Office (PMO) increases efficiency and effectiveness. A PMO should be organized to support organizational goals and to provide benefits such as:

● Enabling project managers to take a consistent, systematic and repeatable approach to project management
● Continuously improving the skill set of project managers
● Providing management visibility and control of projects to reduce the occurrence of troubled projects
● Optimizing project resources, project selection and prioritization
● Performing project, program, and portfolio administrative functions to free up the project manager to focus on project/contract
● Supporting culture shift to project management

Executive focus and commitment is needed in establishing and enforcing the use of the project office. Executives need to sponsor and commit time, attention, and financial resources to making the project office successful. Management must then assign a respected project manager capable of working effectively across multiple projects as the project office manager and also ensure the PMO is staffed with experienced, capable, and full-time individuals. As always, management must effectively communicate and ensure acceptance of the project office's roles and responsibilities across the organization.

Finally, if the project office is well managed, additional benefits result, both at the project and organizational level. Projects see improved delivery times, reduced cost, improved collection of project intellectual materials and historical data for subsequent reuse, and improved estimates. Organizations realize improved communications, improved ability to anticipate rather than react to problems, enhanced "what if" analysis and corrective action planning, and improved business management skills throughout the organization.

Professional Development—Qualification   The project management profession is one of several IBM global professions established to ensure availability and quality of professional and technical skills within IBM. The PM Professional Development initiative includes worldwide leadership of IBM's project management profession, its qualification processes, IBM's relationship with Project Management Institute (PMI), and project management skills development through education and mentoring. These programs are targeted to cultivate project management expertise and to maintain standards of excellence within the PM profession.

What is the context of a profession within IBM? The IBM professions are self-regulating communities of like minded and skilled IBM professionals and managers who do similar work. Their members perform similar roles wherever they are in the organizations of IBM and irrespective of their current job title. Each profession develops and supports its own community, including providing assistance with professional development, career development and skills development. The IBM professions:

- Help IBM develop and maintain the critical skills needed for its business;
- Ensure IBM clients were receiving consistent best practices and skills in the area of project management; and
- Assist employees in taking control of their career and professional development.

All IBM jobs have been grouped into one of several different functional areas, called job families. A job family is a collection of jobs that share similar functions or skills. If data is not available for a specific job, the responsibilities of the position are compared to the definition of the job family to determine the appropriate job family assignment.

Project managers and, for the most part, Program Managers fall into the Project Management Job Family. Project Management positions ensure customer requirements are satisfied through the formulation, development, implementation and delivery of solutions. Project management professionals are responsible for the overall project plan, budget, work breakdown structure, schedule, deliverables, staffing requirements, managing project execution and risk and applying project management processes and tools. Individuals are required to manage the efforts of IBM and customer employees as well as third-party vendors to ensure that an integrated solution is provided to meet the customer needs. The job role demands significant knowledge and skills in communication, negotiation, problem solving, and leadership. Specifically, project management professionals need to demonstrate skill in:

- Relationship management skills with their teams, customers, and suppliers
- Technology, industry, or business expertise
- Expertise in methodologies
- Sound business judgment

Guidance is provided to management on classifying, developing and maintaining the vitality of IBM employees. In the context of the PM profession, vitality is defined as professionals meeting project management skill, knowledge, education, and experience requirements (qualification criteria) as defined by the profession, at or above their current level. Minimum qualification criteria are defined for each career milestone and used as individual's business commitments or development objectives, in addition to business unit and individual performance targets.

Skilled project management professionals are able to progress along their career paths to positions with more and more responsibility. For those with the right blend of skills and expertise, it is possible to move into program management, project executive, and executive management positions. Growth and progression in the profession are measured by several factors:

- General business and technical knowledge required to be effective in the job role;
- Project management education and skills to effectively apply this knowledge;
- Experience that leverages professional and business-related knowledge and skills "on the job"; and
- Contributions to the profession, known as "giveback", through activities that enhance the quality and value of the profession to its stakeholders.

IBM's **project management profession** has established an **end-to-end process to "quality assure" progress through the project management career path**. This process is called "qualification" and it achieves four goals:

- Provides a worldwide mechanism that establishes a standard for maintaining and enhancing IBM's excellence in project management. This standard is based on demonstrated skills, expertise, and success relative to criteria that are unique to the profession.
- Ensures that consistent criteria are applied worldwide when evaluating candidates for each profession milestone.
- Maximizes customer and marketplace confidence in the consistent quality of IBM project management professionals through the use of sound project management disciplines (i.e. a broad range of project and program management processes, methodologies, tools and techniques applied by project management professionals in IBM).
- Recognizes IBM professionals for their skills and experience.

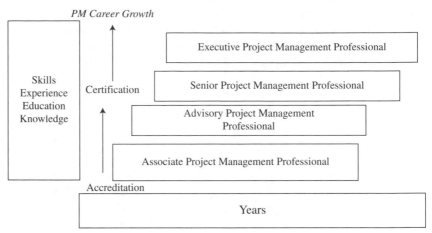

*Career path allows growth from an entry level
to an executive management position*

FIGURE 15–6. IBM's Project Management Career Growth Path.

The IBM project management profession career path allows employees to grow from an entry level to an executive management position. Professionals enter the profession at different levels depending upon their level of maturity in project management. Validation of a professional's skills and expertise is accomplished through the qualification process. The qualification process is composed of accreditation (at the lower, entry levels), certification (at the higher, experienced levels), recertification (to ensure profession currency), and/or level moves (moving to a higher certification milestone). (See Figure 15–6.)

Accreditation is the entry level into the qualification process. It occurs when the profession's qualification process evaluates a project management professional for Associate and Advisory milestones.

Certification is the top tier of the qualification process and is intended for the more experienced project manager. It occurs when the profession's qualification process evaluates a project management professional for Senior and Executive Project Management milestones. These milestones require a more formal certification package to be completed by the project manager. The manager authorizes submission of the candidate's package to the Project Management Certification Board. The IBM Project Management Certification Board, comprised of profession experts, administers the authentication step in the certification process. The Board verifies that the achievements documented and approved in the candidate's certification package are valid and authentic. Once the Board validates that the milestones were achieved, the candidate becomes certified as a Senior or Executive PM.

Recertification evaluates IBM certified project management professionals for currency at Senior and Executive Project Management milestones. Recertification occurs on a three year cycle and requires preparing a milestone package in which a project manager documents what he/she has done in project management, continuing education, and give-back since the previous validation cycle.

IBM continues to be committed to improving its project management capabilities by growing and supporting a robust, qualified project management profession and by providing quality project management education and training to its practitioners.

PM Curriculum The IBM PM Curriculum is a global curriculum. The development and delivery of this **cohesive, world class PM curriculum** was a major element in the establishment of a consistent base of terminology and working knowledge for the more than 25,000 project management professionals worldwide in IBM. A Curriculum Steering Committee (CSC) with global representation manages its development, delivery and deployment. Every classroom course is delivered the same way in every country using the same classroom materials and facilities with local instructors certified to teach the course. The e-learning courses are on servers that are accessible 24 hours a day, 365 days a year by students around the world. In 2008, 64% of IBM's PM education was delivered via e-learning with over 135,000 student days of PM education delivered in 34 countries.

As the needs of the PM professionals and the skills at each project management job level were better understood and matured, the PM curriculum designed a multi-tier architecture (see Figure 15–7) that mapped course content against the skills requirements necessary to develop a project management competence within IBM. The architecture is composed of: PM Basic education that addresses the fundamentals of project management for employees with limited or no project management experience; PM Enabling education that deepens and broadens project management skills with deeper topical content; and Program Management Education that builds on PM skills by enhancing general business skills and providing project based tools and techniques to manage large programs with

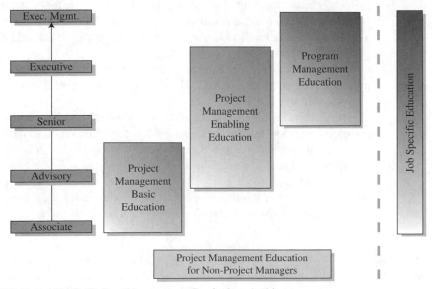

FIGURE 15–7. IBM's Project Management Curriculum Architecture.

multiple projects and business objectives. Once the architecture and curriculum were set, we created roadmaps so students could easily determine what courses to take.

Organization of the Project Management Curriculum

Most IBM project managers begin their IBM project management education by taking courses following the roadmap through the **PM Basic Education**. This education is designed for project managers with limited or no prior project management experience who are working toward IBM project management certification. Within the PM Basic Education curriculum, there is also a roadmap for those experienced project managers who are working towards certification. The Basic Curriculum includes PMI examination preparatory courses including PMI Examination Preparation and PMP Exam Prep Intensive Workshop.

To ensure that students exit the Basic Curriculum with the required IBM PM skills, exams are given at the end of most courses. The content and rigor of our curriculum led to George Washington University declaring that the IBM Basic Curriculum was equivalent to their Masters certificate in project management. Several courses in the curriculum are also accredited by the American Council on Education for both Graduate and Undergraduate credit in North America and by the European National Academic Recognition Information Centre (NARIC) in Europe and Asia.

Project Management Enabling Education is designed to deepen and broaden project management skills. The audience for this education varies by course. In general, the just-in-time courses and short 2-to-4 hour e-learning courses require that students have a fundamental understanding of project management but are not required to have completed the Basic Education. However, the more in-depth classroom or e-learning courses do generally require completion of the Basic program.

Program Management Education is designed to build on project management skills by enhancing general business skills and providing tools and techniques to manage complex projects, programs and project portfolios. IBM certified project managers are the audience for this education as they will generally be tapped for the larger, more significant projects or programs.

Project Management Education for Non-Project Managers contains courses in the project management curriculum to help managers and project team members who support or work on project teams.

In addition to project and program management specific education, all project managers need education that prepares them for the specific environment in which they work. This education may be unique to their specialty, business unit, industry, country, or other job elements. Job specific education is not specifically addressed by the project management curriculum but is recognized as a necessary part of a project manager's training. **Specialty education**, a part of job specific education, is required for IBM project management certification. After certification, credit is given for job specific education for recertification.

Project Management Methodologies

To provide its teams with consistent methods for implementing project management worldwide, IBM developed the Worldwide Project Management Method (WWPMM), which establishes and provides guidance on the best project management practices for defining, planning, executing and controlling a wide variety of projects. IBM's **Worldwide Project Management Method (WWPMM)** is a response to the Corporate Executive Council (CEC) action to establish a single, common project management method for IBM projects worldwide. WWPMM is an implementation of the PMI *PMBOK® Guide* designed for the IBM environment.

Activities within the project management process can be classified into four basic groups: Defining, Planning, Executing and Controlling, and Closing. A project usually consists of a series of phases, known as the project life cycle, and these groups of process activities can be applied to each phase individually or to a set of multiple phases.

In the Defining group, agreement is reached on the objectives of the project, the scope of the project is established, the initial organization is defined, responsibilities are assigned, and the assessment of situational factors is documented.

In the Planning group, detailed work and risk plans are drawn up, the organization is confirmed, and staff assignments are made. No significant amount of resource may be expended on the project (that is, execution does not begin) until clear plans are in place and authorization to proceed has been received at the end of this phase.

In the Executing and Controlling group, the plans and controls are used to execute and manage the project as project development and delivery work is performed. As work proceeds, plans are expanded or refined as necessary.

In the Closing group, the sponsor agrees to close the project, the project is closed down, and the evaluation report (also known as lessons learned) is produced.

Recursively, during the life of a program and that of projects, multiple of these activities can be occurring and recurring concurrently. As programs are defined and planned, projects are identified which, in turn, are defined, planned, executed, controlled, and closed. As the programs are executed and controlled, periodic reassessment of goals can lead to identification of new projects to be defined as well as to the closure of projects that no longer provide value. As a project is executed, it is usual that plans are refined at the end of some phases, to prepare for the execution of the next phases.

In order to be generic and applicable across IBM, the project management method does not describe lifecycle phases, but rather PM activity groups that can be used repeatedly across any lifecycle. This allows the flexibility for the method to be used with any number of technical approaches and lifecycles. (See Figure 15–8.) WWPMM is how projects and programs are managed in IBM and is deployed worldwide through a variety of business unit specific methods and management systems such as Global Services Method and Integrated Product Development to name a few.

WWPMM contains reference documentation (domains), process steps (work patterns), and work products or templates for PM system components as well as support material that explains WWPMM usage. WWPMM describes how to shape and plan a project and then manage its execution through three interrelated views. (See Figure 15–9.)

WWPMM helps define the Project Management System, a collection of plans, procedures and records that direct all PM activities. It describes the current state and history of the project, including the metrics necessary to track and measure the project. Generic templates are provided in downloadable form, from the WWPMM reference page and through various PM tools. When used with appropriate tools and integrated with business and technical management systems, this material provides a comprehensive PM environment. It also

**WWPMM Process Groups can be used
iteratively in the IPD Phases**

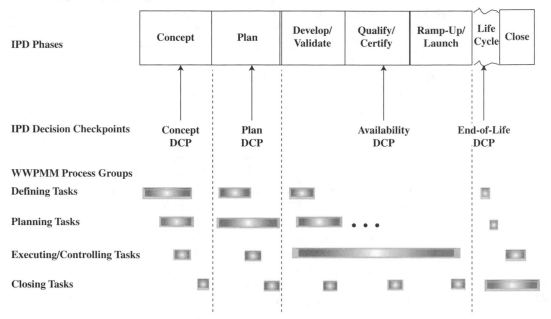

Example: Within the IPD Concept Phase, the majority of the tasks fall within the WWPMM Defining and Planning process groups,
though there are tasks which also fall within the Executing/Controlling and Closing WWPMM process groups

FIGURE 15–8. IBM Worldwide Project Management Method.

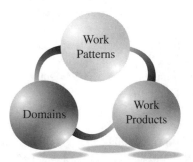

FIGURE 15–9. The Interrelated views of IBM WWPMM.

defines the documentation that will be created and delivered as well as how and where that documentation will be stored.

Keeping with the need to be flexible, the Project Management System templates and work products can be tailored to meet geography, business line, or client specific requirements while still maintaining our commitment for "one common project management method". Managers of project, programs, or portfolio managers can insure their project managers are using a project management system appropriate for the risk and complexity of their specific project and business line needs.

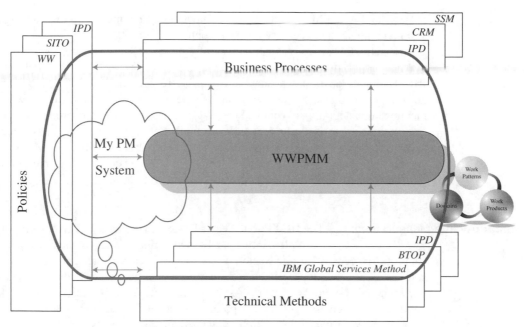

FIGURE 15–10. The Worldwide Project Management Methodology (WWPMM) provide[s] a basis for a comprehensive project management system.

The project management system is unique to each project. An organization's current policies, business and technical methods, and the PM methods of the Worldwide Project Management Methodology (WWPMM) provide a basis for a comprehensive project management system (See Figure 15–10). The project management system should then be tailored to the needs of each individual project to account for complexity of the project and the current business climate of the project environment. Project managers determine what procedures, processes, and tools are best for their projects based on the standards or guidance that has been defined by the business unit. Using a PM system helps ensure that each project team is working in a structured and controlled environment, working towards the same project goals. But, even with our own methodology, IBM stays in sync with what is happening in the industry and with responding to business unit requirements. **Agile** is another technical approach used to develop software and IT solutions within IBM. While the industry has employed it for years, IBM is now expanding its use as clients increasingly request IBM to utilize Agile techniques.

In IBM today, most Agile projects are pieces of larger development or services projects. Thus, a project manager is needed to ensure that the deliverables of the larger project are delivered and the requirements of the business management system are met. For projects where Agile development comprises the entire project, business management requirements and controls must still be followed. To address these business management system requirements, the IBM Worldwide PM Method (WWPMM) is designed to be both scalable and adaptable to meet the needs of all projects, including Agile projects. IBM provides guidance to its project managers on how WWPMM can be streamlined for Agile projects and how PM disciplines and behaviors can be applied to embrace agility while still maintaining the necessary business controls.

A method, however robust, is generally not as effective without a tool to enable its practical application. One effective way to implement WWPMM is through tools such as IBM Rational Portfolio Manager™. This tool provides the capability to deliver virtually all of a project management system in a single web accessible location as well as implementing and integrating a technical method. Through the use of its pre-approved templates and routine monitoring, a project team can implement the appropriate project management system. Rational Project Manager usage is growing dramatically as IBM project managers and their managers realize the value of planning, executing, and monitoring project portfolios in real time through a common and integrated tool.

Earned-Value Measurement The IBM Public Sector Process and Quality Management (PQM) team is implementing an Earned Value Management System (EVMS) to support the needs of Public Sector. The essence of earned value management is that at some level of detail appropriate for the program, a target value (budget) is established for each scheduled element of work. As these elements of work are completed, their target values are "earned." As such, work progress is quantified and the earned value becomes a measure of performance in relation to planned and actual costs.

The Public Sector EVMS emphasizes structured planning and management controls to direct and monitor program functions, and assist program management in making informed decisions in a proactive and timely manner. The current phase includes installation of EVMS tools and establishing pilot projects. The EVMS team is also establishing a greater level of EVMS awareness through communication and training that focuses on educating staff on the value, usage, and applicability of EVMS within Public Sector. A longer term objective of this initiative is to establish a validated EVMS which will support Public Sector in its continued growth as a major competitor for large scale federal contracts, and improve IBM's ability to effectively manage those engagements.

IBM uses an Earned Value Management System (EVMS). The purpose of an EVMS is to help manage projects by defining schedule, resource and scope baselines, controlling changes to those baselines, and measuring performance against those baselines. A byproduct of the EVMS is measurement of results; however the main focus of Earned Value Management is on management, not measurement.

EVMS is documented in the System Description Document and accompanying Desktop Procedures. It is a completely integrated management control system that addresses the basic concepts and requirements for an Earned Value Management system (EVMS) that complies with the most current version of the authoritative industry guidance document, "Earned Value Management Systems" (ANSI/EIA-748). It includes such forms as:

- Responsibility Assignment Matrix
- Work Authorization Document
- Contract Performance Report
- Variance Analysis Report

- Baseline Change Request
- Estimate at Completion Change Request
- Project Baseline Log

Variances are usually analyzed on a monthly basis. The results of the analysis and any impacts and corrective actions are documented in Variance Analysis Reports which are signed by the responsible Control Account Manager and the Project Manager. Occasionally, at a client request, variances are calculated and analyzed on a weekly or bi-weekly basis.

Variances always need corrective action; however, sometimes corrective action is impossible, e.g., if there is no way to mitigate the effect of schedule delays or cost over-runs. Even in that case, corrective action may consist of recognizing an overrun and obtaining additional resources to compensate for it, or of recognizing a schedule delay. These determinations are made on a case by case basis by the responsible Control Account Manager and the Project Manager.

Finally, this approach does use thresholds, based on risk analysis and on client requirements. The thresholds are usually expressed in terms of percent and dollars (e.g., Cumulative variances greater than 10% of budget AND greater than $10,000).

Best Practices
A web search of the keywords "ibm + best practice" yields literally millions of hits. A best practice can be defined as a method, process, or technique, to deliver an outcome that has been shown to provide a more reliable and predictable desired result than other techniques. IBM has throughout its history developed many best practices for many different aspects of information technology. These have been produced in many different levels of rigor and formality. In the technical world, these best practices may be published as white papers or, more formally, within IBM's Redbook series of publications. Within the project management discipline at IBM, it is no different. Since 1996, IBM has developed many project management best practices. The PM/COE has the job of communicating and educating the company's project management stakeholders on these best practices.

- As discussed earlier in this chapter, a set of key project management initiatives was established as the core approach to deploying project management within the enterprise. Together in partnership, the PM/COE and business units throughout IBM worked to deploy eight key initiatives that formed the foundation for IBM's vision of its enterprise project management deployment. The eight key initiatives include (see Figure 15–11):
- Education
- Qualification
- Method
- Select Projects
- Tools
- Portfolio Management
- Maturity
- Policy

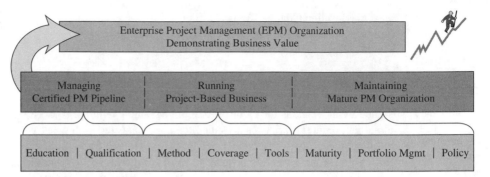

FIGURE 15–11. Within each of the initiatives, best practices were developed to support deployment within the organization.

Within each initiative, best practices were developed to support deployment within the organization. Some were referenced previously but this snapshot provides the basis for the measurements against which IBM's progress to becoming a project based enterprise is measured.

Education

IBM's approach to PM education was outlined in a previous section. The PM curriculum, roadmap, and approach to education have been a very formal best practice within IBM for a number of years. All project managers in IBM are taught the same curriculum and the same messages, which have evolved over time as IBM has matured in project management. This best practice is maintained in a set of documents that are overseen by a small team of staff and a steering committee.

Qualification

IBM's approach to qualifying and certifying project managers through their career has proven to yield value to our clients, our project managers, and the company. This approach is documented as a best practice in the Project Management Profession Guide and maintained by approval of a professional development committee. A small team oversees the administration of this best practice. The material and updates are communicated to the project management community through the PM/COE's communications program to assure that all project managers worldwide are adhering to the same qualification process.

Method

IBM has been very method driven for many years. The challenge within project management was to bring the company together in a unified method of managing projects. While there are many dozens of technical methods maintained by IBM to define the different aspects of performing work, there is only one method for project management used by all parts of IBM in many different types of projects. WWPMM (Worldwide Project Management Method) was described in an earlier section, and is IBM's best practice for managing projects. It is maintained by the PM/COE and updated periodically to reflect changes in requirements. IBM business units have, in most cases, adapted WWPMM into their management processes.

Selected Projects   While Education, Qualification, and Method are IBM best practices with very formal documentation and support mechanisms, Selected Projects is a best practice that is maintained much differently. One of the founding principles put in place when the PM/COE was formed was that "selected" key projects would be managed by the most qualified project managers. Because of the variety in how the different parts of IBM operate and staff their engagements, this best practice is defined in concept at a high level, and the PM/COE works with the different business units to develop a model that works for their business. It is still considered a best practice because the premise is very important to successful project management, yet you will not find detailed process documentation for the practice.

Tools   When IBM started the journey to enterprise project management, many different tools were used to manage our projects. This was very inefficient, both in the cost to maintain many tools, as well as the resource inefficiency with staffing projects with individuals that knew the tools. The PM/COE, working together with the different parts of the business, selected a strategic tool to use for project management. This allowed for better investment in and integration of the tool with key elements such as labor accounting. A team working with the PM/COE has maintained best practices for the implementation of the tool into the business. As new units begin to deploy the tool for their organization, these best practices are followed to minimize problems and streamline the implementation.

Portfolio Management   Project Portfolio Management groups, views, analyzes, and manages projects together to maximize positive business results within an organization's resource constraints. Projects are either included in, or excluded from, the portfolio based on their alignment with the portfolio strategy and their performance against the business objectives. IBM has a formal published policy which is our best practice for portfolio management. The PM/COE works with business units to adapt this policy into their business processes for managing their portfolio of projects.

Maturity   IBM developed a standard tool and best practice, the Project Management Progress Maturity Guide (PMPMG), for measuring project management maturity. PMPMG enables an organization to assess the maturity of its project management capabilities at both the project and organizational levels. By understanding its strengths and weaknesses in these areas, an organization can identify actions for continuous improvement in managing projects/programs and achieving its business objectives. The PMPMG consists of a set of yes/no questions separated into 5 levels that seek information on the PM processes in use. Upon completion, a weighted average of answers is used to compute a relative value for the organization's maturity. Within the guide, the questions apply either to the organization as a whole, or to projects within the organization. Successful execution of the tool will lead to a score of 1 to 5 on maturity, and a list of action items to address deficiencies. The PM/COE maintains the PMPMG as a Lotus Notes tool and database, and assists different business units with the performance of assessments.

Policy

The PM/COE maintains formal corporate practices for project and program management that define the recommended set of management practices that should be used for all IBM projects and programs. These practices are formal corporate level best practice that the whole company is expected to adhere to. In order to deal with the variety of types of businesses in IBM, the PM/COE then works with each unit to develop their own policy statement that adopts the corporate practices and other initiatives and adapts them all to the specifics of their organization. These units' policies are sponsored by the project management executive sponsors for the organizations and published on the PM/COE intranet site so that all project management stakeholders can easily view them.

These eight initiatives comprise a large collection of best practices in project management that have been the pillars of IBM's move to enterprise project management. The PM/COE measures the progress of IBM as a company by monitoring these initiatives using a Red/Amber/Green scorecard that is updated quarterly for each part of the business and each geography around the world. As changes occur, such as a reorganization, market change, or executive leadership move, we see movement in the performance against our scorecard. The PM/COE works with the Executive Steering Team to help correct areas of concern to keep the company on track to maintaining our progress in maturity.

But, the best practices do not stop there. Support of our professionals is also at the core of the project based transformation.

Knowledge Sharing

While the eight initiatives are more formalized best practices, an opportunity exists to leverage lessons learned on engagements by IBM's project managers. The PM Knowledge Network (PMKN) supports IBM's becoming a project-based enterprise by leveraging knowledge through sharing and reusing assets (intellectual capital). The PMKN repository supports the PMKN Community with a wide range of assets that include templates, examples, case studies, forms, white papers, and presentations on all aspects of project management. Practitioners may browse, download, or reuse any of the more than 2,400 entries to aid their projects, their proposals, or their understanding.

Community and Communications

According to Debi Dell, PMP, Manager of Communications and Operations, the Project Management Center of Excellence has driven a strong sense of community for its global project management professionals; this is a best practice among IBM's professions.

Within IBM, a community is defined as a collection of professionals who share a particular interest, work within a knowledge domain, and participate in activities that are mutually beneficial to building and sustaining performance capabilities. Our community focuses on its members and creating opportunities for members to find meaning in their work, increase their knowledge and mastery of a subject area, and feel a sense of membership—that they have resources for getting help, information, and support in their work lives. Knowledge sharing and intellectual capital reuse are an important part of what a community enables but not the only focus. Communities provide value to the business by reducing attrition, reducing the speed of closing sales, and by stimulating innovation.

Communities are part of the organizational fabric but not defined or constrained by organizational boundaries. In fact, communities create a channel for knowledge to cross boundaries created by workflow, geographies, and time and in so doing strengthen the social fabric of the organization. They provide the means to move local know-how to collective information, and to disperse collective information back to local know-how. Membership is totally based on interest in a subject matter and is voluntary. A community is NOT limited by a practice, a knowledge network, or any other organizational construct.

The PM Competency Network (previously referenced as part of the management structure) is not only a community with its own requirements and objectives but also as a source of information for other members of the PM profession. The success of IBM as a project based enterprise is dependent on its ability to effectively integrate project management into the infrastructure, organization and processes used to execute its business. The project management competency community is critical to this integration and has a clear set of responsibilities with which it is charged. These PM competency leaders and teams are responsible to see that PM disciplines, tools, methods and education are implemented within their geography or business unit.

The PM Knowledge Network (PMKN) Community is also sponsored by the PM/COE. Membership is open to all IBM employees with a professional career path or an interest in project management. The PMKN is a self-sustaining community of practice with almost 12,000 members who come together for the overall enhancement of the profession. Members share knowledge and create PM intellectual capital. The PMKN offers an environment to share experiences and network with fellow project managers.

Upon entering the PM community, professional hires into IBM are often asked the question: "What is the biggest cultural difference you have found in IBM compared to the other companies in which you have worked?" The most common answer is that their peers are extremely helpful and are willing to share information, resources and help with job assignments. The culture of IBM lends itself graciously to mentoring. As giveback is a requirement for certification, acting as a mentor to candidates pursuing certification not only meets a professional requirement but also contributes to the community.

Mentoring can be a fragile relationship. A corporate worldwide study revealed that the most common reason that mentoring relationships ended prematurely was due to the lack of a formal structure or goal to keep the momentum going. In addition, a world wide mentoring tool was not available. The Project Management Center of Excellence responded to the study by creating a "Project Management Skills Mentoring Process". In line with the discipline of Project Management, a mentoring process was designed to follow the basic constructs of a project. A mentoring relationship had a definite beginning, a desired project outcome (raise the level of a skill or knowledge area), a plan, a schedule, milestones, tracking and completion. Sample mentoring plans, agreements, milestones, and closing documentation along with a PM Skills Mentoring Guide were provided on the Web site.

The WW IBM mentoring team saw an opportunity to capitalize on using an existing infrastructure—the corporate on-line employee directory (known as Blue Pages) to develop one common corporate-wide mentor matching and tracking tool. This directory requires all employees to list their profiles including skills, career interests, experience, education, internal/external professional group affiliation and contact information. The PM/COE immediately got on board with this corporate wide mentoring tool.

To address the communities and all project management professionals, primary channels of communications include the PM/COE web site and focused newsletters and announcements. Project management can even be added into a PM's corporate web profile. However, as the project management profession grows so do the requirements for projects targeted at specific communities. The Project Management Center of Excellence has developed information for the following communities with the PM profession:

- New to the Profession
- Managers of Project Managers
- Middle Management
- PM Knowledge Network
- Specific geographies' PM communities
- Project and Portfolio Management Communities of Practice
- PM Competency Network

But IBM's best practices are not just recognized within the company. Many have received recognition from industry sources.

According to Debi Dell, the Project Management Center of Excellence tracked its awards and achievements and shared these accomplishments across IBM, the PM community, and with clients in various proposals. This is a partial list of activities that highlight where IBM was recognized for its excellence in project management.

Awards

- 2007 PMI Distinguished Project (IBM Stockholm Congestion Tax Project)
- 2006 PMI Professional Development Provider of the Year Award
- 2006 PMI Education Provider of the Year Award
- 2005 PMI Professional Development Provider of the Year Award
- 2005 PM/COE Director Carol Wright receiving a PMI Distinguished Contribution Award
- 2004 PMI Professional Development Provider of the Year Award
- 2001 Patent award for "Learn How. . . Do It Now. . ." awarded to Dillon Edwards (PM Curriculum Team)
- 2001 International Society for Performance Improvement (ISPI) Award of Excellence for Contracting for Project Management Lotus LearningSpace course (PM Curriculum Team)
- 2000 American Society for Training and Development Excellence in Practice Award recognizing excellence in Training and career development processes. (PM Curriculum Team)

Curriculum

- PM curriculum recognized for its "Contracting for Project Managers" e-learning course by the International Society for Performance Improvement (ISPI) with an Award for Excellence.
- Recognized by American Society for Training & Development (ASTD) with an Excellence in Practice award for its project management curriculum initiative.
- The American Council on Education (ACE) has recommended our Tier 1 courses for credit toward undergraduate/graduate degree programs.

External

- IBM is key participant in PMI Corporate Project Management Council.
- IBM PM Practices cited in Technical Business Review Report in 2007.

Publications

- IBM PM Center of Excellence recognized as one of the top 25 project-based organizations in 2007 by PMI.
- The success of the professional development approach was documented in the PM Best Practices Report.
- The robust implementation of distance learning techniques with the curriculum was cited by Nancy DeViney in Newsweek.
- Article in Nikkei magazine interviewing Carol Wright on the IBM Journey to PM Excellence.
- The Japanese Project Management Forum Journal article on Carol Wright's keynote address at November 2001 international conference.

WW PM Method (WWPMM)

- WWPMM is in the IP catalogue, providing the basis for licensing WWPMM to customers for use on their projects.
- Elements of WWPMM are included in many GS Method Engagement Models that are already licensed as IP.

Figure 15–12 summarizes the value that IBM sees in project management.

Contributors **Deborah (Debi) A. Dell, PMP®** is Manager of Communications and Operations for IBM's Project Management Center of Excellence. In her twenty-nine years experience in IBM, Debi has held positions in project management, business planning, mobile and wireless technologies, and market and product planning. She co-authored ThinkPad: A Different Shade of Blue in 1999. Debi is PMI certified and has a Masters of Science in the Management of Technology.

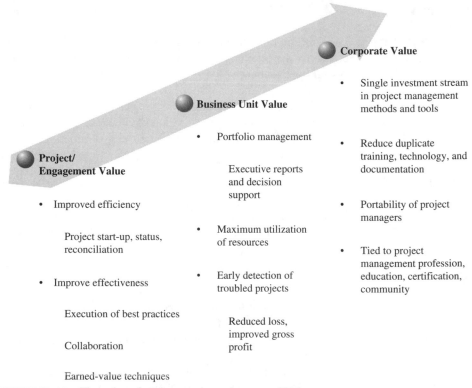

Corporate Value

- Single investment stream in project management methods and tools

- Reduce duplicate training, technology, and documentation

- Portability of project managers

- Tied to project management profession, education, certification, community

Business Unit Value

- Portfolio management

  Executive reports and decision support

- Maximum utilization of resources

- Early detection of troubled projects

  Reduced loss, improved gross profit

Project/ Engagement Value

- Improved efficiency

  Project start-up, status, reconciliation

- Improve effectiveness

  Execution of best practices

  Collaboration

  Earned-value techniques

FIGURE 15–12. The value of project management across IBM.

Rachel Ciliberti is a Project Management Initiatives Consultant in IBM's PM Center of Excellence (PM/COE). She has over fifteen years of experience as a Project Manager and Project Executive. She acquired her project management experience leading various commercial and international IBM accounts in the areas of Systems Integration, Data Center Operations, and Application Development. Rachel is also PMI certified and has a Masters Degree in Project Management.

Mike Collins is currently an Executive Project Manager working in IBM's Project Management Center of Excellence. Mike's current duties include supporting a global business unit's deployment of project management, and leading IBM's Value of Project Management and Worldwide PM Metrics projects. Formerly, Mike worked as a Delivery Manager and Project Executive in IBM Global Business Services managing a portfolio of development projects for NY State. Mike has been an IBM Certified Executive Project Manager and PMP for over 10 years, has 26 years industry experience, and is a member of IBM's PM Certification Board.

Steve DelGrosso, PMP® is Director of IBM's Project Management Center of Excellence. In his thirty years experience in different specialization areas of the IT industry Steve has served as IBM's North American PM Profession Leader, System Engineering manager, developer and instructor for IBM PM curriculum, and also teaches Project

Management to graduate students currently at the University of Miami in Coral Gables, FL. Steve is a member of the PMI College of Performance Management, GovSIG, and PM Journal Editorial Review Board. Steve has published articles in Industrial Engineer magazine, Personal Systems Journal, and was featured on the cover of PMI's April 2004 PM Network magazine.

Kent Demke is a certified Executive Project Manager in the Project Management Center of Excellence (PM/COE). In this capacity, he works with both Systems and Technology Group (STG) and Software Group (SWG) deploying project management initiatives. He is also a member of the North American PM Certification Board. Kent has a Master's Degree in Electrical Engineering, and began his career at IBM designing electronics where he received 9 patents and published 11 technical articles. He later began to manage hardware development projects, and was a product development leader in the Printing Systems Division before joining the PM/COE. He has over 15 years experience in project management.

Jeanne Fraser is the North American Project Management Professions Leader. Jeanne has over 25 years of Project Management Experience, as a project team leader, Project Manager, Project Executive and various senior and executive management positions. Jeanne has been instrumental in the creation of several Project Management Initiatives and has a passion for Project Management and Mentoring. Jeanne is a member of the WW Mentoring team, is PMI certified, a certified Executive Project Manager and certified Life Coach.

John Marrine is the Worldwide Project Management Profession Leader in IBM's PM Center of Excellence (PM/COE). He has been with IBM for 30 years with the last 20 years in Project Management and Project Executive roles. John has also held several management positions during his tenure at IBM. John holds IBM Executive Project Management, IBM Project Executive, and PMI PMP certifications and has achieved Bachelor Degrees in Electrical Engineering and Computer Science.

Sandy Steinruck is a mechanical engineer by degree but has been doing project management for more than 25 years in IBM. She started in product design, managing printer and copier development projects and then moved to FL where she spent 10 years managing software development and integration projects, in the telecommunications industry. For her project management work she was awarded the PM Excellence award in 1998. Sandy has been a member of IBM's PM Center of Excellence as the Program Manager for IBM's award winning PM curriculum for the last 8 years.

John Acton has over 30 years' experience in Federal acquisition management, contracting, proposal preparation and analysis, and Earned Value Management, including 12 years in a Department of Defense Major Weapon System (ACAT-ID) Program Management Office. He is a Certified Public Accountant (CPA), Project Management Professional (PMP), and Earned Value Professional (EVP), and he is certified at Level III in Program Management and Contracting under the Defense Acquisition Workforce Improvement Act (DAWIA).

Emay Hardies has more than 24 years of experience with government contractors in Earned Value, implementing, analyzing, and supporting military and domestic federal contracts, including over 20 years with IBM and 4 years with Lockheed Martin. Also, she has been involved in Earned Value proposal support, surveillance, training, and tool support.

She is currently the editor of the IBM Business Consulting Services, Public Sector EVMS Newsletter and a certified Project Management Professional (PMP).

15.2 COMPUTER ASSOCIATES (CA) SERVICES: SUCCESSFUL PROJECT DELIVERY AND MANAGEMENT[3]

Introduction

Today most companies utilize industry standard project management techniques, yet the success rates for projects remain low. Project success is typically dependent on the skills and sometimes heroism of the people assigned to the project whereas world class companies are able to repeat project success. CA has recognized that in order to achieve repeatable project success we must go beyond the customary project management best practices; project teams need an arsenal of proven tools, processes, and support.

When an organization completes a project successfully the team most likely had solid project management, high quality deliverables, a product that provided value to the stakeholders, and a methodology for delivery that worked. Most organizations do not take advantage of these success stories to recreate for other projects. The project documentation is archived and the team disbands, then the process starts over again from scratch on the next project. Similar challenges apply to failed projects where lessons learned, issues, and risks are not shared or used to improve the execution for later projects. CA Services is very focused on providing tools and techniques to our project teams to provide higher quality solutions and repeatable success that most organizations are unable to match.

CA's Approach

CA is the leading independent software vendor of IT management solutions. Customers look to CA to provide solutions that unify and simplify complex IT environments—in a secure way—across the enterprise for greater business results. CA calls that Enterprise IT Management (EITM)—a clear vision for the future of IT. It's how one can manage systems, networks, security, storage, applications and databases securely and dynamically. One builds on IT investments, rather than replacing them, and does so at one's own pace.

Customers share the EITM vision: comprehensive management and security of the IT environment. They have some solutions in place today and want (need?!) to reach the EITM vision, but how one gets from where one is today to where one needs to be tomorrow isn't clear.

Figure 15–13 depicts a realistic approach to achieving the vision: rather than attempting to define and implement many new business processes and supporting tools at once, one should take [a] phased approach to achieve the vision. This is how CA Services recommends customers plan on reaching the vision; in logical phases.

---

3. Material on Computer Associates was provided by Robert J. Zuurdeeg, Sr. Director, Global Practices, and Mark Elkins, Sr. Principal, Global Services PMO.

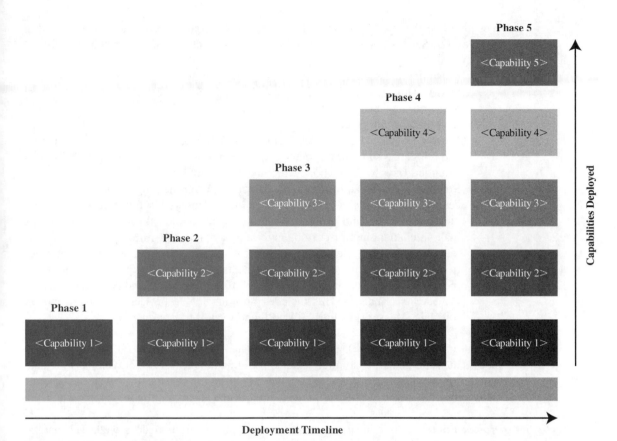

FIGURE 15–13. The CA approach.

CA has defined "Standard Offerings" for many EITM solutions. Standard Offerings define the best practice (proven approach) to implementing EITM solutions. Additional details on these offerings including how they came about, what they contain, and method for continuous improving will be discussed.

Starting Out Right   Many projects are doomed to failure before they even begin. Simply following project management standard practices is not enough to ensure success. CA has created several tools and implemented standard processes to help projects teams begin with a project that is setup from the start to be successful. The following sections outline some of these methods that are utilized prior to a project even kicking off.

Defining Project Success   One obvious question that is usually not asked is how do we define success? The answer that jumps to mind is the project is on budget, on schedule, and the scope statement is completed; however every organization should examine this question more closely. CA did an evaluation of project goals to help define success within Services. CA Services now measures project success based on the following criteria:

- Customer Satisfaction—Short term and long term customer satisfaction is critical to CA. This is measured informally at project closure by meeting with the customer and reviewing project accomplishments. It is also measured more formally through a survey process.
- Customer Time to Value—This includes the ability to get solutions into production and bringing ROI to the customer as quickly as possible. Measures include Time to Contract, Days to Production Deployment, and Solution ROI.
- Solution Standardization and Best Practices—Over the long term, customer success can be negatively impacted if the solution deployed is highly customized and utilizes functions and processes that are out of the norm. Such customizations make it challenging for customers to take advantage of future product capabilities. Therefore long term success for CA includes defining solution best practices and guiding customers to take advantage of these best practices. Measures include deployments of standardized solutions, use cases, and best practices.
- Project Performance—This includes bringing in the project on budget and on schedule. Measures include CPI, SPI, and overall project margin.

CA utilizes a Services Dashboard to help manage the portfolio of projects. These success measurements are captured and displayed within the Services Dashboard to make the data very visible to all groups in CA. This open sharing of data on project success promotes cross-functional support of projects and feedback for continuous improvement.

Project Focus   It is critical to recognize what an organization does well. The natural tendency is to think that we can apply management best practices to execute any project successfully. The reality, though, is most organizations have a sweet spot of projects they do very well; these may be projects that include: specific technologies, are within a particular scope, are within certain industry verticals, solve specific problems, are of a certain size, etc. Organizations need to do an honest evaluation of the projects they do well and the business needs they are able to solve. Failed projects jeopardize business relationships and future business. For CA the focus is on implementing business solutions that utilize CA technology. The results of this evaluation provide the starting point for creating a Services Offering Catalog. This catalog becomes a menu of the project types the organization does well. It also serves as the central repository of project artifacts and best practices for executing each project successfully. The makeup and details of the catalog will later be described in detail.

Standard Solution Offerings   In order to improve Time to Value it is important to understand history and why some projects did not meet success targets. Doing this at CA we discovered a common factor included projects that performed extensive requirements analysis, asking the customer what they wanted, and executing the scope statement perfectly. This may seem like a good formula for project success, but in reality created risk in prolonged analysis and delaying time to value; the customer not receiving optimized solutions and best practices, and too much customization that impacts long term value. A better approach that CA is now utilizing is to define standard offerings that address the majority of what most customers need and allow customers to quickly achieve ROI. Most customers would prefer to receive guidance on what they should be implementing, what works, and solutions being used within their industry. By defining packaged offerings, creating reference architectures, and utilizing standard use cases that can be used for projects with similar goals, CA is able to improve Time to Value for our customers.

Ideally every offering would execute exactly the same scope, project after project. In most situations there is a very large overlap in scope, but there will often be some customizations for each project. But these situations can be controlled by identifying options for customizations or scope changes within each offering. Doing this maintains standardization and increases repeatable success. To support the ability to provide scope options for a common offering, solution calculators are used by CA to support the offering. The calculator provides boundaries to ensure that the scope of the project still falls within the guidelines of the solution offering, but provides options to meet specific customer business needs. Calculators allow for the selection of options and characteristics about the project; the tool can then produce a standard WBS, task list, and estimates by task. These result in standard scope statements and project plans that are executed project after project for each offering.

Feedback and experiences from solution testing, trials and deployments have defined the best practice approach to implementing needed business solutions. That knowledge has been used to define the Standard Offerings. There are 2 primary forms of Standard Offerings: Rapid Implementations and Solution Implementations.

- Standard Implementation Offerings
  - Deliver business value in 60–90 days or less
  - Built on standard components, deployment playbooks and best practices
- Rapid Implementations—good starter package
  - Fixed-priced and functionality, simple scope
  - Pre-approved print and go Statement of Work (SOW)
  - Typically first phase of Solution Implementation
- Solution Implementations
  - Tailor customer requirements to best practices using standard tools
  - Uses Statement of Work Framework
  - May be fixed price or T&M
  - Faster time to contract approval than custom approach
- Additional Standard Offerings include:
  - Assessments
  - Migrations
  - Upgrades

The use of Standard Offerings allows Customers to achieve the EITM vision in the most expeditious, cost-effective manner possible.

Service Offering Catalog   To help the CA Sales organization easily access the Standard Offerings, CA Services created the Service Offering Catalog; a repository where tools are provided to support the positioning, scoping and selling of the Standard Offerings.

The Service Offering Catalog has been recently re-designed; based on feedback from its users. Focus groups provided feedback/input on the tools needed and the repository layout:

- Provide fewer but more comprehensive and focused tools.
- Use as few keystrokes/mouse clicks [as] possible to locate available documents.
- Identify when in the Opportunity Life Cycle each tool is to be used.
- Additionally, provide fast, easy access to the higher level tools in the Sales Enablement Tools section of the Catalog. These include customer presentations and demos, sales and implementation success stories, brochures and references.

The Service Offering Catalog allows a user to enter the catalog (which has been built on Microsoft Sharepoint), quickly locate a Standard Offering, the available tools and download the tools needed at this point of the sales cycle.

Required tools provided in the Service Offering Catalog (SOC) for each standard offering are given in Table 15–1.

TABLE 15–1. REQUIRED TOOLS

SOC Tool	Tool Description
Pocket Guide	The pocket guide is a solution level document that covers one or more offerings. It includes: an overall description and specific deliverables, the value to the customer, target customers, conversation prompters, education components, why CA Services and handling objections.
Service Description	The Service Description is a one page front and back description of the offering, the value and benefits to the customer and specific deliverables. In addition to the catalog, this is available to the customer on ca.com.
Customer Presentation	The Customer Presentation is a set of slides about the specific offering to be incorporated in a larger presentation. It includes a description of the offering, the value and benefits to the customer, specific deliverables and any restrictions that apply.
High Level Design	The High Level Design documents the customer's current business processes, their challenges, proposed new processes and the implementation of new technologies to support the new processes.
Solution Calculator	The Solution Calculator is a spreadsheet which determines the effort (in terms of hours) for the Architect and Consultant delivering a Standard Offering; based on variables unique to the solution.
PM Calculator	The PM Calculator is a spreadsheet which determines the Project Management hours needed for a project, given certain solution and environment variables.
Services Estimating Worksheet	The Services Estimating Worksheet is a spreadsheet used early in the sales cycle to help estimate the approximate price of the Solution Implementation service; this helps set expectations early is the sales cycle.
Project Profiler	This is a tool that evaluates the characteristics of a proposed project during the opportunity phase.
SOW Framework	A SOW Framework (SOWF) is a document with content blocks that contain project specific information based on requirements defined by the Customer to be cut and pasted into the final SOW with the effort and fees.
SOW	For the Rapid Implementation standard offerings the appropriate, completed SOW is provided. The Service Offering Catalog supports CA Services globally; translated and/or localized tools are posted here as well.

Why the Focus on Sales Cycle?

If the focus of this book is Best Practices in Project Management, why are we devoting so much time to what CA Services does during the sales cycle?

That is very simple: a successful services project, which results in the deployment of the right solution and customer satisfaction, comes about as the result of:

1. Understanding the customer's business needs
2. Analyzing existing processes and supporting technologies
3. Customer agreement of the solution (processes and technologies) to be deployed
4. Correctly scoping the project and defining it in the contractual documents
5. Properly setting customer expectations

Benefits to Customers
- Significantly lower implementation risk (than not adhering to best practices)
- Customers get into production quickly using best practices
- Increased satisfaction with defined outcomes delivering value in 60–90 days
- Faster return on investment

Project Execution

Once the services engagement is sold, CA Services prepares to deliver the needed solution. The solution is delivered by following the standard CA Services 9 Stage Deployment Methodology, supported by detailed Deployment Playbooks and managed through CA Services' Project Management methodology.

Deployment Methodology The CA Services 9 stage Deployment Methodology, Figure 15–14, is built on fundamental industry proven systems engineering principles that formalize a consistent, repeatable approach for a rapid implementation or multi-phased solution deployments.

The nine Deployment Methodology stages are:

1. Project Setup and Initiation

Prior to CA arriving at the customer site, a series of joint meetings and/or conference calls are held to prepare both CA and the customer for the beginning of the deployment. Key activities in this Stage are the creation and approval of the Project Management Plan and the Project Schedule, conduct a Risk Assessment and validation of pre-requisites.

2. Solution Requirements Definition

Information on requirements should have been gathered during an assessment or the presales process. In this Stage, a series of requirements gathering sessions with key stakeholders, business managers, customer architects and operations team members are intended to verify and further discover the customer's business drivers, functional and non-functional requirements as necessary. The outcome of this Stage is a Solution Requirements Specification that the customer and CA agree will produce the requested solution.

3. Solution Architecture and Design

Through a series of collaborative workshops and discussions, it is intended that CA analyze the current state of the customer's environment. It is intended that a planned end

Proven Approach:
CA Deployment Methodology

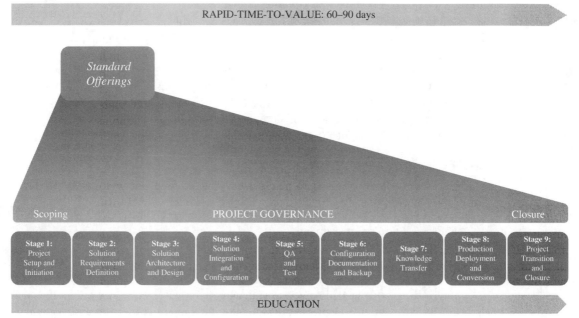

FIGURE 15–14. The nine Deployment Methodology stages.

state be defined, key design decisions and configuration components identified and the rationale for such decisions documented. The outcome of this stage feeds into the Solution Design Specification.

4. Solution Integration and Configuration

The solution is installed and configured into the initial environment (usually development) during this stage. Individual components are deployed to the appropriate environment and then integrated to form a complete solution.

The Solution Integration Specification document is used during this stage. It contains a governance process for the integration process, parameter worksheets for identifying and documenting integration and configuration parameters, and checklists to track the integration of each individual component.

While the Solution is being installed the detailed Solution Test Plan is drafted.

5. Quality Assurance and Test

Upon completion of integration and unit testing, the testing staff (CA and customer) executes a full test suite to validate functional requirements, quality attributes and solution constraints. Results of the testing are recorded in the Test Report. Any significant issues which require remediation will be addressed and then re-tested.

6. Configuration Documentation and Backup

Updates are made to the Solution Integration Specification while the final architecture is documented in the Solution Design Specification. A solution maintenance plan is developed and documented in the Solution Run Book, along with other operational tasks. Finally a backup of the Solution environment is taken and is restored in order to test and validate the process.

7. Knowledge Transfer

A solution demonstration is performed along with a configuration walk-through of key components with designated customer personnel. The Knowledge Transfer Checklist is used to document the knowledge transfer activities with customer personnel.

8. Production Deployment and Conversion

Deployment of the Solution into the customer's production environment is based on the Solution Integration Specification plus any specific procedures developed to convert existing systems to the new solution. A Production Readiness Checklist assists the project team in determining the readiness to move the Solution to production.

9. Project Handoff and Closure

In this stage, the CA and customer project manager conduct a project closure meeting to validate the Solution is in accordance with the requirements identified in the SOW and to discuss proposed next steps and/or deployment phases.

Deployment Playbooks   CA has created deployment playbooks, as seen in Figure 15–15, that align to each of the standard offerings. The playbook provides a roadmap for how the project should be executed. It outlines a staged approach, standard deliverables within each stage, and templates that the delivery team can utilize. Playbooks promote repeatable success by packaging what works and applying lessons learned.

- Methodical instructions that enable CA Services to deliver capabilities in a consistent and repeatable manner
- All of the documents and tools that you will need throughout the 9 stages of a project

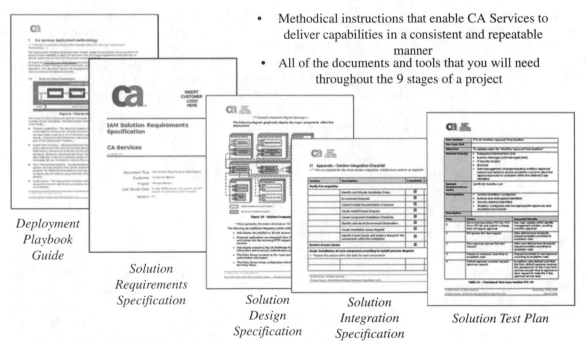

Deployment Playbook Guide

Solution Requirements Specification

Solution Design Specification

Solution Integration Specification

Solution Test Plan

FIGURE 15–15.   The Deployment Playbook.

The playbooks provide tremendous benefits during the execution of the project such as:

- Consistency—As team members move from project to project, they know the project execution will be very similar to other projects they have worked on. The time to bring the team up to speed and bring value to the project is greatly reduced.
- Standard Deliverables—This standardization promotes reuse and reduces time to create.
- Roadmap for the PM—The project manager does not need to create project plans from scratch; there is a predefined roadmap that makes it much easier to manage against.
- Continuous Improvement—The playbooks provide a repository of standard deliverables, tools, and techniques that can continuously be improved from lessons learned and project feedback.

Tools support every stage of the Deployment Methodology. . . . [Table 15–2] lists the tools provided in this Deployment Playbook to support the deployment of the Solution. (Many of the tools were identified in the preceding 9 Stage description.

Some of these tools are for internal CA project team use, others to be used with the customer during the project while others are the basis for the key project deliverables. The tools are applicable to both project managers and team members.

Deployment Playbook Usage by Project Manager

The Deployment Playbooks provide tools for the Project Manager that reinforces project management best practices. (See Figure 15–16) CA Project Managers utilize these tools to provide project governance throughout the project. As the chart (Figure 15–16) shows, governance is a continuous process that crosses the borders from project stage to stage.

The key benefits of project governance to CA projects include:

- Minimize communication gaps
- Efficient use of skilled resources reduces project cost
- Effective financial management and consolidated budget reporting
- Single point of contact for all escalations, financials, resources, and planning
- Management of project success and standards
- Established project quality standards
- Increased customer satisfaction with project results

Global Solutions Engineering-Standard Components to Speed Implementation

As solutions are developed for customers it is not unusual to find some use cases and functional requirements that are not available in the reference architecture or out of the box functionality. CA has created an organization called the Global Solutions Engineering (GSE) team to support the delivery of such use cases. A customer may have a need for a specific report or function

TABLE 15–2. DEPLOYMENT PLAYBOOK TOOLS

Stage	Tool	Tool Description
All Stages	Meeting Minutes Template	A document containing a high level outline for meeting minutes
	Time and Material Project Status Report	A document template for reporting project progress of Time and Material projects
	Fixed Price Project Status Report	A document template for reporting project progress of Fixed Price projects
	Project Issue Log	This document lists any issues that arise during the design or deployment of the Solution. The tracking log keeps a detailed listing of the issue, and the steps for remediation.
	Milestone Acceptance Form	A single page document used for customer sign-off of deliverables
	Acceptance and Release Notice-Doc	A single page document used for customer sign-off of documentation work products; typically used for T&M projects, but may be used in Fixed Price when the deliverable does not map to a defined Milestone
	Acceptance and Release Notice-Non-Doc	A single page document used for customer sign-off of non-documentation work products; typically used for T&M projects, but may be used in Fixed Price when the deliverable does not map to a defined Milestone
	Timesheet Report	A single page document for individual consultants time recording
	Managed Product Identification	A form used whenever the need for an additional (new) product is recognized
1—Project Setup and Initiation	Playbook ReadMe	A document that outlines the playbook concept, the use of the playbook artifacts, known issues and changes to the playbook
	Deployment Playbook Guide	A document (this document) explaining how a project is organized and containing the steps for project completion
	Project Management Plan	A document template for creating a Project Management Plan
	Project Management Plan Workshop Agenda (Kickoff Meeting)	A day long agenda template for kicking off a project
	Risk Management Procedures	A document template for conveying information to the customer about the risk management procedure
	Risk Assessment Spreadsheet	A spreadsheet for assessing project risk
	Risk Inventory and Mitigation Worksheets	A spreadsheet for identifying (inventorying) and mitigating project risk
	Time Cost Expenses Tracker	A spreadsheet for Project Cost Monitoring
	Contact List Template	A template for listing all individuals participating in the project and their contact information
	Project Schedule	This document defines the tasks, milestones and associated resources for the project.

(Continued)

TABLE 15–2. *(Continued)*

Stage	Tool	Tool Description
	Technical Training Evaluation	A presentation template with instructions about how to access information related to customer personnel in CA training classes
	Kickoff Meeting Presentation Template	A presentation template to be used during a project kickoff meeting
	Managed Products Register	A spreadsheet to track the inclusion of managed products
2—Solution Requirements Definition	Requirements Gathering Workshop	A document template to define an outline for a workshop to collect requirements from the customer.
	Requirements Gathering Questionnaire	A questionnaire used for the collection of requirements for the Solution.
	Solution Requirements Specification	This document is intended to completely detail the customer's requirements for the deployment of the Solution.
3—Solution Architecture & Design	Solution Design Specification	This document is intended to outline the proposed solution, and how the Solution is designed to mitigate the current challenges. It is intended to detail the components, the integrations and network configuration of the technology. Additionally, this document is intended to outline any process changes/improvements or personnel re-alignments required to support the new solution.
	Solution Overview Presentation	This presentation outline provides a framework [to] articulate the requirements, current environment and solution to senior management.
4—Solution Integration & Configuration	Solution Integration Specification	This document is intended to outline the procedures to build a solution environment and to deploy the Solution.
	Solution Test Plan	This document is intended to outline the required tests to validate the Solution in each of the deployment environments.
5—Quality Assurance & Test	Test Report	This report is designed to detail the results of the validation tests.
	Testing Register	This report summarizes the tests and test results.
6—Configuration Documentation & Backup	Solution Run Book	This document is intended to identify the procedures to maintain the Solution after deployment.
7—Knowledge Transfer	Knowledge Transfer Checklist	This document contains an outline of the features, functions and responsibilities for the administrator and operator of the Solution.
8—Production Deployment & Conversion	Production Readiness Checklist	This document is intended to list the requirements to proceed from a development or QA environment to the production environment.
9—Project Handoff and Closure	Project Summary and Proposed Next Steps Presentation	This document is intended to outline the "proposed next steps" and additional projects needed to increase the customer's maturity within the Solution parameters.
	Final Project Closure and Acceptance Form	A single page document used for customer sign-off
	Post Project Assessment Report	A simple document form for collecting findings of a post project assessment

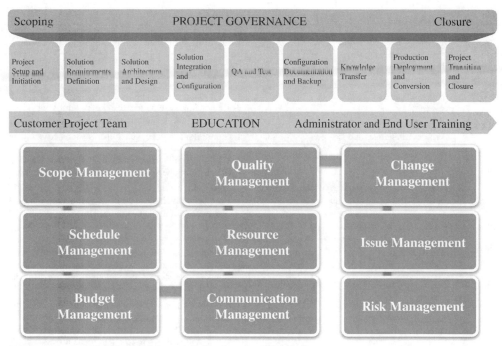

FIGURE 15–16.   The Project Manager's Deployment Playbook.

that is not part of the standard offering; in this situation the development work is performed by GSE in a custom code module. However, a big advantage by using a single group for any custom work is the ability to track customer needs across all projects. It then becomes obvious which use cases are common to multiple customers and where the solution can be expanded to provide the most value. Using this data, the GSE is able to package customizations into pre-built accelerators and components. Ultimately we want to minimize one-off customizations to a single customer and lead customers to take advantage of the pre-built components. This provides faster time to value, reduced project risk, and greater consistency.

Project Closure   One of the stages that CA emphasizes is Project Closure. Many organizations look at this stage as a formality that provides little value. However, CA has found that this stage is critical for long term customer success and sets the stage for additional business. A formal project closure process is published to reinforce best practice. It outlines a project closure meeting with the customer and defines a specific agenda. Table 15–3 outlines the key agenda items during the project closure meeting and how they contribute to project and customer success.

Continuous Improvement   One of the biggest challenges organizations have is their ability to apply lessons learned from a single project to improve planning and execution for future projects. CA has given this responsibility to two primary groups within CA Services—the Global Practices and the PMO. While at any given time a

TABLE 15–3.   PROJECT CLOSURE AGENDA ITEMS

Project Closure Agenda Item	Contribution to Project and Customer Success
Recap of Business Requirements and Goals	Confirmation that the project met the customer objectives and ROI has been created.
Final Scorecard	Final measurements on how well the project executed. Reviewing these metrics can help identify weaknesses for improvement or reinforce execution success.
Project Challenges and Lessons Learned	It is important to get the customer's perspective on any challenges the project had to overcome or lessons learned. Opening up this conversation with the customer brings in a different perspective and potentially fresh ideas for improvement.
Project Documentation CD	All project deliverables and key documentation is packaged and presented to the customer. This helps ensure the customer has copies of the final project work in a consolidated package for easy reference.
Training Plan	A review of education completed and recommendations for additional education can help ensure the customer is ready to take ownership of the deployed solution. This is critical for long term customer success and ROI.
Next Steps	The completed project may be one of many phases or additional opportunities may have been identified for additional ROI. Reviewing next steps helps ensure the customer is aware of recommendations and sets expectations for future activities.
Transition from Services to Support	It's important to have a seamless transition from the team that just completed the project to the support team that will work with the customer long term. This activity reviews with the customer the process for working with Support and transitions key documentation to the Support organization.
Project Closure Signoff	This confirms that the project is ready to close.
Customer Survey	CA surveys customers after project completion to confirm satisfaction and evaluate potential areas for improvement. Reinforcing this during project closure helps improve survey completion rates.
Customer Reference	This activity outlines the CA reference program and determines if the customer is a good fit. Including this activity in project closure helps improve the number and quality of references that can be used for future projects.
Meeting Closure	This activity opens up the discussion to any final comments from the customer and documents closure minutes for distribution.

Project Manager is primarily focused on the successful execution of a single project for a customer, the Global Practices and PMO are responsible for the portfolio of projects and improving our ability to repeat project success.

As projects complete, lessons learned are captured and deltas between the plan and actual execution identified; this feedback is used to update the offerings and Calculators. Most organizations find making use of lessons learned a big challenge, but at CA having

TABLE 15–4.

Lesson Learned	Area for Improvement	Benefit to Project Success
Customer architecture not 100% aligned to reference architecture	Updates to High Level Designer	Ability to more quickly produce architecture and design docs; increased standardization to designs
Additional tasks or deltas in effort to execute project	Updates to Solution Calculator	More accurate WBS and task estimates
Use cases not included in standard offering	Packaging custom components by GSE into accelerators	Reduce time/cost to deploy; standardization of solutions; improved ROI to the customer
Improvements to content in deliverables	Deployment Playbook updates	Higher quality deliverables
Methodology improvements	Deployment Playbook updates	Improve Time to Value
Better alignment of scope to customer business needs	Standard Offerings/Services Offerings Catalog updates	Better solution alignment to customer business needs; Improve Time to Value
Challenges not identified early in life cycle	Updates to Project Profiler and Risk Assessment Templates	Improved mitigation of risks and reinforcement of best practices; Improved Project Success

documented offerings and supporting calculators provides a central repository to take action on lessons learned.

Historically this continuous improvement of project execution and project success has been very challenging for organizations. Often times an organization may capture lessons learned or document results from project reviews, but this information is hidden away and never really used for continuous improvement. At CA, one of the biggest advantages with our repository of standard offerings and playbooks is the ability to improve over time. When a project does something very well or has challenges this information is gathered and improvements made to the repository of tools; thus allowing future projects to be even more successful. The chartin Table 15–4 outlines how some of these lessons learned are used for continuous improvement.

CA utilizes multiple methods for capturing lessons learned and data throughout the execution of the project. Figure 15–17 outlines these methods and when they are utilized for continuous improvement.

Methods for Process Improvement

Project Health Checks: The PMO and Global Practices may conduct a project health check at anytime throughout a project. These are not audits, but rather a method for confirming the tools are providing value and gathering lessons learned. Specific challenges are addressed and additional assistance to the project team provided. The feedback is primarily used to improve the calculators, Deployment Playbooks, and solution offerings.

Architecture Review Board: For large projects with multiple technology integrations, the Architecture Review Board provides additional oversight to the project during the opportunity, requirements, and architecture design stages. The Practice provides experts to the project to review the customer's business needs and ensure the project design will meet

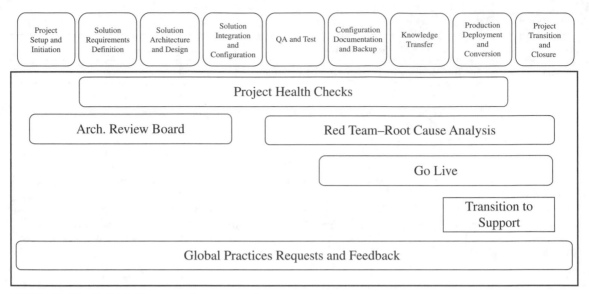

FIGURE 15–17. Methods for process improvement.

these needs. The results of these reviews can then also be used to make further improvements to the high level designer, reference architectures, and solution offerings.

Red Team—Root Cause Analysis: If a project encounters problems during the quality assurance and test or deployment stages, the Red Team may bring additional experts to the project that specialize in troubleshooting and issue resolution. This team's immediate purpose is to resolve the problem, but then a root cause analysis is performed. The results of this analysis can provide insight for areas of improvement; especially for reference architectures.

Go Live: One of the more critical stages of a project is the deployment to production. CA has created a Go Live team that includes representatives from cross-functional groups. This team meets with the project team prior to deploying the solution into production to ensure all requirements have been met, review deployment playbook execution, address any issues, and provide increased support during the deployment. Results of these checks can then provide feedback to continuously improve the Deployment Playbooks.

Transition to Support: It is critical for long term customer success that a project seamlessly transition from the project team that deployed a solution to the support team that will work with the customer long term. During this transition period the customer's architecture and design is shared with the support team along with knowledge transfer. Throughout the post-deployment phase customer success is monitored and feedback used to improve solution offerings.

Global Practices Requests and Feedback: This site allows anyone within CA to record feedback, make suggestions regarding any of the standard offerings, and request Global Practices resources for assistance. Having this data in a centralized tool supports trend analysis, ensures that suggestions are not lost, and tracks the required action items to implement improvements.

15.3 MICROSOFT CORPORATION

There are training programs that discuss how to develop good methodologies. These programs focus on the use of "proven practices" in methodology development rather than on the use of a single methodology. Microsoft has developed a family of processes that embody the core principles of and proven practices in project management. These processes combined with tools and balancing people are called Microsoft Solutions Framework (MSF).[4] What appears in the remainder of this section is just a brief summary of MSF. For more information and a deeper explanation of the topic, please refer to Mike Turner's book in the reference at the bottom of the page.

MSF was created 16 years ago when Microsoft recognized that IT was a key enabler to help businesses work in new ways. Historically, IT had a heritage of problems in delivering solutions. Recognizing this, MSF was created based on Microsoft's experience in solution delivery.

MSF is more than just project management. MSF is about solutions delivery of which project management (aka governance) is a key component. Successful delivery is balancing solution construction with governance. According to Mike Turner:

> At its foundation, MSF is about increasing awareness of the various elements and influences on successful solutions delivery—no one has a methodological silver bullet; it is next to impossible to provide best practice recipes to follow to ensure success in all projects . . . MSF is about understanding your environment so you can create a methodology that enables a harmonious balance between managing projects and building solutions.
>
> Another key point with regard to MSF is that project management is seen as a discipline that all must practice, not just the project managers. Everyone needs to be accountable and responsible to manage their own work (i.e., project manager of themselves)—that builds trust among the team (something very needed in projects with Agile-oriented project management), not so much with formally run projects (still very top-down project management).
>
> The main point that MSF tries to get across is that customers and sponsors want solutions delivered—they frankly see project management as a necessary overhead. Everyone needs to understand how to govern themselves, their team and the work that the project does—not just the project managers.

Good frameworks focus on the understanding of the need for flexibility. Flexibility is essential because the business environment continuously changes, and this in turn provides new challenges and opportunities. As an example, Microsoft recognizes that today's business environment has the following characteristics:

- Accelerating rates of change in business and technology
  - Shorter product cycles
  - Diverse products and services
  - New business models

---

4. M. S. V. Turner, *Microsoft Solutions Framework Essentials*, Microsoft Press. All rights reserved. The author is indebted to Mike Turner for providing the figures for this section in the book.

- Rapidly changing requirements
  - Legislation and corporate governance
  - Growing consumer demand
- New competitive pressures
- Globalization

Typical challenges and opportunities include:

- Escalating business expectations
  - Technology is seen as a key enabler in all areas of modern business
- Increasing business impact of technology solutions
  - Risks are higher than ever before
- Maximizing the use of scarce resources
  - Deliver solutions with smaller budgets and less time
- Rapid technology solutions
  - Many new opportunities, but they require new skills and effective teams to take advantage of them

With an understanding of the business environment, challenges and opportunities, Microsoft created MSF.[5] MSF is an adaptable framework comprising:

- Models (see Figure 15–18)
- Disciplines (see Figure 15–18)
- Foundation principles
- Mindsets

Models

Team
Model

Governance
Model

Disciplines

Project
Management
Discipline

 Risk
Management
Discipline

 Readiness
Management
Discipline

FIGURE 15–18. MSF models and disciplines. *Source:* M. S. V. Turner, *Microsoft Solutions Framework Essentials*, Microsoft Press. All rights reserved.

---

5. MSF is part of a symbiotic relationship between the classic "build it" framework and the "run it'" framework. MSF is the "build it" and Microsoft Operations Framework (MOF) is the "run it." MOF appears in Appendix A. Reproduced by permission of Microsoft Corporation.

● Proven practicesMSF is used for successfully delivering solutions faster, requiring fewer people, and involving less risk, while enabling higher quality results. MSF offers guidance in how to organize people and projects to plan, build and deploy successful technology solutions.

MSF foundation principles guide how the team should work together to deliver the solution. This includes:

● Foster open communications
● Work toward a shared vision
● Empower team members
● Establish clear accountability, shared responsibility
● Deliver incremental value
● Stay agile, expect and adapt to change
● Invest in quality
● Learn from all experiences
● Partner with customers

MSF mindsets orient the team members on how they should approach solution delivery. Included are:

● Foster a team of peers
● Focus on business value
● Keep a solution perspective
● Take pride in workmanship
● Learn continuously
● Internalize qualities of service
● Practice good citizenship
● Deliver on your commitments

With regard to proven practices, Microsoft continuously updates MSF to include current proven practices in solution delivery. All of the MSF courses use two important project management best practices. First, the courses are represented as a framework rather than as a rigid methodology. Frameworks are based upon templates, checklists, forms, and guidelines rather than the more rigid policies and procedures. Inflexible processes are one of the root causes of project failure.

The second best practice is that MSF focuses heavily on a balance between people, process, and tools rather than only technology. Effective implementation of project management is a series of good processes with emphasis on people and their working relationships: namely, communication, cooperation, teamwork, and trust. Failure to communicate and work together is another root cause of project failure.

MSF focuses not only on capturing proven practices but also on capturing the right proven practices for the right people. Mike Turner states:

The main thing that I think sets MSF apart is that it seeks to set in place a common-sense, balanced approach to solutions delivery, where effective solutions delivery is an ever changing balance of people, processes and tools. The processes and tools need to be "right sized" for the aptitude and capabilities of the people doing the work. So often "industry best practices" are espoused to people who have little chance to realize the claimed benefits.

MSF espouses the importance of people and teamwork. This includes:

- A team is developed whose members relate to each other as equals.
- Each team member is provided with specific roles and responsibilities.
- The individual members are empowered in their roles.
- All members are held accountable for the success of their roles.
- The project manager strives for consensus-based decision-making.
- The project manager gives all team members a stake in the success of the project.

The MSF team model is shown in Figure 15–19. The model defines the functional job categories or skill set required to complete project work as well as the roles and responsibilities of each team member. The team model focuses on a team of collaborating advocates rather than a strong reliance on the organizational structure.

On some projects, there may be the necessity for a team of teams. This is illustrated below in Figure 15–20.

Realistic milestones are established and serve as review and synchronization points. Milestones allow the team to evaluate progress and make midcourse corrections where the

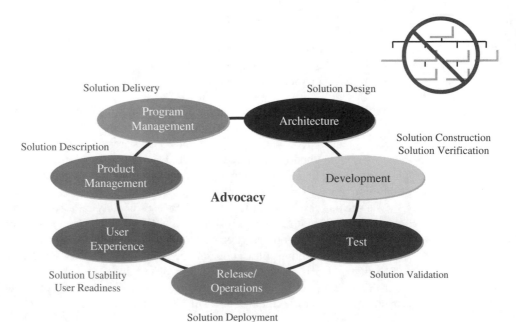

FIGURE 15–19. MSF team model. *Source:* M. S. V. Turner, *Microsoft Solutions Framework Essentials,* Microsoft Press. All rights reserved.

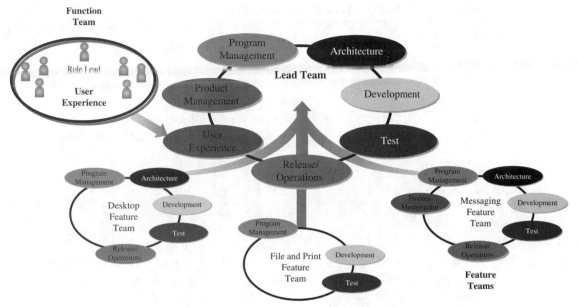

FIGURE 15–20. MSF team of teams. *Source:* M. S. V. Turner, *Microsoft Solutions Framework Essentials,* Microsoft Press. All rights reserved.

costs of the corrections are small. Milestones are used to plan and monitor progress as well as to schedule major deliverables. Using milestones benefits projects by:

● Helping to synchronize work elements
● Providing external visibility of progress
● Enabling midcourse corrections
● Focusing reviews on goals and deliverables
● Providing approval points for work being moved forward

There are two types of milestones on some programs: major and interim. Major milestones represent team and customer agreement to proceed from one phase to another. Interim milestones indicate progress within a phase and divide large efforts into workable segments.

For each of the major milestones and phases, Microsoft defines a specific goal and team focus. For example, the goal of the envisioning phase of a program might be to create a high-level review of the project's goals, constraints, and solution. The team focus for this phase might be to:

● Identify the business problem or opportunity
● Identify the team skills required
● Gather the initial requirements
● Create the approach to solve the problem
● Define goals, assumptions, and constraints
● Establish a basis for review and change

TABLE 15–5.   QUALITY GOALS AND MSF ADVOCATES

MSF Advocate	Key Quality Goals
Product management	Satisfied stakeholders
Program management	Deliver solution within project constraints Coordinate optimization of project constraints
Architecture	Design solution within project constraints
Development	Build solution to specifications
Test	Approve solution for release ensuring all issues are identified and addressed
User experience	Maximize solution usability Enhance user effectiveness and readiness
Release/operations	Smooth deployment and transition to operations

Source: M. S. V. Turner, *Microsoft Solutions Framework Essentials,* Microsoft Press. All rights reserved.

MSF also establishes quality goals for each advocate. This is a necessity because there are natural "opposing" goals to help with quality checks and balances—that way realistic quality is built in the process and not as an afterthought.

This is shown in Table 15–5.

It is often said that many programs can go on forever. MSF encourages baselining documents as early as possible but freezing the documents as late as possible. As stated by Mike Turner:

> The term "baselining" is a hard one to use without the background or definition. When a team, even if it is a team of one, is assigned work and they think they have successfully completed that work, the milestone/checkpoint status is called "Complete" (e.g., Test Plan Complete); whereas "Baseline" is used when the team that is assigned to verify the work agrees that the work is complete (e.g., Test Plan Baselined). After the Baseline milestone/checkpoint, there is no more planned work. At the point when the work is either shipped or placed under tight change control is when you declare it "Frozen"—meaning any changes must be made via the change control process. This is why you want to put off formal change management as late as possible because of the overhead involved.

This also requires a structured change control process combined with the use of versioned releases, as shown in Figure 15–21. What the arrows on the left mean is that as the solution is delivered, the solution completion increases, the knowledge of the solution space increases, and the overall risk to solution delivery goes down. The benefits of versioned releases include:

- Forcing closure on project issues
- Setting clear and motivational goals for all team members
- Effective management of uncertainty and change in project scope
- Encouraging continuous and incremental improvement
- Enabling shorter delivery time

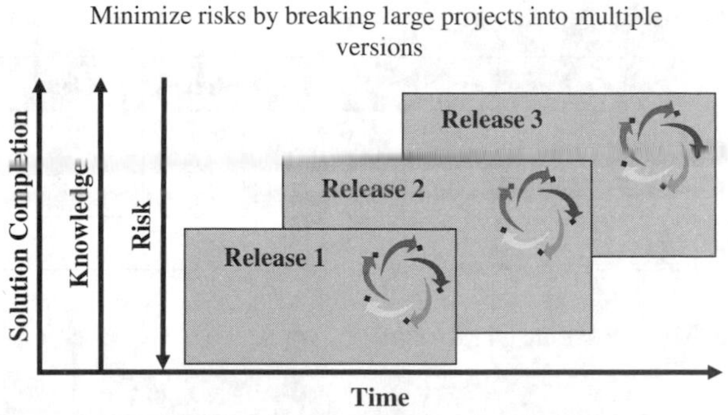

Minimize risks by breaking large projects into multiple
versions

FIGURE 15–21. MSF iterative approach. *Source:* M. S. V. Turner, *Microsoft
Solutions Framework Essentials,* Microsoft Press. All rights reserved.

One of the strengths of MSF is the existence of templates to help create project deliverables in a timely manner. The templates provided by MSF can be custom-designed to fit the needs of a particular project or organization. Typical templates might include:

● Project schedule template
● Risk factor chart template
● Risk assessment matrix template
● Postmortem template

The MSF process for risk management is shown in Figure 15–22. Because of the importance of risk management today, it has become an important component of all project management training programs.

MSF encourages all team members to manage risk, not just the project managers. The process is administered by the project manager.

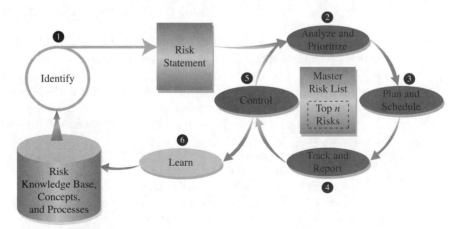

FIGURE 15–22. MSF risk management process. *Source:* M. S. V. Turner, *Microsoft Solutions Framework Essentials,* Microsoft Press. All rights reserved.

Simple Prioritization Approach

Probability/Impact	Low	Medium	High	Critical
High	M	M	H	C
Medium	L	M	M	H
Low	L	L	M	M

Multiattribute Prioritization Approach

Rating	Cost Overrun	Schedule	Technical
Low	Less than 1%	Slip 1 week	Slight effect on performance
Medium	Less than 5%	Slip 2 weeks	Moderate effect on performance
High	Less than 10%	Slip 1 month	Severe effect on performance
Critical	10% or more	Slip more than 1 month	Mission cannot be accomplished

FIGURE 15–23. Risk prioritization examples. *Source:* M. S. V. Turner, *Microsoft Solutions Framework Essentials,* Microsoft Press. All rights reserved.

- *MSF Risk Management Discipline:* A systematic, comprehensive, and flexible approach to handling risk proactively on many levels.
- *MSF Risk Management Process:* This includes six logical steps, namely identify, analyze, plan, track, control, and learn.

Some of the key points in the MSF risk approach include:

- Assess risk continuously.
- Manage risk intentionally—establish a process.
- Address root causes, not just symptoms.
- Risk is inherent in every aspect and at all levels of an endeavor.

There are numerous ways to handle risk and MSF provides the team with various options. As an example, Figure 15–23 shows two approaches for risk prioritization.

In Figure 15–18, we showed that MSF is structured around a team model and a governance model. The governance model is shown in Figure 15–24. This model appears on all of the MSF figures, illustrating that governance is continuously in place.

There are two components to the MSF governance model: project governance and process enactment:

- Project governance
  - Solution delivery process optimization
  - Efficient and effective use of project resources
  - Ensuring that the project team is and remains aligned with:
    - External (strategic) objectives
    - Project constraints
    - Demand for oversight and regulation

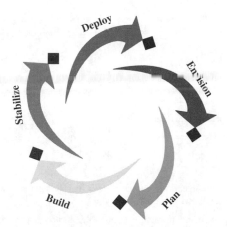

Deliverables

- Vision/scope document
- Project structure document
- Initial risk assessment document

Core Team Organized

Vision/Scope Baselined

Vision/Scope Approved

Goals

- Develop a clear understanding of what is needed and all project constraints
- Assemble the necessary team to envisage possible solution(s) with options and approaches that best meet those needs
- Establish a basis for change for the remainder of the project life cycle

- Process enactment
  - Defining, building, and deploying a solution that meets stakeholders' needs and expectations

The MSF governance model, as shown in Figure 15–24, is represented by five enactment tracks. Figures 15–25 through 15–29 provide a description of each of the enactment tracks.

MSF has established success criteria for each of the tracks:

Envision Track

- Agreement by the stakeholders and team has been obtained on:
  - Motivation for the project
  - Vision of the solution

Deliverables

- Functional specifications
- Master project plan
- Master project schedule

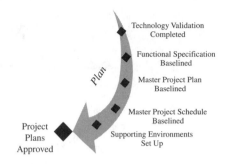

Goals

- Evolve solution concept into tangible designs and plans so it can be built in the developing track
- Find out as much information as possible, as early as possible
- Know when you have enough information to move forward

FIGURE 15–26. MSF plan track. *Source:* M. S. V. Turner, *Microsoft Solutions Framework Essentials,* Microsoft Press.

Deliverables

- Completed solution
- Training materials
- Documentation
    - Deployment processes
    - Operational procedures
    - Support and troubleshooting
- Marketing materials
- Updated master plan, schedule, and risk document

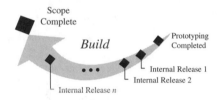

Goals

- Build all aspects of the solution in accordance with deliverables from the plan track (e.g., designs, plans, requirements)
    - Develop solution features and components (code and infra), complete all documentation deliverables and other elements of the solution (training material, etc.)
    - Test all aspects of the solution to assess the state of quality of the solution

FIGURE 15–27. MSF build track. *Source:* M. S. V. Turner, *Microsoft Solutions Framework Essentials,* Microsoft Press.

- Scope of the solution
- Solution concept
- Project team and structure
- Constraints and goals have been identified.
- Initial risk assessment has been done.
- Change control and configuration management processes have been established.
- Formal approval has been given by the sponsors/and or key stakeholders.

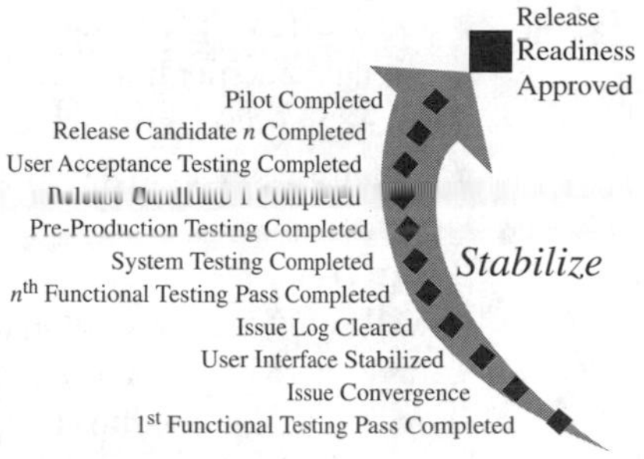

Deliverables

- Pilot review
- Release-ready versions of solution and accompanying collateral
- Test results and testing tools
- Project documents

Release Readiness Approved

Pilot Completed
Release Candidate *n* Completed
User Acceptance Testing Completed
Release Candidate 1 Completed
Pre-Production Testing Completed
System Testing Completed
*n*th Functional Testing Pass Completed
Issue Log Cleared
User Interface Stabilized
Issue Convergence
1st Functional Testing Pass Completed

Stabilize

Goals

- Improve solution quality to meet release criteria for deployment to production
- Validate solution meets stakeholder needs and expectations
- Validate solution usability from a user perspective
- Maximize success and minimize risks associated with solution deployment and operations in its target environment(s)

FIGURE 15–28. MSF stabilize track. *Source:* M. S. V. Turner, *Microsoft Solutions Framework Essentials,* Microsoft Press. All rights reserved.

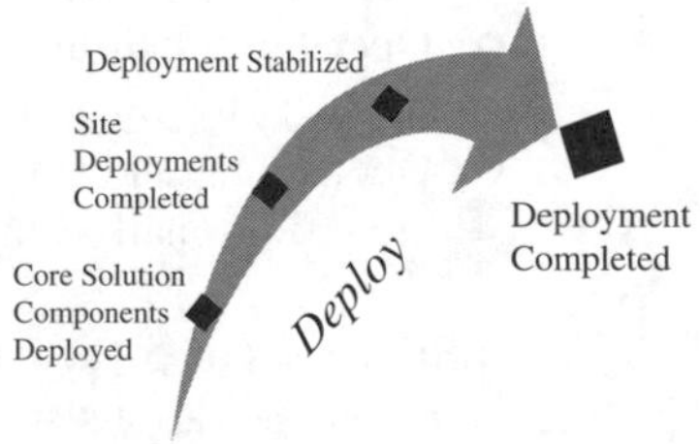

Deliverables

- Operations and support information systems
- Revised processes and procedures
- Repository of all solution collateral

Deployment Stabilized
Site Deployments Completed
Core Solution Components Deployed
Deploy
Deployment Completed

Goals

- Place solution into production at designated environment(s)
- Facilitate smooth transfer of solution from project team to operations team as soon as possible

FIGURE 15–29. MSF deploy track *Source:* M. S. V. Turner, *Microsoft Solutions Framework Essentials,* Microsoft Press. All rights reserved.

Plan Track

- Agreement with stakeholders and team has been obtained on:
  - Solution components to be delivered
  - Key project checkpoint dates
  - How the solution will be built
- Supporting environments have been constructed.
- Change control and configuration management processes are working smoothly.
- Risk assessments have been updated.

- All designs, plans, and schedules can be tracked back to their origins in the functional specifications and the functional specification can be tracked back to envisioning track deliverables.
- Sponsor(s) and/or key stakeholders have signed off.

Build Track
- All solutions are built and complete, meaning:
  - There are no additional development of features or capabilities.
  - Solution operates as specified.
  - All that remains is to stabilize what has been built.
  - All documentation is drafted.

Stabilize Track
- All elements are ready for release.
- Operations approval for release has been obtained.
- Business sign-off has been obtained.

Deploy Track
- Solution is completely deployed and operationally stable.
- All site owners signed off that their deployments were successful.
- Operations and support teams have assumed full responsibility and are fully capable of performing their duties.
- Operational and support processes and procedures as well as supporting systems are operationally stable.

MSF focuses on proactive planning rather than reactive planning. Agreements between the team and the various stakeholder groups early on in the project can make trade-offs easier, reduce schedule delays, and eliminate the need for a reduction in functionality to meet the project's constraints. This is shown in Figure 15–30.

The MSF tradeoff matrix is an early agreement made between the team and stakeholders

	Fixed	Chosen	Adjustable
Resources		✔	
Schedule	✔		
Features			✔

Given a **fixed** schedule, we will **choose** resources,
and **adjust** features as necessary.

FIGURE 15–30. Project trade-off matrix. *Source:* M. S. V. Turner, *Microsoft Solutions Framework Essentials,* Microsoft Press. All rights reserved.

15.4 DELOITTE: ENTERPRISE PROGRAM MANAGEMENT[6]

Organizations are facing increased pressures to "do more with less." They need to balance rising expectations for improved quality, ease of access, and speed of delivery with renewed pressures to demonstrate effectiveness and cost efficiency. The traditional balance between managing the business, that is, day-to-day operations, and transforming the business, that is, projects and change initiatives, is shifting. For most organizations, the proportion of resources deployed on projects and programs has increased enormously in recent years. However, the development of organizational capabilities, structures, and processes to manage and control these investments continues to be a struggle.

Furthermore, there has been a significant increase in project interdependency and complexity. While many projects and programs would have been confined to a specific function or business area in the past, increasingly, we see that there are strong systemic relationships between key initiatives. Most issues do not exist in isolation and resolutions have links and knock-on impacts beyond the scope of one problem. Not only do projects increasingly span people, process, and technology, they also cross functional, geographical, and often organizational boundaries. Without a structured approach to their deployment, projects and programs fail to deliver the expected value. The need for a strategic approach to project, program, and portfolio management has never been greater.

Deloitte's approach to project portfolio management is represented by the guiding enterprise program management (EPM) framework that provides a model within which projects, programs, and portfolios fit into a hierarchy where project execution and program delivery is aligned with enterprise strategy and leads to improved realization of desired benefits. This approach aims to strike a balance between management of results (effectiveness) and management of resources (efficiency) to deliver enterprise value.

In Figure 15–31, *strategy* includes the definition of the organization's vision and mission as well as the development of strategic goals, objectives, and performance measures.

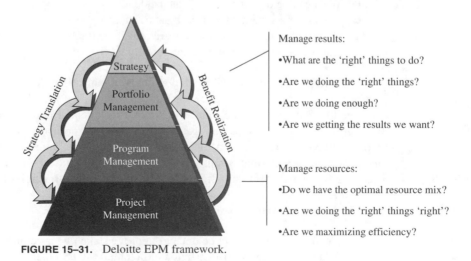

FIGURE 15–31. Deloitte EPM framework.

---

6. Material on Deloitte was provided by Daniel Martyniuk, Manager, Strategy & Operations Consulting.

The *portfolio management* capability translates the organization's enterprise strategy into reality and manages the portfolio to ensure effective program alignment, resource allocation, and benefits realization. *Program management* focuses on structuring and coordinating individual projects into related sets to ensure realization of value that would not have been attained by delivering each project independently in isolation. Disciplined *project management* ensures that defined scope-of-work packages are delivered on time, within budget, and to desired quality standards.

Strategy and Enterprise Value

Today's business leaders live in a world of perpetual motion, running and improving their enterprises at the same time. Tough decisions need to be made every day—setting directions, allocating budgets, and launching new initiatives—all to improve organizational performance and, ultimately, create and deliver value for stakeholders. It is easy to say stakeholder value is important, though much more difficult to make it influence the decisions that are made every day: where to spend time and resources, how best to get things done, and, ultimately, how to win in the competitive marketplace or in the public sector, effectively delivering a given mandate.

Supporting the strategy component of the EPM framework, Deloitte's Enterprise Value Map™ is designed to accelerate the connection between taking action and generating enterprise value. It facilitates the process of focusing on the areas that matter most, identifying practical ways to get things done, and ensuring chosen initiatives deliver their intended business value. The Enterprise Value Map™ makes this process easier by accelerating the identification of potential improvement initiatives and depicting how each can contribute to greater stakeholder value.

- The Enterprise Value Map™, as illustrated at a summary level in Figure 15–32, is powerful and appealing because it strikes a very useful and practical balance between: Strategy and tactics
- What can be done and how it can be done
- The income statement and the balance sheet
- Organizational capability and operational execution
- Current performance and future performance

Overall, the Enterprise Value Map™ helps organizations focus on the right things and serves as a graphic reminder of what they are doing and why. From an executive perspective, the Enterprise Value Map™ is a framework depicting the relationship between the metrics by which companies are evaluated and the means by which companies can improve those metrics. From a functional perspective, the Enterprise Value Map™ is a one-page summary of what companies do, why they do it, and how they can do it better. It serves as a powerful discussion framework because it helps companies focus on the issues that matter most.

The Enterprise Value Map™ is leveraged by Deloitte to:

- Identify things that can be done to improve stakeholder value
- Add structure to the prioritization of potential improvement initiatives
- Evaluate and communicate the context and value of specific initiatives

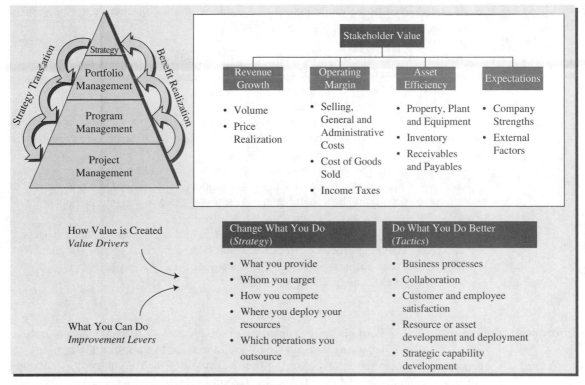

FIGURE 15–32.   Deloitte Enterprise Value Map™ (EVM).

- Deliver insights regarding the organization's current business performance
- Depict how portfolio of project and programs aligns with the drivers of value
- Identify pain points and potential improvement areas
- Depict past, current, and future initiatives

Stakeholder value is driven by four basic "value rivers": revenue growth, operating margin, asset efficiency, and expectations:

1. *Revenue Growth:* Growth in the company's "top line," or payments received from customers in exchange for the company's products and services.
2. *Operating Margin:* The portion of revenues that is left over after the costs of providing goods and services are subtracted—the key measure of operational efficiency.
3. *Asset Efficiency:* The value of assets used in running the business relative to its current level of revenues—the key measure of investment efficiency.
4. *Expectations:* The confidence stakeholders and analysts have in the company's ability to perform well in the future—the key measure of investor confidence.

There are literally thousands of actions companies can take to improve their stakeholder value performance, and the Enterprise Value Map™, in its full version, depicts several hundred of them. While the actions are quite diverse, the vast majority of them revolve around one of three objectives:

- Improve the effectiveness or efficiency of a business process
- Increase the productivity of a capital asset
- Develop or strengthen a company capability

The individual actions in the value map start to identify how a company can make those improvements. Broadly speaking, there are two basic approaches to improvement:

1. Change what you do (*change your strategy*): these actions address strategic changes— altering competitive strategies, changing the products and services you provide and to whom, and changing the assignment of operational processes to internal and external teams.
2. Do the things you do better (*improve your tactics*): these actions address tactical changes—assigning processes to different internal or external groups (or channels), redesigning core business processes, and improving the efficiency and effectiveness of the resources executing those processes.

Portfolio Management   Portfolio management is a structured and disciplined approach to achieving strategic goals and objectives by choosing the right invest- ments for the organization and ensuring the realization of their combined benefits and values while optimizing the use of available resources.

The portfolio management function provides the centralized oversight of one or more portfolios and involves identifying, selecting, prioritizing, assessing, authorizing, manag- ing, and controlling projects, programs, and other related work to achieve specific strategic goals and objectives. Adoption of a strategic approach to portfolio management enables organizations to improve the link between strategy and execution. It allows them to set priorities, gauge their capacity to deliver, and monitor achievement of project outcomes to drive the creation and delivery of enterprise value.

Deloitte's approach to portfolio management allows an organization to link its stra- tegic vision with its portfolio of initiatives and manage initiatives as they progress. It provides the critical link that translates strategy into operational success. As illustrated in Figure 15–33, the portfolio management framework helps to answer the questions of "Are we doing the 'right' things?" "Are we doing enough of the 'right' things?" and "How well are we doing these things?"

Once implemented, the framework helps to transform the business strategy into a coordinated portfolio of initiatives that work together to increase stakeholder value. Additionally, it provides the tools and techniques to keep projects on track, greatly improving an organization's chances of achieving the desired results on time and on budget. It focuses an organization on initiatives that offer the highest value creation opportunities and also provides a structure and discipline to drive performance improve- ment initiatives and aid in the identification of continuous improvement opportunities. Lastly, it confirms that the appropriate resources and budget are made available for critical

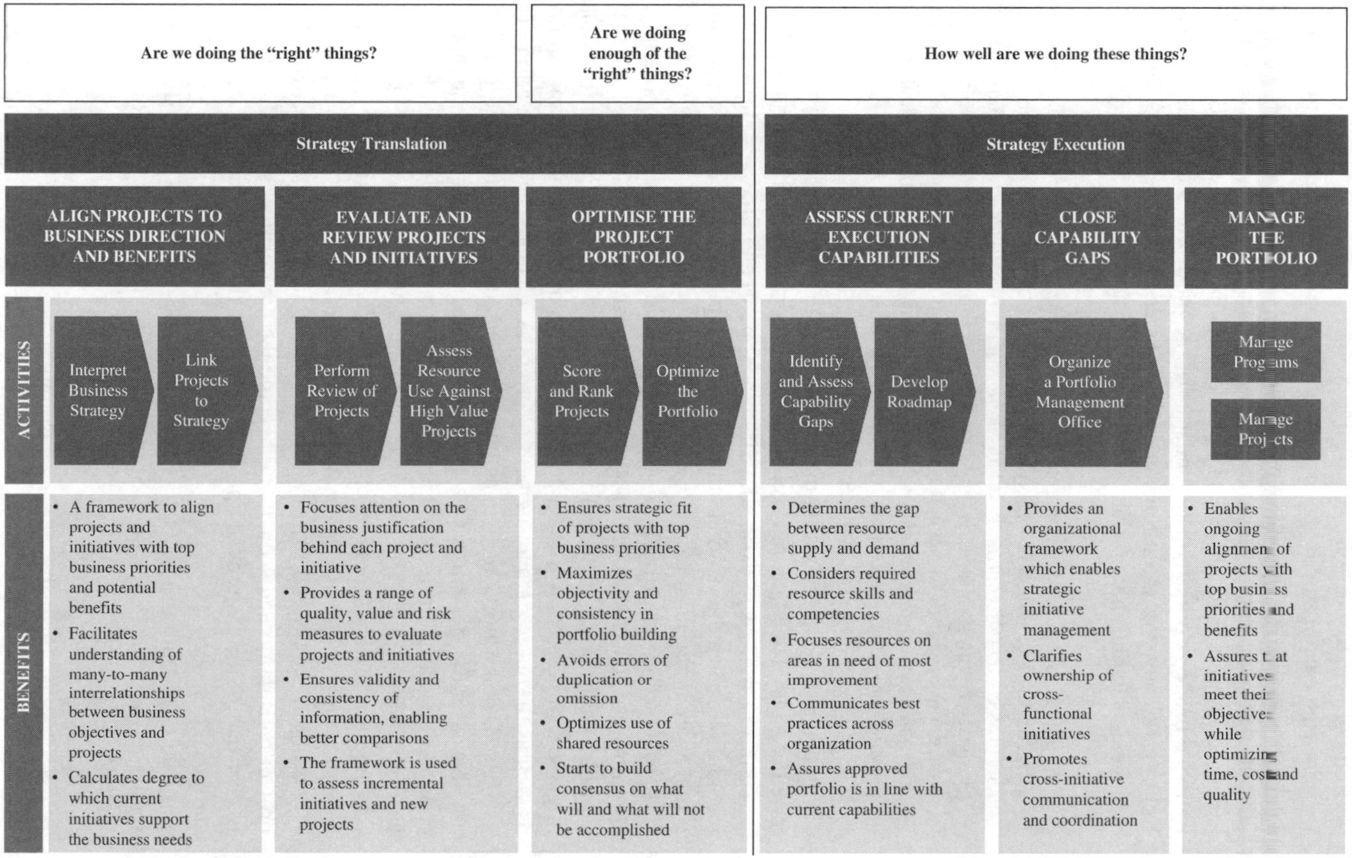

FIGURE 15–33. Deloitte portfolio management framework.

assignments and provides the tools and techniques to effectively manage an organization's portfolio of initiatives.

The first crucial step in the process of developing an effective project portfolio is the establishment of a method for determining which projects will be within and which will fall outside of the scope of the portfolio. A clear definition of what constitutes a "project" is needed, as well as identification of the criteria that will be applied to place a particular initiative inside or outside the boundaries of the portfolio. Daniel Martyniuk, Manager in Deloitte's Strategy & Operations consulting practice that specializes in project portfolio management, highlights:

> While this first step may seem basic insofar as its aim is to provide a basic framework in which to define, sort and categorize projects, the critical component is accurately capturing all projects that are currently undertaken or proposed for approval. Many hard-to-define projects are often missed, as they may take the form of day-to-day activities or may take place 'out-of-sight'. As such, it is essential to define clear boundaries between day-to-day operations and project work—failure to do so may lead to ambiguity and inaccurate representation of the true count of projects in the organization.

A consistent categorization method, such as the Deloitte investment framework illustrated in Figure 15–34, helps to answer the question of "why are we allocating resources to this project?" It aims to define the differences between initiatives allowing for immediate recognition and categorization of projects, and it provides the context for comparing projects that are different in nature or scope. It also facilitates allocation of resources by type first, followed by prioritization of projects within a type. Most importantly, it provides common ground to facilitate dialogue and prioritization discussions. Once the scope of the portfolio has been set, the organization requires a disciplined process to enable continuous alignment of projects to strategic objectives, evaluation, prioritization, and authorization as well as the on-going management of progress, changes, and realization of benefits.

The Deloitte portfolio management process, as illustrated in Figure 15–35, serves as a basis for the definition of a common portfolio management sequence. It allows for coordination across projects to capitalize on synergies and eliminate redundancies. It also helps to outline and identify projects in a comparable format when there are multiple project opportunities and/or organizational pain points to maximize the value created by the organization's initiatives while balancing risk and reward.

When the approved list of projects making up the project portfolio is established, project registration and sequencing become the next critical steps. Just because a project is now part of the "approved" project registry, it does not mean that it should or will be started right away.

There are a number of factors that need to be considered when determining the right sequence for project execution. Some of the key decision criteria for project sequencing include:

● *Strategic priority*—the level of importance placed on this project by stakeholders or organizational leadership; fast track the start of those initiatives that directly contribute to the realization of the stated business objectives.

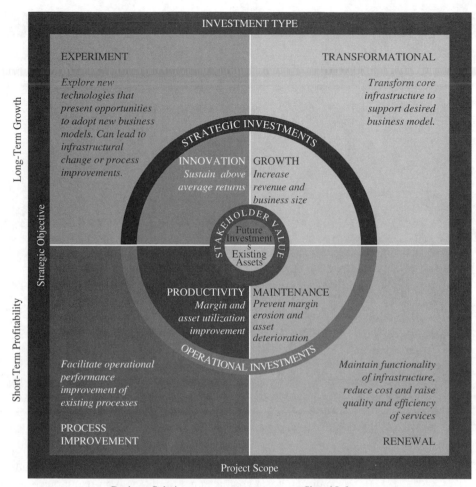

FIGURE 15–34. Deloitte investment framework.

- *Window of opportunity*—some initiatives need to be completed within a certain period of time in order to gain the desired benefits; give those initiatives the required consideration to ensure that the opportunity to provide value is not missed.
- *Project interdependencies*—ensure that all dependencies between related projects have been identified and considered when making project sequencing and initiation decisions; also, consider other dependencies that could impact the start or the successful completion of projects, such as timing of key decisions, budget cycle, etc.
- *Resource availability*—a project cannot be started until the right resources become available to begin working on that particular project; however, remember that "availability" is not a skill, and in addition to getting resources assigned to your projects, make sure that they are the "right" resources in terms of their knowledge, ability, and experience.

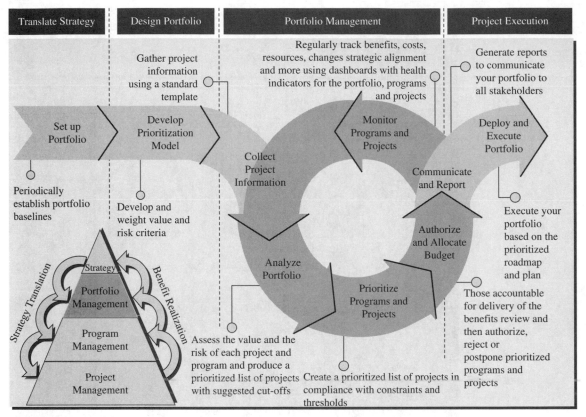

FIGURE 15–35. Deloitte portfolio management process.

● *Risk*—consider the level of risk being taken on as a result of undertaking a given set of projects; it is a good idea not to initiate high-risk projects all at once, all within the same delivery period; high-risk projects should be closely monitored and you should always strive to find a right mix of high-risk and low-risk projects; whenever possible, stagger the execution of high-risk projects and always conduct a full-risk analysis to determine and agree on appropriate risk mitigation strategies.

● *Change*—consider the novelty of the undertaking and the amount of change to be introduced into your organization as a result of implementing the proposed set of projects; ensure that your organization is ready to accept the amount and level of changes being created—there is only so much change that an organization can handle; stagger those projects that introduce significant changes and sequence them accordingly to limit change fatigue in your organization.

The key to proper sequencing, and as a result appropriate portfolio balancing, is having a sound understanding of the organization's capacity to deliver as well as the capabilities of its resources. Organizations need to know who is available to work on projects and what type of skills they have. It is often easy to determine how many people there are—so creating a resource inventory is typically not a problem. The challenge comes when trying to determine what the resources are currently working on and how much availability they have for project work or for additional projects, if they are already working on a project. One of the only methods available to get that accurate picture is to do time tracking of project resources. As a professional services organization, Deloitte leverages time tracking to provide its leadership and managers with an accurate view of who is doing what and for whom.

The expected long-term outcomes and the benefits of adopting a consistent portfolio management framework and process would include but are not limited to having the ability and capability to:

- Make conscious choices in selecting projects for implementation based on accurate and up-to-date information such as strategic alignment to business priorities, expected benefits, estimated costs, and identified risks.
- Determine capacity, that is, the number of concurrent projects, for managing small-, medium-, and large-size projects to enable prioritization and focus on priorities.
- Proactively manage risks associated with small to medium as well as large and complex transformation projects.
- Increase core competencies in project management across the organization and adopt a portfolio management approach to executive decision-making.
- Streamline and standardize processes related to the management of single, as well as the management of related, multiple projects and project portfolios.
- Maintain a current list of all projects, active and inactive, phase the initiation of projects to match capacity, and improve the delivery, that is, on time, within budget, and in accordance to requirements, of approved projects.
- Maximize use of internal resources and rationalize the use of external resources to supplement internal staff, with greater ability to ensure value and efficient completion of the approved projects portfolio.
- Measure actual, real-time performance and track the realization of project and/or program benefits, with the ability to identify actual progress made in the achievement of tangible outcomes and real results.

Program Management   In accordance with Project Management Institute's practice standards, a program is a group of related projects managed in a coordinated way to obtain benefits and controls not available from managing them individually. Programs may include elements of related work (e.g., ongoing operations) outside the scope of the discrete projects in a program. Some organizations refer to large projects as programs. If a large project is split into multiple related projects with explicit management of the benefits, then the effort becomes a program. Managing multiple projects via a program may optimize schedules across the program, deliver incremental benefits, as well as enable staffing to be optimized in the context of the overall program's needs.

As depicted in Figure 15–36, Deloitte's approach to program management highlights four core responsibilities for the program management function: program integration, dependency awareness, standards adherence, and program reporting. The figure further illustrates additional primary and secondary activities that fall within the scope of work for this function.

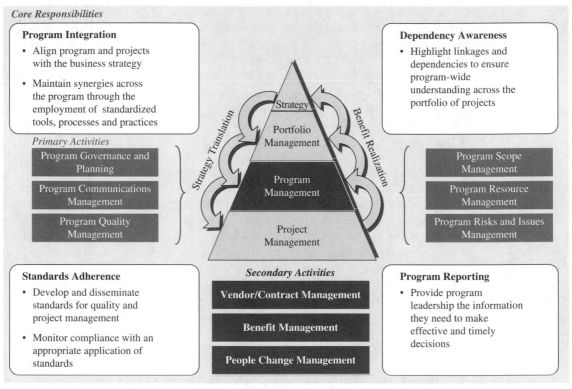

FIGURE 15–36. Deloitte program management framework.

While time, cost, and scope/quality are the key performance measures at the individual project level, coordination, communication and sequencing are the key success factors at the program level. This is because program management involves grouping and managing a series of projects in an integrated manner, and not just completing individual projects. In the end, good project management will definitely help to deliver the program on time, within budget, and according to the planned scope. Good program management will also provide a better understanding of the linkages and dependencies between projects and programs across the overall projects portfolio.

Project Management   Last, but not least, project management is concerned with the definition and delivery of specific work streams within an overall enterprise program management framework. The primary focus of the project management function includes:

- Execution and delivery of planned project deliverables and milestones
- Management of project schedules, budget, issues, risks, and change requests using defined processes and tools
- Escalation of project issues, change requests, and risks, where necessary

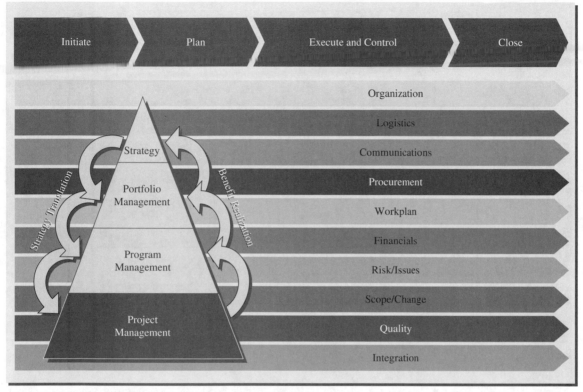

FIGURE 15–37. Deloitte project management method (PMM).

Deloitte's project management method (PMM), as illustrated in Figure15–37, systematically addresses all key components of project management. This method is scalable and flexible; it can be integrated into other Deloitte methods in whole or in part to address relevant project management issues.

Designed to help Deloitte practitioners manage their projects, the PMM is:

- *Scalable*—uses a modular design to maximize its flexibility and fits any project regardless of size or scope and accommodates each project's unique requirements.
- *Deliverables based*—allows for the iterative nature of project management processes.
- *Prescriptive*—includes tools, detailed procedures, templates, and sample deliverables specific to the management of the project that help practitioners initiate, plan, execute, control, and close the project.
- *Rich in information*—houses extensive information about method processes, work distribution, and deliverable creation.
- *Based on experience*—allows practitioners to reuse, rather than replace, the vast experience and knowledge of our practice.

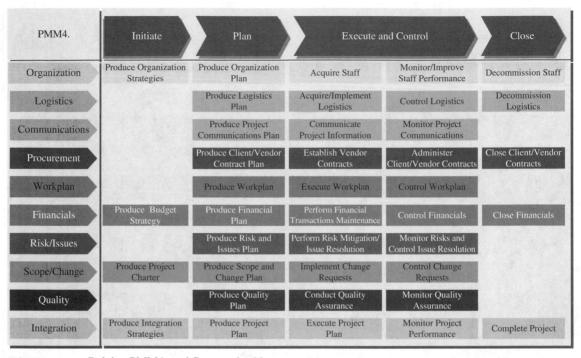

PMM4.	Initiate	Plan	Execute and Control		Close
Organization	Produce Organization Strategies	Produce Organization Plan	Acquire Staff	Monitor/Improve Staff Performance	Decommission Staff
Logistics		Produce Logistics Plan	Acquire/Implement Logistics	Control Logistics	Decommission Logistics
Communications		Produce Project Communications Plan	Communicate Project Information	Monitor Project Communications	
Procurement		Produce Client/Vendor Contract Plan	Establish Vendor Contracts	Administer Client/Vendor Contracts	Close Client/Vendor Contracts
Workplan		Produce Workplan	Execute Workplan	Control Workplan	
Financials	Produce Budget Strategy	Produce Financial Plan	Perform Financial Transactions Maintenance	Control Financials	Close Financials
Risk/Issues		Produce Risk and Issues Plan	Perform Risk Mitigation/ Issue Resolution	Monitor Risks and Control Issue Resolution	
Scope/Change	Produce Project Charter	Produce Scope and Change Plan	Implement Change Requests	Control Change Requests	
Quality		Produce Quality Plan	Conduct Quality Assurance	Monitor Quality Assurance	
Integration	Produce Integration Strategies	Produce Project Plan	Execute Project Plan	Monitor Project Performance	Complete Project

FIGURE 15–38. Deloitte PMM4 workflows and subject areas.

- *Based on best practices*—reflects Deloitte best practices and industry research and experience, allowing Deloitte practitioners to share a common language worldwide.
- *Practical*—provides realistic and useful information, focusing on what truly works.

Currently in its fourth iteration, Figure 15–38 illustrates how PMM4 organizes activities into workflows and subject areas. PMM4 categorizes tasks into four workflows: initiate, plan, execute and control, and close. workflows represent sequential processes that continue throughout the life cycle of a project and are repeated as necessary in each project phase through project completion. The subject area view categorizes tasks into 10 subject areas: organization, logistics, communications, procurement, workplan, financials, risk/issues, scope/change, quality, and integration. Subject areas aggregate tasks based on subject focus, not by when in the project life cycle they occur.

A number of benefits result from consistently applying provided project management workflows and deliverables:

- Helps project managers see the "big" picture and accelerates work
- Provides a consistent approach and a common language
- Includes deliverable templates and tools
- Incorporates quality and risk management, making it easier to improve quality and reduce risk of project deliverables
- Can be used to manage programs as well as projects

Leadership and Governance   There are additional factors that influence an organization's ability to generate value and deliver transformation results that go beyond having the right project, program, and portfolio management processes or templates. "The importance of proper governance, leadership and accountability cannot be underestimated," adds Martyniuk. "In my experience implementing project portfolio management, having the right framework to guide project stakeholders through the myriad of decisions that needs to be made on a constant basis is a critical differentiating factor between a project's success or failure."

The main purpose of governance is to specify decision rights, clarify accountabilities, and encourage desirable behaviors. Governance is about bringing the right individuals to the table to have the right conversation under the right process to make the best decisions given available information. Governance frameworks depict the structures and processes by which decisions are made, and they define sets of principles and practices for managing:

- *What* decisions need to be made
- *Who* has the authority and accountability for making decisions, and with whose input
- *How* decisions get implemented, monitored, measured, and controlled

As illustrated in Figure 15–39, effective governance requires strong executive sponsorship, clear "business" ownership, and sound technical advisory to ensure compliance

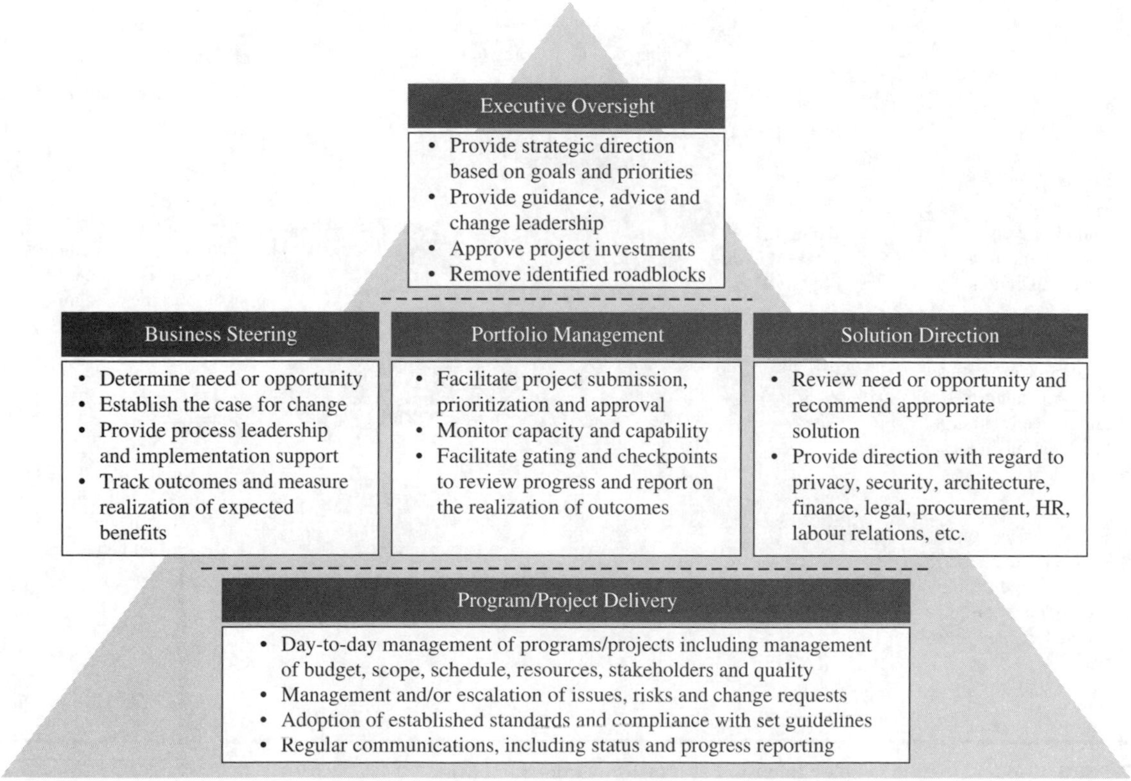

Executive Oversight
- Provide strategic direction based on goals and priorities
- Provide guidance, advice and change leadership
- Approve project investments
- Remove identified roadblocks

Business Steering
- Determine need or opportunity
- Establish the case for change
- Provide process leadership and implementation support
- Track outcomes and measure realization of expected benefits

Portfolio Management
- Facilitate project submission, prioritization and approval
- Monitor capacity and capability
- Facilitate gating and checkpoints to review progress and report on the realization of outcomes

Solution Direction
- Review need or opportunity and recommend appropriate solution
- Provide direction with regard to privacy, security, architecture, finance, legal, procurement, HR, labour relations, etc.

Program/Project Delivery
- Day-to-day management of programs/projects including management of budget, scope, schedule, resources, stakeholders and quality
- Management and/or escalation of issues, risks and change requests
- Adoption of established standards and compliance with set guidelines
- Regular communications, including status and progress reporting

FIGURE 15–39.   Deloitte project portfolio governance framework.

with established regulations, standards, and guidelines. It also requires some type of benefit and value oversight. This can be done through a committee of key stakeholders who understand the qualitative aspects of a project's value. Most importantly, each function of the chosen governance framework needs to be empowered with the required decision-making authority within their area of responsibility.

People and Organizational Change Management

Lastly, though most importantly, it is *people* that are the critical element to achieving project and transformation objectives. They are also the leading cause of transformation results falling short. Integrated people and organizational change management, human resources, and learning services need to be delivered across the portfolio at the program and project levels to drive consistency, alignment, and effective delivery across the overall transformation effort.

Illustrated in Figure 15–40, the Deloitte people dimension of transformation is a comprehensive framework that aligns with the business strategy and addresses everything from risk assessment and leadership alignment to behavioral change, communication, training, organizational design, and more.

One of the major causes of a transformation not achieving its desired objectives is the stakeholders' inability to see and feel the compelling reason for the change. As a

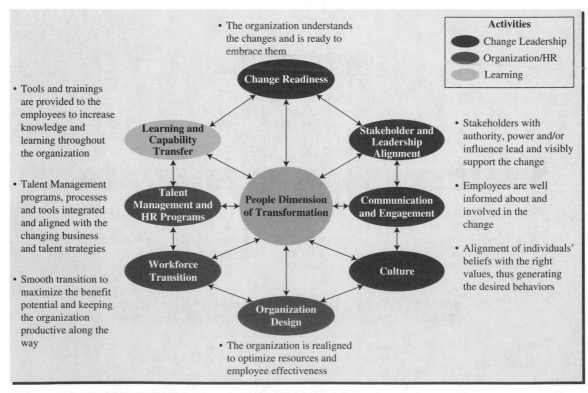

FIGURE 15–40. Deloitte people dimension of transformation framework.

result, fear, anger, or complacency can take root and cause resistance. In cases where change is more successful, individuals have a sense of passion. They create compelling, eye-catching, and dramatic situations to help people see and visualize problems, solutions, or progress in addressing complacency, lack of empowerment, or other key issues.

Sustained transformation also requires deep, personal commitment at every level of the organization. Some stakeholders will be co-creators who help shape the transformation vision and plans. Some will be interpreters. Other stakeholders will be consumers of the transformation. Effective transformation requires contributions and involvement from all types of players. Alignment and internal commitment start at the top—all leaders should be aligned, willing to seek out resistance, and committed to leading the transformation by example.

Transformation projects usually alter structures, work processes, systems, relationships, leadership styles, and behaviors that together create what we know as organizational culture. Creating the culture the organization wants—or preserving the one it already has—requires a deliberate program that aligns with other transformation activities. Without a conscious effort, it is easy to end up with an organization stuck in between new ways of working and old modes of behavior.

To optimize the investment in new business models, technologies, and processes, a formal and deliberate program of education and skills development for all people affected by the transformation is essential. Yet education and training are usually near the bottom of the transformation to-do list.

Selected keys to effectively approaching people and organizational change management on transformation projects include:

- *Get your stakeholders right.* Understand how the transformation can affect each stakeholder's group as well as key individuals.
- *Anticipate risks.* Identify pockets of resistance before they surface, along with any potential business disruptions and risks that might arise.
- *Assess the situation.* Determine whether the magnitude and pace of change are energizing or paralyzing the organization.
- *Set priorities.* Prioritize activities, tackling the most critical barriers first.
- *Influence the influencers.* Identify people within each stakeholder group who command the most respect and then get them involved as champions for the transformation.
- *Strive for real commitment.* Understand people's needs and aspirations—and then make a concerted effort to accommodate them.
- *Equip leaders to drive transformation.* Equip leaders with the unique knowledge and skills needed to help their people get through this challenging and often traumatic period. Make leaders the role models for the desired behavior.
- *Recognize there may be winners and losers.* The impact of transformation varies from one stakeholder group to the next, and some may not be happy with the outcome. Understand, engage, and inform all stakeholders.
- *Focus on the things that really matter.* An effective culture is one that creates sustainable business value, differentiates the organization from its competitors,

supports the unique requirements of the industry, and helps customers get what they really want.

- *Be consistent.* Things that drive behavior and culture must align with one another. Misalignment simply confuses people.
- *Reinforce.* Align all people-related initiatives—particularly rewards and incentives—to help foster the new culture. Establish the right leadership models and introduce new words and vocabulary that highlight the desired behavior.
- *Retain key staff.* Identify top performers and other key staff who are critical to the organization's future results. Let them know they are not at risk.
- *Capture knowledge.* Establish formal processes and systems to transfer and capture organizational knowledge—particularly for sourcing transformations.
- *Be kind but confident.* Decision makers should be gentle but not show any doubt that decisions were necessary, appropriate, and final

Conclusion

The adoption and consistent application of a standard project, program, and portfolio management frameworks, as well as implementation of the right governance along with effective people and organizational change management techniques, can lead the organization to the realization of a number of benefits, including:

- *Improved executive decision-making*—enhanced ability to determine which projects to continue/stop, based on accurate, up-to-date project status/progress information.
- *Financial transparency and accountability*—improved ability to manage budget under- and overruns and shift funds within the portfolio to better manage and respond to unforeseen circumstances and changes in priorities.
- *Enhanced resource capacity management*—availability of needed information and data to make the best use of available resources and ability to shift resources within the portfolio to optimize resource utilization across projects.
- *Proactive issues and risk management*—ability to foresee and respond to challenges before they escalate into major problems; a mechanism for bringing key issue resolution or risk mitigation decision/action requirements to the attention of the executives.
- *Standardization and consistency*—apples-to-apples comparison between projects; improved and more timely internal and external communications with staff, clients, executives, and other stakeholders.
- *Increased collaboration and better results*—optimized realization of benefits through the joint management of initiatives as an integrated portfolio; co-operation and improved removal of roadblocks to success

Although not exhaustive, the topics addressed in this submission outline selected critical success factors that, based on our practical experience, can guide any organization in the "right" direction as it embarks on the road to implement sustained project portfolio management capability to deliver real, tangible enterprise value.

15.5 LESSONS LEARNED FROM JOHNSON CONTROLS AUTOMOTIVE EXPERIENCE'S GLOBAL PROJECTS[7]

Automotive manufacturers today require suppliers that are able to leverage the opportunities of the global marketplace through effective management of multinational, multiregional projects. Johnson Controls Automotive Experience—the seventh largest automotive supplier worldwide—is a leader in global project management.

Like most global companies, Johnson Controls is structured around regional centers: North America, Europe, China, Japan/Korea, and South East Asia. To achieve successful outcomes for programs that span these regions, new ways of thinking are required. Johnson Controls' leaders have learned that in order to manage effectively in the global environment certain core deliverables and processes are critical. The following delves into some of these lessons learned, including planning, project team organization, change management, product definition, communication, and culture sensitivity.

Planning

Robust planning is a foundational element of *PMBOK® Guide* and the project management methodology. For regional projects, weak or poor planning can be transcended by a team's sheer determination and tenacity to succeed. Global teams with similar levels of resolve, who work virtually and whose members are multiculture in nature, have generally not been able to prevail over poor planning in a similar manner. Johnson Controls' automotive teams have found *robust* planning to be particularly critical to achieving successful global project outcomes.

Robust planning includes four elements: critical timing, global resource planning, and product definition planning. None of the elements are new to PMBOK; the best practice, therefore, is in the robustness of the planning activity as well as the flexibility to quickly change and communicate plans across regions.

The critical timing plan, also known at Johnson Controls as the global Product Launch System Action Plan (PAP), utilizes a Gantt-like diagram to communicate critical activities, constraints, and milestones of a project with respect to time (see Figure 15–41). The global PAP comprehends all of a project's multiple launches that occur in multiple regions. Multiregional customer product requirement dates define and drive supply chain part requirement dates and plans, which in turn drive regional PAPs and supporting tactical plans. Careful management of supply chain planning as well as part revision levels is critical to managing global projects.

The global PAP is the baseline by which the global team will operate and only changes to address major scope change. As such, the global PAP provides schedule stability from a global perspective. With this information regional support teams are able to develop their own specific PAP and tactical schedules that support the global PAP. All supporting plans and schedules are designed to be flexible to meet the needs of regional teams.

---

7. Material provided by David B. Kandt, Group Vice President: Quality, Program Management and Continuous Improvement, Automotive Group, Johnson Controls, Inc., and Terri Pomfret, D.M., PMP, Director of Technical Training, AE Leadership Institute, Automotive Group, Johnson Controls, Inc.

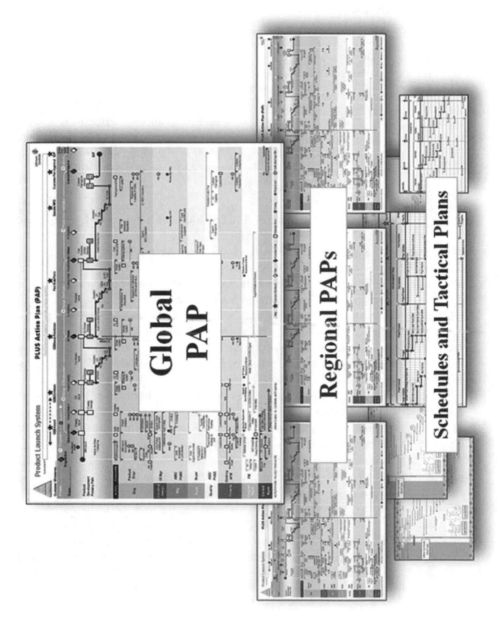

FIGURE 15-41. Cascading global PAPs.

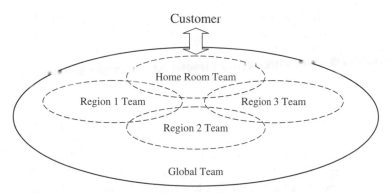

FIGURE 15–42. Johnson Controls AE global team framework.

The global project team—made up of a *home-room team* and *supporting regional teams*— meets regularly to review the status of the various schedules with respect to the global PAP. Adjustments and trade-offs in regional activities are discussed and made during these meetings to ensure the global PAP will be achieved. While a global PAP is not a panacea for global schedule management, it has improved global team effectiveness at Johnson Controls.

Project Team and Organization

Global project teams at Johnson Controls are generally made up of a home-room team and multiple supporting regional teams (see Figure 15–42). The home-room team is first responsible for global leadership as well as primary customer contact and negotiations. The home-room team is secondarily responsible for project tasks and interests concerning their region. This construct precipitates a conflict of interest as home-room members may feel more allegiance to their region than to the global project. To mitigate this potential tendency, the home-room team is aligned with the project sponsors who represent all regions. Multiple supporting teams are assembled in each region where project work must take place. These teams are semiautonomous, responsible for their own plans and activities. The regional teams are obligated to build plans that support the global PAP as well as their regional needs. Regional teams are not permitted to negotiate or work independently with the customer; customer interface is managed from a global perspective.

In order for the home-room and support teams to function harmoniously, their respective regional organizations must work cooperatively. Johnson Controls' global leaders have found that two regions cannot *co-own* project execution and decision making. To address this challenge, a global team organization chart is constructed for each global project. Generally, the organization where the home-room team is located takes the lead in the global organization, but not always. To negotiate these agreements, a leader from the company who can stand apart from the regional alignments is selected and identified as project sponsor. The negotiator's mission is to identify *who is ultimately responsible* for project execution, engineering, finance, purchasing, manufacturing, and primary customer interface; these individuals may or may not be located in the same region. With this information the sponsor builds a global project organization chart that outlines specific roles and responsibilities.

In addition to the global project organization chart, a global responsibility–approval–support–inform–consult (RASIC) chart is created. Developing a robust RASIC chart can be very difficult but critically important for global projects. Johnson Controls' global project leaders have found the RASIC chart especially helpful when disagreements arise with respect to authority and responsibility within the team.

Change Management

Understanding roles and responsibilities is important for effective change management of any type of project, but it is especially critical for global projects. Change occurs throughout the product development cycle; coordinating and communicating change are challenging when multiple regions are involved in product development. For that reason, Johnson Controls' global leaders have found that one region must be responsible for drawing and release control.

Unfortunately, however, customers do not always have clear roles and responsibilities identified and thus drive regional part requirements and/or specifications. This situation can drive up tooling and engineering costs as well as introduce unintended risk into the project. To mitigate this situation, the home-room team works with the customer—at each of its regional locations if necessary—to establish a global product definition that meets all of their needs. This effort often takes significant negotiation skill and leadership support.

Product Definition

Defining product composition for a global project would seem to be similar to that required for strictly regional projects. Global product definition, however, requires significantly more consideration and planning. Two product definition problems global projects commonly face are regional preferences and a "not-invented-here" mindset. These challenges can be driven internally and/or by the customer.

Regions accustomed to product development activity will have pre-established design preferences and practices (e.g., designing to laser weld rather than spot weld or designing a seat frame with a tube structure rather than a stamped structure). As a result of these preferences, regions tend to develop their portion of the product differently, potentially causing redundant work and/or incomparability of products. This situation causes waste as well as frustration among and between team members, sponsors, and customers. A solution to this problem is to define the product such that it must conform to the *best* design practice.

A challenge with this solution is in identifying which region employs the best design practice for a given product requirement. In some instances, the best practice is clear. All too often, however, there are multiple solutions that would achieve the product requirements. Imposing a so-called best design practice on all regions may become problematic if regional team members whose *best* practices were not selected may develop a not-invented-here mindset. Compulsory directives cause teams to be less than enthusiastic about the project and their level of responsibility and ownership.

Johnson Controls' solution to this challenge is in the use of global product centers (GPCs). The centers are strategically located throughout the Johnson Controls sphere of engineering/manufacturing interest with the mission of developing best product practices that reach across regional preferences based on data and rigorous evaluation. These centers are led by both local and global representatives whose mission is to collaborate and build global alliances of product practices. GPC leaders work with global project team leaders during the planning stages of project development to establish a detailed product definition that will meet customer's requirements as well as regional team needs. The GPC leaders also work with regional purchasing specialists to find suppliers with appropriate *global reach*.

Communication   While communication is an inherent challenge for almost any type of project, two aspects of communication have been found to be especially problematic for global projects: virtual communication and English as a second language.

Virtual Communication   People work together better when they have met each other and have developed personal relationships. Moreover, co-located teams that can meet face-to-face communicate more effectively and efficiently than virtual teams. Global projects, however, do not generally have the luxury of co-location and often never meet face-to-face. Tools used to mitigate this communication challenge include eRooms, message boards, same-time voice and visual networking, and conference calls. In addition to these tools, Johnson Controls' global team model (see Figure 15–42) is made up of subteams which are co-located by home-room and region.

Each subteam meets face-to-face on a regular basis as well as virtually with the larger global team. Only documents and communiqués that are global in nature are communicated up to the global team. This approach attempts to minimize virtual communication to only that which is critical to the global team.

To improve communication with project sponsors and stakeholders, regular *project review* meetings are held. Because these meetings are often a combination of face-to-face and virtual, they follow a standard agenda with the purpose of providing information and garnering support from global leaders.

English as a Second Language   Like most companies, Johnson Controls' global teams use English for business communication. English is, however, a second language for many global team members, which can result in miscommunication and/or misunderstanding. Global teams have found two key practices that minimize this problem: use of *simple* English and following up verbal communication with written communication.

Simple English is nothing more than avoiding colloquialisms, slang, jargon, and metaphors. Unfortunately, even with the best intentions to do otherwise, native English speakers do not often recognize the use of a sports metaphor or some type of slang. Changing to simple English takes purposeful recognition and practice.

Team members with English as a second language often find the written word easier to understand than the spoken. For that reason, global teams are encouraged to follow verbal interactions, particularly those involving agreements or direction, with written communiqués. Teams have found that including sketches or pictures along with text further improves communication.

Cultural Sensitivity   Matters involving cultural sensitivity are often forgotten or not considered by global teams. Cultural misunderstanding, however, can be a significant challenge for multicultural project teams. For example, the word *yes* means different things to different groups of people. In some cultures, *yes* means *I agree*, in others it means *I acknowledge* what has been said, not necessarily that I agree; *no* has some of the same challenges. Neither understanding of the word *yes* or *no* is necessarily right or wrong. Multiple meanings will, however, lead to misunderstanding. For that reason, it is important global teams recognize cultural differences among team members and then develop strategies to overcome them.

At the onset of each global project, team members participate in a one- to two-day workshop focused on planning and alignment activities. Recently added to the workshop is a cultural awareness activity where team members identify and discuss cultural topics relative to their team's makeup. The team then discusses approaches they will leverage to mitigate such challenges.

15.6 SIEMENS PLM SOFTWARE: DEVELOPING A GLOBAL PROJECT METHODOLOGY

For decades, large companies have allowed their multinational divisions tremendous autonomy in the way they do business. This works well as long as the various units do not have to interact and work together on projects. But when interaction is required, and each division has a different approach to project management—using different tools and processes—unfavorable results can occur. Today, the trend is toward the development of an enterprisewide methodology. Siemens PLM Software is an example of a company that has successfully developed such a methodology.[8]

About Siemens PLM Software   Siemens PLM Software, a business unit of the Siemens Industry Automation Division, is a leading global provider of product lifecycle management (PLM) software and services with 5.9 million licensed seats and 56,000 customers worldwide. Headquartered in Plano, Texas, Siemens PLM Software works collaboratively with companies to deliver open solutions that help them turn more ideas into successful products.

About Siemens Industry Automation Division   The Siemens Industry Automation Division (Nuremberg, Germany) is a worldwide leader in the fields of automation systems, low-voltage switchgear and industrial software. Its portfolio ranges from standard products for the manufacturing and process industries to solutions for whole industrial sectors that encompass the automation of entire automobile production facilities and chemical plants. As a leading software supplier, Industry Automation optimizes the entire value added chain of manufacturers—from product design and development to production, sales, and a wide range of maintenance services. With around 42,900 employees worldwide Siemens Industry Automation achieved in fiscal year 2008 total sales of EUR 8.7 billion.

---

8. The remainder of this section has been provided by Jan Hornwall, Global Services PMO, Siemens PLM Software. For more information on Siemens PLM Software products and services, visit www.siemens.com/plm. Note: Siemens and the Siemens logo are registered trademarks of Siemens AG.

Abstract

A methodology has been developed by Siemens PLM Software, a business unit of the Siemens Industry Automation Division and a leading global provider of product lifecycle management (PLM) software and services, which includes project and program management, technical activities and project governance. It is based on the use of an internal web site allowing quick access for all employees and has been successfully deployed and taught globally. The remainder of this section describes the background of the project methodology and identifies best practices for similar efforts in the future.

Project Background

Siemens PLM Software develops and deploys enterprise software for Product Lifecycle Management which includes solutions for computer aided design, manufacturung and engineering analysis (CAD/CAM/CAE) as well as for data management, collaboration, and digital factory simulation. The company's global sales and services organization is responsible for configuring and deploying the solutions at customer sites. Engagments can range from small projects over a few man-months to large multi-year global programs with hundreds of people. These projects have been executed against several different methodologies at locations worldwide. Due to the increasingly global nature of manufacturing companies and the increased demand for a variety of subject matter experts from around the world, the initiative to create a single global methodology for Siemens PLM Software was started.

Business Benefits

The following business benefits drove the development of the new methodology:

- Sharing best practices and good examples across projects and geographies
- Accelerating project deployment through quick access to tools, guides, templates and best practices
- Establishing a common methodology "language" that is used across all geographies and projects
- Sharing resources around the world and quickly developing new hires and external employees
- Enabling increased repeatability and predictability resulting in reduced risk and faster delivery times
- Providing a structured project/program governance framework
- Increasing reuse of information in projects; laying the foundation for knowledge management
- Presenting a unified and consistent project management experience for global customers

Methodology Development

Once the initiative was started, the first decision was to either develop our own methodology or procure an off-the-shelf PM methodology. Management, together with the project team, decided on developing our own methodology based on existing in-house experience. The key decision criteria was that the methodology must cover not only the project management activities but also the technical activities specific to our business. It was crucial to work on both the project management and the technical tracks together in each phase since quite a few projects are done in small teams and sometimes even the project manager has the dual role of also being the chief technical solution architect. Another key criteria was that we wanted to leverage what we already had in terms of processes and templates. We also anticipated a much higher adoption rate if the people in the field recognized parts of the methodology.

The project was planned and a project management plan was written and signed off by the project sponsor. The project team consisted of key persons from all zones; Americas, Europe, Asia-Pacific and our in-house offshore team in India who implemented the methodology web site. In total the core team consisted of approximately ten people.

The scope of the project was developed and included the following:

- The methodology must cover the entire lifecycle of a service project, from it's [*sic*] inception to close down. In addition, it must contain program management and governance
- General PM activities will be included in a section which is the same for all phases; "Manage Project"
- Technical activities will cover methods on how to identify out-of-the-box solutions while keeping customer customizations to a minimum; as well as modern techniques such as rapid prototyping and iterative development
- The activities in each phase are structured and had to be described; responsibilities of each task had to be defined; and template guidelines and supporting tools had to be available
- The methodology had to be aligned with surrounding processes such as sales process and post project support process
- Consolidate various existing service delivery methods, leveraging recognized best practices already existing within the company
- The project management track must be aligned with PMI (Project Management Institute) process and terminology
- Start with a comprehensive methodology, then a "small projects" version had to be developed
- Technology would include a web site with content management and a feedback tool with tracking functionality
- A downloadable version must be available for all employees working outside the internal network, such as in defense industry
- Training and deployment globally

The team worked virtually through conference calls but also had three face to face meetings in various locations around the globe, each four to five days long. The project to develop the methodology lasted ten months. If we had been all in the same location the time would have been significantly shorter.

At the start of the project, Siemens PLM Software was known as UGS, an independent and privately owned company. By the end of the project UGS was acquired by Siemens and renamed Siemens PLM Software. It was kept as an intact business unit. Since the methodology is tailored to our PLM business, management decided to continue the project and deploy the methodology. Alignment with the mandatory Siemens project management aspects would take place in consecutive releases.

Resulting Methodology—PLM VDM—Description

The product lifecycle management value delivery methodology (PLM VDM) provides a structured process for delivering a PLM solution. (See Figure 15–43.) PLM VDM emphasizes the unique aspects of delivering an enterprise wide solution using Siemens PLM Software products and has been adopted across the Siemens PLM Software services organization.

PLM VDM encompass both project management and technical delivery work streams. It is structured to allow iterative and flexible project delivery while maintaining "quality gates" and milestones between phases.

The seven methodology phases are:

Prealign

The purpose of the Pre-align phase is to gain a sufficient understanding of customer requirements and the scope of the project, to be able to define the high level solution outline and statement of work.

The pursuit team works with sales and the customer to establish the overall project scope, determine a preliminary project schedule, define the services strategy, conduct an infrastructure assessment and develop the initial project budget.

Align

In the Align phase, the project team works with the customer to transform the solution concepts that were defined during the Pre-align activities into a well defined overall solution architecture.

- The objectives of the Align phase are to establish a common understanding between the customer and the implementation team on all aspects of the project by capturing a complete and accurate project definition through technical workshops, use case definition, rapid prototyping and aligning the solution requirements to the "out of the box" product capabilities.
- This phase is complete when the customer accepts the use cases and requirements and authorizes the work to proceed.

Plan

In the Plan phase the project team works with the customer to develop the remaining documents that are used to execute and control the project and to develop the technical design.

- Depending on complexity of the solution, the team defines detailed plans for scope, schedule, cost, skills, resources, risks, quality and communication.
- In addition to the test environment, the team "baselines" the system infrastructure to create a stable platform for the development, test and training environments.
- This phase is complete when all necessary project management plans and the required functional and design specifications have been reviewed and base lined.

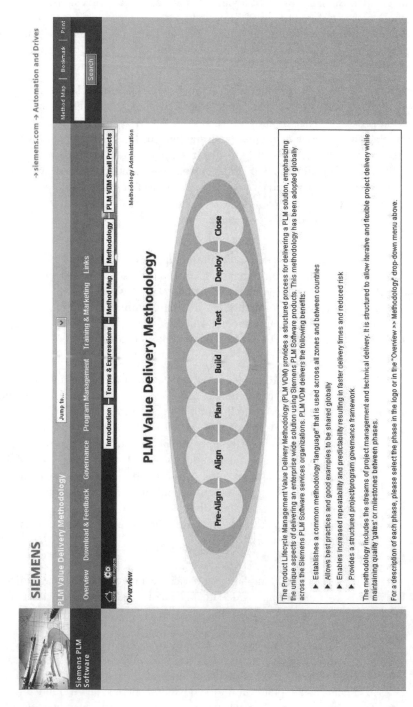

FIGURE 15–43. The PLM VDM internal website.

Build   In the Build phase the team works with the customer to create the defined solution, keeping strict adherence to the requirements.

- During the Build phase the technical team configures and tests the solution, implements the data migration strategy and develops the training materials. The Build phase also includes internal unit and integration testing.
- This phase is complete when the solution is ready for customer testing.

Test   In the Test phase the team validates that the solution is ready for production use.

- During the Test phase, representatives from the user community perform functional and system tests to verify that the system fulfills the requirements.
- This phase is complete when the solution is accepted by the customer and is ready for deployment into a production environment.

Deploy   In the Deploy phase the team delivers the production-ready solution to the end users.

- Deploying the solution consists of ensuring all the data has been migrated to the production environment, the solution is working with all interfaces and the users and helpdesk teams have been trained.
- The Deploy phase completes when the solution has been turned over to the customer for production use.

Close   In the Close phase the team assures that all administrative aspects of the project are complete.

- During the Close phase the project team completes and archives project documents and conducts project retrospective to capture and document lessons learned. The project team is released

The additional sections of the methodology are:

Program Management   Following the PMI standard, the five phases of program management are described including supporting templates:

- Pre-Program Setup
- Program Setup
- Establishing Program Management & Technical Infrastructure Setup
- Delivering Incremental Benefits
- Closing the Program

Small Projects   Projects smaller than US$100,000 in total revenue can select a
                              simplified methodology, with shorter activity descriptions and simpli-
fied templates.

Governance   Project Governance is about the line organization making sure that
                              the project is governed correctly and ensuring efficient and effective
decision-making, steering the project to success. This is done by ensuring that the following is in place:

- Project Charter
- Assigning Project Manager Authority
- Project Steering Board
- Management Review Board (MRB)
- Technical Review Board (TRB)
- Project Healthchecks and Project Retrospectives
- Approval process for new projects

Best practices are described in these sections together with supporting templates. This
is an area where the alignment with the mandatory Siemens processes is done.

Training and Marketing   This important section covers material for training, release updates,
                              links to PLM VDM training on the internal training site, presentations
of the methodology internally and for customers.

Launch and Deployment   The launch and deployment of the methodology globally included the
                              following activities:

- Announcement at conferences in each of the geographies. Management reserved
  time for the methodology to be presented which sent a positive and powerful message to all employees.
- Developing an one hour overview of the methodology as a voice-over presentation.
  This was made available on the in-house training infrastructure and participants
  could view this globally over intranet.
- Developing training material for a two day classroom training. This training was
  then given in many classes on four continents to approximately 600 persons over
  six months. This training covered all roles and included exercises.
- Additional live presentations in conference calls for several hundred additional
  people
- Developing marketing collateral; factsheet for the methodology—same format as
  for our products, developing a logotype and customer presentations
- Follow-up activities; monitoring adoption through KPIs and acting on users'
  feedback

Lessons Learned and Best Practices

- Developing a methodology in a remote/virtual team is possible but plan for several face to face meetings. These meetings were crucial to the success of the project; to bring all together in one room for longer workshops, divide work and make break-out sessions, collect all again and take final decisions. It is also important to get to know each other and have fun in off-hours—this makes working remotely afterwards much more efficient. These meetings also served as a recognition of the contribution made by all.
- Involve key persons in the various geographies in the development early on. It takes much longer than a pure top down approach by a central PMO but these persons became champions in each geographic location during the deployment and significantly increases long-term adoption.
- If you need a pure project management methodology, consider procuring an off-the-shelf product.
- If you need a business specific project methodology, consider developing your own. Most often there is tremendous knowledge in the company, it is just a matter of getting hold of it, put it in writing and making it available globally within the company
- To get management buy-in is crucial; to secure the active participation of key persons, to get slots at conferences for the announcement and to get the logotype of the methodology visible in top management presentations which is essential for adoption.
- Don't underestimate the time it takes to develop training material and giving the training globally. The success of this methodology is thanks to all the hard work done by the dedicated people successfully completing these tasks!

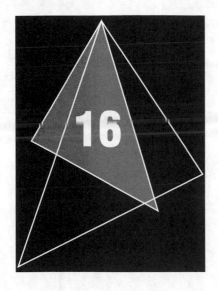

16 Value-Driven Project Management

16.0 UNDERSTANDING VALUE

Over the years, we have come to accept the traditional definition of project success, namely meeting the triple constraint. More recently, we modified our definition of success by stating that there must be a valid business purpose for working on the project. Success was then recognized as having both a business component and a technical component.

Today, we are modifying the definition of success further by adding in a "value" component, as seen in Figure 16–1. In others words, the ultimate purpose of working on the project should be to provide some form of value to both the client and the parent organization. If the project's value cannot be identified, then perhaps we should not be working on the project at all.

Value can be defined as what the stakeholders perceive the project's deliverables as being worth. Each stakeholder can have a different definition of value. Furthermore, the actual value may be expressed in qualitative terms rather than purely quantitative terms. It simply may not be possible to quantify the actual value.

The importance of the value component cannot be overstated. Consider the following statements:

- Completing a project on time and within budget does not guarantee success if you were working on the wrong project.
- Completing a project on time and within budget does not guarantee that project will provide value at completion.
- Having the greatest enterprise project management methodology in the world cannot guarantee that value will be there at the end of the project.

These three statements lead us to believe that perhaps value is now the dominating factor in the selection of a project portfolio. Project requestors must now clearly articulate the value component in the project's business case or run the risk that the project will not be considered.

597

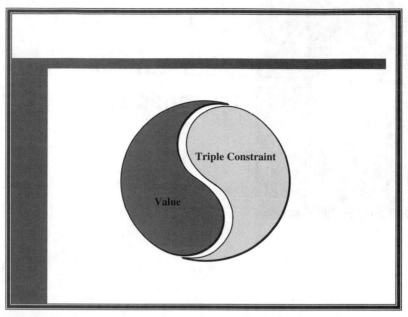

FIGURE 16–1.   Definition of success.

16.1 VALUE OVER THE YEARS

Surprisingly enough, numerous research on value has taken place over the past 15 years. Some of the items covered in the research include:

- Value dynamics
- Value gap analysis
- Intellectual capital valuation
- Human Capital Valuation
- Economic value-based analysis
- Intangible value streams
- Customer value management/mapping
- Competitive value matrix
- Value chain analysis
- Valuation of IT projects
- Balanced scorecard

The evolution of value-based knowledge seems to follow the flowchart in Figure 16–2. Research seems to take place in a specific research area such as calculating the value of software development projects or calculating shareholder value. The output of such research is usually a model that is presented to the marketplace for acceptance, rejection and/or criticism. Soon others follow with similar models but in the same research area such as software development. Once marketplace acceptance concurs on the validity of these models, textbooks appear discussing the pros and cons of one or more of the models.

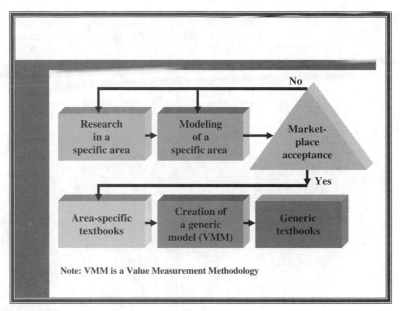

Note: VMM is a Value Measurement Methodology

FIGURE 16–2. Evolution of value-based knowledge.

With the acceptance of modeling in one specific area, modeling then spreads to other areas. The flowchart process continues until several areas have undergone modeling. Once this is completed, textbooks appear on generic value modeling for a variety of applications. Listed below are some of the models that have occurred over the past 15 years:

- Intellectual capital valuation
- Intellectual property scoring
- Balanced scorecard
- Future Value Management™
- Intellectual Capital Rating™
- Intangible value stream modeling
- Inclusive Value Measurement™
- Value performance framework
- Value measurement methodology (VMM)

There is some commonality among many of these models such that they can be applied to project management. For example, Jack Alexander created a model entitled value performance framework (VPF). The model is shown in Figure 16–3. The model focuses on building shareholder value rather than creating shareholder value.[1] The model

---

1. J. Alexander, *Performance Dashboards and Analysis for Value Creation*, Wiley, Hoboken, NJ, 2007, p. 5. Reproduced by permission of John Wiley & Sons.

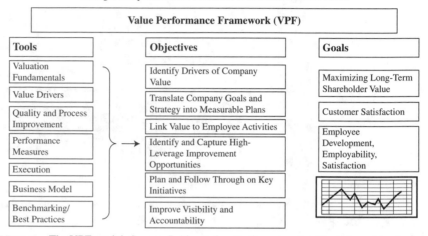

Key to maximizing long-term, sustainable shareholder value is to identify and improve on the critical value performance drivers.

The Value Performance Framework (VPF) integrates fundamental economic valuation principles, process improvement, execution planning and follow-through, and performance measures to build stakeholder value:

FIGURE 16–3. The VPF model. *Source*: J. Alexander, *Performance Dashboards and Analysis for Value Creation*, Wiley, Hoboken, NJ, 2007, p. 5. Reproduced by permission of John Wiley & Sons.

is heavily biased toward financial key performance indicators. However, the key elements VPF can be applied to project management as shown in Table 16-1. The first column contains the key elements of VPF from Jack Alexander's book and the second column illustrates the application to project management.[2]

16.2 VALUES AND LEADERSHIP

The importance of value can have a significant impact on the leadership style of project managers. Historically, project management leadership was perceived as the inevitable conflict between individual values and organizational values. Today, companies are looking for ways to get employees to align their personal values with the organization's values.

Several books have been written on this subject and the best one, in this author's opinion, is *Balancing Individual and Organizational Values* by Ken Hultman and Bill Gellerman.[3] Table 16–2 shows how our concept of value has changed over the years.[4] If you look closely at the items in Table 16–2, you can see that the changing values affect more

2. Ibid., p. 6.
3. K. Hultman and B.l Gellerman, *Balancing Individual and Organizational Values*, Jossey-Bass/Pfeiffer, San Francisco, 2002.
4. Ibid., pp. 105–106.

TABLE 16–1. APPLICATION OF VPF TO PROJECT MANAGEMENT

VPF Element	Project Management Application
Understanding key principles of valuation	Working with the project's stakeholders to define value
Identifying key value drivers for the company	Identifying key value drivers for the project
Assessing performance on critical business processes and measures through evaluation and external benchmarking	Assessing performance of the enterprise project management methodology and continuous improvement using the PMO
Creating a link between shareholder value and critical business processes and employee activities	Creating a link between project values, stakeholder values, and team member values
Aligning employee and corporate goals	Aligning employee, project, and corporate goals
Identifying key "pressure points" (high leverage improvement opportunities) and estimating potential impact on value	Capturing lessons learned and best practices that can be used for continuous improvement activities
Implementing a performance management system to improve visibility and accountability in critical activities	Establishing and implementing a series or project-based dashboards for customer and stakeholder visibility of key performance indicators
Developing performance dashboards with high-level visual impact	Developing performance dashboards for stakeholder, team and senior management visibility

Source: J. Alexander, *Performance Dashboards and Analysis for Value Creation*, Wiley, Hoboken, NJ, 2007, p. 6. Reproduced by permission of John Wiley & Sons.

TABLE 16–2. CHANGING VALUES

Moving Away From: Ineffective Values	Moving Toward: Effective Values
Mistrust	Trust
Job descriptions	Competency models
Power and authority	Teamwork
Internal focus	Stakeholder focus
Security	Taking risks
Conformity	Innovation
Predictability	Flexibility
Internal competition	Internal collaboration
Reactive management	Proactive management
Bureaucracy	Boundaryless
Traditional education	Lifelong education
Hierarchical leadership	Multidirectional leadership
Tactical thinking	Strategic thinking
Compliance	Commitment
Meeting standards	Continuous improvements

Source: Adapted from K. Hultman and B. Gellerman, *Balancing Individual Organizational Values,* Jossey-Bass/Pfeiffer, San Francisco, © 2002.

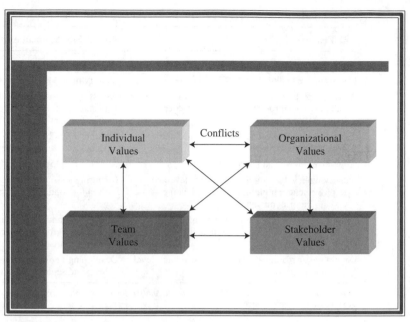

FIGURE 16–4.   Project management value conflicts.

than just individual versus organization values. Instead, it is more likely to be a conflict of four groups, as shown in Figure 16–4. The needs of each group might be: Project manager

- Accomplishment of objectives
- Demonstration of creativity
- Demonstration of innovation
- Team members
  - Achievement
  - Advancement
  - Ambition
  - Credentials
  - Recognition
- Organization
  - Continuous improvement
  - Learning
  - Quality
  - Strategic focus
  - Morality and ethics
  - Profitability
  - Recognition and image
- Stakeholders
  - Organizational stakeholders: job security
  - Product/market stakeholders: quality performance and product usefulness
  - Capital markets: financial growth

There are several reasons why the role of the project manager and the accompanying leadership style have changed. Some reasons include:

- We are now managing our business as though it is a series of projects.
- Project management is now viewed as a full-time profession.
- Projects manager are now viewed as both business managers and project managers and are expected to make decisions in both areas.
- The value of a project is measured more so in business terms rather than solely technical terms.
- Project management is now being applied to parts of the business that traditionally have not used project management.

The last item requires further comment. Project management works well for the "traditional" type of project, which includes:

- Time duration of 6–18 months.
- The assumptions are not expected to change over the duration of the project.
- Technology is known and will not change over the duration of the project.
- People that start on the project will remain through to completion.
- The statement of work is reasonably well defined.

Unfortunately, the newer types of projects are more nontraditional and have the following characteristics:

- Time duration over several years.
- The assumptions can and will change over the duration of the project.
- Technology will change over the duration of the project.
- People that approved the project may not be there at completion.
- The statement of work is ill defined and subject to numerous changes.

The nontraditional types of projects have made it clear why traditional project management must change. There are three areas that necessitate changes:

- New projects have become:
  - Highly complex and with greater acceptance of risks that may not be fully understood during project approval
  - More uncertain in the outcomes of the projects and with no guarantee of value at the end
  - Pressed for speed-to-market irrespective of the risks
- The statement of work (SOW) is:
  - Not always well-defined, especially on long-term projects
  - Based upon possibly flawed, irrational, or unrealistic assumptions
  - Inconsiderate of unknown and rapidly changing economic and environmental conditions
  - Based upon a stationary rather than moving target for final value

- The management cost and control systems [enterprise project management methodologies (EPM)] focus on:
  - An ideal situation (as in the *PMBOK® Guide*)
  - Theories rather than the understanding of the workflow
  - Inflexible processes
  - Periodically reporting time at completion and cost at completion but not value (or benefits) at completion
  - Project continuation rather than cancelling projects with limited or no value

Over the years, we have taken several small steps to plan for the use of project management on nontraditional projects. This included:

- Project managers are provided with more business knowledge and are allowed to provide an input during the project selection process.
- Because of the above item, project managers are brought on board the project at the beginning of the initiation phase rather than the end of the initiation phase.
- Projects managers now seem to have more of an understanding of technology rather than a command of technology.

The new types of projects combined with a heavy focus on business alignment and value brought with it a classification system as shown in Figure 16–5. *Operational Projects:* These projects, for the most part, are repetitive projects such as payroll and taxes.

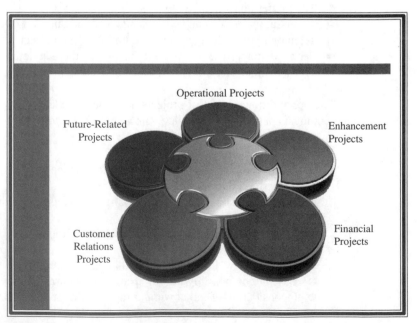

FIGURE 16–5. Classification of projects.

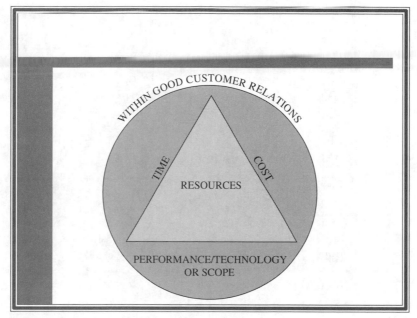

FIGURE 16–6. Traditional triple constraint.

They are called projects but they are managed by functional managers without the use of an enterprise project management methodology.

- *Enhancement or Internal Projects:* These are projects designed to update processes, improve efficiency and effectiveness, and possibly improve morale.
- *Financial Projects:* Companies require some form of cash flow for survival. These are projects for clients external to the firm and have an assigned profit margin.
- *Future-Related Projects:* These are long-term projects to produce a future stream of products or services capable of generating a future cash flow. These projects may be an enormous cash drain for years with no guarantee of success.
- *Customer-Related Projects:* Some projects may be performed, even at a financial loss, to maintain or build a customer relationship. However, performing too many of these projects can lead to financial disaster.

These new types of projects focus more on value than on the triple constraint. Figure 16–6 shows the traditional triple constraint whereas Figure 16–7 shows the value-driven triple constraint. With the value-driven triple constraint, we emphasize stakeholder satisfaction and decisions are made around the four types of projects (excluding operational projects) and the value that is expected on the project. In others words, success is when the value is obtained, hopefully within the triple constraint. As a result we can define the four cornerstones of success using Figure 16–8. Very few projects are completed without some trade-offs. This holds true for both the tradition projects and the value-driven projects. As shown in Figure 16–9, traditional trade-offs result in an elongation of the schedule

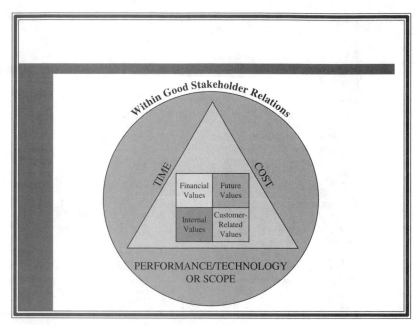

FIGURE 16–7. Value-driven triple constraint.

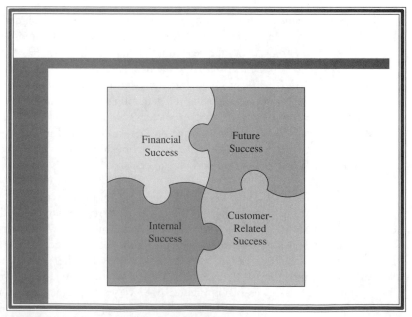

FIGURE 16–8. Four cornerstones of success.

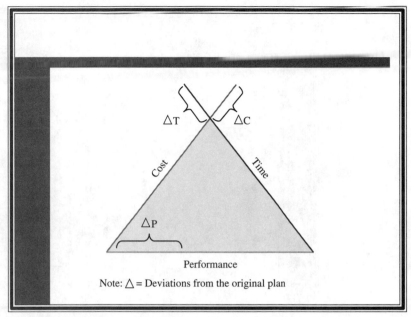

FIGURE 16–9. Traditional trade-offs.

and an increase in the budget. The same holds true for the value-driven projects shown in Figure 16–10. But the major difference is with performance. With traditional trade-offs, we tend to reduce performance to satisfy other requirements. With value-driven projects, we tend to increase performance in hopes of providing added value, and this tends to cause much larger cost overruns and schedule slippages than with traditional trade-offs.Projects managers generally do not have the sole authority for scope/performance increases or decreases. For traditional trade-offs, the project manager and the project sponsor, working together, may have the authority to make trade-off decisions.

However, for value-driven projects all or most of the stakeholders may need to be involved. This can create additional issues, such as:

● It may not be possible to get all of the stakeholders to agree on a value target during project initiation.
● Getting agreement on scope changes, extra costs, and schedule elongations is significantly more difficult the further along you are in the project.
● Stakeholders must be informed of this at project initiation and continuously briefed as the project progresses; that is, no surprises!

Conflicts among the stakeholders may occur. For example:

● During project initiation, conflicts among stakeholder are usually resolved in favor of the largest financial contributors.
● During execution, conflicts over future value are more complex, especially if major contributors threaten to pull out of the project.

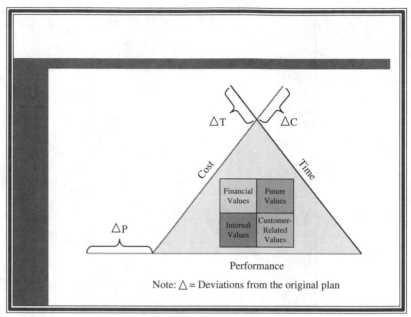

FIGURE 16–10. Value-driven trade-offs.

For projects that have a large number of stakeholders, project sponsorship may not be effective with a single-person sponsor. As such, committee sponsorship may be necessary. Membership in the committee may include:

- Perhaps a representative from all stakeholder groups
- Influential executives
- Critical strategic partners and contractors
- Others based upon the type of value

Responsibilities for the sponsorship committee may include:

- Taking a lead role in the definition of the targeted value
- Taking a lead role in the acceptance of the actual value
- Ability to provide additional funding
- Ability to assess changes in the enterprise environment factors
- Ability to validate and revalidate the assumptions

Sponsorship committees may have significantly more expertise than the project manager in defining and evaluating the value in a project.

Value-driven projects require that we stop focusing on budgets and schedules and instead focus on how value will be captured, quantified, and reported. Value must be measured in terms of what the project contributes to the company's objectives. To do this, an understanding of four terms is essential.

- *Benefits:* An advantage.
- *Value:* What the benefit is worth.
- *Business Drivers:* Target goals or objectives defined through benefits or value and expressed more in business terms than technical terms.
- *Key Performance Indicator (KPI):* Value metrics that can be assessed either quantitatively or qualitatively.

Traditionally, business plans identified the benefits expected from the project. Today, portfolio management techniques require identification of the value as well as the benefits. However, conversion from benefits to value is not easy.[5] Table 16–3 illustrates the benefit-to-value conversion. Also, as shown in Figure 16–11, there are shortcomings in the conversion process that can make the conversion difficult.

We must identify the business drivers and they must have measurable performance indicators using KPI. Failure to do so may make true assessment of value impossible. Table 16–4, illustrates typical business drivers and the accompanying KPIs.

TABLE 16–3. MEASURING VALUE FROM BENEFITS

Expected Benefits	Value Conversion
Profitability	Easy
Customer satisfaction	Hard
Goodwill	Hard
Penetrate new markets	Easy
Develop new technology	Medium
Technology transfer	Medium
Reputation	Hard
Stabilize work force	Easy
Utilize unused capacity	Easy

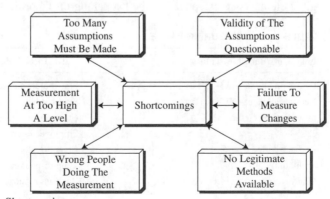

FIGURE 16–11. Shortcomings.

5. For additional information on the complexities of conversion, see J. J. Phillips, T. W. Bothell, and G. L. Snead, *The Project Management Scorecard*, Butterworth Heinemann, Oxford, United Kingdom 2002, Chapter 13.

TABLE 16–4.   BUSINESS DRIVERS AND KPI

Business Drivers	Key Performance Indicators
Sales growth	Monthly sales or market share
Customer satisfaction	Monthly surveys
Cost savings	Earned-value measurement system
Process improvement	Time cards

KPIs are metrics for assessing value. With traditional project management, metrics are established by the enterprise project management methodology and fixed for the duration of the projects life cycle. But with value-driven project management, metrics can change from project to project, during a life-cycle phase and over time because of:

● The way the company defines value internally
● The way the customer and contractor jointly define success and value at project initiation
● The way the customer and contractor come to an agreement at project initiation as to what metrics should be used on a given project
● New or updated versions of tracking software
● Improvements to the enterprise project management methodology and accompanying project management information system
● Changes in the enterprise environmental factors

Even with the best possible metrics, measuring value can be difficult. Some values are easy to measure while others are more difficult. The easy values to measure are often called soft or tangible values whereas the hard values are often considered as intangible values. Table 16–5 illustrates some of the easy and hard values to measure. Table 16–6 shows some of the problems associated with measuring both hard and soft values.

The intangible elements are now considered by some to be more important than tangible elements. This appears to be happening on IT projects where executives are giving

TABLE 16–5.   MEASURING VALUES

Easy (Soft/Tangible) Values	Hard (Intangible) Values
Return on investment (ROI) calculators	Stockholder satisfaction
Net present value (NPV)	Stakeholder satisfaction
Internal rate of return (IRR)	Customer satisfaction
Cash flow	Employee retention
Payback period	Brand loyalty
Profitability	Time-to-market
Market share	Business relationships
	Safety
	Reliability
	Reputation
	Goodwill
	Image

TABLE 16–6. PROBLEMS WITH MEASURING VALUES

Easy (Soft/Tangible) Values	Hard (Intangible) Values
Assumptions are often not disclosed and can affect decision-making	Value is almost always based upon subjective-type attributes of the person doing the measurement
Measurement is very generic	It is more of an art than a science
Measurement never meaningfully captures the correct data	Limited models are available to perform the measurement

significantly more attention to intangible values. The critical issue with intangible values is not necessarily in the end result, but in the way that the intangibles were calculated.[6]

Tangible values are usually expressed quantitatively whereas intangible values are expressed through a qualitative assessment. There are three schools of thought for value measurement:

- *School 1:* The only thing that is important is ROI.
- *School 2:* ROI can never be calculated effectively; only the intangibles are what are important.
- *School 3:* If you cannot measure it, then it does not matter.

The three schools of thought appear to be an all-or-nothing approach where value is either 100 percent quantitative or 100 percent qualitative. The best approach is most likely a compromise between a quantitative and qualitative assessment of value. It may be necessary to establish an effective range, as shown in Figure 16–12, which is a compromise among the three schools of thought. The effective range can expand or contract.

The timing of value measurement is absolutely critical. During the life cycle of a project, it may be necessary to switch back and forth from qualitative to quantitative assessment and, as stated previously, the actual metrics or KPIs can change as well. Certain critical questions must be addressed:

- When or how far along the project life cycle can we establish concrete metrics, assuming it can be done at all?
- Can value be simply perceived and therefore no value metrics are required?
- Even if we have value metrics, are they concrete enough to reasonably predict actual value?
- Will we be forced to use value-driven project management on all projects or are there some projects where this approach is not necessary?
  - Well defined versus ill defined
  - Strategic versus tactical
  - Internal versus external

6. For additional information on the complexities of measuring intangibles, see note 5, Chapter 10. The authors emphasize that the true impact on a business must be measured in business units.

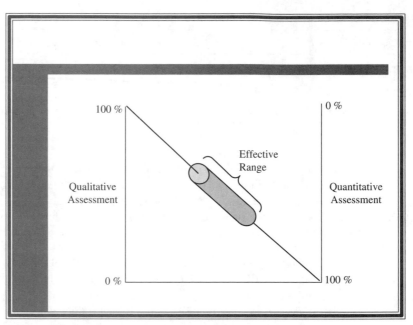

FIGURE 16–12. Quantitative versus qualitative assessment.

● Can we develop a criterion for when to use value-driven project management or should we use it on all projects but at a lower intensity level?

For some projects, assessing value at closure may be difficult. We must establish a time frame for how long we are willing to wait to measure the value or benefits from a project. This is particularly important if the actual value cannot be identified until some time after the project has been completed. Therefore, it may not be possible to appraise the success of a project at closure if the true economic values cannot be realized until some time in the future.

Some practitioners of value measurement question whether value measurement is better using boundary boxes instead of life-cycle phases. For value-driven projects, the potential problems with life-cycle phases include:

● Metrics can change between phases and even during a phase.
● Inability to account for changes in the enterprise environmental factors.
● Focus may be on the value at the end of the phase rather than the value at the end of the project.
● Team members may get frustrated not being able to quantitatively calculate value.

Boundary boxes, as shown in Figure 16–13, have some degree of similarity to statistic process control charts. Upper and lower strategic value targets are established. As long as the KPIs indicate that the project is still within the upper and lower value targets, the project's objectives and deliverables will not undergo any scope changes or trade-offs.

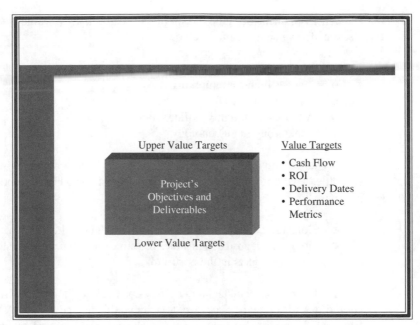

FIGURE 16–13. The boundary box.

Value-driven projects must undergo value health checks to confirm that the project will make a contribution of value to the company. Value metrics, such as KPIs, indicate the current value. What is also needed is an extrapolation of the present into the future. Using tradition project management combined with the traditional enterprise project management methodology, we can calculate the time at completion and the cost at completion. These are common terms that are part of earned-value measurement systems. But as stated previously, being on time and within budget is no guarantee that the perceived value will be there at project completion.

Therefore, instead of using an enterprise project management methodology which focuses on earned-value measurement, we may need to create a VMM which stresses the value variables. With VMM, time to complete and cost to complete are still used, but we introduce a new term entitled value (or benefits) at completion. Determination of value at completion must be done periodically throughout the project. However, periodic re-evaluation of benefits and value at completion may be difficult because:

● There may be no reexamination process.
● Management is not committed and believes that the re-examination process is unreal.
● Management is overoptimistic and complacent with existing performance.
● Management is blinded by unusually high profits on other projects (misinterpretation).
● Management believes that the past is an indication of the future.

An assessment of value at completion can tell us if value trade-offs are necessary. Reasons for value trade-offs include:

- Changes in the enterprise environmental factors
- Changes in the assumptions
- Better approaches found, possibly with less risk
- Availability of highly skilled labor
- Breakthrough in technology

As stated previously, most value trade-offs are accompanied by an elongation of the schedule. Two critical factors that must be considered before schedule elongation takes place are:

- Elongating a project for the desired or added value may incur risks.
- Elongating a project consumes resources which may have already been committed to other projects in the portfolio.

Traditional tools and techniques may not work well on value-driven projects. The creation of a VMM may be necessary to achieve the desired results. A VMM can include the features of earned-value measurement systems (EVMSs) and enterprise project management systems (EPMs), as shown in Table 16–7. But additional variables must be included for the capturing, measurement, and reporting of value.

TABLE 16–7.   COMPARISON OF EVMS, EPM, AND VMM

Variable	EVMS	EPM	VMM
Time	✓	✓	✓
Cost	✓	✓	✓
Quality		✓	✓
Scope		✓	✓
Risks		✓	✓
Tangibles			✓
Intangibles			✓
Benefits			✓
Value			✓
Trade-offs			✓

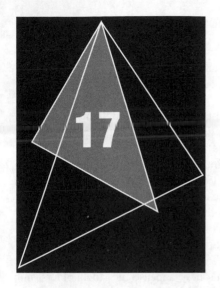

Effect of Mergers and Acquisitions on Project Management

17.0 INTRODUCTION

All companies strive for growth. Strategic plans are prepared identifying new products and services to be developed and new markets to be penetrated. Many of these plans require mergers and acquisitions to obtain the strategic goals and objectives. Yet even the best-prepared strategic plans often fail. Too many executives view strategic planning as planning only, often with little consideration given to implementation. Implementation success is vital during the merger and acquisition process.

17.1 PLANNING FOR GROWTH

Companies can grow in two ways: internally and externally. With internal growth, companies cultivate their resources from within and may spend years attaining their strategic targets and marketplace positioning. Since time may be an unavailable luxury, meticulous care must be given to make sure that all new developments fit the corporate project management methodology and culture.

External growth is significantly more complex. External growth can be obtained through mergers, acquisitions, and joint ventures. Companies can purchase the expertise they need very quickly through mergers and acquisitions. Some companies execute occasional acquisitions, whereas other companies have sufficient access to capital such that they can perform continuous acquisitions. However, once again, companies often neglect to consider the impact on project management. Best practices in project management may not be transferable from one company to another. The impact on project management systems resulting from mergers and acquisitions is often irreversible, whereas joint ventures can be terminated.

This chapter focuses on the impact on project management resulting from mergers and acquisitions. Mergers and acquisitions allow companies to achieve strategic targets at a speed not easily achievable through internal growth, provided that the sharing of assets and capabilities can be done quickly and effectively. This synergistic effect can produce opportunities that a firm might be hard-pressed to develop themselves.

Mergers and acquisitions focus on two components: preacquisition decision-making and postacquisition integration of processes. Wall Street and financial institutions appear to be interested more in the near-term financial impact of the acquisition rather than the long-term value that can be achieved through better project management and integrated processes. During the mid-1990s, companies rushed into acquisitions in less time than the company required for a capital expenditure approval. Virtually no consideration was given to the impact on project management and whether or not the expected best practices would be transferable. The result appears to have been more failures than successes.

When a firm rushes into an acquisition, very little time and effort appear to be spent on postacquisition integration. Yet, this is where the real impact of best practices is felt. Immediately after an acquisition, each firm markets and sells products to each other's customers. This may appease the stockholders, but only in the short term. In the long term, new products and services will need to be developed to satisfy both markets. Without an integrated project management system where both parties can share the same best practices, this may be difficult to achieve.

When sufficient time is spent on preacquisition decision-making, both firms look at combining processes, sharing resources, transferring intellectual property, and the overall management of combined operations. If these issues are not addressed in the preacquisition phase, unrealistic expectations may occur during the postacquisition integration phase.

17.2 PROJECT MANAGEMENT VALUE-ADDED CHAIN

Mergers and acquisitions are expected to add value to the firm and increase its overall competitiveness. Some people define value as the ability to maintain a certain revenue stream. A better definition of value might be defined as the competitive advantages that a firm possesses as a result of customer satisfaction, lower cost, efficiencies, improved quality, effective utilization of personnel, or the implementation of best practices. True value occurs *only* in the postacquisition integration phase, well after the actual acquisition itself.

Value can be analyzed by looking at the value chain: the stream of activities from upstream suppliers to downstream customers. Each component in the value chain can provide a competitive advantage and enhance the final deliverable or service. Every company has a value chain, as illustrated in Figure 17–1. When a firm acquires a supplier, the value chains are combined and expected to create a superior competitive position. Similarly, the same result is expected when a firm acquires a downstream company. But it may not be possible to integrate the best practices.

Historically, value chain analysis was used to look at a business as a whole.[1] However, for the remainder of this chapter, the sole focus will be the project management value-added chain and the impact of mergers and acquisitions on the performance of the chain.

Figure 17–2 shows the project management value-added chain. The primary activities are those efforts needed for the physical creation of a product or service. The primary

---

1. M. E. Porter, *Competitive Advantage*, Free Press, New York, 1985, Chap. 2.

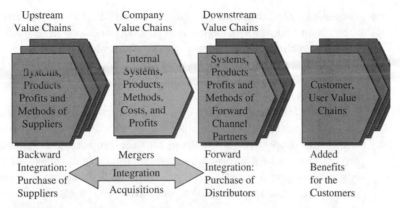

FIGURE 17–1. Generic value-added chain.

activities can be considered to be the five major process areas of project management: project initiation, planning, execution, control, and closure.

The support activities are those company-required efforts needed for the primary activities to take place. At an absolute minimum, the support activities must include:

● *Procurement Management:* The quality of the suppliers and the products and services they provide to the firm.

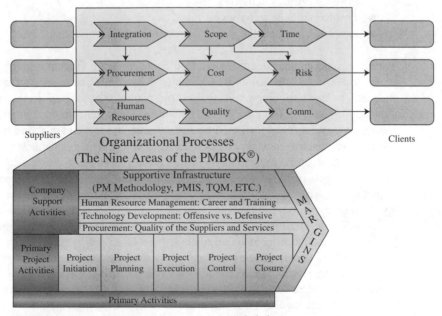

FIGURE 17–2. Project management value-added chain.

- *Technology Development:* The quality of the intellectual property controlled by the firm and the ability to apply it to products and services both offensively (new product development) or defensively (product enhancements).
- *Human Resource Management:* The ability to recruit, hire, train, develop, and retain project managers. This includes the retention of project management intellectual property.
- *Supportive Infrastructure:* The quality of the project management systems necessary to integrate, collate, and respond to queries on project performance. Included within the supportive infrastructure are the project management methodology, project management information systems, total quality management system, and any other supportive systems. Since customer interfacing is essential, the supportive infrastructure can also include processes for effective supplier–customer interfacing.

These four support activities can be further subdivided into the nine process areas of the *PMBOK*® *Guide*. The arrows connecting the nine *PMBOK*® *Guide* areas indicate their interrelatedness. The exact interrelationships may vary for each project, deliverable, and customer.

Each of these primary and support activities, together with the nine process areas, is required to convert material received from your suppliers into deliverables for your customers. In theory, Figure 17–2 represents a work breakdown structure for a project management value-added chain:

- *Level 1:* value chain
- *Level 2:* primary activities
- *Level 3:* support activities
- *Level 4:* nine *PMBOK*® *Guide* process areas

The project management value-added chain allows a firm to identify critical weaknesses where improvements must take place. This could include better control of scope changes, the need for improved quality, more timely status reporting, better customer relations, or better project execution. The value-added chain can also be useful for supply chain management. The project management value-added chain is a vital tool for continuous improvement efforts and can easily lead to the identification of best practices.

Executives regard project costing as a critical, if not the most critical, component of project management. The project management value chain is a tool for understanding a project's cost structure and the cost control portion of the project management methodology. In most firms, this is regarded as a best practice. Actions to eliminate or reduce a cost or schedule disadvantage need to be linked to the location in the value chain where the cost or schedule differences originated.

The "glue" that ties together elements within the project management chain is the project management methodology. A project management methodology is a grouping of forms, guidelines, checklists, policies, and procedures necessary to integrate the elements within the project management value-added chain. A methodology can exist for an individual process such as project execution or for a combination of processes. A firm can also design its project management methodology for better interfacing with upstream or downstream organizations that interface with the value-added chain. Ineffective integration at supplier–customer interface points can have a serious impact on supply chain management and future business.

17.3 PREACQUISITION DECISION-MAKING

The reason for most acquisitions is to satisfy a strategic and/or financial objective. Table 17–1 shows the six most common reasons for an acquisition and the most likely strategic and financial objectives. The strategic objectives are somewhat longer-term than the financial objectives that are under pressure from stockholders and creditors for quick returns.

The long-term benefits of mergers and acquisitions include:

- Economies of combined operations
- Assured supply or demand for products and services
- Additional intellectual property, which may have been impossible to obtain otherwise
- Direct control over cost, quality, and schedule rather than being at the mercy of a supplier or distributor
- Creation of new products and services
- Pressure on competitors through the creation of synergies
- Cost cutting by eliminating duplicated steps

Each of these can generate a multitude of best practices.

The essential purpose of any merger or acquisition is to create lasting value that becomes possible when two firms are combined and value that would not exist separately. The achievement of these benefits, as well as attaining the strategic and financial objectives, could rest on how well the project management value-added chains of both firms integrate, especially the methodologies within their chains. Unless the methodologies and cultures of both firms can be integrated, and reasonably quickly, the objectives may not be achieved as planned.

TABLE 17–1. TYPES OF OBJECTIVES

Reason for Acquisition	Strategic Objective	Financial Objective
Increase customer base	Bigger market share	Bigger cash plow
Increase capabilities	Provide solutions	Wider profit margins
Increase competitiveness	Eliminate costly steps	Stable earnings
Decrease time-to-market (new products)	Market leadership	Earnings growth
Decrease time-to-market (enhancements)	Broad product lines	Stable earnings
Closer to customers	Better price–quality–service mix	Sole-source procurement

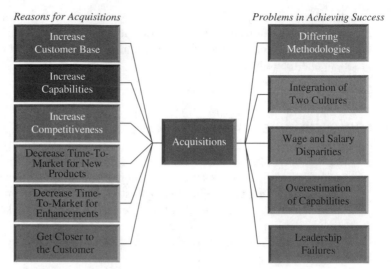

FIGURE 17–3. Project management problem areas after an acquisition.

Project management integration failures occur after the acquisition happens. Typical failures are shown in Figure 17–3. These common failures result because mergers and acquisitions simply cannot occur without organizational and cultural changes that are often disruptive in nature. Best practices can be lost. It is unfortunate that companies often rush into mergers and acquisitions with lightning speed but with little regard for how the project management value-added chains will be combined. Planning for better project management should be of paramount importance, but unfortunately is often lacking.

The first common problem area in Figure 17–3 is the inability to combine project management methodologies within the project management value-added chains. This occurs because of:

● A poor understanding of each other's project management practices prior to the acquisition
● No clear direction during the preacquisition phase on how the integration will take place
● Unproven project management leadership in one or both of the firms
● A persistent attitude of "we–them"

Some methodologies may be so complex that a great amount of time is needed for integration to occur, especially if each organization has a different set of clients and different types of projects. As an example, a company developed a project management methodology to provide products and services for large publicly held companies. The company then acquired a small firm that sold exclusively to government agencies. The company realized too late that integration of the methodologies would be almost impossible because of requirements imposed by the government agencies for doing business with the government. The methodologies were never integrated and the firm servicing government clients was allowed to function as a subsidiary, with its own specialized products and services. The expected synergy never took place.

Some methodologies simply cannot be integrated. It may be more prudent to allow the organizations to function separately than to miss windows of opportunity in the marketplace. In such cases, "pockets" of project management may exist as separate entities throughout a large corporation.

The second major problem area in Figure 17–3 is the existence of differing cultures. Although project management can be viewed as a series of related processes, it is the working culture of the organization that must eventually execute these processes. Resistance by the corporate culture to support project management effectively can cause the best plans to fail. Sources for the problems with differing cultures include:

- A culture in one or both firms that has limited project management expertise (i.e., missing competencies)
- A culture that is resistant to change
- A culture that is resistant to technology transfer
- A culture that is resistant to transfer of any type of intellectual property
- A culture that will not allow for a reduction in cycle time
- A culture that will not allow for the elimination of costly steps
- A culture that must "reinvent the wheel"
- A culture in which project criticism is viewed as personal criticism

Integrating two cultures can be equally difficult during both favorable and unfavorable economic times. People may resist any changes in their work habits or comfort zones, even though they recognize that the company will benefit by the changes.

Multinational mergers and acquisitions are equally difficult to integrate because of cultural differences. Several years ago, a U.S. automotive supplier acquired a European firm. The American company supported project management vigorously and encouraged its employees to become certified in project management. The European firm provided very little support for project management and discouraged its workers from becoming certified, using the argument that their European clients do not regard project management in such high esteem as do General Motors, Ford, and Chrysler. The European subsidiary saw no need for project management. Unable to combine the methodologies, the U.S. parent company slowly replaced the European executives with American executives to drive home the need for a singular project management approach across all divisions. It took almost five years for the complete transformation to take place. The parent company believed that the resistance in the European division was more of a fear of change in its comfort zone than a lack of interest by its European customers.

During the past 40 years, Philip Morris has systematically pursued a diversification strategy to reduce its dependence on cigarettes. In addition to its cigarette business, it owns Clark Chewing Gum, Kraft General Foods, Miller, Miller Lite, Lowenbrau, Jello-O, Oscar Meyer, Maxwell House, Sealtest, Orowheat Baked Goods, and Louis Kemp Seafood. The strategy of Philip Morris was to acquire well-established industry leaders. Each organization acquired had a distinctive corporate culture. Yet even though all of the organizations were successful, some of the expected synergies were not realized because of cultural dissimilarities. Each acquisition was then treated as a stand-alone organization. Although it could be argued that there was no reason to merge these firms into one culture, it does show that even highly successful firms are often resistant to change.

Sometimes there are clear indications that the merging of two cultures will be difficult. When Federal Express acquired Flying Tiger, the strategy was to merge the two into one smoothly operating organization. At the time of the merger, Federal Express employed a younger workforce, many of whom were part time. Flying Tiger had full-time, older, longer-tenured employees. Federal Express focused on formalized policies and procedures and a strict dress code. Flying Tiger had no dress code and management conducted business according to the chain of command, where someone with authority could bend the rules. Federal Express focused on a quality goal of 100 percent on-time delivery, whereas Flying Tiger seemed complacent with a 95–96 percent target. Combining these two cultures had to be a monumental task for Federal Express. In this case, even with these potential integration problems, Federal Express could not allow Flying Tiger to function as a separate subsidiary. Integration was mandatory. Federal Express had to address quickly those tasks that involved organizational or cultural differences.

Planning for cultural integration can also produce favorable results. Most banks grow through mergers and acquisitions. The general belief in the banking industry is to grow or be acquired. During the 1990s, National City Corporation of Cleveland, Ohio, recognized this and developed project management systems that allowed National City to acquire other banks and integrate the acquired banks into National City's culture in less time than other banks allowed for mergers and acquisitions. National City viewed project management as an asset that has a very positive effect on the corporate bottom line. Many banks today have manuals for managing merger and acquisition projects.

The third problem area in Figure 17–3 is the impact on the wage and salary administration program. The common causes of the problems with wage and salary administration include:

- Fear of downsizing
- Disparity in salaries
- Disparity in responsibilities
- Disparity in career path opportunities
- Differing policies and procedures
- Differing evaluation mechanisms

When a company is acquired and integration of methodologies is necessary, the impact on the wage and salary administration program can be profound. When an acquisition takes place, people want to know how they will benefit individually, even though they know that the acquisition is in the best interest of the company.

The company being acquired often has the greatest apprehension about being lured into a false sense of security. Acquired organizations can become resentful to the point of physically trying to subvert the acquirer. This will result in value destruction, where self-preservation becomes of paramount importance to the workers, often at the expense of the project management systems.

Consider the following situation. Company A decides to acquire company B. Company A has a relatively poor project management system in which project management is a part-time activity and not regarded as a profession. Company B, on the other hand, promotes project management certification and recognizes the project manager as a full-time, dedicated position. The salary structure for the project managers in company B was significantly higher than for their counterparts in company A. The workers in company B expressed concern that "We don't want to be like them," and self-preservation led to value destruction.

Because of the wage and salary problems, company A tried to treat company B as a separate subsidiary. But when the differences became apparent, project managers in company A tried to migrate to company B for better recognition and higher pay. Eventually, the pay scale for project managers in company B became the norm for the integrated organization.

When people are concerned with self-preservation, the short-term impact on the combined value-added project management chain could be severe. Project management employees must have at least the same, if not better, opportunities after acquisition integration as they did prior to the acquisition.

The fourth problem area in Figure 17–3 is the overestimation of capabilities after acquisition integration. Included in this category are:

- Missing technical competencies
- Inability to innovate
- Speed of innovation
- Lack of synergy
- Existence of excessive capability
- Inability to integrate best practices

Project managers and those individuals actively involved in the project management value-added chain rarely participate in preacquisition decision-making. As a result, decisions are made by managers who may be far removed from the project management value-added chain and whose estimates of postacquisition synergy are overly optimistic.

The president of a relatively large company held a news conference announcing that his company was about to acquire another firm. To appease the financial analysts attending the news conference, he meticulously identified the synergies expected from the combined operations and provided a timeline for new products to appear on the marketplace. This announcement did not sit well with the workforce, who knew that the capabilities were overestimated and that the dates were unrealistic. When the product launch dates were missed, the stock price plunged and blame was placed erroneously on the failure of the integrated project management value-added chain.

The fifth problem area in Figure 17–3 is leadership failure during postacquisition integration. Included in this category are:

- Leadership failure in managing change
- Leadership failure in combining methodologies
- Leadership failure in project sponsorship
- Overall leadership failure
- Invisible leadership
- Micromanagement leadership
- Believing that mergers and acquisitions must be accompanied by major restructuring

Managed change works significantly better than unmanaged change. Managed change requires strong leadership, especially with personnel experienced in managing change during acquisitions.

Company A acquires company B. Company B has a reasonably good project management system but with significant differences from company A. Company A then decides that "We should manage them like us," and nothing should change. Company A then replaces several company B managers with experienced company A managers. This was put into place with little regard for the project management value-added chain in company B. Employees within the chain in company B were receiving calls from different people, most of whom were unknown to them and were not provided with guidance on who to contact when problems arose.

As the leadership problem grew, company A kept transferring managers back and forth. This resulted in smothering the project management value-added chain with bureaucracy. As expected, performance was diminished rather than enhanced.

Transferring managers back and forth to enhance vertical interactions is an acceptable practice after an acquisition. However, it should be restricted to the vertical chain of command. In the project management value-added chain, the main communication flow is lateral, not vertical. Adding layers of bureaucracy and replacing experienced chain managers with personnel inexperienced in lateral communications can create severe roadblocks in the performance of the chain.

Any of the problem areas, either individually or in combination with other problem areas, can cause the chain to have diminished performance, such as:

- Poor deliverables
- Inability to maintain schedules
- Lack of faith in the chain
- Poor morale
- Trial by fire for all new personnel
- High employee turnover
- No transfer of project management intellectual property

17.4 LANDLORDS AND TENANTS

Previously, it was shown how important it is to assess the value chain, specifically the project management methodology, during the preacquisition phase. No two companies have the same value chain for project management as well as the same best practices. Some chains function well; others perform poorly.

For simplicity sake, the "landlord" will be the acquirer and the "tenant" will be the firm being acquired. Table 17–2 identifies potential high-level problems with the landlord–tenant relationship as identified in the preacquisition phase. Table 17–3 shows possible postacquisition integration outcomes.

The best scenario occurs when both parties have good methodologies and, most important, are flexible enough to recognize that the other party's methodology may have desirable features. Good integration here can produce a market leadership position.

TABLE 17–2. POTENTIAL PROBLEMS WITH COMBINING METHODOLOGIES BEFORE ACQUISITIONS

Landlord	Tenant
Good methodology	Good methodology
Good methodology	Poor methodology
Poor methodology	Good methodology
Poor methodology	Poor methodology

TABLE 17–3. POSSIBLE INTEGRATION OUTCOMES

Methodology		
Landlord	Tenant	Possible Results
Good	Good	Based upon flexibility, good synergy achievable; market leadership possible at a low cost.
Good	Poor	Tenant must recognize weaknesses and be willing to change; possible cultural shock.
Poor	Good	Landlord must see present and future benefits; strong leadership essential for quick response.
Poor	Poor	Chances of success limited; good methodology may take years to get.

If the landlord's approach is good and the tenant's approach is poor, the landlord may have to force a solution upon the tenant. The tenant must be willing to accept criticism, see the light at the end of the tunnel, and make the necessary changes. The changes, and the reasons for the changes, must be articulated carefully to the tenant to avoid cultural shock.

Quite often a company with a poor project management methodology will acquire an organization with a good approach. In such cases, the transfer of project management intellectual property must occur quickly. Unless the landlord recognizes the achievements of the tenant, the tenant's value-added chain can diminish in performance and there may be a loss of key employees.

The worst-case scenario occurs when neither the landlord nor the tenant have good project management systems. In this case, all systems must be developed anew. This could be a blessing in disguise because there may be no hidden bias by either party.

17.5 BEST PRACTICES: CASE STUDY ON JOHNSON CONTROLS, INC.[2]

The Automotive Systems Group (ASG) of Johnson Controls, Inc. (JCI) is one of the best-managed organizations in the world, with outstanding expertise in the management of the project value-added chain. ASG is a global supplier of automotive interior systems and batteries.

2. D. Kandt, Group Vice President, Quality, Program Management and Continuous Improvement, and Alok Kumar of the Automotive Systems Group of Johnson Controls, Inc., provided much of the information in the remainder of this section. Originally published in Harold Kerzner, *Advanced Project Management: Best Practices on Implementation*, Wiley, 2004, Hoboken, NJ, pp. 520–524.

During the 1990s, JCI automotive business expanded more than 10-fold, with some of the growth coming from strategic acquisitions. ASG was increasingly working on development projects that required multiple teams in multiple locations developing products for a single customer. It was clear that ASG needed to have one common global project management methodology or process that would allow *all* ASG employees and teams to communicate better and improve the efficiency of the development process. Without this, the results could have been devastating. ASG was no longer supplying simply products. It was now providing complete solutions for its customers: namely, the interior of the vehicle.

Johnson Controls, Inc. Automotive Systems Group Integration with Prince and Becker Corporation

JCI purchased both Prince and Becker approximately 10 years ago, with the intent of becoming an integrated interior supplier to the automotive industry. The addition of Prince's overhead systems, door panel, and instrument panel capabilities to Becker's interior plastic trim capability and ASG seating products established ASG immediately as an interior supplier. Organizational changes were necessary to truly integrate the companies and their capabilities. The result was a new business model from which the company would be reorganized into vehicle platforms so as to provide complete solutions for its customers rather than merely components.

All three companies had different project management systems, project management position descriptions, and organizational structures. For example, Becker in Europe had project accounters, which were combined project managers and sales personnel. Prince, on the other hand, in addition to project managers, had project coordinators who handled the administrative functions for the project managers. JCI followed a traditional approach, with a well-defined matrix organization and project managers who were fully responsible for all aspects of project execution and success. JCI decided quickly that the three project management systems had to be integrated and commonized. This proved to be a difficult but productive task.

Integrating the Methodologies   Johnson Controls' Automotive Systems Group had a 15-year history of project management with a project management methodology that was well integrated with total quality control. Between 1995 and 2000, ASG had received more than 164 quality awards from its customers and other organizations, and much of the credit was given to the way in which projects were managed. This system had nine life-cycle phases. Prince's system was referred to as a new product development/product development process and had seven life-cycle phases. Becker's system had six life-cycle phases and was called the Projekt Management Handbuch (Project Management Handbook). Not only were the systems different, the languages were different. Each company thought that its system was superior and should be adopted by the entire company. Reaching agreement just on position titles required extensive discussion.

Corporate Cultures   In addition to different organizations and systems, the three companies had different values and all were understandably proud of their own ways of operating. Prince Corporation had a very well integrated culture, with a strong focus on the leadership that had been provided by the founder, Ed Prince. Prince Corp. (located in Holland, Michigan) had never had any real need to consider European interests, especially those of Becker, which had traditionally been a competitor or supplier. JCI had a very strong presence in Europe, with its central office in Burscheid, Germany. The Burscheid office followed North America's lead for project management systems but had strong opinions on how these systems should function in the European culture and followed a much more laissez-faire approach to managing projects. In keeping with the European culture and traditional separation of European countries, the various ASG development centers located in different countries in Europe operated pretty much on their own, with little centralized influence. The end result was a lot of opinions on how to integrate the project management systems, with some common principles and some widely varying values.

Product Focus   In addition to the differences listed above, the purchase of Prince and Becker was intended to bring JCI into an automotive total interior systems position with the vehicle companies. This meant that not only differences in culture, organization, and systems had to be overcome, but also differences in product, equipment, and core manufacturing processes. ASG had to find a way to commonize. In addition, the ASG had to position the new systems to allow the development and launch of total vehicle interiors, a fundamentally different scope for all three of the companies involved. The newly developed integrated project management value-added chain would be a completely new approach for all three companies.

The Integration   A team was created to integrate the various systems, organization, and values of the three companies. Project managers from all three companies, including North America and Europe, were appointed. Also, representatives from all functional departments were represented. Representatives from Quality, Engineering, Manufacturing, and Finance were all part of the team and were able to influence the direction.

The team was challenged to create a project management methodology (and multinational project management value-added chain) that would achieve the following goals:

- Combine best practices from all existing project management methodologies and project management value-added chains
- Create a methodology that encompasses the entire project management value-added chain from suppliers to customers
- Meet the industry standards established by the Automotive Industry Action Group (AIAG), Project Management Institute (PMI), and International Organization for Standardization (ISO)
- Share best practices among all ASG global locations
- Achieve the corporate launch goals of timing, cost, quality, and efficiency
- Accommodate all ASG products
- Optimize procedures, deliverables, roles, and responsibilities
- Provide clear and useful documentation

The system that evolved, called PLUS (Product Launch System), incorporated all three companies' ideas and best practices.[3] PLUS had five phases (plus an initial phase called phase 0 for ideation, which was Prince's product creation phase). All three companies were able to identify with this new system and provided the basic project management structure for JCI to pursue its strategic target of providing vehicle interiors. PLUS repackaged and improved the three existing systems into a new process with optimized procedures, roles, responsibilities, and deliverables. The new system was also painstakingly mapped to the Quality System Requirements QS9000 (AIAG, March 1998), Advanced Product Quality Planning (APQP) and Control Plan (AIAG, June 1994), the Automotive Project Management Guide (AIAG, 1997), and *PMBOK® Guide* (2000).

There were still some problems that had to be resolved. This became evident when teams with more than one year of development time were required to convert to PLUS. PLUS was introduced to the workers through an online PLUS Overview Course that was not sufficient to prepare the teams for using PLUS. Additional training programs were then put in place with customization for particular audiences of workers. The training program provided guidance on why PLUS was developed, how to apply it, and procedures for suggested improvements.

The Outcome   PLUS was well received by all three companies. More than 350 programs were using PLUS. The team that created the new system was given the Chairman's Award for Customer Satisfaction. ASG now had a basic system that allowed everyone in the company to utilize a common language for product development and launch. In fact, this development paved the way for an organizational change that came one year later, referred to as the New Business Model, which reorganized the company into vehicle platform teams, emphasizing JCI's strategic direction to become a total vehicle interior supplier.

This change allowed ASG to function as an integrated company. The platform teams (project offices) had become true teams and felt a sense of identity based on the entire vehicle rather than on doors, seats, and cockpits. The major metrics, such as timing, profitability, and customer relations, were now viewed from the perspective of the entire vehicle. The customers were happier because they now had a single point of contact for their vehicle instead of multiple points of contact.

Follow-on Work   The benefits of PLUS were clear: integrated system, common roles and responsibilities, and the ability to manage projects as integrated platforms. The downside was that PLUS was designed by a committee. Because everyone had a vote, the system was somewhat more bulky and less elegant than it could be. The next step was obvious—PLUS needed to be simplified and streamlined. Out of respect for the strong opinions and cultures of the three companies, PLUS was allowed to remain as it was for approximately 1 to 1.5 years. After that, a central group of project management directors took a minimalist approach to PLUS, removed some unnecessary content, focused on the critical deliverables, and made it much simpler and user friendlier. The revised system,

---

3. The complexities in the development and roll-out of PLUS and the mapping of PLUS against PMBOK® can be found in R. Spigarelli and C. Allen, "Implementation of an Automotive Product Launch System," *Proceedings of the Project Management Institute Annual Seminars and Symposiums*, November 1–10. 2001, Nashville, Tennessee. The latest version of their methodology, called PAP, appears in Section 15.5 of this book.

launched in October 2001, focused on the critical deliverables and on a one-deliverable, one-owner principle. The results were well accepted in Europe and North America.

The system is revised twice a year at an interval of six months. The focus continues on making it intuitive and keeping it flexible, driven by critical deliverables and having clear accountability. Proper integration of methodologies, along with possibly the entire project management value-added chain, can provide excellent benefits. At JCI, the following benefits were found:

- Common terminology across the entire organization
- Unification of all companies
- Common forms and reports
- Guidelines for less experienced project managers and team members
- Clearer definition of roles and responsibilities
- Reduction in the number of procedures and forms
- No duplication in reporting
- Reduction in the number of timeline items from 184 to 110

Recommendations for Other Companies   The following recommendations can be made:

- *Use a common written system for managing programs.* If new companies are acquired, bring them into the basic system as quickly as is reasonable.
- *Respect all parties.* You cannot force one company to accept another company's systems. There has to be selling, consensus, and modifications.
- *It takes time to allow different corporate cultures to come together.* Pushing too hard will simply alienate people. Steady emphasis and pushing by management are ultimately the best way to achieve integration of systems and cultures.
- *Sharing management personnel among the merging companies helps to bring the systems and people together quickly.*
- *There must be a common "process owner" for the project management system.* A person on the vice-presidential level would be appropriate.

17.6 INTEGRATION RESULTS

The best prepared plans do not necessarily guarantee success. Reevaluation is always necessary. Evaluating the integrated project management value added after acquisition and integration is completed can be done using the modified Boston Consulting Group Model (BCG), shown in Figure 17–4. The two critical parameters are the perceived value to the company and the perceived value to customers.

If the final chain has a low perceived value to both the company and the customers, it can be regarded as a "dog." The characteristics of a dog include:

- There is a lack of internal cooperation, possibly throughout the entire value-added chain.
- The value chain does not interface well with the customers.

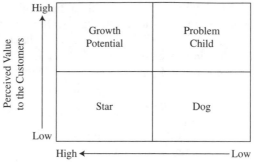

FIGURE 17–4.   Project management system after acquisition.

- The customer has no faith in the company's ability to provide the required deliverables.
- The value-added chain processes are overburdened with excessive conflicts.
- Preacquisition expectations were not achieved, and the business may be shrinking.

Possible strategies to use with a dog include:

- Downsize, descope, or abandon the project management value-added chain.
- Restructure the company to either a projectized or departmentalized project management organization.
- Allow the business to shrink and focus on selected projects and clients.
- Accept the position of a market follower rather than a market leader.

The *problem child* quadrant in Figure 17–4 represents a value-added chain which has a high perceived value to the company but is held in low esteem by customers. The characteristics of a problem child include:

- The customer has some faith in the company's ability to deliver but no faith in the project management value-added chain.
- Incompatible systems may exist within the value-added chain.
- Employees are still skeptical about the capability of the integrated project management value-added chain.
- Projects are completed more on a trial-by-fire basis rather than on a structured approach.
- Fragmented pockets of project management may still exist in both the landlord and the tenant.

Possible strategies for a problem child value chain include:

- Invest heavily in training and education to obtain a cooperative culture.
- Carefully monitor cross-functional interfacing across the entire chain.

- Seek out visible project management allies in both the landlord and the tenant.
- Use of small breakthrough projects may be appropriate.

The *growth-potential* quadrant in Figure 17–4 has the potential to achieve preacquisition decision-making expectations. This value-added chain is perceived highly by both the company and its clients. The characteristics of a growth potential value-added chain include:

- Limited, successful projects are using the chain.
- The culture within the chain is based upon trust.
- Visible and effective sponsorship exists.
- Both the landlord and the tenant regard project management as a profession.

Possible strategies for a growth potential project management value-added chain include:

- Maintain slow growth leading to larger and more complex projects.
- Invest in methodology enhancements.
- Begin selling complete solutions to customers rather than simply products or services.
- Focus on improved customer relations using the project management value-added chain.

In the final quadrant in Figure 17–4 the value chain is viewed as a *star*. This has a high perceived value to the company but a low perceived value to the customer. The reason for the customer's low perceived value is because you have already convinced the customer of the ability of your chain to deliver, and your customers now focus on the deliverables rather than the methodology.

The characteristics of a star project management value-added chain include:

- A highly cooperative culture exists.
- The triple constraint is satisfied.
- Your customers treat you as a partner rather than as a contractor.

Potential strategies for a star value-added chain include:

- Invest heavily in state-of-the-art supportive subsystems for the chain.
- Integrate your PMIS into the customer's information systems.
- Allow for customer input into enhancements for your chain.

17.7 VALUE CHAIN STRATEGIES

At the beginning of this chapter the focus was on the strategic and financial objectives established during preacquisition decision-making. However, to achieve these objectives, the company must understand its competitive advantage and competitive market after acquisition

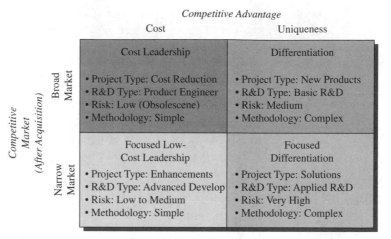

FIGURE 17–5.   Four generic strategies for project management.

integration. Four generic strategies for a project management value-added chain are shown in Figure 17–5. The company must address two fundamental questions concerning postacquisition integration: Will the organization now compete on cost or uniqueness of products and services?

● Will the postacquisition marketplace be broad or narrow?

The answer to these two questions often dictates the types of projects that are ideal for the value-added chain project management methodology. This is shown in Figure 17–6. Low-risk projects require noncomplex methodologies, whereas high-risk projects require complex methodologies. The complexity of the methodology can have an impact on the time needed for postacquisition integration. The longest integration time occurs when a company wants a project management value-added chain to provide

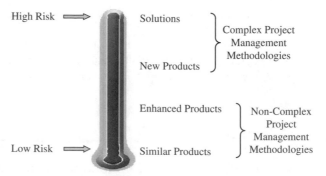

FIGURE 17–6.   Risk spectrum for type of project.

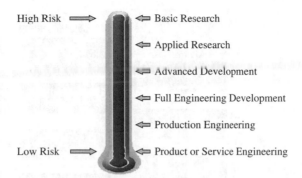

High Risk ⟹   ⟸ Basic Research

⟸ Applied Research

⟸ Advanced Development

⟸ Full Engineering Development

⟸ Production Engineering

Low Risk ⟹   ⟸ Product or Service Engineering

FIGURE 17–7. Risk spectrum for the types of R&D projects.

complete solution project management, which includes product and service development, installation, and follow-up. It can also include platform project management, as was the case with Johnson Controls, Inc. Emphasis is on customer satisfaction, trust, and follow-on work.

Project management methodologies are often a reflection of a company's tolerance for risk. As shown in Figure 17–7, companies with a high tolerance for risk develop project management value-added chains capable of handling complex R&D projects and become market leaders. At the other end of the spectrum are enhancement projects that focus on maintaining market share and becoming a follower rather than a market leader.

17.8 FAILURE AND RESTRUCTURING

Great expectations often lead to great failures. When integrated project management value-added chains fail, the company has three viable but undesirable alternatives:

- Downsize the company.
- Downsize the number of projects and compress the value-added chain.
- Focus on a selected customer business base.

The short- and long-term outcomes for these alternatives are shown in Figure 17–8.
Failure often occurs because the preacquisition decision-making phase was based on illusions rather than fact. Typical illusions include:

- Integrating project management methodologies will automatically reduce or eliminate duplicated steps in the value-added chain.
- Expertise in one part of the project management value-added chain could be directly applicable to upstream or downstream activities in the chain.

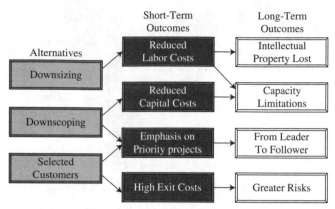

FIGURE 17–8.   Restructuring outcomes.

- A landlord with a strong methodology in part of its value-added chain could effectively force a change on a tenant with a weaker methodology.
- The synergy of combined operations can be achieved overnight.
- Postacquisition integration is a guarantee that technology and intellectual property will be transferred.
- Postacquisition integration is a guarantee that all project managers will be equal in authority and decision-making.

Mergers and acquisitions will continue to take place regardless of whether the economy is weak or strong. Hopefully, companies will now pay more attention to postacquisition integration and recognize the potential benefits.

APPENDIX A

SOLUTION*ACCELERATORS*

Act faster. Go further.

Microsoft® Operations Framework

Version 4.0

MOF Overview

Published: April 2008

For the latest information, please see
microsoft.com/technet/**SolutionAccelerators**

Microsoft

Contents

Introduction to MOF

Microsoft® Operations Framework (MOF) consists of integrated best practices, principles, and activities that provide comprehensive guidelines for achieving reliability for IT solutions and services. MOF provides question-based guidance that allows you to determine what is needed for your organization now, as well as activities that will keep the IT organization running efficiently and effectively in the future.

The guidance in the Microsoft Operations Framework encompasses all of the activities and processes involved in managing an IT service: its conception, development, operation, maintenance, and—ultimately—its retirement. MOF organizes these activities and processes into Service Management Functions (SMFs), which are grouped together in phases that mirror the IT service lifecycle. Each SMF is anchored within a lifecycle phase and contains a unique set of goals and outcomes supporting the objectives of that phase. An IT service's readiness to move from one phase to the next is confirmed by management reviews, which ensure that goals are achieved in an appropriate fashion and that IT's goals are aligned with the goals of the organization.

Goal of MOF

The goal of MOF is to provide guidance to IT organizations to help them create, operate, and support IT services while ensuring that the investment in IT delivers expected business value at an acceptable level of risk.

MOF's purpose is to create an environment where business and IT can work together toward operational maturity, using a proactive model that defines processes and standard procedures to gain efficiency and effectiveness. MOF promotes a logical approach to decision-making and communication and to the planning, deployment, and support of IT services.

How to Use MOF

The MOF guidance consists of a series of phase overviews and SMF guides. These describe the activities that need to occur for successful IT service management—from the assessment that launches a new or improved service, through the process of optimizing an existing service, all the way to the retirement of an outdated service.

The guidance is written for a number of audiences—Corporate Information Officers (CIOs), IT managers, and IT professionals:

- Overview guides are directed toward CIOs who need to see the big picture.

- Overview and workflow information in function-specific guides is geared toward IT managers who need to understand the IT service strategies.

- Activities in function-specific guides are meant for the IT professionals who implement MOF within their work.

The MOF guidance is available on Microsoft TechNet at http://www.microsoft.com/technet/solutionaccelerators/cits/mo/mof/default.mspx.

The IT Service Lifecycle

The IT service lifecycle describes the life of an IT service, from planning and optimizing the IT service to align with the business strategy, through the design and delivery of the IT service, to its ongoing operation and support. Underlying all of this is a foundation of IT governance, risk management, compliance, team organization, and change management.

The Lifecycle Phases

The IT service lifecycle is composed of three ongoing phases and one foundational layer that operates throughout all of the other phases. They are:

- The Plan Phase.
- The Deliver Phase.
- The Operate Phase.
- The Manage Layer.

Figure 1. IT service lifecycle

Your IT organization is probably managing many services at any given time, and these services may be in different phases of the IT service lifecycle. Therefore, you will get the most benefit from MOF if you understand how all phases of the lifecycle operate and how they work together.

- The Plan Phase is generally the preliminary phase. The goal of this phase is to plan and optimize an IT service strategy in order to support business goals and objectives.
- The Deliver Phase comes next. The goal of this phase is to ensure that IT services are developed effectively, are deployed successfully, and are ready for Operations.

- Next is the Operate Phase. The goal of this phase is to ensure that IT services are operated, maintained, and supported in a way that meets business needs and expectations.
- The Manage Layer is the foundation of the IT service lifecycle. Its goal is to provide operating principles and best practices to ensure that the investment in IT delivers expected business value at an acceptable level of risk. This phase is concerned with IT governance, risk, compliance, roles and responsibilities, change management, and configuration. Processes in this phase take place during all phases of the lifecycle.

Following the guidelines contained in MOF can help:

- Decrease risks through better coordination between teams.
- Recognize compliance implications when policies are reviewed.
- Anticipate and mitigate reliability impacts.
- Discover possible integration issues prior to production.
- Prevent performance issues by anticipating thresholds.
- Effectively adapt to new business needs.

When an IT service is first released, it is generally the result of a new initiative: IT-driven or business-driven. Throughout the lifecycle of the service, whenever changes—minor, major, and significant—are made, the MOF IT service lifecycle phases should be applied.

Use the MOF IT service lifecycle phases regardless of the size or impact of a change. The formality with which you apply the lifecycle is proportionate to the risk of the change. You need to do just what is required—no more and no less. For example, a major new initiative, such as a new service on which a business function depends, should go through in-depth analysis and review in the Plan Phase, a formal project plan in the Deliver Phase, and preparation for and review of how IT will implement, support, and monitor this service in the Operate Phase.

A smaller change that does not have as much risk should also go through the lifecycle phases, but can do so in a more nimble way. For example, a change to a Web site is requested. It is not defined as a standard change, so the MOF IT service lifecycle needs to be applied. In the Plan Phase, the requirements are defined with the business, policies are checked, and reliability tradeoffs are made. In the Deliver Phase, the change is designed and tested. In the Operate Phase, the change is made in the production environment, the service is then monitored and adjustments are made as needed, and customer support is provided to assist users with issues that may arise from the change. All of this can be done very quickly for this lower-risk change by using a minimum set of MOF processes.

Service Management Functions Within the Phases

Each phase of the IT service lifecycle contains service management functions (SMFs) that define the processes, people, and activities required to align IT services to the requirements of the business. Each SMF has its own guide that explains the flow of the SMF and details the processes and activities within it.

Figure 2 shows the IT service lifecycle phases and the SMFs within each phase.

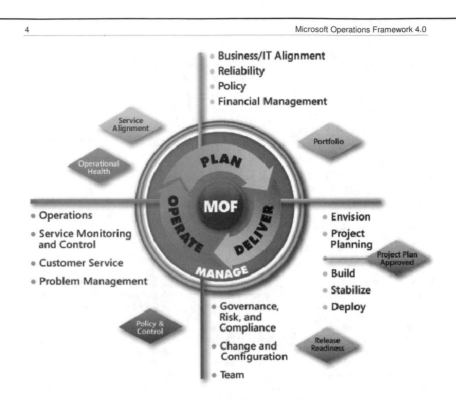

Figure 2. The IT service lifecycle phases and SMFs

Although each SMF can be thought of as a stand-alone set of processes, it is important to understand how the SMFs in all of the phases work to ensure that service delivery is at the desired quality and risk level. In some phases (such as Deliver), the SMFs are performed sequentially, while in other phases (such as Operate), the SMFs may be performed simultaneously to create the outputs for the phase.

Management Reviews

For each phase in the lifecycle, management reviews (MRs) serve to bring together information and people to determine the status of IT services and to establish readiness to move forward in the lifecycle. MRs are internal controls that provide management validation checks, ensuring that goals are being achieved in an appropriate fashion, and that business value is considered throughout the IT service lifecycle. The goals of management reviews, no matter where they happen in the lifecycle, are straightforward:

- Provide management oversight and guidance.
- Act as internal controls at the phase level of the IT lifecycle.
- Assess the state of activities and prevent premature advancement into the next phases.
- Capture organizational learning.
- Improve processes.

During a management review, the criteria that a service must meet to move through the lifecycle are reviewed against actual progress. The MRs make sure that business objectives are being met, and that IT services are on track to deliver expected value.

The MRs, their locations in the IT service lifecycle, and their inputs and outputs are shown in the following table.

Table 1. MOF Management Reviews

MR	Owned by Phase	Inputs	Outputs
Service Alignment	Plan	• Results of the Operational Health Review • Service level agreements (SLAs) • Customer input	• Opportunity for a new or improved project • Request for changes to SLA
Portfolio	Plan	• Project proposals	• Formation of a team • Initial project charter
Project Plan Approved	Deliver	• Business requirements • Vision statement	• Formation of the project team • Approved project plan
Release Readiness	Deliver	• Documentation showing that the release meets requirements • Documentation showing that the release is stable • Documentation showing that the release is ready for operations	• Go/no go decision about release
Operational Health	Operate	• Operating level agreement (OLA) documents • OLA performance reports • Operational guides and service-solution specifications	• Request for changes to the OLA documents • Request for changes to the IT services • Configuration changes to underlying technology components

MR	Owned by Phase	Inputs	Outputs
Policy and Control	Manage	• Operational and security policies • Policy violations, compliance incidents • Policy change requests • Changes in regulations, standards, or industry best practices	• Requests for changes to policies and controls • Requests for changes to policy and control management

Figure 3 illustrates the IT service lifecycle phases and the MRs that connect them.

Figure 3. The IT service lifecycle with MRs

The MRs are described in more detail in the phase overview documents.

Appendix A   **645**

Goals and Functions of the IT Service Lifecycle Phases

The following sections discuss the goals and functions of each phase and how the particular SMFs within that phase achieve their objectives.

Plan Phase

Figure 4 illustrates the Plan Phase.

Figure 4. Plan Phase

During the Plan Phase business and IT work together to determine how IT will deliver valuable services that enable the organization to succeed. Doing that requires:

- Understanding the business strategy and requirements and how the current IT services support the business.

- Understanding what reliability means to this organization and how it will be measured and improved, as well as reviewing and taking action to improve the current state where needed.

- Understanding the organization's policy requirements and how they affect the IT strategy.

- Providing the financial structure to support the IT work and drive the right decisions.

- Creating an IT strategy that provides value to the business strategy and making portfolio decisions accordingly.

The goal of the Plan Phase is to make the right decisions about IT strategy and the project portfolio, ensuring that the delivered services have the following attributes and outcomes:

- Are valuable and compelling in terms of business goals.

- Are predictable and reliable.

- Are cost-effective.
- Are in compliance with policies.
- Can adapt to the changing needs of the business.

The following SMFs support the primary activities of the Plan Phase.

Table 2. Plan Phase SMFs

SMF	Deliverable/Purpose	Outcomes
Business/IT Alignment	**Deliverable:** IT service strategy **Purpose:** - Deliver the right set of services as perceived by the business	- IT Service Portfolio that is mapped to business processes, functions, and capabilities - Services that support the business needs - Knowledge of service demand and usage - Customer satisfaction
Reliability	**Deliverable:** IT standards **Purpose:** - Ensure that service capacity, service availability, service continuity, and data integrity are aligned to business needs in a cost-effective manner	- Reliability plans - Reliability performance reports - Predictable services

Policy	**Deliverable:** IT policies
	Purpose:
	- Efficiently define and manage IT policies required
		- Documented IT polices that are mapped to business policies
		- IT policies required for the effective management of IT
		- Policies documented for the following areas:
		  - Security
		  - Privacy
		  - Appropriate use
		  - Partner and third-party management
		  - Asset protection

SMF	Deliverable/Purpose	Outcomes
Financial Management	**Deliverable:** IT financial planning and measurement **Purpose:** • Accurately predict, account for, and optimize costs of required resources to deliver end-to-end IT services	• Accurate accounting of IT expenditures • Costs mapped to IT services • Budget that supports IT-required investments • Model to determine IT investment opportunities and predict lifecycle costs

Deliver Phase

Figure 5 illustrates the Deliver Phase.

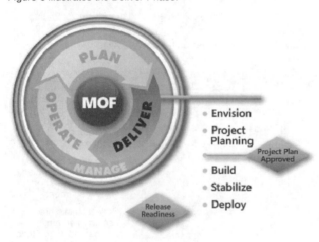

Figure 5. Deliver Phase

Once you have a solid plan for IT service strategy in place, you can begin to create new or updated IT services. The goal of the Deliver Phase is to help IT professionals work within a project management discipline to build, stabilize, and deploy IT services, applications, and infrastructure improvements in the most efficient way possible.

Think of the IT service lifecycle as a continuum: it begins with the efforts of IT to understand the services that the business needs and ends with those services operating in a production environment. The Deliver Phase, then, is the part of the continuum where changes to the services are planned, designed, built, and deployed.

Goals of the Deliver Phase

The primary goals of the Deliver phase are to ensure that IT services, infrastructure projects, or packaged product deployments are envisioned, planned, built, stabilized, and deployed in line with the organization's requirements and the customer's specifications.

Specifically, that means ensuring that the project team:

- Captures the business needs and requirements prior to planning a solution.

- Prepares a functional specification and solution design.

- Develops work plans, cost estimates, and schedules for the deliverables.

- Builds the solution to the customer's specification, so that all features are complete, and the solution is ready for external testing and stabilization.

- Releases the highest-quality solution by performing thorough testing and release-candidate piloting.

- Deploys a stable solution to the production environment and stabilizes the solution in production.

- Prepares the operations and support teams to manage and provide customer service for the solution.

The following SMFs support the primary activities of the Deliver Phase.

Table 3. Deliver Phase SMFs

SMF	Deliverable/Purpose	Outcomes
Envision	**Deliverable:** Vision document **Purpose:** • Clearly communicate the project's vision, scope, and risk	• Vision and scope of the project are clearly documented and understood by the team and the customer • Conceptual design of the proposed solution is recorded as part of the vision document • Project's risks are documented and understood by the team and the customer
Project Planning	**Deliverable:** Project plan document **Purpose:** • Obtain agreement from the project team, customer, and stakeholders that all interim milestones have been met, that the project plans reflect the customer's needs, and that the plans are realistic	• The design and features of the solution are clearly documented in the functional specification • The design and features of the solution are traceable to business, user, operational, and system requirements

MOF Overview   11

SMF	Deliverable/Purpose	Outcomes
Build	**Deliverable:** Developed solution **Purpose:** • Build a solution that meets the customer's expectations and specifications as defined in the functional specification	• A final design that meets business, user, operational, and system requirements • A solution that meets the customer's expectations and specifications as defined in the functional specification
Stabilize	**Deliverable:** Tested and stable solution **Purpose:** • Resolve all issues found by testing and through pilot feedback, and release a high-quality solution that meets the customer's expectations and specifications as defined in the functional specification	• All issues found by testing and through pilot feedback are resolved • A high-quality solution that meets the customer's expectations and specifications as defined in the functional specification
Deploy	**Deliverable:** Service in operation **Purpose:** • Deploy a stable solution that satisfies the customer, and successfully transfer it from the project team to the operations and support teams	• A stable solution deployed into the production environment • A customer who is satisfied with and accepts the deployed solution • A solution successfully transferred from the project team to the operations and support teams

Operate Phase

Figure 6 illustrates the Operate phase.

Figure 6. Operate Phase

The Operate Phase of the IT service lifecycle represents the culmination of the two phases that precede it. The Operate Phase focuses on what to do after the services are in place.

After an IT service has been successfully deployed, ensuring that it operates to meet business needs and expectations becomes the top priority. This is the focus of the Operate Phase, which depends on four primary endeavors:

- Effective ongoing management of the service

- Proactive and ongoing monitoring of its health

- Effective and readily available help to assist with use of the service

- Restoration of a service to health when things go wrong

The primary goal of the Operate Phase is to ensure that deployed services are operated, maintained, and supported in line with the service level agreement (SLA) targets that have been agreed to by the business and IT.

Specifically, that means ensuring:

- That IT services are available by improving IT staff use and better managing workload.

- That IT services are monitored to provide real-time observation of health conditions, and ensuring that team members are trained to handle any problems efficiently and quickly.

- That IT services are restored quickly and effectively.

The SMFs described in the following table support the primary activities of the Operate Phase.

Table 4. Operate Phase SMFs

SMF	Deliverable/Purpose	Outcomes
Operations	**Deliverable:** Operations guide **Purpose:** • Ensure that the work required to successfully operate IT services has been identified and described • Free up time for the operations staff by reducing reactive work • Minimize service disruptions and downtime • Execute recurring IT operations work effectively and efficiently	• Improved efficiency of IT staff • Increased IT service availability • Improved operations of new/changed IT services • Reduction of reactive work
Service Monitoring and Control	**Deliverable:** IT health monitoring data **Purpose:** • Observe the health of IT services • Initiate remedial actions to minimize the impact of service incidents and system events	• Improved overall availability of services • Reduced number of SLA and OLA breaches • Improved understanding of the infrastructure components responsible for delivery of the service • Improvement in user satisfaction with the service • Quicker and more effective responses to service incidents
Customer Service	**Deliverable:** Effective assistance for service users **Purpose:** • Provide a positive experience to the users of a service • Address complaints or issues	• Maintained levels of business productivity • Increased value added by IT • Improved business functionality, competitiveness, and efficiency

SMF	Deliverable/Purpose	Outcomes
Problem Management	**Deliverable:** Effective problem resolution process **Purpose:** • Provide root cause analysis to find problems • Predict future problems	• Reduced number of incidents and problems that occur, and lessened impact of those that do occur • Increased number of workarounds and permanent solutions to identified problems • More problems resolved earlier or avoided entirely

Manage Layer

Figure 7 illustrates the Manage Layer.

• Governance, Risk, and Compliance

• Change and Configuration

• Team

Policy & Control

Figure 7. Manage Layer

The Manage layer integrates the decision making, risk management, and change management processes that occur throughout the IT service lifecycle. It also contains the processes related to defining accountability and associated roles.

The Manage Layer represents the foundation for all phases of the lifecycle. The Manage Layer promotes consistency in planning and delivering IT services and provides the basis for developing and operating a resilient IT environment.

The primary goal of the Manage Layer is to establish an integrated approach to IT service management activities. This approach helps to coordinate processes described throughout the three lifecycle phases: Plan, Deliver, and Operate. This coordination is enhanced through:

- Development of decision making processes.
- Use of risk management and controls as part of all processes.
- Promotion of change and configuration processes that are controlled.
- Division of work so that accountabilities are clear and do not conflict.

Specific guidance is provided to increase the likelihood that:

- The investment in IT delivers the expected business value.
- Investment and resource allocation decisions involve the appropriate people.
- There is an acceptable level of risk.
- Controlled and documented processes are used.
- Accountabilities are communicated and their ownership is apparent.
- Policies and internal controls are effective and reliable.

Meeting these goals is most likely to be achieved if IT works toward:

- Explicit IT governance structures and processes.
- The IT organization and the business organization sharing a common approach to risk management.
- Regularly scheduled management reviews of policies and internal controls.

The SMFs described in the following table support the primary activities of the Manage Layer.

Table 5. The Manage Layer SMFs

SMF	Deliverable/Purpose	Outcomes
Governance, Risk, and Compliance	**Deliverable:** IT objectives achieved, change and risk managed and documented **Purpose:** Support, sustain, and grow the organization while managing risks and constraints	IT services are seamlessly matched to business strategy and objectives
Change and Configuration	**Deliverable:** Known configurations and predictable adaptations **Purpose:** Ensure that changes are planned, that unplanned changes are minimal, and that IT services are robust	IT services are predictable, reliable, and trustworthy

SMF	Deliverable/Purpose	Outcomes
Team	**Deliverable:** Clear accountabilities, roles, and work assignments **Purpose:** Agile, flexible, and scalable teams doing required work	IT solutions are delivered within specified constraints, with no unplanned service degradation Service operation that is trusted by the business

The following list describes how each of the Manage Layer SMFs can be applied to the phases of the IT services lifecycle:

- The Plan Phase purpose is getting business and IT to productively work together. In this phase, the focus of GRC is on the clear communication of strategy, IT decisions being made by the desired stakeholders, and an overall consideration of risk and benefits in terms of the service portfolio. Change management is oriented to issues related to program plans and approaches to initiatives. Plan accountabilities described in the Team SMF are Service and Architecture.

- The Deliver Phase is focused on building the desired solution in the right way. Here, GRC is focused on project scope decisions, project stakeholder involvement, and project risks. Change management is also project-focused, often driving risk management activity for each project. The Deliver accountability described in the Team SMF is Solution,

- The Operate Phase is primarily focused on the daily running and service delivery tasks of IT. In this phase, GRC focuses on capturing information from processes and applications, as well as on demonstrating compliance to policy and regulations. Change management is applied to minimize unplanned changes, preserving service performance and availability, and smoothly introducing standard changes. Operate accountabilities described in the Team SMF are Operations and Support.

There is also an interrelationship among the Manage SMFs themselves. The Change and Configuration SMF processes provide information needed for the Governance, Risk, and Compliance (GRC) SMF processes. In turn, the GRC SMF helps determine who is involved in change management, defines the process for identifying, evaluating, and managing risk associated with the change, ensures that changes reflect policy, and documents these changes appropriately. The Manage accountabilities described in the Team SMF are Management and Compliance.

Objectives, Risks, and Controls

To ensure that the work done in each phase of the IT service lifecycle meets its key objectives, MOF identifies internal controls to minimize the risks to those objectives. Controls are processes and procedures put in place in each phase to ensure that the tasks are performed as expected and that management objectives can be achieved. Many of the controls are driven by the management reviews.

Further information about controls can be found in the *Manage Layer Overview*. The specific controls to ensure that the work performed in each phase is completed as agreed are described in the individual phase overview documents.

MOF Applied

This scenario uses a fictional company, Woodgrove Bank, to demonstrate the use of MOF 4.0. The scenario shows how Woodgrove Bank IT uses each MOF component in meeting a new business objective. For more information on any of these processes, see the MOF Phase Overviews and SMFs.

Woodgrove Bank is a small, regional U.S. bank. It has a new business opportunity that involves expanding operations to California. This puts the company under state laws that they have not had to comply with in the past. These state laws require that they make some changes to their existing security and privacy policies and systems.

Janet is the CEO of Woodgrove Bank. She and other members of the senior staff are going to drive the California expansion project, and they need to identify the actual work that needs to be done, assess the risks, and ensure compliance with the new state regulations. She knows that Woodgrove must greatly improve the ways in which it protects the security of the bank's data and the privacy of its customers' personally identifiable information (PII).This involves thoroughly understanding the California laws, defining the necessary security and privacy controls, assessing the risk and impact of the changes, planning for the changes, building, testing, and deploying the changes, and preparing Customer Service and Operations to support the changes.

Governance, Risk, and Compliance SMF

Janet arranges a series of meetings with the management team to determine how to provide the necessary security and privacy controls for the California expansion. During their meetings, they review pertinent California regulations and define the following security and privacy controls:

- Standards, policies, and procedures
- Data access controls, including encryption
- Controls that protect the bank's computer systems

The bank already has an IT governance policy in place, but Janet knows that they must assess the risks to security and privacy in their new endeavor and then design, document, and implement the proper controls. They begin by examining Woodgrove's:

- Mission statement.
- Risk tolerance and approach to risk management.
- IT portfolio.
- IT service maps.
- Incident reports and security events.
- Regulatory environment and past non-compliance events.

The result is an IT services risk characterization report. This and an awareness of possible threats and vulnerabilities to the bank's data, allows the team to prioritize the risks and identify the necessary controls—the most important of which is the encryption of data—and then implement them. The team also recommends systems for monitoring, tracking, and reporting risks and controls.

The team knows that the most critical aspect of its preparations for the expansion concerns compliance with the California state regulations and law. The team identifies all pertinent California policies, laws, and regulations, creates a compliance plan, and sets up compliance auditing.

Policy SMF

Janet (CEO), Charles (CIO), Kevin (Security Manager), and Neil (IT Manager) need to update the bank's policies to incorporate the new state laws with which they must comply and to increase security and privacy through the use of data encryption.

First, they determine the areas requiring policy changes. Next, they create security, privacy, and appropriate use policies. After the policies have been created, they perform a policy review and incorporate any changes.

The team then establishes controls for enforcing the policies and corrective actions for infractions, sets up a system for recording infractions, and then releases the policies to bank employees who need to know about them. After the policies have been released, the team needs to analyze enforcement of the policies and evaluate their effectiveness, so it sets up a structure for periodically reviewing policies and changing them if necessary.

Reliability SMF

Neil (IT Manager), Ray (Infrastructure Manager), Linda (Development Manager), and Jamie (Operations Manager) tackle ensuring the reliability, dependability, and trustworthiness of the changes that must be made to comply with the California laws.

First, they define the business and service requirements for the changes by examining the new laws, internal bank policies, risk analysis, and reliability parameters. This results in service level requirements, data classification, and data handling policies. Next, they plan and develop a reliability specification model, project plan, and responsibilities for development activities.

The team reviews and updates the following documents:

- The availability plan, which describes the plans for ensuring high availability for the service and which addresses the hardware, software, people, and processes related to the service.

- The capacity plan, which outlines the strategy for assessing overall service and component performance and uses this information to develop the acquisition, configuration, and upgrade plans.

- The information security plan, which describes how the service will be brought to acceptable levels of security. It details existing security threats and how implementing security standards will mitigate those threats. The information security plan addresses data confidentiality, data integrity, and data availability.

- The disaster recovery plan, which specifies the actions and activities that IT will follow in the event of a disaster. The purpose of this plan is to ensure that critical IT services are recovered within required time frames.

- The monitoring plan, which defines the process by which the operational environment will monitor the service, the information being sought, and the ways in which the results will be reported and used.

The team reviews the plans with Janet (CEO), Charles (CIO), and Kevin (Security Manager), make appropriate changes, and then approve the plans.

Financial Management SMF

Working with its Finance team, the team creates an IT cost model and sets up the budget model covering the proposed changes. As a result of this process the team:

- Determines maintenance and operations costs.
- Defines how to track the project and operations costs.
- Adds the tracking requirements to the project plan.
- Estimates the project costs and expected value realization.

The team was satisfied that it now had a methodology to forecast, track, and eventually evaluate the business value realized from the decision to move into the California market.

Business/IT Alignment SMF

The team decides to create a service map for the changes to clarify the relationships between IT and the business and to determine requirements for the work that needs to be done.

Service mapping requires that the team:

- Identify the specific services to be created and their owners.
- Identify the key customers and users.
- Classify and categorize the components of the new services.

The team then takes this information and creates a draft of the service mapping illustration in Microsoft Visio® and a draft of the service mapping table in Microsoft Excel®, reviews the drafts, and publishes them.

Change and Configuration SMF

Kevin (Security Manager) and Neil (IT Manager) baseline the bank's system configuration and do an initial assessment of the impact of the changes. After this is complete, they initiate a Request for Change (RFC) and route it to the bank's change advisory board (CAB) for approval. Since this is not a standard change, the CAB analyzes the impact of the changes and then approves the RFC.

The teams will use change and configuration management throughout the rest of the IT service lifecycle.

Portfolio Management Review

The team presents the project proposal at a Portfolio MR to review and approve the proposed project, and then forms teams to design and deliver the new services. The ultimate outcome of this MR is the initial project charter. With the charter in hand, the team can now begin the process of creating and deploying the new services.

Team SMF

The management team analyzes the scope of the project, determines the roles that will be needed, and then organizes the core project team. The team then creates the project structure document that describes the project team's organization, and specifies roles and responsibilities of each project team member. Phil, the IT Services Manager, also regularly participates in the team meetings to ensure that the customer is well represented. The management team now hands off the work to the project team.

Envision SMF

The project team identifies a number of projects that will bring Woodgrove Bank into compliance with California laws, but it decides to start with the most critical component of the solution. The team decides to call this project the Data Encryption Project, and its goal is to implement Internet Protocol Security (IPSec) to encrypt all of the bank's data including customer personal data. IPSec is a suite of cryptography-based protection services and security protocols.

The team then writes the vision/scope document, which provides a clear vision of what it wants to accomplish. The vision/scope document defines the deliverables for the project:

- IPSec deployment plan
- Definition and assignment of IPSec policies and rules

After the vision/scope document is approved by the team and the stakeholders, the project team completes the milestone review report for the Vision/Scope Approved Milestone. The team and stakeholders sign off on this report, indicating that the project is ready to proceed to planning.

Project Planning SMF

Now that the vision/scope document has been approved, the project team is ready to do the planning work for the project.

The team documents the project requirements and usage scenarios and creates the functional specification.

The project team also creates the design documents (conceptual design, logical design, and physical design) that record the results of the design process. These documents are separate from the functional specification and are focused on describing the internal workings of the solution.

Now the team begins the detailed planning of the project. The team leads prepare project plans for the deliverables in their areas of responsibility and participate in team planning sessions.

After the project plans are complete and approved, the project team rolls up the individual project plans into the master project plan. It then creates individual project schedules and rolls them up into the master schedule, which includes the release date for the solution.

Project Plan Approved Management Review

The project team, management team, and stakeholders review the functional specification, the master plan, and the master schedule and agree that the project team has met the requirements of the Project Plan Approved MR. The team is ready to move on to the development of the solution.

Build SMF

Now the project team begins developing the solution: the IPSec deployment plan, policies, and documentation.

The developers begin by developing the IPSec deployment plan. This plan addresses the following considerations:

After it has created the deployment plan, the project team starts the IP Security Policy Management snap-in in the test lab and begins defining the IPSec policies.

The developers implement the policies and rules in interim builds, and the testers review each build, providing feedback to the developers through the bug-tracking database. To obtain realistic performance data in the test environment, they run standard workloads on programs.

At the same time, the technical writers begin developing the user and operations documentation. This work continues through a number of interim milestones until all of the solution's features are completed. After this happens, the project team can begin to prepare for the solution's release. It prepares for deployment of the solution by developing a pilot plan, a deployment plan, and a deployment infrastructure, including hardware and software.

The team also prepares a training plan, so that the bank's employees will understand the new requirements and learn how to use the new software. Developing ends with the Scope Complete Milestone.

Stabilize SMF

After the development of the solution's features is complete, stabilizing the solution begins. During the initial parts of stabilizing, Test writes the test specification document, and Test and Development work together to find and resolve bugs. They hold regularly scheduled bug meetings to triage bugs, and team members report and track the status of each bug by using the bug-tracking procedures developed during planning.

As stabilizing progresses, the team begins to prepare release candidates. Testing after each release candidate indicates whether the candidate is fit to deploy to a pilot group. After Test and Development have a pilot-ready release candidate, the stakeholders, including Operations and Support, perform user acceptance testing of the solution. When the users accept the solution as pilot-ready, the project team archives the solution in the definitive software library (DSL).

The project team then performs the pilot test, which is a deployment of the solution in a subset of the live production environment. During the pilot test, the team collects and evaluates pilot data, such as user feedback. The team continues to fix reported bugs and other issues until it is satisfied that the solution is stable.

Release Readiness Management Review

Stabilizing culminates in the Release Readiness MR. This review occurs after a successful pilot has been conducted, all outstanding issues have been addressed, and the solution is released and made available for full deployment in the production environment. This review is the opportunity for customers and users, operations and support personnel, and key project stakeholders to evaluate the solution and identify any remaining issues that they must address before deployment.

When the review is complete, the project team approves the release signoff.

Deploy SMF

The solution is now ready to be deployed in the production environment. The project team deploys the infrastructure that supports the solution—new domain controllers and remote access servers. Then, the team deploys the solution to all targeted users and computers at each site.

Stabilization continues during deployment as customer and user feedback from the deployed sites reveals bugs with the solution. The development team fixes the bugs that are approved through the project team change control process.

The Post-Implementation Review finalizes the project and signifies that the project team has fully disengaged and transferred the solution to Operations and Support personnel. At the completion of this review, the project team signs off on the project. Finally, the project team members, management team members, and stakeholders perform a post-project analysis, sign off on the project, and then close it.

Service Monitoring and Control SMF

Jamie (Operations Manager) knows that she needs to determine what is required to monitor the health of the new service. She creates a health model and defines the monitoring requirements.

Next, she determines that the new service is aligned to existing processes and functions, specifies which Operations groups will do the monitoring, and defines the service monitoring requirements in the bank's service monitoring and control (SMC) tool. She also ensures that the service will be continuously monitored.

Operations SMF

Jamie now leads a team that identifies the operational requirements for the new services—for instance how the service is backed up and restored, the tasks needed for disaster management, and the tasks needed for the security of the services.

The team builds the operations plan and then uses it to write the operational work instructions. These instructions identify resources and operational guidance for the new services, and specify the operational work instructions. The team then plans the operational work, assigns the resources, and identifies dependencies. Finally, the team executes the operational work and performs maintenance on the work instructions.

Operational Health Management Review

The new service is added to the regularly scheduled Operational Health MR with the Operations team. It is during this review that requests for changes to the OLA documents, IT services, and configuration of the underlying technology components are made.

Customer Service SMF

Even though the software has been deployed automatically to all bank employee computers, the project team knows that customer service representatives (CSRs) will have to respond to questions and requests for information. In response to these needs, the project team writes knowledge base articles and procedural documentation and trains CSRs to deal with the new IPSec policies and rules. When incidents are recorded, Kate (a CSR) reviews the knowledge base articles about IPSec to find answers needed to resolve the incidents.

Problem Management SMF

Kate noticed in the incident trend reports that there were many incidents regarding intermittent problems with connections refused. She escalates this to the Problem Management team. Jerry on the Problem Management team researches the incidents related to the problem and finds that one server was not updated correctly. He submits a standard change request to update this server and the problem is fixed. He further finds that a root cause of the problem is that the update to the servers should have been done using Auto Enrollment rather than a manual configuration of the certificates. Jerry updates the knowledge base article with this information.

Service Alignment Management Review

The new service is added to the regularly scheduled Service Alignment MR with the account manager and the business customers. It is during this review that requests for changes to the SLA documents and IT services are made.

Policy and Control Management Review

After operations had been running for several months, Woodgrove Bank's management assessed the policies and controls that were put in place to support the bank's move into the California market. Taking a view across the lifecycle, they could see how well risk is being managed and the likelihood that management's objectives were on track to be achieved. They found that one policy around access to customer credit card information was not effectively being enforced. Their outsourced storage solution provider was allowing shared administrator passwords that violated Woodgrove's policy for tracking the specific person that had access to the data. They initiated a change request to address the situation with the service provider both in terms of their contract and their SLA.

Conclusion

Woodgrove Bank's management and project teams have successfully deployed IPSec to provide data encryption for the bank's data. With this important step completed, they can now begin work on other security and privacy solutions that will allow them to expand their operations to California and take advantage of new and exciting opportunities.

Next Steps

This document provides a broad overview of the Microsoft Operations Framework 4.0 and its related SMFs, management reviews, and controls. The next step in putting MOF 4.0 into practice is to consider your organization's needs, and then read and use the relevant SMFs. Their step-by-step guidance will be of value to IT organizations whose goal is reliable, efficient, and compelling IT services.

Feedback

Please direct questions and comments about this guide to mof@microsoft.com.

Attributions

The MOF 4.0 Core Team would like to express our deep appreciation to the many people who participated in the development and review of this latest version of MOF. Contributions have been made at many different levels, from individuals who provided subject matter expertise, to beta participants who gave us much food for thought and valuable suggestions for improvements, to those who reviewed and commented on the content throughout the development process.

The following people are from Microsoft, unless otherwise noted.

Subject Matter Experts:

Alex Broekarts	Tony Noblett (Socair Solutions)
Chuck Chemis (ExamineOps)	Kurt Skjødt Pedersen
Edwin Griffioen	Michael Polzin
Jerry Honeycutt (Studio B)	Gary Roos
Lasse Wilén Kristensen	Rune Ungermann
Morten Lauridsen	Rob van der Burg
Paul Leenards (Getronics)	Hans Vriends (Getronics)

Contributors:

Khalid AlHakim	Karl Grunwald
Michelle Arney	Valentina Haack
Rick Baker	Kelly Hengesteg
Dave Beers	Gerald Herbaugh
Mary Anne Blake	Kevin Hite
Tom Bondi	Byron Holder
Simon Boothroyd	Betty Houser (Volt)
Nigel Cain	Michael Kaczmarek
Derick Campbell	Peter Kendon
Bill Canning	Larry Killingsworth (Air Products)
Chase Carpenter	Kevin Klein (Xtreme Consulting Group Inc.)
Andre Carter	David Krogh
Bret Clark	Brent Kronenberg (Avenade)
Shiloh Cleofe	Shawn LaBelle
CyBOOK, Inc. for graphics	Peter Larsen
Brian DeZell	Matthew Lehman
Laurie Dunham	Lex Liao
Flicka Enloe	David Lymburner

Robin Maher	Paul Ross
Cleber Marques	Gerard Roth
Luis Martinez	Patricia Rytkonen (Volt)
Susan McEver	Terri Snider
Steve McReynolds	Mitch Sonnen
Daniel Rubiolo Mendoza	Jennifer Stevens
Linda Moschell (Volt)	Jim Stuart
Shane Muffley	Robert Sympson
Rich Nardi	Kendall Tieck
Jeff Newfeld	Kathleen Troy
Baldwin Ng	Jan van Bon (Inform-IT)
Michael Oppenheimer (Volt)	Shane van Jaarsveldt (brightenup)
Catherine Paoletti	Meera Venkatesh
Mark Pohto	Mike Walker
Ruth Preston (Volt)	Jeffrey Welton
David Pultorak (Pultorak & Assoc., Ltd.)	Jeff Yuhas

Our thanks to each of you for making this contribution to the IT service management community. We look forward to ongoing collaboration through our MOF Community Portal!

The MOF Core Team
Betsy Norton-Middaugh, Jerry Dyer, Clare Henry, Don Lemmex, and Jason Osborne

Index